Coastal Research Library

Volume 22

Series editor
Charles W. Finkl
Coastal Education and Research Foundation (CERF)
Fletcher, NC, USA

Department of Geosciences
Florida Atlantic University
Boca Raton, FL, USA

The aim of this book series is to disseminate information to the coastal research community. The Series covers all aspects of coastal research including but not limited to relevant aspects of geological sciences, biology (incl. ecology and coastal marine ecosystems), geomorphology (physical geography), climate, littoral oceanography, coastal hydraulics, environmental (resource) management, engineering, and remote sensing. Policy, coastal law, and relevant issues such as conflict resolution and risk management would also be covered by the Series. The scope of the Series is broad and with a unique cross-disciplinary nature. The Series would tend to focus on topics that are of current interest and which carry some import as opposed to traditional titles that are esoteric and non-controversial. Monographs as well as contributed volumes are welcomed.

More information about this series at http://www.springer.com/series/8795

Marcos R. Rossi-Santos • Charles W. Finkl
Editors

Advances in Marine Vertebrate Research in Latin America

Technological Innovation and Conservation

Springer

Editors
Marcos R. Rossi-Santos
Laboratório de Ecologia Acústica e Comportamento Animal/Acoustic Ecology and Animal Behavior Laboratory
Centro de Ciências Agrárias, Ambientais e Biológicas - Universidade Federal do Recôncavo da Bahia (UFRB)
Cruz das Almas, Bahia, Brazil

Programa de Pós-Graduação em Ecologia e Biomonitoramento
Universidade Federal da Bahia (UFBA)
Salvador, Bahia, Brazil

Charles W. Finkl
Coastal Education and Research Foundation (CERF)
Fletcher, NC, USA

Department of Geosciences
Florida Atlantic University
Boca Raton, FL, USA

ISSN 2211-0577 ISSN 2211-0585 (electronic)
Coastal Research Library
ISBN 978-3-319-56984-0 ISBN 978-3-319-56985-7 (eBook)
DOI 10.1007/978-3-319-56985-7

Library of Congress Control Number: 2017946079

Printed on acid-free paper

This Springer imprint is published by Springer Nature
The registered company is Springer International Publishing AG
The registered company address is: Gewerbestrasse 11, 6330 Cham, Switzerland

Preface

This volume in the Coastal Research Library (CRL) gathers the most recent research findings on the ecology and conservation of marine vertebrates in Latin America, making use of advanced technological methods to show readers the diversity of the marine research that has been conducted in these countries over the last decades. The book brings together authors from more than 25 institutions of different countries developing the most diverse research aiming at ocean conservation through the ecology of different vertebrate animals, such as whales, dolphins, manatees, turtles, seabirds, and fish.

This eclectic collection is broad in scope but provides detailed summaries of new methods that are deployed in the study of marine environmental conservation. Key issues revolve around the development and application of educational methodologies in the field of marine vertebrate research, which provide a rational basis for better management of marine environments using modern techniques associated with GIS, satellite tracking, aerial systems, bioacoustics, biogeochemistry, genetics, underwater videography, species photoidentification, molecular biology, trophic ecological methods, ethological methods, arts awareness, and behavioral ecology, among others. Discussion and elucidation of these kinds of techniques are aimed at university-level students and postgraduate researchers. The scope of this volume includes whales, sharks, rays, dolphins, tropical fishes, turtles, and manatees as well as aspects of Latin American marine ecosystem conservation.

The 17 chapters in this book are organized in three parts, with Part I focusing on the historical development of aquatic mammal research in Latin America, Part II dealing about technological innovation for marine vertebrate research in Latin America, and Part III bringing studies on the conservation ecology of marine vertebrate research in Latin America.

Part I includes three chapters, the first being "Mastozoology History and Perspectives in Latin America: Marine and Freshwater Mammals in Spotlight," where Paulo Simões-Lopes introduces the historical context of the first studies in Latin America as well as the potential perspectives for the future of this field. In Chap. 2, "Long-Term Monitoring of Dolphins in a Large Estuarine System of

Southeastern Brazil," Emygdio Monteiro Filho, Lucimary Deconto, Caio Louzada, Rebeca Wanderley, Daniela Godoy, and Eric Medeiros describe one of the first in situ long-term studies for a marine vertebrate species (the Guiana dolphin) in Latin America. Chapter 3, "An Integrated Framework to Assess the Carrying Capacity of Whale-Watching Tourism in Praia do Forte, Northeastern Brazil," by Luena Fernandes and Marcos Rossi-Santos, represents an important effort to understand the whale-watching operation toward its sustainability in Brazilian waters.

Part II contains six chapters that generally deal with technological innovation for marine vertebrate research. In Chap. 4, "Land-Based Studies of Aquatic Mammals in Latin America: Understanding Behavior for Conservation," Maria Morete, Daniela Abras, and Cristiane Martins relate their use of theodolites to monitor whale behavior from land stations and use the acquired information to support conservation actions. Chapter 5, "Marine Mammal Bioacoustics Using Towed Array Systems in the Western South Atlantic Ocean," by Artur Andriolo, Franciele de Castro, Thiago Amorim, Gustavo Miranda, Juliana Di Tulio, Juliana Moron, Bruna Ribeiro, Gabriela Ramos, and Raíssa Mendes, presents the advanced technique of acoustic towed arrays to monitor marine mammals in oceanic waters. Also about acoustic approaches, Mariana Melcón, Vanesa Reyes, and Miguel Iñíguez in Chap. 6, "Bioacoustic Techniques Applied to Odontocete Conservation and Management in Argentina," discuss on the application of such methods in Argentinean waters. In order to contribute with the present knowledge of the marine turtle species of Latin America, Maria Angela Marcovaldi, Guy Marcovaldi, Alexsandro Santos, Paulo Lara, and Milagros López-Mendilaharsu, in Chap. 7, "Novel Research Techniques Provide New Insights to the Sea Turtle Life Cycle," bring an interesting view of their pioneering study and rewarded research program in Brazil. In Chap. 8, "Using Opportunistic Platforms to Study Dolphin Occurrence and Distribution in the North Coast of Bahia State, Northeastern Brazil," Mateus Carvalho, Marcos Rossi-Santos, Elitieri Santos-Neto, and Clarêncio Baracho describe the important findings of dolphin occurrence, with the use of tourism boats as a research platform, minimizing costs for the marine fauna monitoring. Chapter 9, "Using Chemical Elements to the Study of Trophic and Spatial Ecology in Marine Mammals of the Southwestern Atlantic Ocean," by Elisa Seyboth, Silvina Botta, and Eduardo Secchi, contributes with biochemistry methods to study the trophic ecology of marine mammals as well as their conservation.

Part III consists of seven chapters that deal with the conservation ecology of marine vertebrate research in Latin America. In Chap. 10, "Elasmobranchs Consumption in Brazil: Impacts and Consequences," Hugo Bornatowski, Raul Braga, and Rodrigo Barreto tell readers about the increasing consumption of sharks and rays and what this implies for marine ecology and conservation. Chapter 11, "West Indian Manatee (*Trichechus manatus*) in South America: Distribution, Ecology and Health Assessment," by Ana C. O. de Meirelles, Vitor L. Carvalho, and Miriam Marmontel, constitutes a great overview of the research with this endangered marine vertebrate species, enriching the species diversity of this volume. Representing some of the advanced research conducted in Central America, Chap. 12, "Ecology and Conservation of Cetaceans of Costa Rica and Panama," by Laura

May-Collado, Marisol Amador-Caballero, José Julio Casas, Mónica Gamboa-Poveda, Frank Garita-Alpízar, Tim Gerrodette, Rocío González-Barrientos Gabriela Hernández-Mora, Daniel Palacios, José Palacios-Alfaro, Betzi Pérez, Kristen Rasmussen, Lissette Trejos, and Javier Rodríguez-Fonseca, presents the state of art of the cetacean research in these two countries. Another high topic is debated in Chap. 13, "Anthropogenic Noise and Guiana Dolphins (*Sotalia guianensis*) in Brazil: Ecological and Conservation Concerns," by Fábio de Sousa Pais, Rafaela Cardoso, Leonardo Wedekin, Dalila Martins, Flavio Silva, Emygdio Monteiro Filho, and Marcos Rossi-Santos, discussing the occurrence of human disturbance on the daily life of a marine top predator. Another great contribution to enlarge the vertebrate diversity of this book is presented in Chap. 14, "The Atlantic Goliath Grouper: Conservation Strategies for a Critically Endangered Species in Brazil," by Maurício Hostim-Silva, Áthila Bertoncini, Maíra Borgonha, Jonas Leite, Matheus Freitas, Felippe Daros, Leonardo Bueno, Ana Paula Farro, and Christopher Koenig, bringing the efforts for the conservation of one of the most endangered fish species of the world. In Chap. 15, "Spreading Message in a Noisy World: Song Behavior and Acoustic Ecology in Humpback Whales (*Megaptera novaeangliae*), from the Southwestern Atlantic Ocean," Marcos Rossi-Santos presents the results of a long-term study of one of the most complex singers in the animal world. Chapter 16, "Conservation Advances for the Southern Right Whales in Brazil," by Karina Groch, relates the victories and challenges of the efforts to understand and conserve whales in Latin America since the 1980s. The last Chap. 17, "Behavior and Ecology of Endangered Species Living Together: Long-Term Monitoring of Resident Sympatric Dolphin Populations in South Brazil," by Marta Cremer, Annelise Holz, Camila Sartori, Beatriz Schulze, Renan Paitach, and Paulo Simões-Lopes, brings the unique research that studies Franciscana dolphins in the wild along to Guiana dolphins, in an important Brazilian bay ecosystem.

With this book, we would like to present to the world the efforts from Latin American researchers in the search for high technological methods to achieve enhanced results in the marine ecology and conservation science, and we hope that it may be taken as an example for the future generations of investigators in the developing countries worldwide. Come and join us in this amazing journey!

Praia do Forte, BA, Brazil — Marcos R. Rossi-Santos
Fletcher, NC, USA — Charles W. Finkl

Contents

About the Editors

Marcos R. Rossi-Santos is a biologist, with a MSc in Zoology from the Universidade Federal do Paraná and PhD in Animal Behavior from the Universidade Federal do Rio Grande do Norte. He has been investigating the ecology and conservation of aquatic mammals and other vertebrates for the past 23 years, developing a broad experience about behavioral ecology and anthropogenic impacts on natural ecosystems, mainly through the acoustic studies. Along this time, he has collaborated with many Brazilian research institutions, exploring different areas such as the Brazilian coast and islands, Amazon, Galapagos Islands, and South Georgia Island, in Antarctica. He is a professor at the Universidade Federal do Recôncavo da Bahia, Brazil, coordinating the Acoustic Ecology and Animal Behavior Laboratory, and also associated with the Graduation Program in Ecology of the Universidade Federal da Bahia.

Charles W. Finkl has edited and/or contributed to more than eight volumes in the Encyclopedia of Earth Sciences Series. For the past 33 years, he has been the executive director of the Coastal Education and Research Foundation (CERF) and Editor-in-Chief of the international *Journal of Coastal Research*. In addition to these duties, he is Professor Emeritus at Florida Atlantic University in Boca Raton, Florida, USA. He is a graduate of the University of Western Australia (Perth) and previously worked for a wholly owned Australian subsidiary of the International Nickel Company of Canada (INCO). During his career, he acquired field experience in Australia; the Caribbean; South America; SW Pacific islands; Southern Africa; Western Europe; and the Pacific Northwest, Midwest, and Southeast USA.

Part I
Historical Development of Aquatic Mammal Research in Latin America

Chapter 1
Mastozoology History and Perspectives in Latin America: Marine and Freshwater Mammals in Spotlight

Paulo C. Simões-Lopes

Abstract This chapter will introduce readers to the mastozoology history of the marine and freshwater mammals of Latin America. Its inconspicuous beginning in the nineteenth century, its emergence in the mid-twentieth century and the retrospective of the last 32 years based on an evaluation of 16 bi-annual congresses of experts in aquatic mammals of Latin America. Brazil, Argentina and Chile have covered about 75% of all Latin American scientific production (from 29 countries), and studies about cetaceans covered 51–67% of all scientific production. Behavioral and population ecology studies have grown continuously and works focused on conservation, tourism, environmental education, the development of new research methods and new equipment are growing markedly.

Keywords Cetaceans • Pinnipeds • Sirenians • Mustelids

1.1 Introduction

1.1.1 Beginnings

The nineteenth century marked the beginning of the large foreign expeditions through South America as *Voyage Amérique Méridionale, Castelnau Expedicion Amérique Sud,* and later, the *Expedicion Antartic Belgique* (the latter took place at the beginning of the twentieth century). Famous personalities such as Étienne Geoffroy Saint-Hilaire, Alexander von Humboldt, Hermann von Ihering and Charles Darwin himself devoted their efforts in cataloging the "tropical nature" as it was called. However, the aquatic mammals were not the object of their immediate

P.C. Simões-Lopes (✉)
Aquatic Mammals Lab (LAMAQ), Departamento de Ecologia e Zoologia, Universidade Federal de Santa Catarina, P.O. Box, 5102, Florianópolis 88040-970, Brazil
e-mail: lamaqsl@ccb.ufsc.br

M.R. Rossi-Santos, C.W. Finkl (eds.), *Advances in Marine Vertebrate Research in Latin America*, Coastal Research Library 22,
DOI 10.1007/978-3-319-56985-7_1

interest. On the other hand, direct competition between various eminent naturalists served as a trigger for the first discoveries.

The Brazilian naturalist Alexandre Rodrigues Ferreira, who was unfortunately forgotten, had a vital importance due to his work. In many ways, he was the pioneer to unravel the Amazon's tributaries in a famous expedition, and also, the specimens collected by him were sent to Coimbra and Lisbon in the eighteenth century. In 1790, he collected the type specimen of the Amazon River dolphin, *Inia geoffrensis* (de Blainville, 1817). Originally sent to the *Museu da Ajuda*, in Lisbon, the specimen was taken to Paris by Napoleon looters, during the Peninsular Wars. Geoffroy Saint-Hilaire himself was responsible for the "transfer" and ended up being honored as the specific epithet, while Alexandre Rodrigues Ferreira remained anonymous. Therefore, this is the time to pay tribute for Ferreira's pioneering in the Latin American aquatic mastozoology.

The itinerant French naturalist Alcide d'Orbigny (1802–1857) and his colleagues from the *Musée d'Histoire Naturelle* in Paris and the *Académie Royale de Sciences de Belgique* in Brussels were the first to devote more time to the aquatic mammals. They also collected material and illustrated color plates that served for the description of the Franciscana, *Pontoporia blainvillei* (Gervais & d'Orgigny, 1844). He also described the genus *Inia* d'Orbigny, 1834.

Paul Gervais and P.J. van Bénéden made use of that precious material nominating valid species as the Amazonian tucuxi, *Sotalia fluviatilis* (Gervais, 1852) and the estuarine dolphin or Guiana dolphin, *S. guianensis* (P.J. van Bénéden, 1864). Descriptions of *S. pallida* (van Bénéden & Gervais, 1880) and *S. brasiliensis* (van Bénéden, 1875) are currently junior synonyms of the related species mentioned above.

The Baron Georg Heinrich von Langsdorff, a recognized entomologist, was in Brazil, in December 1803 (Berger 1984). He reported the whaling in the state of Santa Catarina, southern Brazil, citing the sperm whales, *Physeter macrocephalus* Linnaeus, 1756. His apparently glaring mistake was due to the hunting occur in winter and spring, but never in the summer when he stayed there. Moreover, the only target species during whaling in southern Brazil was the southern right whales, *Eubalaena australis* (Desmoulins, 1822), which was a species not scientifically described at that time yet. That mistake was fixed only with a full assessment of significant amounts of bone material from the whaling stations that showed the southern right whales was the only species hunt on the southern coast of Brazil (Simões-Lopes and Ximenez 1993).

The greatest contribution of that period was that of another entomologist, the German Herman Karl Conrad Burmeister. He described small and large cetaceans like the strange Burmeister's porpoise, *Phocoena spinipinnis* Burmeister, 1865, the Antarctic minke whale, *Balaenoptera bonaerensis* Burmeister, 1867, and founded the first cetology center in the *Museo Argentino de Ciencias Naturales*, in Buenos Aires. Such facts gave the front in this field of knowledge to Argentina and absorbed other foreign and autochthonous scientists.

1.1.2 The Twentieth Century: An Emerging Science

The first half of the twentieth century was discrete about changes. The French naturalist Fernando Lahille (often regarded as Argentinean) and the Argentinean Francisco Pascasio Moreno, also known as Perito Moreno, were based in the *Museo de La Plata* and contributed to the description of species such as the bottlenose dolphin, *Tursiops ghephyreus* (Lahille, 1908) and the spectacle porpoise, *Phocoena dioptrica* (Lahille, 1912).

In Brazil and Argentina there was the outstanding participation of the German naturalist Hermann von Ihering, and later, his Brazilian son Rodolpho von Ihering who published the record of the *Stenodelphis blainvillei* (= *Pontoporia blainvillei* (Gervais & d'Orgigny, 1844) for the Lagoa dos Patos, in southern Brazil.

Almost in the middle of the twentieth century happened the first significant leap in mastozoology with the works of Angel Cabrera in Argentina, and Alipio Miranda Ribeiro, Carlos Cunha Vieira and later Cory Texeira de Carvalho, in Brazil (Avila-Pires 2009).

In the second half of the century, there was a true "quantum leap" in the volume and quality of scientific production in the area. Anélio Aguayo Lobo of the *Universidad Nacional Autónoma de México* went to Chile and with his experience leveraged the development in that field. Angel Crovetto†, Jorge A. Oporto, Luis Pastene, Walter Sielfeld, Doris Oliva and Jose Yanez, among others, continued the challenging work seeking information between the jagged fjords and rocky islets of the Chilean coast. In Mexico, the exchange among experts started early due to the creation of SOMEMMA (Mexican Society of Marine Mammals), at the end of the 1970s. Jorge Urban was one of the main supporters and the first one to seek integration with the South American colleagues.

The Uruguayans Raul Vaz Ferreira†, Isaias Ximenez and Alberto Ponce de León studied the behavior and the killing of sea lions and fur seals, *Otaria flavescens* (= *O. byronia*) and the *Arctocephalus australis*, in the breeding colonies. Also, Ricardo Praderi† and the North American Robert Brownell were pioneers in the studies of incidental mortality of the Franciscana, on shark fishing nets. The conservation timeline of what is today the most endangered species owes them that auspicious beginning.

The publication of Cabrera's catalog allowed Argentina to lead a significant progress in the study of marine mammals. The foreigners Bernd Würsig and Roger Payne studied the bottlenose dolphins and right whales, in *Península de Valdés, Patagonia.* The Argentineans Ricardo Bastida, Enrique Crespo, Claudio Campagna, Juan Carlos Lopez, Julio Loureiro, Luiz Capozzo and Mirtha Lewis, among others, deepened and gave maturity to a large research program of the Argentine Mastozoological Fauna. In Tierra del Fuego, Argentina, there was an extraordinary collaboration of the English Natalie Pros Goodall† who arrived there to photograph lichens and ended up collecting a myriad of marine mammal carcasses mummified by the cold.

Hugo Patrício Castello had a remarkable influence both in Argentina and Brazil. Moreover, he included the South American researchers with the organization of the *"Primera Reunión de Trabajos de Espertos en Mamíferos Acuáticos de América del Sur"* in 1984, in Buenos Aires. These bi-annual conferences became known as RTs and remain uninterrupted until now.

The research development in Brazil owes its influence to the Admiral Ibsen de Gusmão Câmara† who was a pioneer in whaling assessment. In southern Brazil, Hugo P. Castello, Maria Cristina Pinedo and the Uruguayan Alfredo Ximenez† deserve to be highlighted because they organized the first research laboratories. The Canadian Robin Best† and the Brazilian Vera Maria Ferreira da Silva, based in Manaus, did the same in the Amazon Basin and also encouraged many other young researchers. Fernando W. Rosas and Miriam Marmontel belong to the subsequent generation of researchers responsible for leveraging that area of knowledge and develop research programs that included dolphins, whales, manatees and pinnipeds.

Other South American countries such as Peru, Colombia and Ecuador also contributed to the study of humpback whales and Amazonian otters. The Galapagos archipelago was and still is one of the most extraordinary open-air laboratories of the world, highlighting Ecuador in the international arena.

1.2 Results

1.2.1 Retrospective of the Last 32 Years

In an attempt to reconstruct a timeline that minimally expresses what happened in three decades, in Latin America, we are going to start with the first RT (Buenos Aires, 1984) and the preceding year. Thus, 16 bi-annual congresses of experts in aquatic mammals were sampled, covering 32 uninterrupted years and with the presentation of 3019 abstracts (Table 1.1). From 1996 (Viña del Mar, Chile), these symposia started to be organized by SOLAMAC (The Latin American Society of Specialists in Aquatic Mammals, Fig. 1.1) and the number of submitted abstracts has grown substantially.

Twenty-nine countries were represented in these scientific events, and many of them did not represent Latin America (Table 1.2). Many of these outside interests were related to postgraduate courses taken abroad by Latin-American researchers, especially in the United States. Also, many international partnerships contributed with their logistics and technology in more expensive studies. The three countries with the highest participation (Brazil, Argentina and Chile) have covered about 75% of all Latin American scientific production over these 32 years (Table 1.2). Certainly, the volume of papers presented by Brazil, which is almost half of the production in these 32 years, has a direct relation to its large population and a coastline of about 7500 km.

Table 1.1 Meeting of Specialists on Aquatic Mammals in Latin America (RTs) and number of scientific presentations (abstracts)

Year	RT	City	Country	N^0. Abstracts
1984	I	Buenos Aires	Argentina	51
1986	II	Rio de Janeiro	Brazil	54
1988	III	Montevideo	Uruguay	83
1990	IV	Valdivia	Chile	158
1992	V	Buenos Aires	Argentina	155
1994	VI	Florianópolis	Brazil	156
1996	VII	Viña del Mar	Chile	183
1998	VIII	Recife	Brazil	233
2000	IX	Buenos Aires	Argentina	205
2002	X	Valdivia	Chile	170
2004	XI	Quito	Ecuador	235
2006	XII	Mérida	Mexico	258
2008	XIII	Montevideo	Uruguay	238
2010	XIV	Florianópolis	Brazil	373
2012	XV	Puerto Madryn	Argentina	270
2014	XVI	Cartagena	Colombia	115
Total	**16**	**11**	**7**	**3019**

The shaded area identifies the symposium organized by SOLAMAC – The Latin American Society of Specialists in Aquatic Mammals

The six most active countries that presented over 100 abstracts, in that period, traced different trajectories (Table 1.2, Fig. 1.2). The leadership of Argentina until 1992, supported by the Southern Cone countries, allowed the consolidation of that area of knowledge and influenced generations of new scientists. After 1994, there was a hegemony alternation. A profusion of Non-Governmental Organizations (NGOs) and an efficient system of academic post-graduate studies allowed Brazil to qualify young scientists, renew the team of researchers and strengthen research programs and laboratories already established. Mexico developed a different trajectory since it had an Independent Society, but promoted the union of these initiatives in Merida (2006).

Fig. 1.1 Founding members of the SOLAMAC – The Latin American Society of Specialists in Aquatic Mammals. From *left* to *right*: seated (Luiz Capozzo, Miriam Marmontel, Mario Cozzuol, Jose Yanez, Mirtha Lewis, Vera Maria Ferreira da Silva and Lilian Flores-Gonzáles); standing (Paulo C. Simões-Lopes, Marcela Junin, Hugo P. Castello, Claudio Campagna, Anélio Aguayo, Jorge A. Oporto, Maria Cristina Pinedo, Monica Muelbert and Jorge Urban)

Productivity fluctuations between the participating countries, over the years, do not reflect immediate changes in their research efforts; however, they are a mere consequence of each country's distances to host the events. There is also an indirect and very harmful consequence, the international economic crisis that has started in recent years. SOLAMAC and local organizing committees are surviving with limited resources and, in most cases, non-existent ones.

Palacios et al. (2012) used a different array to measure the scientific production among Latin American countries. Therefore, they considered the Latin American Journal of Aquatic Mammals (LAJAM), but they reached to similar results with the dominance of Brazil, Argentina, Mexico, Uruguay and the USA. In our assessment, Chile appears in third place while Colombia and Peru are in sixth and seventh places, respectively. These two readings can help future research programs to guide the relevance of scientific production in Latin-America.

The importance given to each Taxa has also changed over the years. The pinnipeds (sea lions, fur seals and seals) were losing ground, gradually, over the advancement of studies with cetaceans (Fig. 1.3). In the case of sirenians and mustelids, despite the small number of species, have been occupying, proportionally, the same space.

Behavioral and population ecology studies involving cetaceans and pinnipeds in the natural environment have grown continuously (Fig. 1.4). In general, they are related to acoustic issues or abundance and density estimates. The interest in skeleton anatomy, comparative morphology, zooarchaeology and paleontology have also gradually increased, although less impressively. In this group, there are taxonomic and evolutionary studies. The focus on health or pathologies as well as in parasites has not increased as we expected.

Table 1.2 Scientific production by country based on the number of abstracts presented at the Meeting of Specialists on Aquatic Mammals in Latin America (RTs) (1984–2014)

Order	Country	Continents	%	N⁰. Abstracts
1	Brazil	SA	48.7	1471
2	Argentina	SA	17.3	524
3	Chile	SA	8.5	258
4	Mexico	NA	5.7	174
5	Uruguay	SA	4.6	141
6	Colombia	SA	4.6	139
7	Peru	SA	2.4	73
8	United States	NA	1.9	61
9	Ecuador	SA	1.8	56
10	Venezuela	SA	0.9	28
11	Spain	EU	0.5	17
12	Puerto Rico	SA	0.4	13
13	England	EU	0.4	13
14	Canada	EU	0.3	11
15	Costa Rica	SA	0.3	9
16	New Zealand	OC	0.2	6
17	Bolivia	SA	0.1	5
18	China	AS	0.1	5
19	Germany	EU	0.1	4
20	Denmark	EU	0.0	2
21	French Guiana	SA	0.0	1
22	Panama	CA	0.0	1
23	Guatemala	CA	0.0	1
24	Dominican Rep.	CA	0.0	1
25	Cuba	CA	0.0	1
26	Italy	EU	0.0	1
27	Portugal	EU	0.0	1
28	Australia	OC	0.0	1
29	Philippines	AS	0.0	1
Total	**29**	**6**	**100.0**	**3019**

The shaded area identifies countries outside Latin America.

The shaded area identifies countries outside Latin America
Abbreviations: *AS* Asia, *CA* Central America, *EU* Europe, *NA* North América, *OC* Oceania, *SA* South America

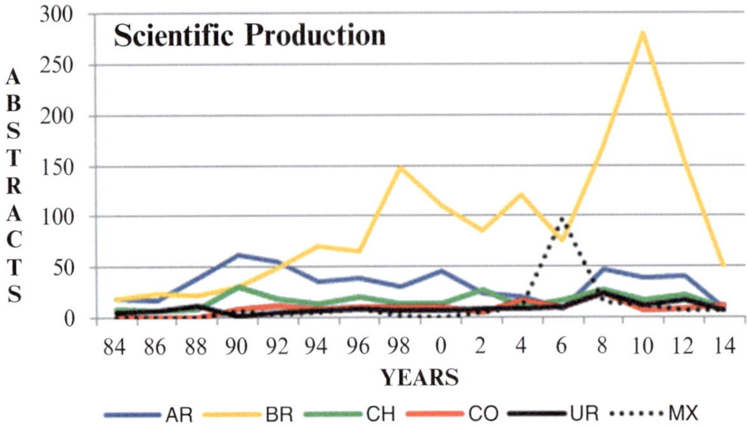

Fig. 1.2 Number of abstracts submitted by each country over 32 years, during the Meeting of Specialists on Aquatic Mammals in Latin America (RTs). Abreviations: *AR* Argentina; *BR* Brazil; *CH* Chile; *CO* Colombia; *UR* Uruguay and *MX* – Mexico

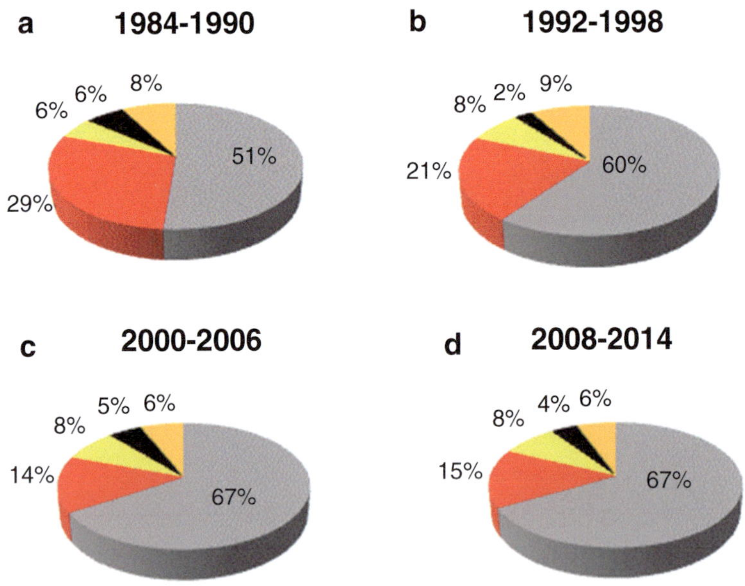

Fig. 1.3 Percentage of abstracts presented at the Meeting of Specialists on Aquatic Mammals in Latin America (RTs). Captions: *gray* (cetaceans); *red* (pinnipeds); *yellow* (sirenians); *black* (mustelids); *orange* (marine mammals). (**a**) 1984–1990; (**b**) 1992–1998; (**c**) 2000–2006; (**d**) 2008–2014

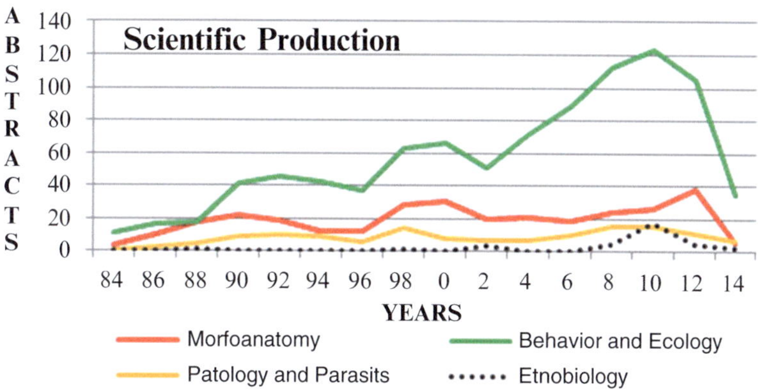

Fig. 1.4 Abstracts covering ethnobiological, behavioral, ecological and morphoanatomical studies and pathologies or parasites, over 32 years, in the Meeting of Specialists on Aquatic Mammals in Latin America (RTs)

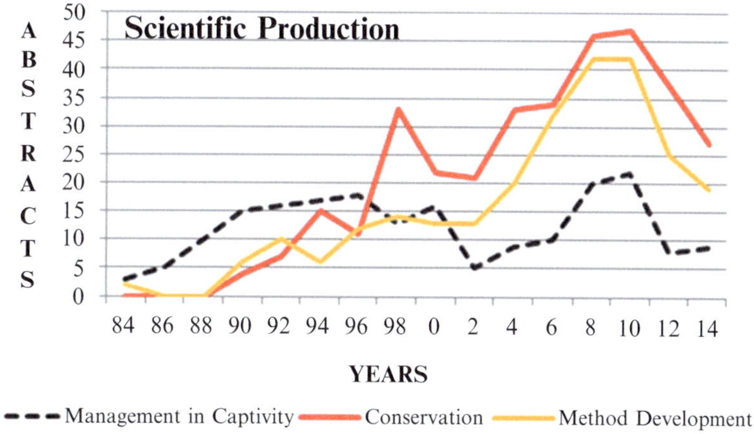

Fig. 1.5 Abstracts covering captive management studies, conservation and development of new methods, during 32 years, in the Meeting of Specialists on Aquatic Mammals in Latin America (RTs)

Issues related to captive management have lost ground in the Latin American context (Fig. 1.5). However, works directed to conservation, tourism, environmental education, development of new research methods and new test equipment are on the rise and increasingly occupying the researchers' interest.

Ten years after the first Latin American meeting of experts in aquatic mammals, the first panorama seemed to be changing. Rosas et al. (1994), Corcuera and Cornejo (1994) and Simões-Lopes (2006) performed an assessment of the period and found that the most immediate topics such as stranding records, casual sightings at sea and distribution expansions were giving way to more elaborate studies both in the field and in the laboratory. However, if we consider a more extended period as the reported

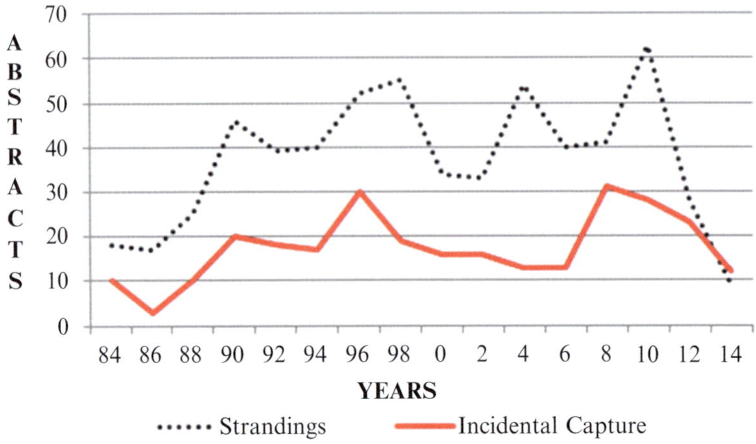

Fig. 1.6 Abstracts covering studies of strandings and accidental captures in fishing nets, over 32 years, in the Meeting of Specialists on Aquatic Mammals in Latin America (RTs)

here, it is possible to realize that there is no decrease and the subject is still in vogue (Fig. 1.6). Instead, the so-called accidental mortality in fishing nets and entanglements of marine mammals in cables and 'ghost nets' at sea – an issue of extreme importance in the early twenty-first century – is not getting the necessary attention.

1.2.2 Perspectives in Latin America

Somehow, there was a growing involvement of new researchers, expansion of expertise areas and consolidation of several research groups in Latin-America. On the other hand, some countries failed to conduct a proper replacement of human resources devoted to advanced studies. It may lead them to a collapse of human resources, especially in international economic crisis times.

Studies with stable isotopes, satellite tagging on migratory species, reviews of environmental-organic contamination and oxidative stress, bone DNA extraction, microsatellite locus, molecular eschatology, animal health ratings, use of computed tomography (CT scans), improvements in imaging quality and sound and complex population modeling have received more attention, but it is necessary to improve the overall research efforts. The incorporation of new analytical techniques and modeling was remembered by Oliva (2012), but the challenge, she said, is to move from a population-based approach to an ecosystem approach. Such perception seems to be quite adequate; however, there is no stimulating panorama on that aspect yet.

Campagna (1998) already anticipated that we were going through a descriptive-naturalistic start to a technological-instrumental stage, and the future of our science seemed to be dictated by the technological development. He ruled that we were part of a settled science in content, but juvenile in its influences, and it delegates the own trend to computers rather than taking the helm of the situation.

Thus, we are at this crucial point. There is still a lot of repetition and "reinventing the wheel" and, therefore, there is a lot of effort, human resources and money wasted. It is more evident just in areas of knowledge on the rise as environmental education, conflicts with tourism and conservation programs. Scientific meetings aimed at protecting threatened species and populations are represented by biologists, oceanographers and veterinarians, but there is a lack of economists, politicians and diplomats. For that reason we cannot pass from the "population vision to the ecosystem vision," and thus, we are still part of a "juvenile science in its influences."

It is necessary to develop a philosophy for this area of knowledge, a research program and financial support that values the tradition and maturity of groups already established. It could solve some problems about "reinventing the wheel," but such philosophy should not allow the established "fiefdoms" to dictate the performance limits and control the specialized personnel training. The formation of new research groups, qualifications, engagement and self-determination are fundamental to "pass the ecosystem vision" in our science.

References

Ávila-Pires F (2009) Mamíferos aquáticos: águas passadas e perspectivas futuras. Boletim da Sociedade Brasileira de Mastozoologia 55:6–10

Berger P (1984) Ilha de Santa Catarina – Relatos de Viajantes Estrangeiros nos séculos XVIII e XIX, 2ª edn. Editora da Universidade Federal de Santa Catarina, Florianópolis, 335p

Campagna C (1998) La Investigación en mamíferos acuáticos: desde ayer y hacia mañana. In: Lima RP (eds) Anais da 8° Reunião de Trabalhos de Especialistas em Mamíferos Aquáticos da América do Sul e 2⁰ Congresso SOLAMAC. Recife, 244 p

Corcuera J, Cornejo I (1994) La investigacion en mamíferos marinos en America del Sur. In: Ximenez A, Simões-Lopes PC (eds) Anais da 6ª Reunião de Trabalho de Especialistas em Mamíferos Aquáticos da América do Sul. Universidade Federal de Santa Catarina, Florianópolis, p 41. 142 p

Oliva D (2012) Desarrollo de la Investigación sobre mamíferos acuaticos en América del Sur: avances y perpectivas. In: Anales de la 15° Reunion de Trabajo de Expertos em Mamíferos Acuaticos de America del Sur. 9° Congresso SOLAMAC. Puerto Madryn, p 11

Palacios D, Martins C, Gómez-Salazar C, Olavarria C (2012) La mastozoologia acuática en América-Latina: un análisis bibliométrico y de autorias para los primeiros ocho años del Latin American Journal of Aquatic Mammals (2002–2010). In: 15° Reunion de Trabajo de Expertos em Mamíferos Acuaticos de America del Sur. 9° Congresso SOLAMAC. Puerto Madryn, p 12

Rosas FCW, Marmontel M, Castello HP (1994) Reuniões de Trabalho de Especialistas em Mamíferos Aquáticos da América do Sul: histórico e evolução dos 10 primeiros anos. In: Ximenez A, Simões-Lopes PC (eds) Anais da 6ª Reunião de Trabalho de Especialistas em Mamíferos Aquáticos da América do Sul. Universidade Federal de Santa Catarina, Florianópolis, pp 1–4. 142 p

Simões-Lopes PC (2006) Mamíferos Aquáticos em Retrospectiva: a breve história de uma ciência emergente no Brasil, Cap 18. In: Monteiro-Filho ELA, Aranha JMR (eds) Revisões em Zoologia – Volume Comemorativo dos 30 anos do Curso de Pós-Graduação em Zoologia da UFPR. SEMA/PR, Curitiba, pp 331–338. 391 p

Simões-Lopes PC, Ximenez A (1993) Annotaded list of the cetacens of Santa Catarina coastal waters, southern Brazil. Biotemas 6(1):67–92

Chapter 2
Long-Term Monitoring of Dolphins in a Large Estuarine System of Southeastern Brazil

Emygdio L.A. Monteiro-Filho, Lucimary S. Deconto, Caio N. Louzada, Rebeca P. Wanderley, Daniela F. Godoy, and Eric Medeiros

Abstract Considering long-term study as one that through constant efforts subsidizes future studies, an example in Brazil is "Projeto Boto-Cinza" adding up more than 36 continued years of research and data application on Guiana dolphin biology, ecology and conservation. The first effort we dedicated was toward understand the fishing strategies used by this species, lasting about 10 years. Latter, as we gained knowledge in behavior study, the next step was to access parental care, youngling behavior, learning process of fishing, acoustic repertoire, and recently social networks. Throughout this time, it was possible to access also the mortality, and based on carcasses that gave such information, access parameters of life history such as reproductive anatomy of males and females, pregnancy period, birth size, and even age. Population density estimates could be done in different periods and areas; photo-identification enabled the detection of individual spatial fidelity. Modeling permitted to observe which parameters seem to be associated with the spatial distribution of population units inside the estuary. All these data contributed to a greater knowledge about the species, creation of public policies by conservation units' chiefs, education purposes, and conservation of the species.

E.L.A. Monteiro-Filho (✉)
Department of Zoology, Centro Politécnico, Universidade Federal do Paraná, Caixa postal 19020, 81.531-980, Curitiba, PR, Brazil

Instituto de Pesquisas Cananéia,
Rua Tristão Lobo, 199, Centro, 11.990-000, Cananéia, SP, Brazil
e-mail: kamonteiro@uol.com.br; elamf@ufpr.br

L.S. Deconto • C.N. Louzada
Programa de Pós-Graduação em Zoologia, Departamento de Zoologia, Centro Politécnico, Universidade Federal do Paraná, Caixa postal 19020, 81.531-980, Curitiba, PR, Brazil

Instituto de Pesquisas Cananéia,
Rua Tristão Lobo, 199, Centro, 11.990-000, Cananéia, SP, Brazil

R.P. Wanderley • D.F. Godoy • E. Medeiros
Instituto de Pesquisas Cananéia,
Rua Tristão Lobo, 199, Centro, 11.990-000, Cananéia, SP, Brazil

M.R. Rossi-Santos, C.W. Finkl (eds.), *Advances in Marine Vertebrate Research in Latin America*, Coastal Research Library 22,
DOI 10.1007/978-3-319-56985-7_2

Keywords *Sotalia guianensis* • Cananéia estuary • Guiana dolphin • Paranaguá estuary • Projeto Boto-Cinza

2.1 Historic Background

Although not being common, researchers developed in African ecosystems several natural history long-term studies – those in which constant efforts subsidizes future. Some examples of these studies are Jane Goodall's with chimpanzees (*Pan troglodytes*), Craig Packer's with lions (*Panthera leo*), Holly Dublin's with elephants (*Loxodonta africana*), Robert Sapolsky's with baboons (*Papio cynocephalus* and *Papio anubis*), Laurence Frank's with hyenas (*Crocuta crocuta*), among others.

Despite the challenges that arise in marine environment long-term studies, a handful of good examples can here be listed. Regarding these studies, we point out efforts conducted by Roger Paine as of 1967, initially dedicated to Humpback whales (*Megaptera novaeanglia*) in the United States, then to cetacean conservation, and more recently to Right whales (*Eubalaena australis*) in Argentina coast. Also, Paul Spong and collaborators since the 1970s with Killer whales (*Orcinus orca*) in Vancouver Island, Canada, and by John Reynolds with Manatees in (*Trichecus manatus*) in Florida also since the 1970s.

Particularly in the Brazilian coast, different studies have been made in long-term projects such as the ones conducted by Flávio J. Lima Silva and José Martins Silva Júnior with Spinner dolphins (*Stenella longirostris*) in Fernando de Noronha Archipelago, the ones by Leonardo Flach and Elaine Ferreira with Guiana dolphin (*Sotalia guianensis*) in Sepetiba Bay, and the ones by Marta J. Cremer with Franciscanas (*Pontoporia blainvillei*) in Babitong Bay. All of the above lasting more than 15 years of integrated research. Adding up more than 36 years of uninterrupted studies in the Cananéia estuary in Southeastern Brazil, the "Projeto Boto-Cinza" studies gave support for the presentation here made about aspects of biology, ecology, and conservation studies about this species.

2.2 Target Species

Sotalia guianensis (Fig. 2.1) is a Delphinidae with wide coastal distribution between Honduras (Edwards and Schnell 2001) and Baía Norte, southern Brazil (Simões-Lopes 1988). Since its description, this species received different denominations, having been originally described as *Delphinus guianensis* van Bénéden, 1864 (Hershkovitz 1966). With a review by Gray in 1866, the species started to belong to *Sotalia*, not *Delphinus*, remaining for about 100 years. By this time, the *Sotalia* had three species: *S. fluviatilis* (Gervais and Deville 1853) in Amazon

Fig. 2.1 A Guiana dolphin, *Sotalia guianensis*, photographed at Cananéia Lagoon Estuarine Complex, State of São Paulo, Brazil (Photo: Caio N. Louzada)

Basin, *S. guianensis* van Bénéden, 1864 in the northern South America, and *S. brasiliensis* van Bénéden, 1875 strictly at Guanabara Bay in southeastern Brazil (Hershkovitz 1966).

During the 1960s, some doubts arose about the validity of such names and in 1963, Carvalho reports for the municipalities of Santos and Cananéia the presence of a common dolphin known as the boto-da-baía-de-guanabara (Guanabara Bay dolphin) (Monteiro-Filho et al. 2008a). In the following years, the doubt remained on which name should be given to animals observed in several spots of the Brazilian coast. After short considerations about the species, Rice (1977) started to consider the existence of a single species (*Sotalia fluviatilis*), with two subspecies being one restricted to the Amazon Basin (*S. f. fluviatilis*) and the other to the marine coast (*S. f. guianensis*).

For 25 years the classification proposed by Rice (1977) remained, even with some authors arbitrarily refusing the possibility of the subspecies. In 2002, Monteiro-Filho et al. (2002) published a paper questioning the name given to marine and fresh water animals of *Sotalia*. Authors used geometric morphometric to access the skull shape of 104 individual from different locals of the Brazilian coast and Amazon fresh waters. In addition to biological knowledge already published about the species (see Monteiro-Filho et al. 2008a), the results point to two distinct biological units, and therefore repurposing the use of *S. fluviatilis* to describe fresh water animals and *S. guianensis* for marine ones (Monteiro-Filho et al. 2002, 2008a).

Latter, molecular studies (Cunha et al. 2005) converged to the same conclusion. These morphometric and molecular studies surely contribute so that researchers can designate the results of their own studies to specific biological units. However, despite of the academic importance, such studies are no news since researchers described the species as separate units more than 150 years before. This fact brings up the fantastic capacity of natural observation of such researchers of the nineteenth and twentieth century.

2.3 Study Area

Studies performed by this project were conducted mainly in a large estuarine system called Estuarine Lagoon Complex of Iguape Cananéia Paranaguá. The huge biodiversity of the region attracted the attention of researchers to this area at the beginning of 1980s. They looked to main species that inhabit that place and one received particular attention, the Guiana dolphin. This region presents an environmental complexity, composed by several river that flow into the estuary and creates many lagoons, mangrove islands, flooded areas and sand beaches. At many of these places is possible to watch plenty of coastal and estuarine species that here to forage, breed, and raise offspring.

The estuary is part of the biggest remaining patch of Atlantic rain forest (SOS Mata Atlântica 2013), located between boundaries of States of São Paulo and Paraná (Fig. 2.2). Belongs to one of the most important Brazilian coastal ecosystem

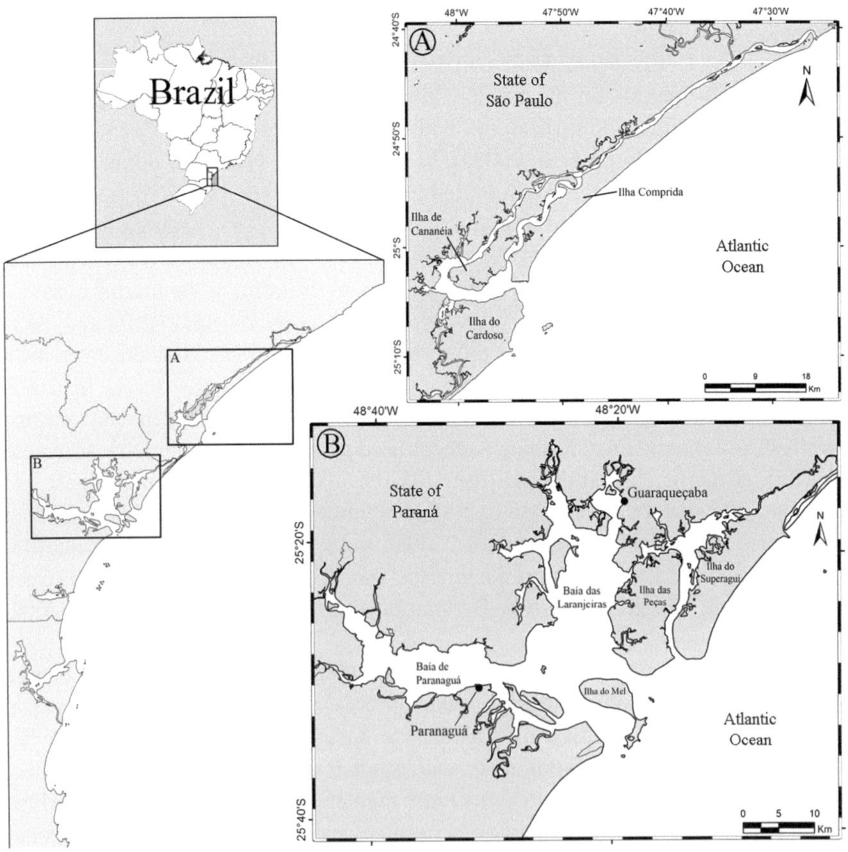

Fig. 2.2 Map of at Estuarine Lagoon Complex of Iguape Cananéia Paranaguá, located at boundaries of States of São Paulo and State of Paraná, Brazil. (**a**) Detailed map of Cananéia estuary, State of São Paulo, Brazil. (**b**) Detailed map of Paranaguá estuary, State of Paraná, Brazil

(Diegues 1987) and is considered as an internationally sanctuary that needs to be preserved (Mendonça and Katsuragawa 2001). According to IUCN (1984), this regions is one of the five most productive and less disturbed estuary in the world and is a priority to conservation according to the World Conservation Strategy. The estuary is legally protected by several federal, state and municipal protected areas, which together create a mosaic of conservation units that aims the conservation of local biodiversity (Schaeffer-Novelli et al. 1990). Besides that, is stated as "The Mata Atlântica Biosphere Reserve" and "World Heritage Site" by UNESCO at 1991 and 1999 (UNESCO 1992, 1999).

Cananéia estuary (24°40′S; 47°25′W and 25°05′S; 48°02′W; Fig. 2.2a) has 110 km of extension, consisting of a large protected channel (Mar Pequeno), one bay (Baía de Trapandé) and three islands (Ilha Comprida at east, Ilha do Cardoso at south and Ilha de Cananéia at west). Which communicate with the Atlantic Ocean by two inlets (Barra do Icapara at north and Barra de Cananéia at south). The channels are composed by brackish water, bordered by mangrove vegetation. The three main species of mangroves are *Rhizophora mangle* (red mangrove), *Laguncularia racemosa* (white mangrove) and *Avicennia schaueriana* (black mangrove) (Schaeffer-Novelli et al. 1990).

Paranaguá estuary (25°13′S; 48°06′W and 25°45′S; 48°44′W; Fig. 2.2b) shows two distinct components. The first one is represented by sandbars and many lagoons. The second one is more affected by the fluvial system. The flow of the water inside the estuary is regulated mainly by tidal currents, commonly found in estuarine systems. The fresh water and salt wedge interacts with each other creating a great distinctness during low and high tide (Bigarella 1978). The vegetation is mainly composed by mangroves, salt marshes and "restingas" (Lana 1986).

Both regions show a high variation of the bathymetry. Sand and mud banks are commonly found, which increases heterogeneity of underwater landscape. This ground depth variation builts many different coastal formations, flooded marshals and floodplains. The formation of vegetation islands enforces the water currents to narrow channels, creating a strong flow of water during tidal currents, which impacts water dynamics (Miyao and Sarti 1986). All complexity of this system receives high influence of tides, fluvial flow, and ocean winds (Miranda et al. 1995). The salinity patterns change seasonally, according to fresh water outflow, being higher during the Austral winter (dry season) and lower at Austral summer (rain season) (Miyao and Sarti 1986). Consequently, all these features are suitable to build a high heterogeneity habitat, which supports the presence of several species, including the Guiana dolphin.

2.4 Morphology

As part of our efforts, some studies started almost at the same time because of the necessity of collecting carcasses along the beach shores. One example of carcass-based study is described above, with the use of geometric morphometric (Monteiro-Filho et al. 2002), counting with great support of other institutions' scientific

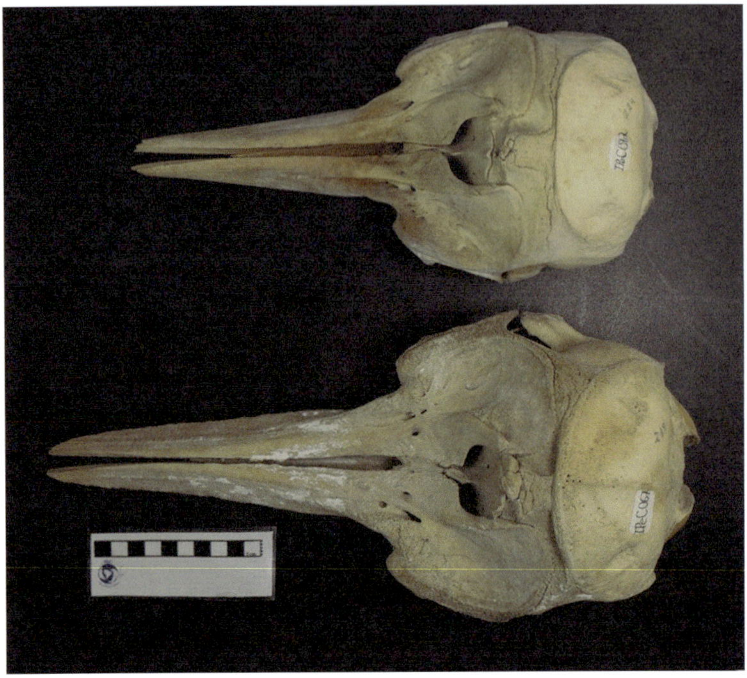

Fig. 2.3 Skulls of two specimens of *Sotalia guianensis* used in cranial morphometric studies. Upper skull is from a calf specimen and below skull is from an adult. The skull of tucuxi (*S. fluviatilis*) is very similar to the skull of a calf *S. guianensis*, due to minor development of temporal region and minor projection of condyles occipitals. Specimens are deposited at Instituto de Pesquisas Cananéia (IPeC) scientific collection, municipality of Cananéia, State of São Paulo, Brazil (Photo: IPeC)

collections [Projeto Mamirauá (State of Amazonas), Museu Nacional (State of Rio de Janeiro), Museu de Zoologia da USP (State of São Paulo), Museu de História Natural da UNICAMP (State of São Paulo), Laboratório de Mamíferos Aquáticos da UFSC (State of Santa Catarina)] besides the collection from Instituto de Pesquisas Cananéia (IPeC), responsible for the "Projeto Boto-Cinza". Besides the discussion about the group systematics, the study points out to some other aspects. Specific adaptive characteristics in skulls were linked to the difference in the environment as well as different needs for each species. The authors discuss about differences in size of both species, showing that fresh water individuals have skull shape similar to younglings (paedomorphosis, Fig. 2.3.) and end up by making a reflection on the possibility of both species being more similar if fresh water individuals grew up to proportions of marine individuals. It seems not to be the case, since the divergence would tend to be even greater.

Concomitantly, Rosas and Monteiro-Filho (2002) evaluated the species' gonads shapes and sizes, describing testis as cylindrical with conical tips with average length of 7.36 cm (±1.82) and average width of 2.28 cm (±0.90) in immature and up to 25.52 cm (±3.48) and 9.65 cm (±1.14), respectively, in mature animals. Just like

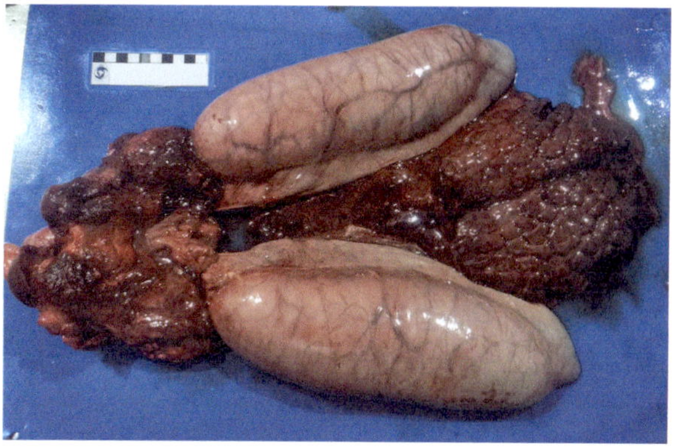

Fig. 2.4 Testis of *Sotalia guianensis* of a sexually mature specimen dissected after necropsy. Specimen was found during beach monitoring activities at Cananéia estuary (Photo: IPeC)

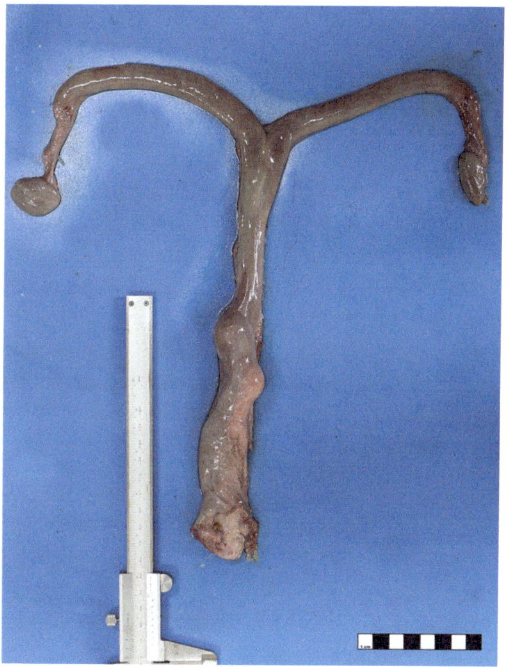

Fig. 2.5 Ovaries and bipartite uterus of *Sotalia guianensis* of a sexually mature specimen dissected after necropsy. Specimen was found during beach monitoring activities at Cananéia estuary (Photo: IPeC)

the size, testis mass vary largely between immature, with 52.3 g (±33.5), and mature animals, with 2.270 g (±0.86) (Fig. 2.4). The testis mass correspond to 0.11% of body mass in immature and 2.52% in mature individuals, keeping these proportions throughout the year, with no apparent seasonality in testis activity detected. Female, just as with other Delphinidae, have bipartite uterus (Fig. 2.5) with the shape of a small almond (Rosas and Monteiro-Filho 2002; Rosas and Barreto 2008). The authors detected little variation in ovary morphometric, ranging from 2.3 cm (±0.5) long and 1 cm (±0.2) wide in immature to 3 cm (±0.4) long and 1.7 cm (±0.3) wide in mature animals. Differently, ovary masses increase significantly, from 1.6 g (±1)

Fig. 2.6 Photography of two specimens of *Sotalia guianensis*. Dolphin on left is an adult, while the dolphin on right side is an infant, which is recognized by pinkish color of dorsal fin, back and flanks (Photo: Caio N. Louzada)

in immature to 7.9 g (±4.2) in mature animals (Rosas and Monteiro-Filho 2002; Rosas and Barreto 2008).

The newborn average size was estimated in 92 cm. The estimated gonadal maturity occurs between 165 and 170 cm in females and between 170 and 175 cm in males (Rosas and Monteiro-Filho 2002). In relation to the cetacean morphology, the skin is mostly evaluated regarding the blubber (e.g. Sokolov 1960; Giacometti 1967; Simpson and Gardner 1972; Pfeifer and Jones 1993). The skin thickness changes according to age in *S. guianensis*, with expressive ontogenetic variation in pigmentation (Randi et al. 2008). Therefore, the pink color that appear in infants (Fig. 2.6) results of intense vascularization and a thin blubber layer rather than by red pigmentation on the skin. The only detected pigment was melanin, scarce in infants and increases with aging. Such increase tend to cover the pink tones of vascularization, making adults mostly grey in the dorsal portion of the body, reducing towards the flanks.

2.5 Age Evaluation

Successful age estimates were made with *S. guianensis* carcasses through counting of dentin growth layers deposited in teeth (Schmiegelow 1990; Ramos 1997, Rosas 2000; Rosas et al. 2003; Sydney and Monteiro-Filho 2011; Wanderley 2013; Fig 2.7). The term “Growth Layer Group” (GLG) describes the distinct lines formed by deposition of dentin and cementum in layers. Although different possible techniques may be used to access age in cetacean, Rosas et al. (2003) describe some guidelines for age estimates with teeth since there was no specific model for dentin deposition or GLG reading for the species. Researches in Cananéia (SP) and Paranaguá (PR) registered a variation between 0 (newborn) and 30 years, being female the eldest representation (Rosas et al. 2003; Wanderley 2013) and immature (up to 6 years old) present greater mortality frequency, indicating a possible greater vulnerability to natural and/or anthropic threats due to inexperience of younger animals (Wanderley 2013).

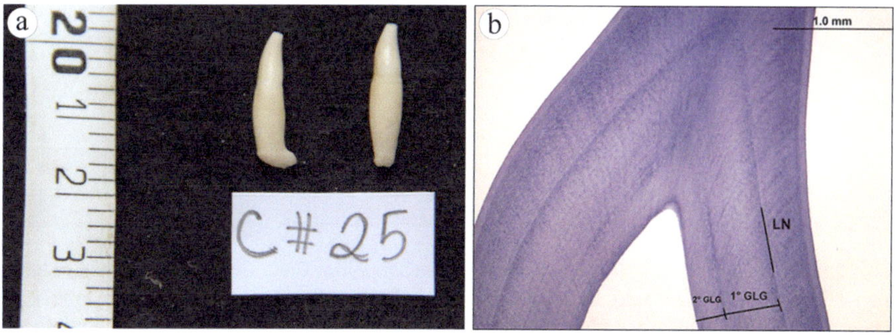

Fig. 2.7 (**a**) *Sotalia guianensis* teeth deposited at Instituto de Pesquisas Cananéia (IPeC) scientific collection, municipality of Cananéia, State of São Paulo, Brazil. (**b**) Photomicrography of apical region of *Sotalia guianensis* teeth, cut at 30 μm and colored with Harris' Hematoxylin. *LN* Neonatal line, 1° GLG – first growth layer group referring to the first year of life, *2° GLG* second growth layer group referring to an incomplete second year of life (Photos: Rebeca P. Wanderley)

Beyond the importance of GLG in age estimation in *S. guianensis*, there is strong evidence that the end of milk-feeding period is also registered. Rosas et al. (2003) described an accessory layer deposited in the dentin between the neonatal layer and first GLG as being possibly linked to the end of milk-feeding of the infants, which varied between 6.7 and 10.3 months. According to the authors, the decrease of calcium caused by ablactation could reflect in a hipomineralized layer. Rosas and Monteiro-Filho (2002) described the milk composition, where protein (9.5 g/100g) and fat (17.1 g/100g) values were similar to those observed in other delphinids' milk.

2.6 Population Parameters and Habitat Index

Monteiro-Filho (2000) conducted the first studies related to Guiana dolphin group organization in Cananéia, defining three categories: individual (Fig. 2.8a), family (Fig. 2.8b), and school (Fig. 2.8c). The family organization is the most commonly seen in the study region, registered in more than 80% of the field observations (Monteiro-Filho 2000). The basic structure is composed by a female and its calf, or even by an adult, a female and its calf. For the first months after birth, while the calf remains close to its mother (Fig. 2.9), the associated adult may aid the mother in feeding. Schools happen when there is an association of several families (Fig. 2.8c), which can happen temporary (from 5 to 60 min) for feeding or displacement purposes.

Beside studies about social organization, it was also necessary to estimate dolphin density in the region. For cetaceans, researchers use diverse techniques to assess population density such as census, mark-recapture method, land observations and line transect. Bisi (2001) and Havukainen et al. (2011) made density estimates for Cananéia region based on line transect, dividing the area in sectors and obtaining values reaching 12.41 individuals/km^2. The division of the area in sectors enabled results to show a non-uniform use within sectors.

Fig. 2.8 Examples of most frequent group structure found at Cananéia estuary Guiana dolphin population, State of São Paulo, Brazil. (**a**) Solitary individual. (**b**) Family, composed by one adult and one calf, and frequently a plus one adult. (**c**) School, which represents an association among several families (Photos: Caio Louzada)

Fig. 2.9 A calf of Guiana dolphin (*Sotalia guianensis*) is frequently observed swimming side-by-side with its parental (Photo: Caio N. Louzada)

The distribution pattern of animals may also be investigated in an individual basis and to assess such information the capture-mark-recapture method is the most usual. Since the physical capture of the Guiana dolphin would not be an adequate procedure, we made use of individual characteristics, captured on film by camcorder (de Oliveira and Monteiro-Filho 2008) or photographic camera. The photo-identification (Fig. 2.10) well known among cetacean researches, were originally based on film revelation (Würsig and Würsig 1977). Being non-invasive and therefore unlikely to alter natural behavior of the target species, the individual identification through images can answer questions related to individuals. De Oliveira and Monteiro-Filho (2008) reported the same individuals captured in image repeatedly in the same place: two individuals seen frequently close to a traditional local fishing trap, cerco-fixo, and a mother-calf association observed in two beaches in different islands. Medeiros (2014) presented similar results for both beaches (Fig. 2.11). In one of these areas, the Ponta da Trincheira, the author identified several individuals using the area, with no significant difference between them, while in the other, Praia do Pereirinha, fewer individuals were captured and all but two individuals were captured only once. These two individuals used this site significantly more than any other individual did, constantly accompanied by calves. Thereby, it is pos-

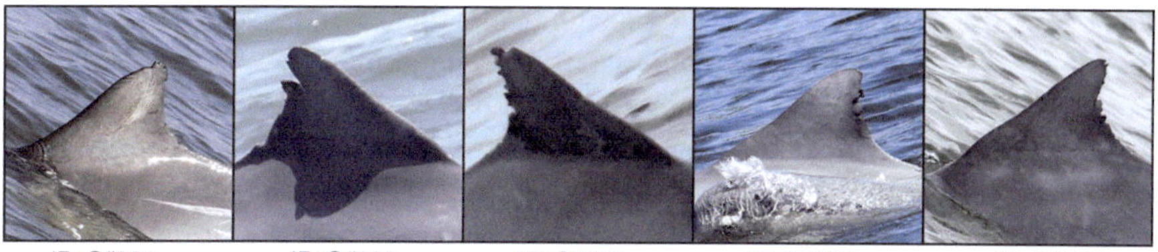

Fig. 2.10 Individuals of photo-identification catalogue of Instituto de Pesquisas Cananéia, used in studies of population dynamics, special patterns and social organization of Guiana dolphin's population of Cananéia estuary, State of São Paulo, Brazil

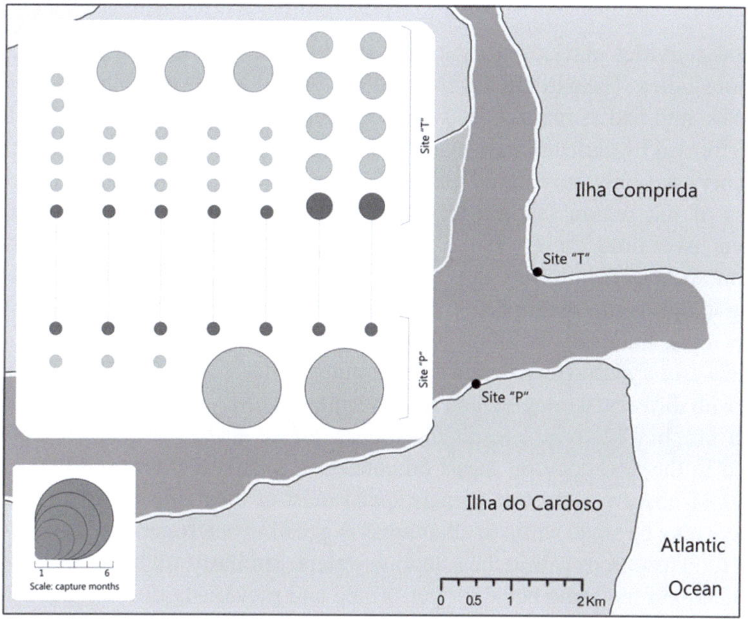

Fig. 2.11 Individual photo-identification in two close, but geomorphologically distinct sites; Site "T", known as "Ponta da Trincheira", and Site "P", known as "Praia do Pereirinha". *Light shaded circle* and *dark shaded diads* represents single individual captured exclusively in one or in both sites, respectively. Circle size refers to the number of months individuals were seen in each site, regardless the capture frequency within each month

sible to observe that dolphins use the area irregularly and some individuals may use more frequently one sector. Knowledge such as this may indicate different degree of fidelity for the area as a whole as well as for different sectors, besides assisting in the creation of management plans for conservation units according to local need of a better understanding of the area use of the Guiana dolphin.

Further, recent studies seek to understand what factors influence the occurrence and habitat use of Guiana dolphins in Cananéia region. Godoy et al. (2015) observed that dolphins concentrate mainly in the region close to the adjacent ocean. The authors verified that the generalized linear model with best fit presented salinity and depth with a positive correlation with the Guiana dolphin occurrence in the region. For a better understanding about the correlations between the species occurrence and environmental factors, the former researcher is developing additive generalized model to evaluate the use of habitat surrounding Cananéia Island, using several environmental variables and fish captured in the cerco-fixo fishing gear (Godoy et al. unpublished data). This study aims to evaluate anthropic interference in the distribution of Guiana dolphins through a predictive modelling.

2.7 Behavior

The first activities carried out with this species in the Cananéia region were the behavior studies. The naturalistic observations of the Guiana dolphin began early, in the 1980s, and had as main objective to document and describe the various behaviors performed by these animals, even facing various difficulties at the time. Initially, researchers did only land-based observations, since the geomorphological characteristics of the region favored this method of observation from some beaches. However, over time, there was the need to expand the observations of behavior in different areas of the estuary, thus scientists started to use different types of boats, making it possible to obtain data from different perspectives.

The more than 600 h of observations of different behaviors carried out in Cananéia estuary allowed a detailed description of the main behavioral events associated with different strategies, and measure them. The first set of descriptions of the Guiana dolphin behaviors resulted from an effort of 10 years of observations, detailed in the first doctoral thesis on cetaceans in Brazil (Monteiro-Filho 1991). The author observed that the animals spend most of their time foraging activities, and there may be small shifts to other areas suggesting heterogeneity in the distribution of food resources within the estuarine waters. Similarly, night observations also have identified the same behavior performed and previously described for the day periods (Atem and Monteiro-Filho 2006).

Monteiro-Filho (1991) described some basic behaviors such as swimming, shallow diving, deep diving, and rest, as well as a number of hunting strategies, individuals and those carried out in groups. Fishing behaviors described for Guiana dolphin range from more general behaviors such as pursuit (Fig. 2.12a), to more complex behaviors such as pursuit with bubbles (Fig. 2.12b). Beside, there are cooperative fishing activities, as circle formation (Fig. 2.12c), which may involve a large number of individuals fishing cooperatively on a shoal.

The descriptions of such behaviors opened the door to further studies to identify details in each behavior initially observed. Variations of these main behaviors received emphasis and then compared with behavior of populations from other estuaries in order to identify possible variations due to changes in environmental characteristics (Domit 2006). Monteiro-Filho (1992) observed interspecific interactions of the Guiana dolphin with at least five different species of marine and estuarine birds commonly spotted in the region: boobies (*Sula leucogaster*, Boddaert, 1783),

Fig. 2.12 Large variations of feeding behaviors were described for the population of Guiana dolphin at Cananéia estuary and Paranaguá estuary. (**a**) An individual pursuits a shoal of fishes and finalizes its attack with prey capture (a mullet, *Mugil* sp.) at water surface (Photo: Eric Medeiros). (**b**) Another more complex strategy used by dolphins during its foraging is the pursuit with bubbles, whereas the dolphin exhale air from lungs during a dive and creates a bubble barrier that encloses the shoal to opposite direction, where dolphins are waiting to prey on (Photo: Clarissa R. Teixeira). (**c**) Cooperative fishing is also observed as the circle formation in the picture (Photos: Eric Medeiros)

frigatebird (*Fregata magnificens*, Mathews, 1914), terns (*Sterna hirundinacea*, Lesson, 1831), gulls (*Larus dominicanus*, Lichenstein, 1983), and cormorants (*Phalacrocorax brasilianus*, Gmelin 1789; Fig. 2.13).

One notable feature of the Guiana dolphin behavior in Cananéia estuary is the interaction with the artisanal fishing traps, regionally called cerco-fixo (for more information on the operation of the cerco-fixo see Dias 1990; Oliveira and Hanazaki 2011). The dolphins use the fishing apparatus to aid in hunting strategies (Monteiro-Filho 1995; Louzada 2013; Fig. 2.14). In this positive interaction between local fishermen and dolphins, both sides benefit, as the Guiana dolphin is able to catch fish more easily and fishermen get higher fish catch in the presence of animals hunting near the cerco-fixo (Louzada 2010).

Over the years, the accumulation of hours of behavior recordings within the estuary found that dolphins used different sets of fishing behaviors in different areas. Such remarks oriented future studies to search which factors influence the behavioral patterns observed in the Guiana dolphin population in Cananéia estuary. After quantifying different fishing strategies in estuarine regions with different geomorphological characteristics, Louzada (2014) noted that dolphins adapt the fishing strategies to fit better in each distinct areas of the estuary. This finding helped to understand the reasons why dolphins presented some behaviors with particular characteristics at specific locations (adaptations due to site geomorphology) and different one across the estuary (adaptations due to the prey characteristics). Moreover, the different associations between individuals and the group composition, while foraging, was also a factor influencing the fishing strategies used. Therefore, the author suggested that the Guiana dolphin has a behavioral plasticity large enough that allows it to adapt their foraging strategies to different fishing locations and different group structures. This behavioral plasticity had already been suggested to explain small differences in descriptions presented by different authors in different geographic areas (Monteiro-Filho 2008).

Louzada (2014) also assessed whether the fishing strategies varied according to the different types of prey consumed, once that prey exert a great influence on many ecological patterns of their predators (Würsig 1986). By grouping the Guiana dolphin main prey in the region according to their ecological guilds (i.e. resident estuarine species, marine visitors, and juvenile marine species), vertical guilds (i.e. benthic, pelagic, nektonic demersal, and benthic-pelagic demersal), and shoal formation habit (i.e. forming small shoal, forming large shoals, and not forming shoals) and comparing them to the foraging strategies, no relationship was found between both factors. The author suggested, once again, that behavioral plasticity of the Guiana dolphin allows the development of foraging strategies specific to several types of prey and under different environmental conditions. Thus, the set of behavioral studies of dolphins in the region, has allowed us to show that the animal has an extensive repertoire of foraging strategies, leading it to fish effectively, on different guilds, in different regions of the estuary, and in different group structures. If such behaviors were not efficient in the process of prey capture and intake, the same would not be part of this vast repertoire and would be discarded long ago.

Studies such as this, that evaluate the types of prey consumed by the Guiana dolphin, were only possible thanks to years of beach monitoring in the Cananéia region in search of stranded carcasses. Carcasses collected serve as osteological material and as samples for natural history studies of the species. By analyzing the stomach contents (to study methods and analysis of stomach contents see Fitch and Brownell 1968), researchers determined that the Guiana dolphin feeds on a wide variety of prey, which together add up to 34 different species (29 teleost species, two species of cephalopods, and three species of crustaceans). However, even the dolphins being able to consume a large variety of prey, some particular item in their

Fig. 2.13 An interspecific interaction among Guiana dolphin and several species of seabirds is frequently observed at Cananéia estuary. At this picture is possible to see a large group of brown boobies (*Sula leucogaster*) foraging in association with a Guiana dolphin group (Photo: Caio N. Louzada)

Fig. 2.14 At Cananéia estuary, Guiana dolphins have a positive interaction with artisanal fishermen. At the estuarine waters, fixed fishing traps are built, which benefit dolphins foraging due to a limitation of escape routes for shoals during foraging. Groups of Guiana dolphins are found frequently using these traps during fishing activities (Photo: Caio N. Louzada)

diet have a greater relative importance of the total ingested items. These items possibly represent prey species that are abundant in the region during some periods, since the richness of teleost, crustaceans and cephalopods available for consumption is almost twice as much as the number of species ingested by the Guiana dol-

Fig. 2.15 One of most executed parental care behavior is the echelon swimming, in which calf swim side-by-side of its parental during whole dive and surfacing time. Most of times the behavior is synchronic and the calf keeps a physical contact with the adult (Photos: Caio N. Louzada)

phin in the region, showing that the dolphins are able to select some species for consumption (Louzada 2014).

One of the factors that strongly influences the fishing strategies is the presence of calves in the group (Monteiro-Filho 1991, 2000). Calves require great attention of parents, especially during the first months of life, when they are more vulnerable to environmental adversities. Rautenberg and Monteiro-Filho (2008) described in detail the key behaviors of parental care observed for the Guiana dolphin in the Cananéia estuary. These authors identified one of the key behaviors, as already described for other species of cetaceans, was the echelon swimming (Fig. 2.15). Other behaviors identified were the alternation, escort, meeting among families and crèche.

Later, it was possible to deepen the studies to understand how the quality and quantity of parental care varies according to the calves' ontogeny and the variations of parental care in different sites of the estuary. Teixeira et al. (*unpublished data*) observed that while offspring developed, contrary to what was initially expected, there is no decrease in the quantity of parental care effort, but a change in the quality. In the same way, extrinsic factors, such as changes in geomorphological characteristics of the site, are able to influence the frequency of care for the calves executed by parents (Teixeira 2013).

Parallel to the study of Guiana dolphin parental care behavior, some studies focused their observations on the behavior of infants and juveniles. The fact that births occur throughout the whole year in the Cananéia region (Schmiegelow 1990; Rosas and Monteiro-Filho 2002) favored behavioral studies of calves, rarely done in other regions of the country by the time. The calves are born after a gestation period ranging from about 11.5 to 12 month (Rosas and Monteiro-Filho 2002), with an average size of 88–98 cm long (Monteiro-Filho et al. 2008b). Through the observations, researchers characterized the behavior of infants, as has been observed for other species, by a high frequency of contacts between calves and parents (Domit 2002). The author also noted that the young are involved in many events of play, consisting of jumps, chases, and capture of possible prey. Many of the observations of these plays are associated with fishing learning periods (Domit 2002, 2006). Markedly, calves also perform various activities near the surface to perform a visual recognition of the area as to spot other individuals in the surroundings (Monteiro-Filho et al. 2008b). Finally, through the behavior, one can recognize four age classes for the Guiana dolphin calves classified as newborn, F1 infant, F2 infant, and juvenile (Fig. 2.16).

Fig. 2.16 The color patterns of calves skins are used to identify different ages classes. (**a**) A newborn shows a strong pinkish color the whole body. (**b**) F2 has an light pinkish color mainly at dorsal fin and flanks. (**c**) In calves, the skin pigmentation is almost complete and only a small region of dorsal fin remains pink, while the whole body is a dark-grey, like other adults (Photos: Eric Medeiros)

2.8 Acoustic

Our acoustic studies with the Guiana dolphin began in the 1980s and after 10 years of sampling, Monteiro-Filho and Monteiro (2001) described the specie's sound repertoire, categorizing sounds in whistles, calls, clicks and gargle (emitted by calves; Fig. 2.17), and even some complex sequences, as possible phrases. This was the first study to assess sounds produced by the species in natural environment, as prior studies only investigated Guiana dolphin in captivity (e.g. Caldwell and Caldwell 1970; Kamminga et al. 1993; Sauerland and Dehnhardt 1998).

Following the sound characterization for the species, studies have assessed the influence of noise generated by vessels, concluding that anthropogenic noise forces dolphins to raise the frequency range of the sounds used in social communication, seeking a new acoustic niche (Rezende 2008; Fig. 2.18). Also in relation to the noise produced by ships, Monteiro-Filho et al. (2008c) found that wooden boats and inboard engines (fishing vessels) generate low intensity noise and in frequencies below those used by dolphins. On the other hand, vessels with outboard engines (fishing sports, passenger transport, and tourism) generate high intensity noise with frequency equivalent to the lowest frequencies produced by dolphin, which may cause the phenomenon of interference (Fig. 2.18).

Despite advances in knowledge of the sound emission of the species and the context in which emission happen such as social, fishing, and nocturnal activities (Monteiro-Filho and Monteiro 2001, 2008; Rodrigues 2003; Atem and Monteiro-Filho 2006; Wanderley 2006; Oliveira et al. 2008; Rossi-Santos and Monteiro-Filho 2008), samples were specific and limited to 8 kHz and then to 20 kHz. Since such frequencies did not allowed the interpretation of some features of dolphins' acoustic universe, from 2011, with availability of better equipment and systematic sampling, we began evaluating sounds that reach sampling frequencies up to 48 kHz (Deconto and Monteiro-Filho 2013, 2016; Fig. 2.19).

The most recent studies have advanced in relation to the characterization of the sounds used in daytime and nighttime activities (Deconto and Monteiro-Filho 2016). The authors brought up a the new classification of dolphin sound emission in the region, braking down the calls, originally described by Monteiro-Filho and Monteiro (2001), into burst pulse sounds and low frequency narrow-band sounds

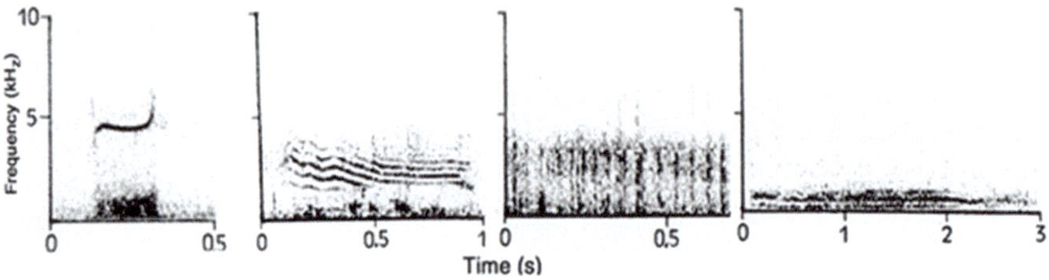

Fig. 2.17 Sounds produced by *Sotalia guianensis* and categorized by Monteiro-Filho and Monteiro (2001). From left to right: whistles, screams, clicks, gargle. Y axis represents frequency in kHz; X axis I time in seconds

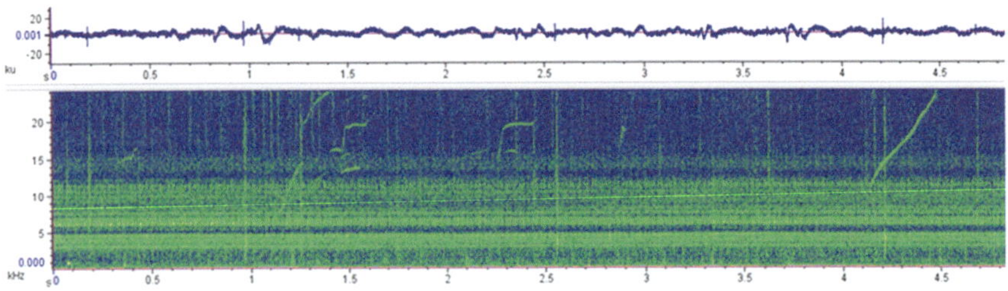

Fig. 2.18 A spectrogram highlighting the "noise" produced by a vessel equipped with an outboard motor, represented by the *green color* horizontal band, which reaches up to 15 kHz
Note that whistles and clicks produced by *Sotalia guianensis* are masked by noise. Y axis represents frequency in kHz; X axis time in seconds

(LFN sounds; Fig. 2.20). Moreover, authors demonstrated a greater amount of sounds emitted, excluding the clicks, with lower frequencies during the day than the night. Thus, the sound repertoire of the species is daylight dependent, with influence on their social communication.

Another result of the study phase that began in 2011 was the observation of high initial and minimum frequencies of Guiana dolphin whistles in the region compared to previous studies for the species in Brazil. Deconto and Monteiro-Filho (2013) recorded sounds with these parameters very close to those recorded in Costa Rica (May-Collado and Wartzok 2009). Until then, the most accepted hypothesis was that the frequencies of the Guiana dolphin whistles increased from higher latitudes toward areas of lower latitudes (Azevedo and Van Sluys 2005; Rossi-Santos and Podos 2006). Thus, reinforcing a hypothesis initially raised by Rossi-Santos and Podos (2006), Deconto and Monteiro-Filho (2013) present new evidence that the latitudinal variations reflect adaptations to the environment, a fact recently confirmed (Leão et al. 2015).

Continuing the acoustic study focusing on environmental influence on the spread of Guiana dolphin sounds, IPeC researchers have conducted experiments in natural environment both in the Northeast (Praia da Pipa) and in Southeastern Brazil

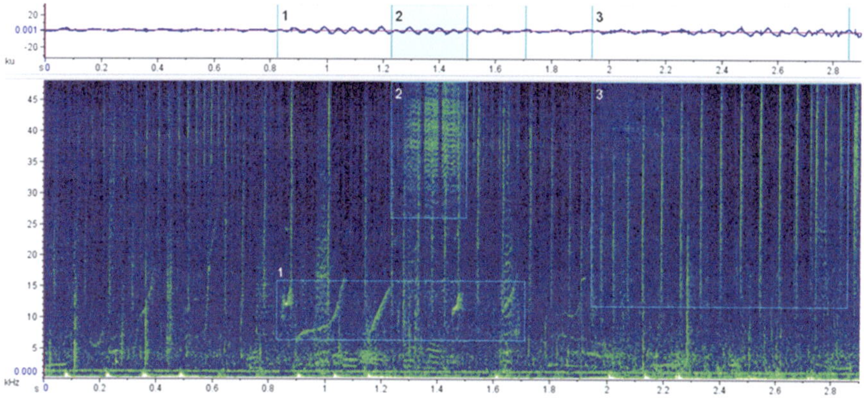

Fig. 2.19 A spectrogram showing the sampling frequency of 48 kHz, highlighting different sounds produced by *Sotalia guianensis*, which could occupy this acoustic niche. *1* Whistles; *2* burst pulses; *3* clicks. Y axis represents frequency in kHz; X axis time in seconds

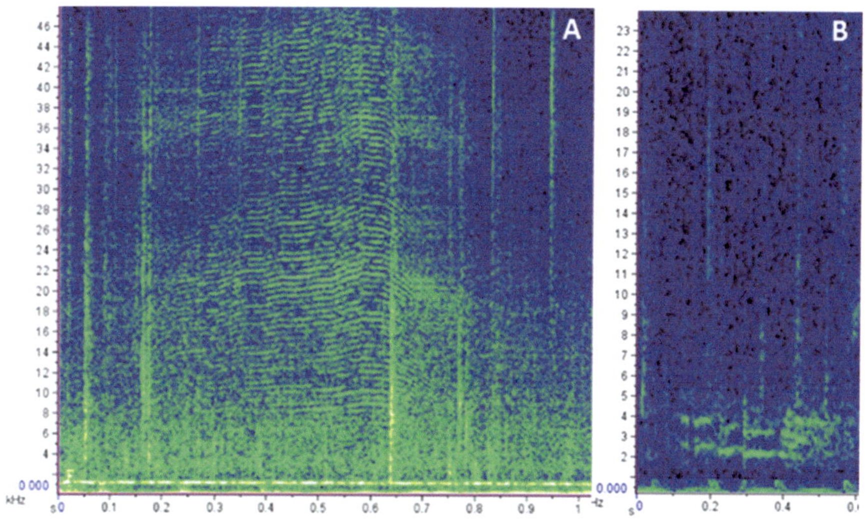

Fig. 2.20 The calls produced by *Sotalia guianensis* described by Monteiro-Filho and Monteiro (2001) were subdivided into two categories by Deconto and Monteiro-Filho (2016), burst-pulse sounds (**a**) and low frequency narrow-band sounds (**b**)

(Cananéia). The two study areas were chosen because they are two different habitats with respect to environmental and geomorphological parameters, which we expect that after the analysis, will recognize the influence of latitude, temperature, salinity, depth, turbidity and the sound source of the distance in the repertoire Guiana dolphin sound.

2.9 Conservation

Throughout this period of study, information on the biology and ecology of the estuarine dolphin allowed us to understand some of its needs in Cananéia estuary as well as the various interactions of this species with humans and their technologies.

Cananéia is within an area of great biological and scenic abundance ["World Natural Heritage Site" and "Atlantic Forest Biosphere Reserve" (UNESCO 1992, 1999)]. Here, two activities are widely developed: artisanal fishing and tourism. Both activities involve the use of vessels with vary propeller engines. Even considering that the estuary is large and the number of vessels is not very big, we could detect different reactions of the Guiana dolphin in relation to wooden vessels with inboard engine, aluminum vessels with outboard motors, and fiber sport vessels (Fig. 2.21). The study showed that the central engine of wooden vessels produce low intensity noise in the water, because wood is a great sound conductor and disperses noise produced into the air. Thus, boats used on local artisanal and some wooden boats used in tourism, produce low intensity noise and are not fast, allowing Guiana dolphin to control proximity (Monteiro-Filho et al. 2008c). However, the same can not be said for vessels with outboard engines, which sometimes alter the behavior of the dolphins (Monteiro-Filho et al. 2008c). Aiming to minimize the possible impact of noise produced by vessels, a series of tests were done and so generated a simple proposal, by reducing the speed when the conductor sighted the dolphins, noise was also reduced and dolphins were able to manage the approach of vessels.

After these studies, new questions began to arise. How many vessels could operate simultaneously with tourism activities? How close these vessels could approach? How should captains proceed before dolphin groups? In search of these answers, studies were conducted for about 3 years evaluating possible interactions between vessels and Guiana dolphin, how vessels were conducted, and periods and distances that these vessels should remain without interfering with the dolphins' activities (Fig. 2.22).

Fig. 2.21 Different kinds of vessels navigate at Cananéia and Paranaguá estuary waters, Brazil. (**a**) Fishing boats are made of wood equipped with inboard engine. (**b**) The dolphin-watching tourism are performed mainly by small aluminum boats equipped with outboard motor (Photos: Eric Medeiros)

Fig. 2.22 An informative poster was created as a public politic activity developed by Projeto Boto-Cinza in order to advertise the population about the best way to approach and observe dolphins with minimum impact created by vessels

The results produced with the studies on the production of noise (Monteiro-Filho et al. 2008c), added to onboard tourism activities (Filla and Monteiro-Filho 2009 a, b, c), on the estuarine dolphin economic valuation in the estuary of Cananéia (Filla et al. 2012), subsidized municipal legislation that established procedures and restrictions of nautical activities in the region (municipal Law No. 2129/2011), without, however, interfering with commercial activities.

Currently, our data is constantly requested by conservation unit managers in the region for the proposal of nautical activities establishment and standardization. In the year 2015, during the development of the Management Plan of the Cananéia-Iguape-Peruibe Environmental Protection Area (ICMBio 2016), the results of works on distribution and space use by the estuarine dolphin in Cananéia estuary were presented to sustain the creation of a Sustainable Use and Protection of Cetaceans Estuarine Zone (ZEUSPC). Through a technical feedback by some of the project's researchers, the total area of the special zoning was expanded to include areas with considerable concentrations of groups, previously not been taken into account.

References

Atem ACG, Monteiro-Filho ELA (2006) Nocturnal activity of the Estuarine dolphin, *Sotalia guianensis* (CETACEA, DELPHINIDAE), in the region of Cananéia, São Paulo State, Brazil. Aquat Mamm 32:236–241

Azevedo AF, van Sluys M (2005) Whistles of tucuxi dolphins (*Sotalia fluviatilis*) in Brazil: comparisons among populations. J Acoust Soc Am 117:1456–1464

Bigarella JJ (1978) A Serra do Mar e a porção do Estado do Paraná: Um problema de segurança ambiental e nacional (contribuição à geografia, geologia e ecologia regional). Secretaria de Estado do Planejamento, Associação de Defesa e Educação Ambiental (ADEA), Curitiba, p 248

BISI TL (2001) Estimativa da densidade populacional do boto-cinza *Sotalia guianensis* (CETACEA, DELPHINIDAE) na região estuarina lagunar de Cananéia, SP. Undergraduation thesis. Universidade Estadual Paulista. Rio Claro, SP

Caldwell DK, Caldwell MC (1970) Echolocation-type signals by two dolphins, genus *Sotalia*. Q J Florida Acad Sci 33:124–131

Carvalho CT (1963) Sobre um boto comum no litoral do Brasil. (Cetacea:Delphinidae). Rev Bras Biol 23:263–276

Cunha HÁ, da Silva VMF, Lailson-Brito J Jr, Santos COM, Flores PAC, Martin AR, Azevedo AF, Fragoso ABLZ, Zanelatto RC, Solé-Cava AM (2005) Riverine and marine ecotypes of *Sotalia* dolphins are different species. Mar Biol 148:449–457

de Oliveira FC, Hanazaki N (2011) Ethnobotany and ecological perspectives on the management and use of plant species is the traditional fishing trap, southern coast of São Paulo, Brazil. J Environ Manage 92:1783–1792

De Oliveira LV, Monteiro-Filho ELA (2008) Individual identification and habitat use of the estuarine dolphin *Sotalia guianensis* (Cetacea:Delphinidae) in Cananéia, south-eastern Brazil, using video images. J Mar Biol Assoc UK 88:1199–1205

Deconto LS, Monteiro-Filho ELA (2013) High initial and minimum frequencies of *Sotalia guianensis* whistles in the southeast and south of Brazil. J Acoust Soc Am 134:3899–3904

Deconto LS, Monteiro-Filho ELA (2016) Day and night sounds of the Guiana dolphin, *Sotalia guianensis* (Cetacea:Delphinidae) in southeastern Brazil. Acta Ethol 19:61–68

Dias ERA (1990) Produtividade de um cerco–fixo no Parque Estadual da Ilha do Cardoso, na região lagunar– estuarina de Cananéia, SP. In Livro de Resumos. II Simpósio de Ecossistemas da Costa Sul e Sudeste Brasileira: estrutura, função e manejo Águas de Lindóia, SP: Academia de Ciências de São Paulo, p 400

Diegues AC (1987) Conservação e desenvolvimento sustentado de ecossistemas litorâneos no Brasil. São Paulo, Secretaria Estadual do Meio Ambiente de São Paulo, p 46

Domit C (2002) Comportamento de filhotes de *Sotalia guianensis* (Cetacea: Delphinidae), na região do Complexo Estuarino Lagunar de Cananéia, São Paulo. Undergraduation thesis, Universidade Estadual de Londrina, Londrina, PR, Brasil

Domit C (2006) Comportamento de pesca do boto-cinza, *Sotalia guianensis* (van Bénéden, 1864). Master thesis, Universidade Federal do Paraná, Curitiba, PR, Brasil

Edwards H, Schnell G (2001) Status and ecology of *Sotalia fluviatilis* in the Cayos Misquitos Reserve, Nicaragua. Mar Mamm Sci 17:445–472

Filla GF, Monteiro Filho ELA (2009a) A contribuição do olhar e da percepção do turista na conservação de áreas naturais: o turismo de observação do boto-cinza na região de Cananéia (SP). OLAM – Ciênc Tecnol 9(1):243–269

Filla GF, Monteiro Filho ELA (2009b) Monitoring tourism schooners observing estuarine dolphins (*Sotalia guianensis*) in the Estuarine complex of Cananéia, south-east Brazil. Aquat Conserv Mar Freshw Ecosyst 19:772–778

Filla GF, Monteiro Filho ELA (2009c) O desenvolvimento do turismo náutico e a sua ligação com a observação do Boto-cinza (*Sotalia guianensis*) na região de Cananéia, litoral sul do Estado de São Paulo. Turismo em Análise 20(2):282–301

Filla GF, de Oliveira CB, Gonçalves JM, Monteiro-Filho ELA (2012) The economic evaluation of estuarine dolphin (*Sotalia guianensis*) watching tourism in the Cananéia region, south-eastern Brazil. Int J Green Econ 6:95–116

Fitch JE, Brownell RJ Jr (1968) Fish otoliths in cetacean stomachs and their importance in interpreting feeding habits. J Fish Res B Can 25:2561–2574

Gervais FLP, Deville E (1853) Sur les mammiféres marins qui fréquentent les côtes de la France et plus particulìerement sur une novelle espéce de dauphins propre a la Méditerranés. Bulletin Sociéte Centrale d'Agriculture et des Comices Agricoles du Département de l'Herault, Montpellier, pp 140–155

Giacometti L (1967) The skin of the whale (*Balaenoptera physalus*). Anat Rec 159:69–76

Godoy DF, Andriolo A, Filla GF (2015) The influence of environmental variables on estuarine dolphins (*Sotalia guianensis*) spatial distribution and habitat used in the Estuarine Lagunar Complex of Cananéia, southeastern Brazil. Ocean Coastal Mange 106:68–76

Havukainen L, Monteiro-Filho ELA, Filla GF (2011) Population density of *Sotalia guianensis* (Cetacea: Delphinidae) in the Cananéia region, Southeastern Brazil. Rev Biol Trop 59:1275–1284

Hershkovitz P (1966) Catalog of living whales. Smithsonian Institution Press, Washington, DC

ICMBio (2016) Plano de Manejo da Área de Proteção Ambiental Cananéia-Iguape-Peruíbe, SP. Available in: http://www.icmbio.gov.br/portal/unidadesdeconservacao/biomas-brasileiros/marinho/unidades-de-conservacao-marinho/2241-apa-de-cananeia-iguape-peruibe. Accessed in 1 Oct 2016

IUCN (1984) Threatened protected areas of the world. IUCN Commission on National Parks and Protected Areas, Gland, Switzerland

Kamminga C, van Hove MT, Engelsma FJ, Terry RP (1993) Investigations on cetacean sonar X: a comparative analysis of underwater echolocation clicks of *Inia* spp. and *Sotalia* spp. Aquat Mamm 19:31–43

Lana PC (1986) Macrofauna bêntica de fundos sublitorais não consolidados da Baía de Paranaguá (Paraná). Nerítica 1:79–89

Leão DT, Monteiro-Filho ELA, Silva FJL (2015) Acoustic parameters of sounds emitted by *Sotalia guianensis*: dialects or acoustic plasticity? J Mammal 97:611–618

Louzada CN (2010) Interação entre *Sotalia guianensis* (van Bénéden, 1864) e a arte pesqueira artesanal de cerco-fixo no complexo estuarino-lagunar de Cananéia. Undergraduation thesis, Universidade Estadual de Maringá, Maringá

Louzada CN (2013) How do Guiana dolphin (*Sotalia guianensis*), from the Cananéia estuary in state of São Paulo, use cerco-fixo fish traps in their fishing activities? Revista de Etologia 12:18–24

Louzada CN (2014) Características das estratégias de pesca do boto-cinza (*Sotalia guianensis*) em diferentes regiões do complexo Estuarino-lagunar de Cananéia. Master thesis, Universidade Federal do Paraná, Curitiba

May-Collado LJ, Wartzok D (2009) A characterization of Guyana dolphin (*Sotalia guianensis*) whistles from Costa Rica: the importance of broadband recording systems. J Acoust Soc Am 125:1202–1213

Medeiros E (2014) Different types of individual habitat use in two close sites by Guiana dolphin (*Sotalia guianensis*) in the Cananéia Estuary- Brazil. Master thesis, Universidade Federal do Paraná, Curitiba

Mendonça JT, Katsuragawa M (2001) Caracterização da pesca artesanal no complexo estuarino-lagunar de Cananéia-Iguape, Estado de São Paulo, Brasil (1995–1996). Acta Sci Biol Sci 23:535–547

Miranda LB, Mesquita AR, França CAS (1995) Estudo da circulação e dos processos de mistura no extremo sul do mar de Cananéia: condições de dezembro de 1991. Bol Inst Ocean 43:153–164

Miyao SY, Sarti CC (1986) Características Físicas e Químicas Do Sistema Estuarino-Lagunar De Cananéia-Iguape. Bol Inst Ocean 34:23–36

Monteiro-Filho ELA (1991) Comportamento de caça e repertório sonoro do golfinho *Sotalia brasiliensis* (Cetacea:Delphinidae) na região de Cananéia, Estado de São Paulo. PhD thesis, Universidade de Campinas, Campinas

Monteiro-Filho ELA (1992) Pesca associada entre golfinhos e aves marinhas. Rev Bras Zool 9:29–37

Monteiro-Filho ELA (1995) Pesca interativa entre o golfinho *Sotalia fluviatilis guianensis* e a comunidade pesqueira da região de Cananéia. Bol Inst Pesca 22:15–23

Monteiro-Filho ELA (2000) Group organization of the dolphin *Sotalia fluviatilis guianensis* in an estuary of southeastern Brazil. Ciên Cult 52:97–100

Monteiro-Filho ELA (2008) Comportamento de Pesca. In: Monteiro-Filho ELA, Monteiro KDKA (eds) Biologia. Ecologia e Conservação do Boto-cinza. Páginas & Letras, São Paulo, pp 77–89

Monteiro-Filho ELA, Monteiro KDKA (2001) Low-frequency sounds emitted by *Sotalia fluviatilis guianensis* (Cetacea: delphinidae) in an estuarine region in southeastern Brazil. Can J Zool 79:59–66

Monteiro-Filho ELA, Monteiro KDKA (2008) Repertório Sonoro. In: Monteiro-Filho ELA, Monteiro KDKA (eds) Biologia, Ecologia e Conservação do Boto-cinza. Páginas & Letras, São Paulo, pp 67–74

Monteiro-Filho ELA, Monteiro LR, Reis SF (2002) Skull shape and size divergengein the genus *Sotalia*: a tridimensional morphometric analysis. J Mammal 83:125–134

Monteiro-Filho ELA, Monteiro LR, Reis SF (2008a) Sistemática e distribuição de Sotalia. In: Monteiro-Filho ELA, Monteiro KDKA (eds) Biologia, Ecologia e Conservação do Boto-cinza. Páginas & Letras, São Paulo, pp 3–10

Monteiro-Filho ELA, Neto MMS, Domit C (2008b) Comportamento de infantes. In: Monteiro-Filho ELA, Monteiro KDKA (eds) Biologia, Ecologia e Conservação do Boto-cinza. Páginas & Letras, São Paulo, pp 51–65

Monteiro-Filho ELA, Gonçalves MLRL, Keinert AC, Sasaki G (2008c) Comportamento como bioindicador das interações entre Boto-Cinza e o homem (Tráfego de embarcações). In: Rossi-Santos MA, Reis MSS (eds) II Workshop do Nordeste: Pesquisa e Conservação de *Sotalia guianensis*, pp 129–136

Oliveira F, Rodrigues RG, Monteiro-Filho ELA (2008) Atividade noturna. In: Monteiro-Filho ELA, Monteiro KDKA (eds) Biologia, Ecologia e Conservação do Boto-cinza. Páginas & Letras, São Paulo, pp 157–163

Pfeiffer CJ, Jones FM (1993) Epidermal lipid in several cetacean species: ultrastructural observations. Anat Embryol 188:209–218

Ramos R (1997) Determinação de idade e biologia reprodutiva de *Pontoporia blainvillei* e de *Sotalia fluviatilis* (Cetacea: Pontoporiidae e Delphinidae) no Norte do Rio de Janeiro. Tese de Mestrado. Universidade Estadual do Norte Fluminense, Campos dos Goytacazes, Rio de Janeiro

Randi MAF, Rassolin P, Rosas FCW, Monteiro-Filho ELA (2008) Padrão de cor da pele. In: Monteiro-Filho ELA, Monteiro KDKA (eds) Biologia, Ecologia e Conservação do Boto-cinza. Páginas & Letras, São Paulo, pp 11–16

Rautenber M, Monteiro-Filho ELA (2008) Cuidado Parental. In: Monteiro-Filho ELA, Monteiro KDKA (eds) Biologia, Ecologia e Conservação do Boto-cinza. Páginas & Letras, São Paulo, pp 67–74

Rezende F (2008) Alterações acústico-comportamentais. In: Monteiro-Filho ELA, Monteiro KDKA (eds) Biologia, Ecologia e Conservação do Boto-cinza. Páginas & Letras, São Paulo, pp 155–176

Rice DW (1977) A list of the marine mammals of the world. 3rd ed. NOAA technical report, National Marine Fisheries Services 711:1–15

Rodrigues RG (2003) Análise comparativa das emissões sonoras do boto-cinza (*Sotalia guianensis*) nos períodos diurno e noturno. Undergraduation thesis, Universidade Federal do Paraná. Curitiba, Paraná

Rosas FCW (2000) Interações com a pesca, mortalidade, idade, reprodução e crescimento de *Sotalia guianensis* e *Pontoporia blainvillei* (Cetácea, Delphinidae) no litoral sul do Estado de São Paulo e litoral do Estado do Paraná, Brasil. PhD thesis, Universidade Federal do Paraná. Curitiba, Paraná

Rosas FCW, Barreto AS (2008) Reprodução e crescimento. In: Monteiro-Filho ELA, Monteiro KDKA (eds) Biologia, Ecologia e Conservação do Boto-cinza. Páginas & Letras, São Paulo, pp 51–65

Rosas FCW, Monteiro-Filho ELA (2002) Reproduction of the estuarine dolphin (*Sotalia guianensis*) on the coast of Paraná, Southern Brazil. J Mammal 83(2):507–515

Rosas FCW, Barreto AS, Monteiro-Filho ELA (2003) Age and growth of the estuarine dolphin (*Sotalia guianensis*) (Cetacea, Delphinidae) on the Paraná coast, southern Brazil. Fish Bull 101:377–383

Rossi-Santos MA, Monteiro-Filho ELA (2008) Panorama atual de bioacústica com o boto-cinza. In: Rossi-Santos MA, Reis MSS (eds) II Workshop do Nordeste: Pesquisa e Conservação de *Sotalia guianensis*, pp 144–150

Rossi-Santos MR, Podos J (2006) Latitudinal variation in whistle structure of the estuarine dolphin *Sotalia guianensis*. Behaviour 143:347–364

Sauerland M, Dehnhardt G (1998) Underwater audiogram of tucuxi (*Sotalia fluviatilis guianensis*). J Acoust Soc Am 103:1199–1204

Schaeffer-Novelli Y, Mesquita HSL, Cintrón-Molero G (1990) The Cananéia lagoon estuarine system, São Paulo, Brazil. Estuaries 13:193–203

Schmiegelow JMM (1990) Estudo sobre cetáceos odontocetes encontrados em praias da região entre Iguape (SP) e Baía de Paranaguá (PR) (24° 42′S – 25o 28°S) com especial referência a *Sotalia fluviatilis* (Gervais, 1853) (Delphinidae). Dissertação de Mestrado, Universidade de São Paulo, São Paulo, SP, Brasil

Simões-Lopes PC (1988) Ocorrência de uma população de *Sotalia fluviatilis* Gervais 1853, (Cetacea, Delphinidae) no limite sul de sua distribuição, Santa Catarina, Brasil. Biotemas 1:57–62

Simpson JG, Gardner MB (1972) Comparative microscopic anatomy of selected marine mammals. In: Ridgeway T (ed) Mammals of the sea. Eds. Springfield, pp 298–418

Sokolov W (1960) Some similarities and dissimilarities in the structure of the skin among the members of the suborders Odontoceti and Mystacoceti (Cetacea). Nature 185:745–747

SOS Mata atlântica (2013) Atlas dos remanescentes florestais da Mata Atlântica período 2011–2012 Relatório Técnico. Fundação SOS Mata Atlântica e Instituto Nacional de Pesquisas Espaciais, São Paulo

Sydney NV, Monteiro-Filho ELA (2011) Efficiency of wear and decalcification technique for estimating the age of estuarine dolphin *Sotalia guianensis*. J Biosci 36(1):117–121

Teixeira CR (2013) Cuidado parental em boto-cinza, *Sotalia guianensis*, na região do complexo estuarino lagunar de Cananéia, Estado de São Paulo. Master thesis, Universidade Federal do Paraná, Curitiba

UNESCO (1992) United Nations Educational, Scientific and Cultural Organization – man and the Biosphere (MAB) programme. Available in: http://www.Unesco.Org/mabdb/br/brdir/directory/biores.Asp?Mode=gen&code=BRA+01. Accessed in 1 Oct 2016

UNESCO (1999) United Nations Educational, Scientific and Cultural Organization – world Heritage convention. Available in: http://whc.Unesco.Org/en/list/892. Accessed in 1 Oct 2016

Wanderley RP (2006) Análise das emissões sonoras de *Sotalia guianensis* (CETACEA: DELPHINIDAE) em alto mar, no litoral norte do Estado do Paraná, Brasil. Undergraduation thesis, Universidade Católica do Paraná, Curitiba, Paraná

Wanderley RP (2013) Estrutura etária do boto-cinza, *Sotalia guianensis*, no sul de sua distribuição: Uma avaliação dos encalhes e interação com redes de pesca. Master thesis, Universidade Federal do Paraná, Curitiba

Würsig B (1986) Delphinid foraging strategies. In: Schusterman RJ, Thomas JA, Wood FG (eds) Dolphin Cogn. Behav. A Comp. approach, 1st ed. Lawrence Erlbaum Associates, Hillsdale, pp 347–359

Würsig B, Würsig M (1977) The photographic determination of group size, composition, and stability of coastal porpoises (*Tursiops truncatus*). Science 198:755–756

Chapter 3
An Integrated Framework to Assess the Carrying Capacity of Humpback Whale-Watching Tourism in Praia do Forte, Northeastern Brazil

Luena Fernandes and Marcos R. Rossi-Santos

Abstract Whale-watching tourism has been growing in recent years in Praia do Forte, northeastern Brazil and to manage this activity in a sustainable manner it is important to determine its carrying capacity. However, an assessment of whale-watching carrying capacity requires an understanding of the biological, social and economic dimensions, as well as consideration of the several stakeholders involved and the current management capacity in place at a specific locality. Here we analyse the development of this activity in Brazil and Praia do Forte in recent years, the current state of knowledge about the potential impacts of whale-watching on humpback whales and the concept of tourism carrying capacity as it applies to whale-watching. We then bring all of these components together to propose an integrated framework to begin its assessment for Praia do Forte, northeastern Brazil. We conclude that whale-watching in Praia do Forte is currently a low-impact activity, with no evident deleterious biological effects on the target population although we still lack all necessary information to establish its carrying capacity. Therefore, we suggest limiting the activity to its current level, until more data and management capacity become available, and make recommendations to improve our understanding of the impacts of this activity and its sustainable management in this locality.

L. Fernandes (✉)
Projeto Baleia Jubarte/Instituto Baleia Jubarte, Av. do Farol, 2907, Praia do Forte, Mata de São João, Bahia 48287-000, Brazil

Programa de Pos-Graduacao em Ecologia e Biomonitoramento, Universidade Federal da Bahia, Rua Barao de Jeremoabo, 147, Ondina, Salvador, Bahia 44170-110, Brazil
e-mail: luena.fernandes@baleiajubarte.org.br

M.R. Rossi-Santos

Programa de Pos-Graduacao em Ecologia e Biomonitoramento, Universidade Federal da Bahia, Rua Barao de Jeremoabo, 147, Ondina, Salvador, Bahia 44170-110, Brazil

Laboratório de Ecologia Acústica e Comportamento Animal, Centro de Ciências Agrárias, Ambientais e Biológicas, Universidade Federal do Recôncavo da Bahia, Rua Rui Barbosa, 710, Centro, Cruz das Almas, Bahia 48380-000, Brazil

M.R. Rossi-Santos, C.W. Finkl (eds.), *Advances in Marine Vertebrate Research in Latin America*, Coastal Research Library 22,
DOI 10.1007/978-3-319-56985-7_3

3.1 Introduction

Whale-watching is defined by the International Whaling Commission (IWC) as "any commercial enterprise which provides for the public to see cetaceans (whales, dolphins and porpoises) in their natural habitat" (IWC 1994). This activity started in California in the 1950s (Hoyt 1984, 2009) but has been expanding rapidly worldwide since the 1980s (Hoyt 1995, 2001), when the International Whaling Commission called a moratorium on all commercial whaling (IWC 2014). Since then, whale-watching has been largely promoted by businesses and NGO's as an economically viable and sustainable use of cetaceans (Woods-Ballard et al. 2003), recognized as having significant social, economic and educational benefits (e.g. Duffus and Dearden 1990; Forestell 1993; Hoyt 2001; Stamation et al. 2007; Brumatti 2013).

The most recent statistics (O'Connor et al. 2009) revealed that 119 countries worldwide develop whale-watching, involving 13 million tourists, and generating $2.1 billion year^{-1} and an estimated 13,200 jobs. This activity has been growing globally at an average 3.7% year^{-1}, however, on a regional scale, for example in South America, whale-watching has been increasing an average 10% year^{-1} and is now developed by the majority of countries in the continent, with Brazil alone receiving 200,000 tourists in 2008.

This rapid development, however, has been raising concerns over its impacts on cetacean populations (e.g. Parsons 2012; Rossi-Santos 2016) and its long-term sustainability, as its growth has outpaced management in several locations (Forestell 2007; Higham et al. 2015; New et al. 2015).

Although legislation, guidelines and codes-of-conduct have been developed around the globe in an attempt to reduce the potential negative impacts of whale-watching (see Carlson 2012), these are often voluntary, and thus largely ineffective (Allen et al. 2007; Wiley et al. 2008), or unenforced, resulting in non-compliance (Scarpaci et al. 2003; Lusseau 2005; Kessler and Harcourt 2013).

In Brazil, there has been specific legislation to avoid the intentional molesting of cetaceans and regulate whale-watching tourism since 1990. However, these rules are generally unenforced and cases of non-compliance have been documented (Instituto Baleia Jubarte, unpublished data).

To be genuinely sustainable whale-watching must be ecologically, socially and economically sustainable, and this requires an integrated approach to its management, based on the precautionary principle and considering it as a non-lethal consumptive activity (Berrow 2003; Higham et al. 2015). In this context, when the number of operators and vessels is limited through a permit system, there is greater potential for the sustainable management of whale-watching (Berrow 2003). As suggested by Higham et al. (2009), a permit system must be based on the scientific assessment of the carrying capacity for a specific location, and include regular assessment and reapplication. However, obtaining adequate scientific data for this purpose is not an easy task (Ahn et al. 2002).

In Brazil, most studies on humpback whale behaviour (e.g. Petta et al. 2002; Morete 2003a; Bisi 2006; Morete 2007) and interaction with whale-watching tourism (e.g. Morete et al. 2000; Morete 2003b; Morete et al. 2007) have been carried out in the Abrolhos National Marine Park, southern Bahia state, for its ecological importance (Martins et al. 2013). In fact, whale-watching in Brazil started in the Abrolhos region but it is on the state's northern coast, specifically in Praia do Forte, that this activity has shown the greatest increase in recent years (Cipolotti et al. 2005; Brumatti 2008). Therefore, there is an urgent need for more studies to subsidize the management of this activity in this region.

To begin to understand and address the issue of whale-watching carrying capacity in Praia do Forte, we will look at the development of the activity in this location and its impacts on the whales, the tourists and the local operators. Then we analyse the concept of carrying capacity as it applies to whale-watching.

By bringing all these components together, we propose an integrated framework to assess the carrying capacity of commercial humpback whale-watching tourism in Praia do Forte, northeastern Brazil, taking into account the biological, social and economic aspects involved.

3.2 From Humpback Whaling to Whale-Watching in Brazil

Modern commercial whaling during the twentieth century killed over 200,000 humpback whales (*Megaptera novaeangliae*) in the southern hemisphere (Findlay 2001; Rocha et al. 2014). The southwestern Atlantic humpback whale population, classed as southern Atlantic breeding stock A (IWC 2005), was depleted to less than 4% of its pre-exploitation size in the late 1950s (Zerbini et al. 2011). In 1966, the IWC prohibited commercial humpback whaling. The currently available global humpback whale population estimate is 60,000 individuals (Reilly et al. 2008).

By the time Brazil permanently banned whaling in 1987 (Brasil 1987), humpback whales were considered extinct from Brazilian waters, as a direct result of decades of commercial whaling along the Brazilian coast and the Southern Ocean. However, shortly after the whaling prohibition, a small remnant humpback population was 'discovered' by scientists working in the Abrolhos National Marine Park, southern Bahia state (IBAMA/NEMA 1990), and, in 1988, the Projeto Baleia Jubarte (Humpback Whale Project) was created to study this population and promote its recovery, setting up its headquarters in the former whaling town of Caravelas. Subsequently, Abrolhos was recognised as the most important humpback whale breeding ground in the southwestern Atlantic Ocean (Engel and Ades 1996; Martins et al. 2001; Andriolo et al. 2006, 2010). In the early 1990s, humpback whale-watching started in the Abrolhos region as an opportunistic activity carried out by dive operators taking tourists to the archipelago during the winter months (Morete et al. 2000) (Fig. 3.1).

Since the whaling moratorium this population has been recovering from exploitation at an estimated 7.4–15.2% annual rate (Ward et al. 2011; Wedekin 2011;

Fig. 3.1 Humpback whale-watching started in the Abrolhos region, the species' main breeding ground in the southwestern Atlantic ocean (Photos: Enrico Marcovaldi/Instituto Baleia Jubarte)

Zerbini et al. 2011). As a result of this recovery and reoccupation of former breeding range along the Brazilian coast (Mas-Rosa et al. 2002; Pacheco de Godoy et al. 2005; Baracho et al. 2006; Rossi-Santos et al. 2008), from the early 2000s sightings of this species became increasingly frequent along the northern coast of Bahia state. In order to widen its study area and monitor the humpbacks in this region the Projeto Baleia Jubarte installed its second station in the coastal village of Praia do Forte, northern Bahia, in 2001.

Although it was recently down-listed to Near Threatened category in Brazil (ICMBio 2014), this population remains a fraction (17,000 individuals; Instituto Baleia Jubarte, unpublished data) of its hypothesized size prior to commercial whaling (22,000–28,000; Zerbini et al. 2011) and, therefore, demands continued conservation efforts.

Fig. 3.2 Praia do Forte is an excellent whale-watching destination because humpback whales can be watched very close to shore, due to a narrowing of the continental shelf in this region and the whales' preference for shallower waters (Photo: Luena Fernandes/Instituto Baleia Jubarte)

3.2.1 Humpback Whale-Watching in Praia do Forte

The first whale-watching operation in Praia do Forte began in 2001, in a partnership between the Projeto Baleia Jubarte and a local tourism operator (Cipolotti et al. 2005). Praia do Forte has since become a strong whale-watching destination. This is partly due to its proximity to the state capital Salvador (55 km), one of the most sought after tourist destinations in Brazil, and partly because the whales occur very close to the shore, due to a narrowing of the continental shelf in this region (Rossi-Santos et al. 2008; Wedekin et al. 2010; Baracho-Neto et al. 2012) (Fig. 3.2). Since then, the activity has spread along the coast of Bahia and is now carried out in more than ten localities, however, Praia do Forte is where the activity has shown the fastest growth in recent years (Brumatti 2008).

Additionally, the northern coast of Bahia is considered to have the greatest tourism development potential in the state (Limonad 2007). Increasing visitation to Praia do Forte in the coming years is thus very likely to increase the demand for whale-watching and, consequently, the number of operators, creating the need to manage the number of whale-watching boats operating in the area in order to avoid biological damage to the whale population and maintain tourist satisfaction.

In 2003, the North Coast Continental Shelf Protected Area (Área de Proteção Ambiental da Plataforma Continental do Litoral Norte) was created to safeguard the region's rich marine biodiversity, promote the development of economic activities within acceptable limits of change of the ecosystems and improve the quality of life of the coastal communities using the area (Brasil 2003). This Protected Area (PA)

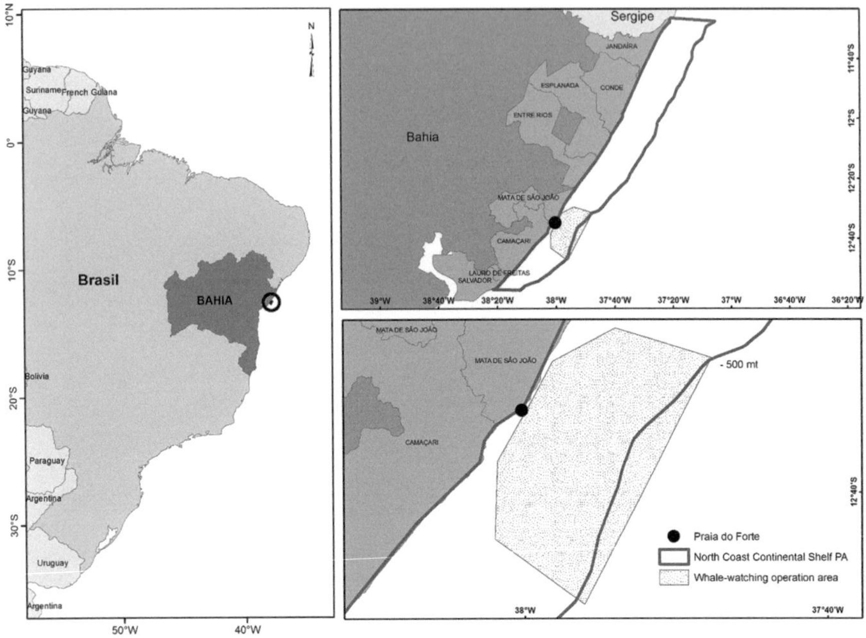

Fig. 3.3 Map showing the state of Bahia, northeastern Brazil, and the port of Praia do Forte, within the North Coast Continental Shelf Protected Area. The *shady polygon* represents the operation area of the whale-watching vessels departing from the port of Praia do Forte (Adapted from Cipolotti 2013)

covers approximately 3,622 km^2 from Itapuã (Salvador) to the border with Sergipe state, from the high tide line to the 500 m isobaths (Fig. 3.3). The main characteristic of this PA is a narrow continental shelf, approximately 15 km wide, with average depths of 50 m (DHN 1995).

Whale-watching vessels departing from the port of Praia do Forte operate within the North Coast Continental Shelf PA (Fig. 3.4). The vessels depart at different times during the day, usually between 8 am and 2 pm and typically navigate for 2–3 h. Trips do not normally exceed 4 h in duration and vessels tend to navigate out in a straight line for a maximum 15 km and then turn north or southwards, depending on wind conditions, stopping to observe any whale groups spotted on the way, before returning to port. The shady polygon in Fig. 3.3 comprises the area covered by the whale-watching vessels operating from Praia do Forte recorded between 2003 and 2012 (Cipolotti 2013). The polygon corresponds to 475 km^2, is 15 km wide and 30 km long parallel to the coastline between Porto de Sauípe to the north (12° 50′) and Guarajuba to the south (12° 60′), and water depths within this area vary between 10 and 500 m, but 37% of the area is within 40–50 m depth (DHN 1995).

Fig. 3.4 The port of Praia do Forte, northern Bahia, is the departure point for whale-watching vessels during the humpback whale season (July–October) and an important tourist destination in northeastern Brazil. The creation of the North Coast Continental Shelf Protected Area in 2003 aimed to protect the region's rich ecosystems and promote sustainable economic activities, such as whale-watching tourism (Photo: Enrico Marcovaldi/Instituto Baleia Jubarte)

Between 2001 and 2004, only one operator carried out commercial whale-watching from Praia do Forte. From 2005 onwards, the number of operators gradually increased but varied over the years, oscillating between two and five operators. In order to characterize the present whale-watching activity in Praia do Forte, operators were asked to answer questionnaires about their operation, results obtained over the years and their expectations for the future of the activity. During 2016, four operators offered the activity in Praia do Forte, using a variety of vessels (Fig. 3.5) and engine types (Table 3.1). The schooners tend to operate one daily trip and the speedboats two, although, depending on demand, some speedboats make up to three trips daily. Multiplying the number of boats conducting whale-watching daily (fleet size) by the usual number of daily trips per boat, the tourism intensity in Praia do Forte is 13 boat trips per day. According to data provided by the four whale-watching companies, 284 whale-watching trips were carried out with 3,117 tourists during the 2015 humpback whale season.

Fig. 3.5 Examples of different boat types operating humpback whale-watching in Praia do Forte, northeastern Brazil: catamaran (**a**), wooden schooner (**b**) and fibreglass speedboat (**c** and **d**). (Photos: **a** and **c** – Luena Fernandes, **b** and **d** – Instituto Baleia Jubarte)

Table 3.1 Summary of whale-watching boats operating from Praia do Forte, northeastern Brazil, during the 2016 humpback whale season (operator, vessel name, body type, size, engine type and power (hp) and tourist capacity).

Operator	Vessel name	Boat type	Size (m)	Engine type	Engine power (hp)	Capacity (tourists)
Portomar	Parasail	Fibreglass speedboat	8.90	Outboard	170	5
	Cannes	Wood schooner	17.75	Inboard	140	65
EmyTour	Fishing	Fibreglass speedboat	8.6	Outboard	300	7
Carcará Ecotours	Água viva	Fibreglass catamaran	8	Outboard	2 × 150	10
Base Náutica	Fenix	Fibreglass speedboat	8	Outboard	200	6
	Catamaran	Fibreglass catamaran	8	Outboard	2 × 150	14

3.3 The Negative Impacts of Whale-Watching on Humpback Whales

An increasing body of evidence suggests that whale-watching can negatively impact individual cetaceans and populations (e.g. IWC 2004; Parsons 2012; Scarpaci and Parsons 2014; Senigaglia et al. 2016), though most studies have only been able to show short-term behavioural responses to observation by whale-watching vessels. Understanding the long-term effects of these responses on individuals and populations, in terms of non-lethal or sub-lethal impacts, is essential to assess the long-term sustainability of whale-watching (Berrow 2003; Lusseau 2004, 2014; Parsons et al. 2015).

However, cetaceans are long-lived and hard to study and thus only three studies to date have been able to demonstrate long-term effects of whale-watching on cetacean vital rates, mainly a decrease in female reproductive success, and all on odontocetes (Bejder et al. 2006; Fortuna 2006; Lusseau et al. 2006). For mysticetes, the long-term effects of whale-watching disturbance are still unknown (Christiansen and Lusseau 2014). Weinrich and Corbelli (2009) looked at calving rate and survival of humpback whales exposed to whale-watching boats in New England and found no long-term effects. Nonetheless, several studies have detected whale-watching induced alterations in the behaviour of humpback whales exposed to observation. Responses include changes in respiration rate, direction of travel, time spent on the surface and dive behaviour (Baker and Herman 1989; Corkeron 1995).

In their breeding grounds, humpback whales have been shown to react to the approach of whale-watching boats by increasing speed significantly, reducing blows per minute, moving more erratically, increasing breach frequency, decreasing resting behaviour and adopting a more direct path after the boat's departure (Scheidat et al. 2004; Morete et al. 2007; Avila et al. 2015). A study of the behavioural reactions of wintering humpback whales to biopsy sampling in Abrolhos, northeastern Brazil, found that groups involved in social and resting activities usually change these states to travel in response to disturbance from vessel approach (Cantor et al. 2010). Close vessel approach during biopsy sampling and whale-watching also result in behavioural reactions such as 'fluke slaps' (personal observation) (Fig. 3.6).

A study in New Caledonia detected significant behavioural changes in over 80% of humpback whales approached by boats at less than 335 m, mostly changes in path predictability, and increased swim speed and dive time (Schaffar et al. 2013).

During migration, groups with calves in Hervey Bay, Australia, also had their behaviour altered in the presence of vessels within 300 m of them, particularly a reduction in surface behaviour and increased diving (Corkeron 1995). Stamation et al. (2010) observed horizontal avoidance and longer dive times in the presence of vessels, as well as a reduction on some surface behaviours. They also found pods with calves to be more sensitive to the presence of vessels than non-calf pods.

In feeding grounds, behavioural responses of humpbacks to tour boats include horizontal avoidance, higher frequencies of some surface activities and changes in blow intervals, indicating possible shifts of behavioural patterns (from foraging to

Fig. 3.6 A 'fluke slap' reaction from a humpback whale (*Megaptera novaeangliae*) in response to close vessel approach during biopsy sampling in Abrolhos (**a**) and whale-watching in Praia do Forte (**b**), northeastern Brazil (Photos: Enrico Marcovaldi/Instituto Baleia Jubarte)

travelling) (Corbelli 2006). Baker and Herman (1989) found that humpback whales responded to vessels operating at distances of up to 4,000 m. Changes in behaviour were correlated with the speed, size, distance, and numbers of vessels and included shorter blow intervals, longer dive times and swimming away from the vessel's path. They also found a reduction in aerial behaviours in response to closer vessel approaches.

Whale-watching vessels also produce underwater noise that can interfere with whale communication. Sousa-Lima and Clark (2009) studied the acoustic interactions between tourism boats and humpback whales in Abrolhos and found the noise produced by approaching vessels caused male singers to stop singing and abandon the area. If, as suggested by Bejder et al. (2006), only the individuals less sensitive to boat disturbances would remain in the area, female choice and consequently the distribution of breeding success among the males of this population could be affected, with unknown long-term effects on the structure of this population (Sousa-Lima and Clark 2009).

Au and Green (2000) studied the effects of whale-watching boat noise on humpback whales in west Maui, Hawaii, and found that boats with outboard engines produced sounds at a greater level than the ambient background noise produced by chorusing humpback whales. Similarly, Rossi-Santos (2016) described the underwater noise originated from whale-watching vessels in Praia do Forte (Fig. 3.7) and presented evidence that the noise produced by the whale-watching industry can impact the reproductive behaviour of humpback whales, either by masking some humpback song components, or by causing the animals to change their behaviour upon vessel approach or sudden departure. A masking effect was also suggested by Dunlop (2016).

Although all types of vessels produce underwater noise, which may cause alterations or induce avoidance behaviour in whales, studies with humpback whales reported only minor behaviour changes to larger whale-watching boats when compared to smaller vessels with outboard engines, which cause greater cavitations and produce more sound than the inboard engines of larger boats (Au and Green 2000).

Apart from singing, humpback whales also communicate through vocal sounds and non-vocal sounds originating from surface behaviours, for example breaching and tail or pectoral slapping. Studies have shown that when wind noise increases humpbacks respond by increasing the level of vocal sounds and switch to using surface-generated sounds, but this response does not occur in reaction to increased vessel noise, suggesting humpback whales might not be as prepared to cope with anthropogenic noise as they are to deal with natural noise (Dunlop 2016).

Fig. 3.7 A long-term study on the acoustic ecology of humpback whales (*Megaptera novaeangliae*) in their breeding ground off northeastern Brazil collected behavioural and acoustic data during whale-watching boat approaches in Praia do Forte and indicated impacts on the whales' surface and underwater behaviour, as well as potential masking of important biological sounds (Photos: Enrico Marcovaldi/Instituto Baleia Jubarte)

3.4 The Sustainable Management of Whale-Watching

A variety of regulations, voluntary and legislative, have been developed around the world in an attempt to manage whale-watching in a sustainable manner (see Wiley et al. 2008; Carlson 2012). There is a general view that voluntary guidelines and codes of conduct may not be enough to safeguard the whales (e.g. Garrod and Fennell 2004; Allen et al. 2007; Wiley et al. 2008; Kessler and Harcourt 2013; Tyne et al. 2014). Instead, regulations and permit systems provide the greatest level of management certainty and control (IWC 2004). For many places, therefore, ensuring that commercial whale-watching can only be undertaken by a limited number of permitted (licensed) operators is the essential first step to sustainable management. Ideally, the number of permits available should be based on a scientific assessment of the carrying capacity for that location and the system should include regular evaluation and renewal of permits (Higham et al. 2009).

Different locations have established varying guidelines for limiting the number of whale-watching permits issued (see Carlson 2012). In Kaikoura, New Zealand, permits are limited by tourist activity, so the company running swim-with-dolphin programmes cannot also run whale-watching trips. Hervey Bay, in Australia, implemented, within 2 years of the industry's start in 1987, a spatially explicit management scheme, with defined management goals and that started with a mathematical model for the number of permits issued (Jeffery et al. 1994; Smith et al. 2006). The management scheme also provided for the transfer of licences (individual transferable quotas) and enforcement.

One of the most advanced permit systems for whale-watching operators is in South Africa, where permits are generally limited to one per location. The system is founded on the principles of 'continuous criteria' and 'permit re-application', whereby operators are required to provide evidence of their high standard of responsible whale and dolphin watching in order to successfully retain a permit on reapplication every 5 years (Schalkwyk 2008). Examples of the evidence provided include cooperation with researchers, equality within the workforce, and involvement in educational programmes.

The guidelines for commercial cetacean-watching activities in the ACCOBAMS area (Black and Mediterranean Sea) recommend permits to be granted by the appropriate authorities and issued by area, to restrict the number of commercial vessels operating in an area on any given day. Issuing of permits should be based on a strict set of criteria to avoid conflicts with the conservation, management or protection of the cetaceans concerned and guarantee the proposed operators have sufficient experience with cetacean tourism (Carlson 2012).

In Peninsula Valdés, Argentina, the current whale-watching regulations resulted from a series of workshops organised by the Tourism Board (see Sironi et al. 2005). The main aspects of the resulting Provincial Law 5714 from Chubut include the issuing of official whale-watching permits by the Enforcement Authority through public bidding for a period of no less than 6 years, and determines a tax that whale-watching companies must pay to the Province. The Enforcement Authority can grant between four and six official permits to commercial whale-watching operations and each whale-watching company is allowed to operate with only one boat at any one time with a maximum capacity of 70 passengers. A whale-watching Advisory Committee chaired by the Enforcement Authority with members from the operating companies was created to suggest improvements to the activity and guarantee the principles for resource conservation and sustainable development. Fines are also predicted for infractions to the Law.

In Chile, commercial cetacean-watching activities can only be carried out under a permit granted by the appropriate authority, with a restricted number of one commercial vessel operating in any given area and permits must be renewed every 3 years after a compliance review process (Carlson 2012).

In the Azores (Portugal) whale-watching companies must also apply for permits to operate. The number of permits issued for the different zones is based on the carrying capacity calculated as a relationship between the relative abundance of cetaceans and the number of vessels operating in the area as such: $N_{\text{whale sightings/day}} \times N_{\text{vessels allowed with each group}}$ (Magalhães et al. 2002). If an average ten groups are sighted in an area on a daily basis and the number of vessels allowed simultaneously with each group is two, then the carrying capacity for that area would be 20 vessels. This particular study used a combination of land station and in water observations to determine cetacean abundance.

The first legislation designed specifically to prohibit any form of intentional harassment to all species of cetacean in Brazilian waters was ordinance IBAMA N° 2.306, of November 22nd 1990, which established the first whale-watching rules. In 1996, this was reformulated by ordinance IBAMA N° 117, of December 26th. The new regulation imposed, among other rules: a minimum distance of 100 m from any species of whale for boats; a minimum altitude of 100 m from cetaceans for any aircraft; the maximum limit of 30 min to observe any group of whales; a minimum distance of 50 m for diving or swimming in the vicinity of any species of whale; and standards for the operation of tourism vessels within Protected Areas (PAs). These standards include the registering of all vessels operating within the PA, the determination of the maximum number of vessels allowed to operate within the PA simultaneously and of the routes and speeds allowed, in addition to the permanent provision of interpretative information on cetaceans and their conservation (Brasil 1996). In 2002, ordinance IBAMA N° 24, from February 8th, altered this ordinance, specifically to establish that boats must keep their engines idle when observing humpback whales, and turned off or idle for other species of cetaceans, and establish the limit of two vessels allowed to approach an individual or group of whales simultaneously.

In 2000, the Right Whale Environmental Protection Area (Área de Proteção Ambiental da Baleia Franca) was created in the state of Santa Catarina, southern Brazil, to protect the main concentration area of a southern right whale (*Eubalaena australis*) wintering ground off Brazil; to promote the sustainable use of the natural resources; and to manage the tourist and research activities and the local vessel and aircraft traffic (Brasil 2000). Whale-watching activities have been conducted in this region since 1999 and, in 2005, in accordance with ordinance IBAMA N° 117, this

PA established a methodology for registering and monitoring all boats involved in this activity. Additionally, six area closures were established in the PA, taking into account scientific and management issues, to allow the monitoring of the short and long-term effects of boat-based tourism in the PA and the design and implementation of adequate management measures to ensure both the species' survival and the sustainability of the whale-watching industry (Palazzo and Groch 2007).

The Abrolhos National Marine Park, in southern Bahia state, is the primary marine protected area (MPA) for the protection of Brazilian humpbacks. Within the Park, all vessels are registered and its Management Plan (MP) determines the carrying capacity for some areas of the Park. For example, in the Abrolhos archipelago and Parcel, the carrying capacity was established at 15 medium-sized boats (35 ft/10 passengers) at any one time, based on the protection offered to the boats by the southerly and northerly wind anchoring sites (IBAMA 1991). The MP also establishes that each vessel can carry up to 15 adult passengers plus 3 children under the age of 10 into the Park's interior. Concerning whale-watching, the MP determines a carrying capacity of up to three vessels approaching the same group of cetaceans simultaneously, based on these animals' susceptibility to stress, particularly of nursing females with calves. However, according to the ordinance IBAMA N° 24, only two vessels are allowed to approach any group of whales at the same time. Therefore, the park's Public Use Plan, currently under discussion, when implemented, will contain specific rules for the practice of whale-watching (Krob 2003).

The North Coast Continental Shelf Protected Area, in northern Bahia state, was created in 2003 to promote sustainable economic activities, such as whale-watching tourism, however, to date, this PA remains unimplemented and without a Management Plan.

In Brazil, any individual with a certified vessel and appropriate boat handling qualifications may offer commercial whale-watching tours. There are no restrictions to new entrants and no regulations relating to the types or numbers of boats used. This means the range of vessels used and the experience and knowledge of the operators are highly variable. A study in Australia found that new operators have a more limited understanding of how to run a 'good' whale-watching operation and are less concerned about the whales themselves than the more experienced operators (Lalime 2005). Similarly, the collaborative relationships that exist in Praia do Forte between the operators and the Instituto Baleia Jubarte are shaken up every time a new operator begins whale-watching activities. As a result, the operators move to a more competitive approach.

With the increasing number of commercial whale-watching operators in Praia do Forte, management measures need to be implemented in order to mitigate the potential impacts on whales (Fig. 3.8). One such measure is the requirement for permits to operate whale-watching commercially. The introduction of a permitting system could be used to manage cumulative effects of increasing vessel numbers and ideally would include training in boat handling around whales for new operators and refreshment courses for existing ones (Kessler and Harcourt 2013).

To ensure that the current regulations are successful in minimizing harm to the whales, it is necessary to enhance their enforcement, which is virtually non-existent

Fig. 3.8 Whale-watching vessel approach in Praia do Forte has been observed to cause whales to change swimming course (**a**). Additionally, groups of whales are often observed by more than one vessel simultaneously and for long periods of time (**b**): the legislation stipulates the maximum observation time for each individual vessel but does not prohibit vessels from immediately approaching the group as soon as others depart (Photos: **a** – Luena Fernandes and **b** – Enrico Marcovaldi/Instituto Baleia Jubarte)

in Praia do Forte, besides the monitoring carried out by the Instituto Baleia Jubarte, which is a non-governmental organization with no enforcement authority and that can only report infractions at best.

Because the long-term consequences of whale-watching interactions are unknown, management of the humpback whale-watching industry should adopt a conservative approach. Improved knowledge of the long-term impacts of multiple exposures to vessels is required to inform its management (Higham et al. 2009; Stamation et al. 2010). Additionally, determining the carrying capacity of whales to whale-watching is one of the greatest challenges for its sustainable management, although little information is available on how to calculate it (Berrow 2003).

3.5 Limiting Growth and the Concept of Carrying Capacity

The concept of carrying capacity has its origins in the pastoral and wildlife management sectors, where it was used to calculate the maximum number of individuals an area could sustain without compromising their reproductive potential or to determine the size of an animal population a specific habitat could support (Berrow 2003; Curtin 2003). Applied to tourism, the concept of carrying capacity assumes a positive relation between the number of tourists and the impact of tourism on a destination and has been criticized in the tourism literature for being subjective, misleading and ineffective as a management tool (e.g. McCool and Lime 2001).

Determining carrying capacity can be a complex process and researchers seem to agree that there is no single number but rather a combination of factors such as management objectives, the biological nature of the resource, and the preference and tolerances of the users, all of which must be considered together to determine a carrying capacity for a specific area (Heberlein 1977). Several methodologies have been proposed to regulate visitor management in protected areas, including Visitor Impact Management (Graefe et al. 1990), Limits of Acceptable Change (Stankey et al. 1985) and Tourism Carrying Capacity (Cifuentes 1992).

The concept of Tourism Carrying Capacity (TCC) includes mainly ecological and social parameters and can be defines as: "*The maximum number of visitors an area can sustain without unacceptable deterioration of the physical environment and without considerably diminishing user satisfaction*" (Salerno et al. 2013). TCC has been challenged by the idea of *Sustainable Tourism* but many of the issues outlined for this new concept are similar and, clearly, the basic element remains the need to limit the growth of the tourist activity and determine which social and environmental changes can be accepted (Butler 1999; Liu 2003). Nonetheless, however criticized this concept may be (McCool and Lime 2001) it remains in use and has evolved to consider ecological, economic, social and perceptual aspects of tourism (Needham et al. 2011; Filby et al. 2015).

However, according to Heberlein (1977), resource managers tend to overemphasize biological capacity. The problem is that, as is the case for whale-watching impacts, although demonstrable, numbers are often limited to trivial biological impact, or no impact at all can be documented. As such, in the absence of information on the biological carrying capacity, in order to maintain the activity's sustainability, important decisions about other sorts of capacity, such as the social capacity for example, should not be delayed.

An interesting approach to the theme of tourism carrying capacity has been presented in a study conducted by IBAMA and WWF-Brazil in the Fernando de Noronha National Marine Park, northeastern Brazil (Mitraud 2001). In this study, the concept of TCC (Cifuentes 1992) is used in conjunction with the Visitor Impact Management (Graefe et al. 1990) and the Limits of Acceptable Change (Stankey et al. 1985). According to Cifuentes (1992), the TCC in protected areas should also take into account the conditions available for the adequate management of the area, which will ultimately determine its real carrying capacity.

Similarly, whale-watching carrying capacity must be assessed with basis on a set of variables related to the three spheres of sustainable management, namely the impact on whales (Biological), on tourist experience (Social) and whale-watching companies (Economic) (Chion et al. 2012), as well as management capacity.

3.5.1 Social Carrying Capacity – The Tourists

The social carrying capacity relates to visitor experience on whale-watching boats, including the time spent observing whales, time spent alone with the whales and satisfaction (Lime and Stankey 1971; Chion et al. 2012).

There are many studies that measured the satisfaction of whale-watching tourists (e.g. Avila-Foucat et al. 2013; Filby et al. 2015; Sitar et al. 2015). Generally, tourists prefer when whale-watching trips have educational elements (Bentz et al. 2016a), licensed boat operators, and boat operators that follow whale-watching regulations (Orams 2000) and are concerned about minimizing harm to whales (Kessler et al. 2014).

The Instituto Baleia Jubarte has been monitoring visitor profile and satisfaction on-board whale-watching trips in Praia do Forte since 2003. This is achieved by asking tourists to answer questionnaires upon return from their whale-watching trips (Fig. 3.9).

Crowding is another potential indicator of social carrying capacity that can influence the probability of tourists returning to a site and has been increasingly applied to assess marine wildlife tourism carrying capacity (Finkler and Higham 2004; Needham et al. 2011; Avila-Foucat et al. 2013; Bentz et al. 2015).

Fig. 3.9 The Instituto Baleia Jubarte monitors visitor satisfaction on-board the whale-watching vessels in Praia do Forte in partnership with the local operators since 2003 (Photos: Enrico Marcovaldi/Instituto Baleia Jubarte)

3.5.2 Biological Carrying Capacity – The Whales

Animals perceive tourism as a risk and may respond to interactions in the same way they would respond to predators (New et al. 2014). This can result in disruptions of their activity budget, displacement from the area and other sub-lethal effects that can lead to long-term consequences for the populations (Pirotta et al. 2014). Methods of boat approach, boat behaviour around the animals and the number of vessels present have all been shown to change humpback whale behaviour (Corbelli 2006; Stamation et al. 2010).

Avoidance of disturbance is one of the most commonly used behaviour measures of disturbance, but there are other forms of behavioural responses to disturbance, for example, increased levels of vigilance, reduction in parental care or amount of time spent resting, that could potentially affect survival or fecundity (Gill et al. 2001).

Short-term effects of individual interactions between whale-watching boats and cetaceans are unlikely to result in any long-term effects for the animals, since they may be able to compensate the effects of the disturbance once it is absent (Christiansen and Lusseau 2014). The problem is that the effects from whale-watching disturbance may be cumulative when they become more frequent or target individuals within a population, certain areas, behaviours or age classes more frequently.

Responding to disturbances has associated energetic consequences. One way to evaluate the energy cost of human disturbances is to analyse the amount of time the animals dedicate to different energy-relevant activities, such as swimming, feeding and resting, in the presence or absence of tourist vessels (Christiansen and Lusseau 2015).

Lusseau (2004) suggests population estimates, structure, reproductive rates and behaviour budgets should be included in a whale-watching monitoring program. To assess the carrying capacity of a population to whale-watching the most sensitive parameter must be determined and its limits assessed to manage the activity within them.

The most biologically significant impact of whale-watching, when performed in a breeding ground, is the interruption of reproductive behaviour, including courtship and parental care. Therefore, if disturbance to female and calf pairs is the most sensitive variable, the number of vessels or time they spend with these groups should be limited to avoid impacts (Berrow 2003).

Future research should consider how soon whales come to rest after disturbance, as an additional indication of the strength of avoidance and whether this is likely to be biologically significant (Kessler et al. 2013).

3.5.3 Economic Carrying Capacity – The Whale-Watching Companies

This aspect of whale-watching carrying capacity refers to the economic contribution the activity makes to coastal economies around the world (Parsons et al. 2003; O'Connor et al. 2009; Cisneros-Montemayor et al. 2010; Mustika et al. 2012). It creates jobs and income for local communities. However, with the increase in numbers of operators in a specific locality, competition may lead to decreased economic viability, for example. So, in analysing this aspect of carrying capacity, factors such as number of jobs created as a result of whale-watching tourism, total income generated from tourism in the region and for the whale-watching operators can be considered (Parsons et al. 2003).

If tourist demand exceeds the capacity of the existing operators, then the activity is below carrying capacity. If, on the other hand, there are more operators than tourists wanting to go whale-watching, then the economic carrying capacity is exceeded. An evaluation of the demand for whale-watching in Praia do Forte region could contribute to the assessment of the economic carrying capacity (Brumatti 2008).

3.5.4 Management Carrying Capacity – The Enforcement Authorities

The management carrying capacity was proposed by Cifuentes (1992) and refers to the existence of policies, equipment, personnel, funding, infrastructure and facilities available in a Protected Area to support the sustainable and effective management of the area and activities carried out within it. This author defines optimal management carrying capacity as "*the best state or conditions the administration of a Protected Area should have to develop its activities and reach its goals*". In Praia do Forte, within the North Coast Continental Shelf PA, northeastern Brazil, the management carrying capacity can be considered one of the main limiting factors of whale-watching carrying capacity.

3.6 An Integrated Framework to Assess Whale-Watching Carrying Capacity

Based on the concepts presented herein we suggest an integrated framework to guide the assessment of whale-watching carrying capacity that takes into account the different elements of sustainable management and the main stakeholders involved in any commercial whale-watching activity (Fig. 3.10). As proposed by Higham et al. (2009), four key stakeholder groups were included: the commercial tourism operators, the social sciences (visitors), the natural sciences (whales) and

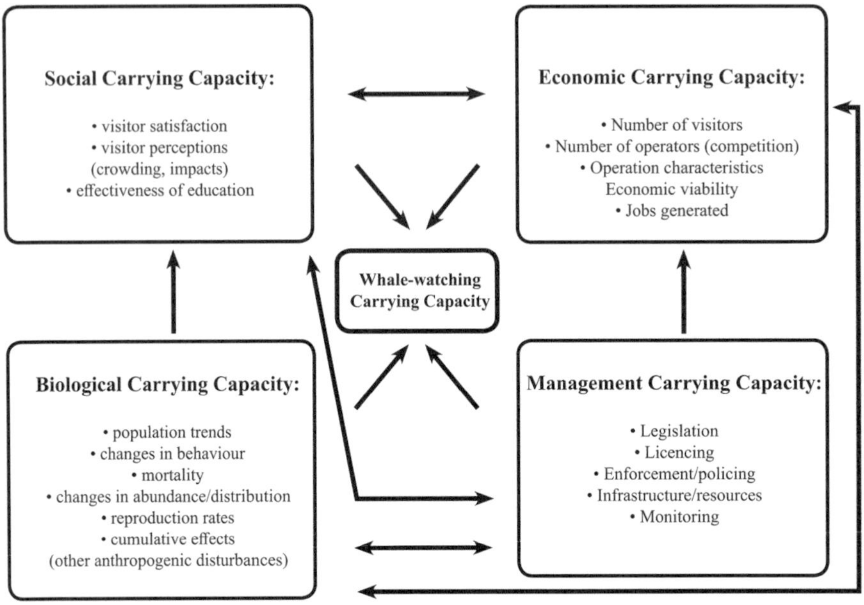

Fig. 3.10 An integrated framework model for the assessment of whale-watching carrying capacity

planning/management agencies. The use of such frameworks, illustrated as models to facilitate understanding and interpretation, is common throughout the whale-watching management literature (e.g. Duffus and Dearden 1990; IFAW 1995; Curtin 2003; Higham et al. 2009; New et al. 2015).

This framework intends to guide the assessment of whale-watching carrying capacity (CC) in a specific location. Four elements of carrying capacity are included. The biological CC refers to the level of interactions the target population can sustain before significant behavioural alterations or population effects are observed. The biological CC influences the social CC as a reduction in numbers of whales in an area will affect visitor satisfaction. It also influences the economic CC since less whales, for example, will reduce profits and the economic viability of the activity, and is influenced by the economic CC since more operators, for example, can result in more competition amongst operators for access to less individuals, resulting in unsustainable pressure on the animals. It also influences and is influenced by the management carrying capacity since adequate policies, based on scientific data and adequate management, will maintain the biological CC and, on the other hand, changes in the populations will demand active management decisions. The social CC, translated into visitor experience and perceptions, is directly influenced by the economic CC, as the services and products provided by the operators, such as naturalist guides, and the conduct of the operator, often influenced by increased competition, all influence visitor experience. On the other hand, social CC, translated into numbers of tourists, demand and willingness to pay, all influence economic

CC. Finally, the management CC, translated into a permit scheme, limited permits or permit conditions, for example, affect the economic CC and the social CC. All four elements of CC and the interactions amongst them must be considered when assessing the carrying capacity of whale-watching in a specific location.

3.7 Discussion

Currently all five companies operating whale-watching in Praia do Forte have voluntary cooperation agreements with the Instituto Baleia Jubarte (IBJ) for the provision of specific training for their staff, interpretative information for their tourists, monitoring and collection of scientific data on-board their vessels (Fig. 3.11). This agreement also involves the operators making a financial contribution to support the research and conservation activities developed by IBJ.

The Instituto Baleia Jubarte has been monitoring visitor profile and feedback since 2003 by means of questionnaires answered voluntarily by the tourists on the return from their whale-watching trips. In 2015, 50% of tourists interviewed stated the experience fully satisfied their expectations and 96% said they would go whale-watching in Praia do Forte again in the future. Additionally, when asked to rate different aspects of the operations, 76% rated the boat conduct excellent, 69% rated the explanations provided boat staff excellent and 69% rated the information provided by the IBJ staff excellent.

For the purpose of this study, in order to characterise the whale-watching activity in Praia do Forte, the operators were also asked to answer questionnaires. The five whale-watching operators in Praia do Forte employ a total of 39 permanent staff and an additional 14 temporary staff during the whale-watching season, of which 26 employees are from the local community. One operator stated that the income from the whale-watching operation provides full employment year round for all the company's staff, some of which would be only temporary otherwise.

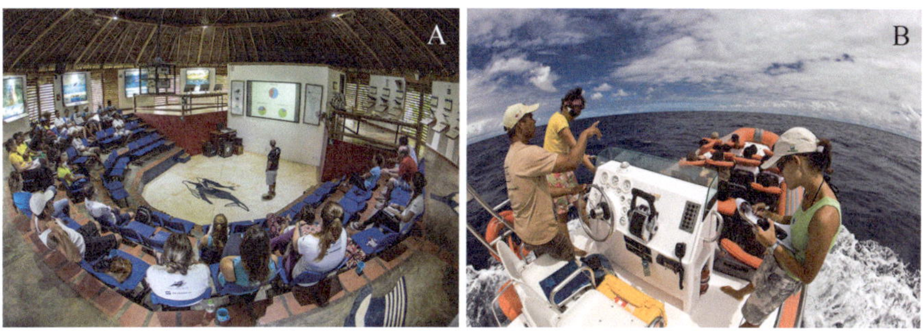

Fig. 3.11 Instituto Baleia Jubarte provides specific training for whale-watching operator staff (**a**), interpretative information for tourists, and monitoring and collection of scientific data on-board whale-watching vessels (**b**) (Photos: Enrico Marcovaldi/Instituto Baleia Jubarte)

An in depth analysis of the social impacts and benefits of whale-watching for the local community of Praia do Forte is beyond the scope of this study. Previous studies on this subject, though, suggested a greater articulation between the local authorities and the tourism operators is needed to promote community engagement and the better distribution of the economic and social benefits generated by the whale-watching tourism industry (Brumatti 2008).

When asked whether their tourists were satisfied with their whale-watching experience, four out of five operators answered they were fully satisfied and one answered the vast majority of tourists were satisfied. When asked whether they consider the whale-watching activity as financially viable and profitable for their companies, all operators answered positively and declared they expect the activity to expand in future years.

When asked what would be an acceptable limit for whale-watching in Praia do Forte, three out of five operators believe there could be more boats operating in Praia do Forte and only one believes it should remain at its current level (6 boats). Suggested levels were two large boats for up to 50 passengers and three to five speedboats for up to 10 passengers operating from Praia do Forte. Two operators mentioned the growth of the activity, and the number of boats, would ultimately be limited by tourist demand, which determines the financial viability for boat owners.

Another factor pointed out by all operators is the absence of involvement or support from public authorities in the development, marketing or management of this activity. This situation has not changed over the years as, previously, Brumatti (2008) also noted the absence of participation by the public authorities and the local communities in the planning and implementation of whale-watching in Bahia. The local governments, environmental agencies and tourism bodies recognise the whale-watching activity and its potential for local tourism development but do not participate or take an active role in its management. Garrod and Fennell (2004) discuss the fact that whale-watching management measures are more likely to be followed if participants have been involved in their development.

To date, the Instituto Baleia Jubarte (IBJ), a non-governmental organisation, has taken responsibility for the promotion, management and monitoring of the whale-watching activity in Bahia. The operators in Praia do Forte see the collaboration with the IBJ as a means of contributing to whale conservation and adding value to their product, attributing a social and environmental responsibility image to their business.

In Praia do Forte, humpback whale-watching is performed on whales involved in breeding ground activities, such as courtship, mating and nursing. Cipolotti (2013) investigated the distribution patterns and habitat use of humpback whales in relation to depth, based on data collected by the Projeto Baleia Jubarte on-board whale-watching vessels operating from Praia do Forte during the breeding season (July to October) between the years 2003–2012. He found that 52% of the whale groups were sighted within 40–50 m depths. Additionally, groups with calves were found to concentrate on shallower waters, closer to shore and therefore easier to reach during whale-watching trips, making them more likely to be observed and thus suffer greater pressure from this activity. He also found the occurrence of groups with calves peaked during September and October and groups without calves and competitive groups decreased towards the end of the breeding season (September–October), when compared with the beginning of the season (July–August). This can be explained by the fact that these whales' migration is age structured, females with new calves being the final cohort (Chittleborough 1965; Dawbin 1966), so in the later stages of the whale-watching season the activity concentrates on females with calves.

However, for boat interactions to affect the vital rates of individual baleen whales in Praia do Forte, they would have to significantly disrupt the breeding or nursing process. In the case of females with calves, any lost nursing opportunities or decreased lactation due to the energy cost of displacement in response to whale-watching disturbance, can be easily compensated at a later moment (Lusseau 2014), since the legislation only allows 30 min of observation and the number of vessels is low. Additionally, humpback whales in the area are able to avoid disturbance from whale-watching vessel approach because they have plenty of alternative sites to go to, since the whale-watching activity currently operates over a limited area (Fig. 3.1).

However, calf growth tends to peak towards the end of the breeding season (Whitehead and Payne 1981), time during which calves need the most energetic investment from their mothers in order to increase lactation (Lusseau 2014), coinciding with a peak in the proportion of time females spend resting (Thomas and Taber 1984), the activity state in which nursing takes place.

As such, this concentration of whale-watching interactions with female and calf groups towards the end of the breeding season, a critical period in calf development, could compromise calf growth, leading to increased mortality or decreased survival probability as they undertake the long migration to the feeding grounds (McMahon et al. 2000).

However, the northern coast of Bahia is not as important a breeding ground as the Abrolhos Bank, in southern Bahia, since a smaller proportion of individuals observed is of new-born calves (Pacheco de Godoy et al. 2005; Lunardi et al. 2008). Additionally, the northern coast is the area along Bahia that shows the smallest annual population growth rate (Wedekin 2011). As such, the whale-watching activity is less likely to impact calf birth or survival rates here than in Abrolhos, which is one of the main ways in which whale-watching can have any long-term impacts on cetacean populations (Christiansen and Lusseau 2014). Additionally, data collected on-board whale-watching vessels for over 10 years in Praia do Forte show an increase in numbers of groups with calves sighted in the region, also a consequence of population growth, suggesting the activity is not having a negative impact on calving rate (Cipolotti 2013).

Another noteworthy aspect is that the whales show low site fidelity and occupancy rates on the north coast of Bahia, remaining in the same area for an average 5.5 days only, contrasting with 15 days in the Abrolhos Bank (Baracho-Neto et al. 2012). There are also significant differences in the concentration and distribution patterns of different social groups between the northern coast of Bahia and Abrolhos. Although these two regions comprise distinct concentrations of humpback whales, with intrinsic environmental differences, behaviour patterns are quite similar, suggesting these are not sub-populations and that differences in behaviour are due to the groups that use each area, groups with calves preferring the shallow and protected waters around Abrolhos (Lunardi et al. 2008).

The presence of coral reefs around the Abrolhos archipelago provide greater protection for calves, different from the straight and narrow continental shelf along Praia do Forte, where the female and calf groups use more extensive areas along this coast (Rossi-Santos et al. 2008). Zerbini et al. (2006) showed humpback whales to be extremely transient, moving around 42 km per day within their breeding grounds and Wedekin et al. (2010) found whales off the Brazilian coast to travel up to 600 km within the same breeding season.

As such, it can be anticipated that the pressure from whale-watching observation on individual groups of females and calves will be less in Praia do Forte than in Abrolhos, where the activity concentrates around the archipelago.

Another way in which whale-watching disturbance can affect humpback whale behaviour and reproduction is through acoustic interferences. Rossi-Santos (2016) observed good compliance by the local operators with the limits imposed by the Brazilian whale-watching regulations (minimum distance of 100 m and maximum 30 min permanence time with each group). However, he also observed that whale-watching operations produce a frequency range similar to the song of the humpback whale, which may result in masking of this important breeding display during boat approach.

The strong growth in the humpback whale population, utilising the waters off Bahia, including Praia do Forte, however, suggests that any changes to behaviour resulting from interactions with whale-watching vessels to date have not been biologically significant, i.e. population growth is not inhibited. However, the increasing popularity of whale-watching means assessment and management of interactions must be ongoing and responsive to changes.

From a conservation perspective, human disturbances are only of concern if they affect survival or fertility, consequently decreasing population size (Gill et al. 2001). As previously mentioned, the Brazilian humpback whale population is demonstrating a consistent growth. Thus, in the absence of evidence of these long-term impacts, and bearing in mind all the important educational, social and economic benefits of whale-watching in Praia do Forte, and the important opportunities for scientific research it offers, the minor short-term levels of disturbance incurred from whale-watching may be considered acceptable (Fig. 3.12). As argued by Gill et al. (2001), without knowing whether behavioural changes do in fact result in population decline or distribution changes, there is a risk in limiting human access to wildlife, such as whale-watching, with all its associated benefits, in order to protect species which have their behaviour altered by human presence but suffer no other impacts.

Fig. 3.12 Whale-watching in Praia do Forte, as a partnership between the IBJ and local operators, offers important awareness raising and education opportunities for the local community (**a**), tourists (**c**) and through the media (**b**), since whales are charismatic animals, capable of inspiring people to care about the oceans (**d**) (Photos: Enrico Marcovaldi/Instituto Baleia Jubarte)

3.8 Recommendations

Based on the information presented in this study, we consider the current level of knowledge of both the impacts on the whales and the quality and effectiveness of the services provided for the whale-watching tourists in Praia do Forte is insufficient to allow the calculation of carrying capacity for the activity. We can, however, make recommendations on aspects that should be taken into consideration and specific studies that should be developed in order to gather adequate information for the future determination of whale-watching carrying capacity in Praia do Forte, the North Coast Continental Shelf PA and other localities along the coast of Bahia:

- Maintaining the long-term monitoring of this activity, implementing specific studies to assess the impacts of whale-watching on the whales, in particular: behavioural studies from a land station (speed changes, path linearity, inter-breath intervals, activity budgets) (e.g. Lusseau et al. 2009; Christiansen et al. 2013a, b), cumulative exposure (encounter rate) (e.g. Christiansen et al. 2015) and bioacoustics studies to assess the potential effects of whale-watching vessel noise on the whales communication and reproduction (e.g. Au and Green 2000; Dunlop 2016)

- Maintaining the long-term monitoring of visitor feedback, implementing specific studies to evaluate visitor perception, educational value and attitudinal changes (e.g. Needham et al. 2011; Bentz et al. 2015, 2016a, b)
- Implementing studies to assess the cumulative anthropogenic impacts of whale-watching and other human activities in the area (see Wright and Kyhn 2015)
- Revising the existing national whale-watching legislation to include: a reduced observation time of up to 15 min for groups containing calves; a 300 m buffer zone within which vessel speed must be gradually reduced and no sudden changes of speed or direction must be made; make compulsory the presence of a trained naturalist guide and the provision of interpretative information on whale biology and conservation on all whale-watching operations; make compulsory specific training for all staff involved in whale-watching activities
- Developing and implementing a Management and Public Use Plan for the North Coast Continental Shelf PA, involving different stakeholders, such as NGO's, tourism operators, local community and government authorities. The Management Plan, according to ordinance N° 117, must: determine the maximum number of vessels allowed to operate simultaneously and the routes and speed these vessels can operate in inside the PA. We also suggest the delimitation of areas where whale-watching is allowed, taking into account where the activity is already in place, such as Praia do Forte, and where it has potential to develop, and establishing closure areas to allow for long-term monitoring through control sites. Additional management measures could include the use of specific boat and engine types, as well as regular maintenance requirements, to reduce the amount of noise around the whales
- Establishing a whale-watching operators association could improve the activity's self-regulation and promote shared management, with initiatives such as the implementation of standard procedures and prices, permit and passenger sharing to reduce the amount of vessels in the water when demand is below capacity and the adoption of well-established certification schemes such as Blue Flag or Whale Heritage Sites
- Implementing data collection in areas where the whale-watching activity can potentially develop in the near future along the Brazilian coast, such as Vitoria/ES, to obtain baseline data and to allow the observation of changes over time as the activity develops (New et al. 2015)

3.9 Conclusions

In this study, we examined the current literature on the sustainable management of whale-watching tourism and on tourism carrying capacity, in an attempt to assess the whale-watching carrying capacity for Praia do Forte, northeastern Brazil. We proposed an integrated framework to assess whale-watching carrying capacity, taking into account biological, social and economic factors, as well as management considerations.

By analysing the whale-watching industry in Praia do Forte under the light of this integrated framework, we concluded that: whale-watching in Praia do Forte is currently a low-impact activity, with no evident deleterious biological effects on the target population. It is economically viable and produces income and jobs for the local community. It has educational value and visitor satisfaction levels are high.

However, there is a lack of involvement from the government authorities in the management and enforcement of this activity, which relies mostly on self-regulation and on the collaboration with a non-governmental organisation for its management. Although we currently lack all necessary information to establish its carrying capacity, we make recommendations to improve our understanding of the impacts of this activity and its sustainable management in this locality and suggest limiting it to its current level until more data and management capacity become available.

Acknowledgements This paper is part of the first author's Master's degree in Ecology Applied to Environmental Management at Universidade Federal da Bahia (UFBA), Brazil. We are most grateful to the entire Instituto Baleia Jubarte team, especially to Sergio Cipolotti and Márcia Engel, for her support and encouragement. We are grateful to Milton Marcondes, José Truda Palazzo and David Lusseau for constructive conversations. We thank Guilherme Maurutto for his help with the map and Enrico Marcovaldi for the use of his photographs. We are grateful to PETROBRAS for the continued financial support of Projeto Baleia Jubarte.

References

Ahn BY, Lee B, Shafer CS (2002) Operationalizing sustainability in regional tourism planning: an application of the limits of acceptable change framework. Tour Manag 23:1–15. doi:10.1016/S0261-5177(01)00059-0

Allen S, Smith H, Waples K, Harcourt R (2007) The voluntary code of conduct for dolphin watching in Port Stephens, Australia: is self-regulation an effective management tool? J Cetacean Res Manage 9:159–166

Andriolo A, Kinas PG, Engel MH, Martins CCA (2006) Monitoring humpback whale (*Megaptera novaeangliae*) population in the Brazilian breeding ground, 2002 to 2005. International Whaling Commission Saint Kitts paper present to Int Whal Comm Sci Comm SC/58/SH15 1–12

Andriolo A, Kinas PG, Engel MH et al (2010) Humpback whales within the Brazilian breeding ground: distribution and population size estimate. Endanger Species Res 11:233–243. doi:10.3354/esr00282

Au WWL, Green M (2000) Acoustic interaction of humpback whales and whale-watching boats. Mar Environ Res 49:469–481

Avila IC, Correa LM, Parsons ECM (2015) Whale-watching activity in Bahía Málaga, on the Pacific coast of Colombia, and its effect on humpback whale (*Megaptera novaeangliae*) behavior. Tour Mar Environ 11:19–32. doi:10.3727/154427315X14398263718394

Avila-Foucat VS, Sánchez Vargas A, Frisch Jordan A, Ramírez Flores OM (2013) The impact of vessel crowding on the probability of tourists returning to whale watching in Banderas Bay, Mexico. Ocean Coast Manag 78:12–17. doi:10.1016/j.ocecoaman.2013.03.002

Baker CS, Herman LM (1989) Behavioral responses of summering humpback whales to vessel traffic: experimental and opportunistic observations. Alaska, USA

Baracho CG, Cipolotti SR, Lima F, et al (2006) Resultados preliminares sobre o estudo de fotoidentificação da baleia jubarte (*Megaptera novaeangliae*) no litoral norte do estado da Bahia,

Brasil, entre 2000–2005. In: Primera Reunión Internacional sobre el Estudio de Mamíferos Acuáticos SOMEMMA–SOLAMAC; Reunión de Trabajos de Especialistas en Mamíferos Acuáticos de América del Sur, 12; Congreso de la Sociedad Latinoamericana de Especialista en Mamíferos Acuáticos—SOLAM. Mérida, México

Baracho-Neto CG, Neto ES, Rossi-Santos MR et al (2012) Site fidelity and residence times of humpback whales (*Megaptera novaeangliae*) on the Brazilian coast. J Mar Biol Assoc UK 92:1783–1791. doi:10.1017/S0025315411002074

Bejder L, Samuels A, Whitehead H et al (2006) Decline in relative abundance of bottlenose dolphins exposed to long-term disturbance. Conserv Biol 20:1791–1798. doi:10.1111/j.1523-1739.2006.00540.x

Bentz J, Rodrigues A, Dearden P et al (2015) Crowding in marine environments: divers and whale watchers in the Azores. Ocean Coast Manag 109:77–85. doi:10.1016/j.ocecoaman.2015.03.001

Bentz J, Lopes F, Calado H, Dearden P (2016a) Enhancing satisfaction and sustainable management: whale watching in the Azores. Tour Manag 54:465–476. doi:10.1016/j.tourman.2015.11.016

Bentz J, Lopes F, Calado H, Dearden P (2016b) Managing marine wildlife tourism activities: analysis of motivations and specialization levels of divers and whale watchers. Tour Manag Perspect 18:74–83. doi:10.1016/j.tmp.2016.01.004

Berrow S (2003) An assessment of the framework, legislation and monitoring required to develop genuinely sustainable whalewatching. In: Garrod B, Wilson JC (eds) Marine ecotourism: issues and experiences. Channel View Publications, Clevedon

Bisi TL (2006) Comportamento de filhotes de baleia-jubarte, *Megaptera novaeangliae*, na região ao redor do arquipélago dos Abrolhos, Bahia (Brasil). Universidade de São Paulo

Brasil (1987) Lei N° 7.643, de 18 de Dezembro de 1987. Casa Civil, Brasília, Brasil

Brasil (1996) Portaria IBAMA N° 117 de 26 de dezembro de 1996. IBAMA

Brasil (2000) Decreto de 14 de setembro de 2000. Casa Civil, Brasília, Brasil

Brasil (2003) Decreto N° 8.553. Secretaria de Meio Ambiente e Recursos Hídricos, Salvador, Brasil

Brumatti PNM (2008) Análise das potencialidades do desenvolvimento sustentável do turismo de observação de baleias, whale watching, na costa da Bahia, Brasil. Universidade Estadual de Santa Cruz – UESC

Brumatti PNM (2013) O papel do turismo de observação da vida selvagem para a conservação da natureza. Rev Bras Ecoturismo 6:191–208

Butler RW (1999) Sustainable tourism: a state-of-the-art review. Tour Geogr 1:7–25

Cantor M, Cachuba T, Fernandes L, Engel MH (2010) Behavioural reactions of wintering humpback whales (*Megaptera novaeangliae*) to biopsy sampling in the western South Atlantic. J Mar Biol Assoc UK 90:1701–1711. doi:10.1017/S0025315409991561

Carlson C (2012) A review of whale watch guidelines and regulations around the world version 2012. Maine

Chion C, Cantin G, Dionne S et al (2012) Spatiotemporal modelling for policy analysis: application to sustainable management of whale-watching activities. Mar Policy 38:151–162. doi:10.1016/j.marpol.2012.05.031

Chittleborough RG (1965) Dynamics of two populations of the humpback whale, Megaptera novaeangliae (Borowski). Aust J Mar Freshwat Res 16:33–128

Christiansen F, Lusseau D (2014) Understanding the ecological effects of whale-watching on cetaceans. In: Higham J, Bejder L, Williams R (eds) Whale-watching, sustainable tourism and ecological management. Cambridge University Press, Cambridge, pp 177–192

Christiansen F, Lusseau D (2015) Linking behavior to vital rates to measure the effects of non-lethal disturbance on wildlife. Conserv Lett 8:424–431. doi:10.1111/conl.12166

Christiansen F, Rasmussen M, Lusseau D (2013a) Whale watching disrupts feeding activities of minke whales on a feeding ground. Mar Ecol Prog Ser 478:239–251. doi:10.3354/meps10163

Christiansen F, Rasmussen MH, Lusseau D (2013b) Inferring activity budgets in wild animals to estimate the consequences of disturbances. Behav Ecol 24:1415–1425. doi:10.1093/beheco/art086

Christiansen F, Bertulli CG, Rasmussen MH, Lusseau D (2015) Estimating cumulative exposure of wildlife to non-lethal disturbance using spatially explicit capture recapture models. J Wildl Manag 79:311–324. doi:10.1002/jwmg.836
Cifuentes M (1992) Determinación de capacidad de carga turística en áreas protegidas. Turrialba, Costa Rica
Cipolotti SRC (2013) Composição dos Grupos Sociais da Baleia Jubarte (*Megaptera novaeangliae*) e sua distribuição por profundidade observada a bordo das embarcações de turismo, na Praia do Forte, Bahia – Brasil. Universidade Estadual de Santa Cruz
Cipolotti SRC, Morete ME, Basto BL et al (2005) Increasing of whale-watching activities on humpback whales in Brazil: implications, monitoring and research
Cisneros-Montemayor AM, Sumaila UR, Kaschner K, Pauly D (2010) The global potential for whale watching. Mar Policy 34:1273–1278. doi:10.1016/j.marpol.2010.05.005
Corbelli C (2006) An evaluation of the impact of commercial whale watching on humpback whales, *Megaptera novaeangliae*, in Newfoundland and Labrador, and the effectiveness of a voluntary code of conduct as a management strategy
Corkeron PJ (1995) Humpback whales (*Megaptera novaeangliae*) in Hervey Bay, Queensland: behaviour and responses to whale-watching vessels. Can J Zool 73:1290–1299. doi:10.1139/z95-153
Curtin SC (2003) Whale-watching in Kaikoura: sustainable destination development? J Ecotour 2:173–195. doi:10.1080/14724040308668143
Dawbin WH (1966) The seasonal migratory cycle of humpback whales. In: Norris KS (ed) Whales, dolphins and porpoises. University of California Press, Berkeley.
DHN (1995) Tábua da Marés, pp 90–92
Duffus DA, Dearden P (1990) Non-consumptive wildlife-oriented recreation: a conceptual framework. Biol Conserv 53:213–231. doi:10.1016/0006-3207(90)90087-6
Dunlop RA (2016) The effect of vessel noise on humpback whale, *Megaptera novaeangliae*, communication behaviour. Anim Behav 111:13–21. doi:10.1016/j.anbehav.2015.10.002
Engel M, Ades C (1996) Comportamento reprodutivo da baleia jubarte (*Megaptera novaeangliae*) em Abrolhos. An Etol
Filby NE, Stockin KA, Scarpaci C (2015) Social science as a vehicle to improve dolphin-swim tour operation compliance? Mar Policy 51:40–47. doi:10.1016/j.marpol.2014.07.010
Findlay K (2001) A review of humpback whale catches by modern whaling operations in the Southern Hemisphere. Mem Queensl Museum 47:411–420
Finkler W, Higham J (2004) The human dimensions of whale watching: an analysis based on viewing platforms. Hum Dimens Wildl 9:103–117. doi:10.1080/10871200490441757
Forestell PH (1993) If Leviathan has a face, does Gaia have a soul?: incorporating environmental education in marine eco-tourism programs. Ocean Coast Manag 20:267–282. doi:10.1016/0964-5691(93)90070-F
Forestell P (2007) Protecting the ocean by regulating whalewatching: the sound of one hand clapping. In: Higham J, Lück M (eds) Marine wildlife and tourism management: insights from the natural and social sciences. CABI Publishing, Oxfordshire, pp 272–293
Fortuna CM (2006) Ecology and conservation of bottlenose dolphins (*Tursiops truncatus*) in the north-eastern Adriatic Sea. University of St. Andrews, Andrews
Garrod B, Fennell DA (2004) An analysis of whalewatching codes of conduct. Ann Tour Res 31:334–352. doi:10.1016/j.annals.2003.12.003
Gill JA, Norris K, Sutherland WJ (2001) Why behavioural responses may not reflect the population consequences of human disturbance. Biol Conserv 97:265–268
Graefe AR, Kuss FR, Vaske JJ (1990) Visitor impact management: a planning framework. Washeington, DC
Heberlein TA (1977) Density, crowding and satisfaction: sociological studies for determining carrying capacity. In: The national symposium on river recreation management and research, pp 67–76

Higham JES, Bejder L, Lusseau D (2009) An integrated and adaptive management model to address the long-term sustainability of tourist interactions with cetaceans. Environ Conserv 35:294–302. doi:10.1017/S0376892908005249
Higham JES, Bejder L, Allen SJ et al (2015) Managing whale-watching as a non-lethal consumptive activity. J Sustain Tour 24:73–90. doi:10.1080/09669582.2015.1062020
Hoyt E (1984) The whale watchers's handbook. Doubleday, Garden City
Hoyt E (1995) The worldwide value and extent of whalewatching 1995, Bath, UK
Hoyt E (2001) Whale watching 2001. Worldwide tourism numbers, expenditures, and expanding socioeconomic benefits. London, UK
Hoyt E (2009) Whale-watching. In: Perrin WF, Wursig B, Thewissen JGM (eds) Encyclopedia of marine mammals. Academic Press, San Diego, pp 1219–1223
IBAMA (1991) Plano de Manejo Parque Nacional Marinho dos Abrolhos. Brasília, Brasil
IBAMA/NEMA (1990) Relatório Técnico Temporada 1989
ICMBio (2014) Diagnóstico do Risco de Extinção de Espécies da Fauna: 2012–2014. Brasília, DF
IFAW (1995) Report of the workshop on the scientific aspects of managing whale-watching. Montecastello di Vibio, Italy
IWC (1994) Report of the forty-fifth annual meeting, appendix 9. IWC Resolution on whalewatching
IWC (2004) Report of the workshop on the science for sustainable whalewatching
IWC (2005) Report of the scientific commitee. Annex H. Report of the sub-commitee on other Southern Hemisphere whale stocks
IWC (2014) International convention for the regulation of whaling, 1946 – Schedule. 16
Jeffery A, Postle D, Simmons M (1994) Beyond the breach – managing for whale conservation and whale-watching in Hervey Bay Marine Park, Queensland. In: Postle D, Simmons M (eds) Encounters with whales '93 – a conference to further explore the management issues relating to human/whale interactions. Townsville, Queensland, pp 91–107
Kessler M, Harcourt R (2013) Whale watching regulation compliance trends and the implications for management off Sydney, Australia. Mar Policy 42:14–19. doi:10.1016/j.marpol.2013.01.016
Kessler M, Harcourt R, Heller G (2013) Swimming with whales in Tonga: sustainable use or threatening process? Mar Policy 39:314–316. doi:10.1016/j.marpol.2012.12.013
Kessler E, Harcourt R, Bradford W (2014) Will whale watchers sacrifice personal experience to minimize harm to whales? Tour Mar Environ 10:21–30. doi:10.3727/154427314X14056884441662
Krob AJD (2003) Plano de Uso Público para o Parque Nacional Marinho dos Abrolhos, pp 1–158
Lalime J (2005) Operator compliance and attitudinal compliance to aspects of Australian whale watch management conditions. James Cook University
Lime DW, Stankey GH (1971) Carrying capacity: maintaining outdoor recreation quality. In: Recreation symposium proceedings. Northeast Forest Experiment Station, Forest Service, USDA, pp 174–184
Limonad E (2007) O Fio da Meada. Desafios ao Planejamento e à Preservação Ambiental na Costa dos Coqueiros (Bahia)
Liu Z (2003) Sustainable tourism development: a critique. J Sustain Tour 11:459–475
Lunardi DG, Engel MH, Macedo RHF (2008) Behavior of humpback whales, *Megaptera novaeangliae* (Cetacea: Balaenopteridae): comparisons between two coastal areas of Brazil. Rev Bras Zool 25:159–164
Lusseau D (2004) The hidden cost of tourism: detecting long-term effects of tourism using behavioral information
Lusseau D (2005) The state of the scenic cruise industry in doubtful sound in relation to a key natural resource: bottlenose dolphins. In: Hall M, Boyd S (eds) Nature-based tourism in peripheral areas: development or disaster? Channel View Publications, Clevedon, pp 246–260
Lusseau D (2014) Ecological constraints and the propensity for populations consequences. In: Higham J, Bejder L, Williams R (eds) Whale-watching, sustainable tourism and ecological management, 1st edn. Cambridge University Press, Cambridge, pp 229–241

Lusseau D, Slooten L, Currey RJC (2006) Unsustainable dolphin-watching tourism in Fiordland, New Zealand. Tour Mar Environ 3:173–178. doi:10.1017/CBO9781107415324.004
Lusseau D, Bain DE, Williams R, Smith JC (2009) Vessel traffic disrupts the foraging behavior of southern resident killer whales *Orcinus orca*. Endanger Species Res 6:211–221. doi:10.3354/esr00154
Magalhães S, Silva MA, Prieto R et al (2002) Desenvolvimento do "Whale-Watching" nos Açores. Relatório Final. Arquivos do DOP. Série Estudos, nº5/2002
Martins CCA, Morete ME, Engel MH et al (2001) Aspects of habitat use patterns of humpback whales in the Abrolhos Bank, Brazil, breeding ground. Mem Qld Mus 47:563–570
Martins CCA, Andriolo A, Engel MH et al (2013) Identifying priority areas for humpback whale conservation at Eastern Brazilian Coast. Ocean Coast Manag 75:63–71. doi:10.1016/j.ocecoaman.2013.02.006
Mas-Rosa S, Baracho CG, Marcovaldi E, Engel ME (2002) Dados preliminares sobre a reocupação de uma antiga área de reprodução de baleia jubarte (*Megaptera novaeangliae*) no litoral norte do estado da Bahia, Brasil. In: 10a Reunião de Trabalho de Especialistas em Mamíferos Aquáticos da América do Sul e 4º Congresso da Sociedade Latino Americana de Especialistas em Mamíferos Aquáticos. Viña del Mar, Chile, p 103
McCool SF, Lime DW (2001) Tourism carrying capacity: tempting fantasy or useful reality? J Sustain Tour 9:372–388. doi:10.1080/09669580108667409
McMahon CR, Burton HR, Bester MN (2000) Weaning mass and the future survival of juvenile southern elephant seals at Macquarie Island. Antarct Sci 12:149–153
Mitraud S (Org. (2001) Uso recreativo do Parque Nacional Marinho Fernando de Noronha: um exemplo de planejamento e implementação. Brasília, Brasil
Morete ME (2003a) A novel behavior observed in humpback whales on wintering grounds at Abrolhos Bank (Brazil). Mar Mamm Sci 19:694–707
Morete ME (2003b) Observação de ponto-fixo para estudo comportamental das baleias jubarte e a interação com embarcações ao redor do Arquipélago de Abrolhos, Bahia. In: XXI Encontro Anual de Etologia e III Simpósio de Ecologia Comportamental e de Interações, Uberlândia, MG
Morete ME (2007) Caracterização temporal da estrutura de grupos e do comportamento de baleias jubarte (*Megaptera novaeangliae*) na área de reprodução da região do Arquipélago dos Abrolhos (Bahia, Brasil). Universidade de São Paulo
Morete ME, Freitas AC, Engel ME, Glock L (2000) Tourism characterization and preliminary analyses of whale watching on humpback whales (*Megaptera novaeangliae*) around Abrolhos Archipelago, Southeastern Bahia, Brazil. Paper present to science committee 52nd meet, International Whaling Commission, , Adelaide, Aust, 3–6 July 2000, pp 1–10
Morete ME, Bisi TL, Rosso S (2007) Mother and calf humpback whale responses to vessels around the Abrolhos Archipelago, Bahia, Brazil. J Cetacean Res Manag 9:241–248
Mustika PLK, Birtles A, Welters R, Marsh H (2012) The economic influence of community-based dolphin watching on a local economy in a developing country: implications for conservation. Ecol Econ 79:11–20. doi:10.1016/j.ecolecon.2012.04.018
Needham MD, Szuster BW, Bell CM (2011) Encounter norms, social carrying capacity indicators, and standards of quality at a marine protected area. Ocean Coast Manag 54:633–641. doi:10.1016/j.ocecoaman.2011.06.004
New LF, Clark JS, Costa DP et al (2014) Using short-term measures of behaviour to estimate long-term fitness of southern elephant seals. Mar Ecol Prog Ser 496:99–108. doi:10.3354/meps10547
New LF, Hall AJ, Harcourt R et al (2015) The modelling and assessment of whale-watching impacts. Ocean Coast Manag 115:10–16. doi:10.1016/j.ocecoaman.2015.04.006
O'Connor S, Campbell R, Knowles T, Cortez H (2009) Whale watching worldwide: tourism numbers, expenditures and expanding economic benefits, a special report from the International Fund for Animal Welfare, Yarmouth MA

Orams MB (2000) Tourists getting close to whales, is it what whale-watching is all about? Tour Manag 21:561–569. doi:10.1016/S0261-5177(00)00006-6

Pacheco de Godoy ML, Baracho C, Ferreira S et al (2005) Humpback whale (*Megaptera novaeangliae*) photo-identification in Brazil: aspects of habitat use in two different breeding areas. In: Biennial conference on the biology of marine mammals, 16. The Society for Marine Mammology, San Diego, p 216

Palazzo JT Jr, Groch KR (2007) Áreas restritas como ferramenta de ordenamento do turismo de observação de baleias na Área de Proteção Ambiental da Baleia Franca, SC. In: Anais V Congresso Brasileiro de Uinidades de Conservação. Foz do Iguaçú, Brasil, pp 1–9

Parsons ECM (2012) The negative impacts of whale-watching. J Mar Biol 2012:1–9. doi:10.1155/2012/807294

Parsons ECM, Warburton CA, Woods-Ballard A et al (2003) The value of conserving whales: the impacts of cetacean-related tourism on the economy of rural West Scotland. Aquat Conserv Mar Freshw Ecosyst 13:397–415. doi:10.1002/aqc.582

Parsons E, Baulch S, Bechshoft T et al (2015) Key research questions of global importance for cetacean conservation. Endanger Species Res 27:113–118. doi:10.3354/esr00655

Petta CB, Andriolo A, Morete ME, Martins CCA (2002) Comportamento respiratório das baleias jubarte (*Megaptera novaeangliae*) em seu sítio reprodutivo no Banco dos Abrolhos, Bahia, Brazil. In: Reunión de trabajo de especialistas en mamíferos acuáticos de América Del Sur – 5º Congreso de la Sociedad Latinoamericana de Epecialistas en Mamíferos Acuáticos

Pirotta E, New L, Harwood J, Lusseau D (2014) Activities, motivations and disturbance: an agent-based model of bottlenose dolphin behavioral dynamics and interactions with tourism in doubtful sound, New Zealand. Ecol Model 282:44–58. doi:10.1016/j.ecolmodel.2014.03.009

Reilly SB, Bannister JL, Best PB et al (2008) *Megaptera novaeangliae*. In: IUCN red list threat. Species. Version 2015.2. http://www.iucnredlist.org/details/13006/0. Accessed 14 July 2015

Rocha RC, Clapham PJ, Ivashchenko YV (2014) Emptying the Oceans: a summary of industrial whaling catches in the 20th century. Mar Fish Rev 76:37–48. doi: 10.7755/MFR.76.4.3

Rossi-Santos MR (2016) Whale-watching noise effects on the behavior of humpback whales (*Megaptera novaeangliae*) in the Brazilian breeding ground. Proc Meet Acoust 27:12

Rossi-Santos MR, Neto ES, Baracho CG et al (2008) Occurrence and distribution of humpback whales (*Megaptera novaeangliae*) on the north coast of the state of Bahia, Brazil, 2000–2006. ICES J Mar Sci 65:667–673. doi:10.1093/icesjms/fsn034

Salerno F, Viviano G, Manfredi EC et al (2013) Multiple carrying capacities from a management-oriented perspective to operationalize sustainable tourism in protected areas. J Environ Manag 128:116–125. doi:10.1016/j.jenvman.2013.04.043

Scarpaci C, Parsons ECM (2014) Recent advances in whale-watching research: 2013–2014

Scarpaci C, Dayanthi N, Corkeron PJ (2003) Compliance with regulations by "swim-with-dolphins" operations in Port Phillip Bay, Victoria, Australia. Environ Manag 31:342–347. doi:10.1007/s00267-002-2799-z

Schaffar A, Madon B, Garrigue C, Constantine R (2013) Behavioural effects of whale-watching activities on an endangered population of humpback whales wintering in New Caledonia. Endanger Species Res 19:245–254. doi:10.3354/esr00466

Scheidat M, Castro C, Janira G, Williams R (2004) Behavioural responses of humpback whales (*Megaptera novaeangliae*) to whalewatching boats near Isla de la Plata, Machalilla National Park, Ecuador. J Cetacean Res Manag 6:1–6

Senigaglia V, Christiansen F, Bejder L et al (2016) Meta-analyses of whalewatching impact studies: comparisons of cetacean responses to disturbance. Mar Ecol Prog Ser 542:251–263. doi:10.3354/meps11497

Sironi M, Schteinbarg R, Losano P, Carlson C (2005) Sustainable whale watching at Península Valdés, Argentina: an assessment by owners and captains of local whale watch companies. J Cetacean Res Manag SC/57/WW2:1–9

Sitar A, May-Collado LJ, Wright A, et al (2015) Tourists' perspectives on dolphinwatching in Bocas del Toro, Pamana, support sustainable and educational tourism. Paper presented to the

Scientific Committee at the 66th meeting of the International Whaling Commission, 19 May–3 June 2015, San Diego, USA
Smith A, Newsome D, Lee D, Stoeckl N (2006) Case studies: monkey mia dolphins and Hervey Bay whale watching, Queensland
Sousa-Lima RS, Clark CW (2009) Whale sound recording technology as a tool for assessing the effects of boat noise in a Brazilian marine park. Park Sci 26:59–63
Stamation KA, Croft DB, Shaughnessy PD et al (2007) Educational and conservation value of whale watching. Tour Mar Environ 4:41–55. doi:10.3727/154427307784835660
Stamation KA, Croft DB, Shaughnessy PD et al (2010) Behavioral responses of humpback whales (*Megaptera novaeangliae*) to whale-watching vessels on the southeastern coast of Australia. Mar Mamm Sci 26:98–122. doi:10.1111/j.1748-7692.2009.00320.x
Stankey GH, Cole DN, R.C. Lucas Margaret, et al (1985) The limits of acceptable change (LAC) system for wilderness planning. General technical report INT-176. Ogden, UT
Thomas PO, Taber SM (1984) Mother-infant interaction and behavioural development in southern right whales, *Eubalaena australis*. Behaviour 88:42–60
Tyne J, Loneragan N, Bejder L (2014) The use of area–time closures as a tool to manage cetacean-watch tourism. In: Higham J, Bejder L, Williams R (eds) Whale-watching, sustainable tourism and ecological management: sustainable management: insights and cases. Cambridge University Press, Cambridge, pp 242–256
van Schalkwyk M (2008) Policy on boat-based whale and dolphin watching. Department of Environmental Affairs and Tourism, Republic of South Africa
Ward E, Zerbini AN, Kinas PG et al (2011) Estimates of population growth rates of humpback whales (*Megaptera novaeangliae*) in the wintering grounds off the coast of Brazil (Breeding Stock A). J Cetacean Res Manag (Special Issue) 3:145–149
Wedekin LL (2011) Ecologia populacional da baleia-jubarte (*Megaptera novaeangliae* Borowski, 1871) em sua área reprodutiva na Costa do Brasil, Oceano Atlântico Sul. Universidade Federal do Paraná
Wedekin LL, Neves MC, Marcondes MCC et al (2010) Site fidelity and movements of humpback whales (*Megaptera novaeangliae*) on the Brazilian breeding ground, southwestern Atlantic. Mar Mamm Sci 26:787–802. doi:10.1111/j.1748-7692.2010.00387.x
Weinrich M, Corbelli C (2009) Does whale watching in southern New England impact humpback whale (*Megaptera novaeangliae*) calf production or calf survival? Biol Conserv 142:2931–2940. doi:10.1016/j.biocon.2009.07.018
Whitehead H, Payne R (1981) New techniques for assessing populations of right whales without killing them
Wiley DN, Moller JC, Pace RM, Carlson C (2008) Effectiveness of voluntary conservation agreements: case study of endangered whales and commercial whale watching. Conserv Biol 22:450–457. doi:10.1111/j.1523-1739.2008.00897.x
Woods-Ballard AJ, Parsons ECM, Hughes AJ et al (2003) The sustainability of whale-watching in Scotland. J Sustain Tour 11:40–55. doi:10.1080/09669580308667192
Wright AJ, Kyhn LA (2015) Practical management of cumulative anthropogenic impacts with working marine examples. Conserv Biol 29:333–340. doi:10.1111/cobi.12425
Zerbini AN, Andriolo A, Heide-Jorgensen MP et al (2006) Satellite-monitored movements of humpback whales, *Megaptera novaeangliae*, in the south-west Atlantic Ocean. Mar Ecol Prog Ser 313:295–304
Zerbini AN, Ward EJ, Kinas PJ et al (2011) A Bayesian assessment of the conservation status of humpback whales (*Megaptera novaeangliae*) in the western South Atlantic Ocean. J Cetacean Res Manag (Special Issue) 3:131–144

Part II
Technological Innovation for Marine Vertebrate Research in Latin America

Chapter 4
Land-Based Station Studies of Aquatic Mammals in Latin America: Understanding Behavior for Conservation

Maria Emilia Morete, Daniela Abras, and Cristiane C.A. Martins

Abstract Behavioral studies are the key to understand animals and to build up the knowledge of their ecology playing a key role in investigating consequences of anthropogenic activities. Land-based station (LBS) enables broad view of animals in their habitat, not interfering in their natural behavior. This chapter presents an overview of LBS research comprehensiveness, limitations, methodologies and equipment, guiding readers to the main literature in the field. In Latin America LBS studies have been identified in 11 countries: Argentina, Brazil, Chile, Colombia, Ecuador, French Guiana, Mexico, Nicaragua, Peru, Uruguay and Venezuela. At least 27 species of aquatic mammals have been studied from land, comprising six Mysticetes, 18 Odontocetes, two Pinnipeds and a Sirenian. LBS research enhanced the knowledge of several species behavior, movement patterns, distribution, and habitat use. They have provided valuable support to conservation measures and coastal management in Latin America countries. Because of their relative low cost, simple logistics and diverse possibilities, LBS studies should be encouraged in all areas where aquatic mammals are present and observations from land are feasible.

Keywords Land-based station • Animal behavior • Behavioral study • Aquatic mammals • Observational methodologies • Theodolite • Total station

---

M.E. Morete (✉) • D. Abras
VIVA Baleias, Golfinhos e cia,
Estr. Ana Procópio de Moraes 783, Santana Do Parnaiba 06528--551, Brazil
e-mail: miamorete@gmail.com

C.C.A. Martins
Tryphon Oceans, 4 Rue du Moulin-à-Baude, CP 451, Tadoussac, QC G0T2A0, Canada

M.R. Rossi-Santos, C.W. Finkl (eds.), *Advances in Marine Vertebrate Research in Latin America*, Coastal Research Library 22,
DOI 10.1007/978-3-319-56985-7_4

4.1 Introduction

The attraction and curiosity of human beings to aquatic mammals dates prehistorical ages. Due to their grace and mystery, aquatic mammals have been observed, described and portrayed in cave paintings, carved in rocks and bones, drawn in vases, mosaics, totems. They were part of Greeks mythology, ancient and autochthonous legends from all over the world. They have been seeing as gods and beasts, have been hunted, kept in captivity, cared, protected and loved. While humans would not venture out to explore the oceans, they would observe aquatic mammals from land (Ellis 2009) and land-based lookouts were probably the first platform used to observe live aquatic mammals.

Aquatic mammals inhabit almost every aquatic habitat on this planet. Study these animals however, is not an easy task, since some have offshore distribution, or pass by the coast only during a period of the year, while several never approach. Some seasonally migrate long distances, inhabit inner regions in estuaries and rivers, many have crypt and/or inconspicuous surface behavior. Additionally, since they spend most of the time diving, being out of view, aquatic mammals are observable for only a small proportion of time.

All research platforms and methods used to study aquatic mammals have advantages and disadvantages. The decision of which use will be mostly guided by the objectives and goals of the research, within the available budget. Nowadays, new technologies, allow animal monitoring for short and long periods, revealing details about their behavior, dive and swimming patterns, distribution, among others. However, with few exceptions, most research methods are in a certain manner invasive. In fact, just the observer's presence may introduce subtle bias even when animals appear to be well habituated (Martin and Bateson 1993). It has been shown that a simple boat approach may interfere in animal behavior (e.g., Corkeron 1995; Bejder et al. 1999; Heckel et al. 2001; Failla et al. 2004; Scheidat et al. 2004; Morete et al. 2007a; Pavez et al. 2011; Lundquist et al. 2013; Vermeulen et al. 2012; Avila et al. 2015) and that most of the new technologies have the capacity to cause pain, injuries, or modify the behavior of the animal being observed (Gales et al. 2003).

Behavioral studies are the key to understanding animals (Mann 2000) and to build up the knowledge of their ecology, playing a role in investigating consequences of anthropogenic activities (e.g., Bejder et al. 2006). However, link short-term behavioral responses to long-term population-level consequences presents difficulties, a fact that can lead to the false, or at least premature conclusion that human activities have no biologically significant effect on the targeted animals (Williams et al. 2006). Behavior can be affected in many different manners but it will only be perceived by the observer when the animal or a group of animal stop the activity they were engaged in, or react in some way to respond to an external stimuli. A punctual behavior change might be meaningless, but they usually have cumulative effects. For example, if an animal stops its feeding activity, it will have to compensate extending its feeding period, which might reduce its socializing or resting time. Thus, what seems to be a simple behavioral short-term change may affects life functions, influencing vital rates, leading to population scale effects (National Research Council 2005). Aquatic mammals are slow breeding animals and the consequences of any disturbance may take some time to be detected, once population level responses are long-term effects (Tyack et al. 2003). Therefore, behavioral studies enable the design of adequate management actions to ensure species' protection (Alcock 1998), addressing important conservation problems by monitoring the health of the target population, showing unrecognized risks (Tyack et al. 2000).

In many places of the Earth, it is possible to observe from a remote place aquatic mammals that inhabit near shore waters, occupy a certain coastal area during a period of the year, migrate by the coastline or approach river banks. A little mountain by the sea, a sand dune by the shore, or a high place at an island, a rock, a tower, a lighthouse, a tree, even a house's roof, or an apartment building balcony, may be a belvedere, an observatory for contemplating and admire the animals. Advantages places like these may also serve a touristic purpose, allowing the development of land-based whalewatching activities, for example. In the other hand, if the objective is to study these animals, these look-outs points can serve as a land-based station (LBS), for scientifical purposes (Fig. 4.1).

Studies from LBS have the great advantage of not interfere in animal natural behavior (e.g., Würsig et al. 1991) being not invasive. LBS enables for a broad view of animals in their habitat, allowing to address different and relevant studies as: in-depth description of behavior, development of an ethogram, characterization of habitat use and distribution, estimation of species density, assessment of the impact of anthropogenic activities, evaluation of the impact of climate changes, etc. Therefore forming the core information for many conservation measures, as baseline information from species and areas, development of management plan and creation of protected areas (i.e., State and National Parks), establishment of tourism and whale/dolphin watching guidelines, artisanal and/or industrial fisheries legislation, vessel and ships traffic, improvement of coastal management, etc.

4.2 Land-Based Studies Overview

Land-based station studies can be a "simple natural history" behavioral pattern study of a species within a certain area. The study area may encompass a core region for the species, or a small part of its distribution, or even just a passage area. These observations can be performed on daily basis, or monthly, seasonally and annually, becoming a short, middle or long term research. Details about the way one individual animal behave, and how it behaves in relation to the group, or how several groups behave within a time frame. How individuals behave according to age, or social status, time of the day, or period of the season. How social aspects change throughout the day, or even during a season. Do they have a typical movement pattern? What speed can they reach? Does calf behavior differ if the mother is accompanied by another individual? Is there a preferred habitat area among different group categories? Do bathymetric features, water and air temperature, tide

Fig. 4.1 Examples of land-based stations diversity in Latin America: (**a**) West Indian Manatee at Praia do Pico, Ceará (Brazil); (**b**) Humpback whale at Bahia Málaga (Colombia); (**c**) Humpback Whales at Abrolhos, Bahia (Brazil); (**d**) Southern Right Whale at Morro do Farol Imbituba, Santa Catarina (Brazil); (**e**) West Indian Manatee at Barra do Mamanguape, Paraiba (Brazil); (**f**) South American fur seal at Isla de Lobos (Uruguay); (**g**) Spinner dolphin at Fernando de Noronha (Brazil);

variation, moon phase play a role in their distribution? What is the density of the species within an area? Does it fluctuate through time? Perhaps the presence of the animals is related to environmental variations... Can an animal react to a playback of a social sound? Or the sound of a predator? What are the consequences of the approach of a vessel? How the behavior of a mother and calf change in the presence of a tourism boat? Does the presence of a boat alter animal activity leading to energetic costs? Does a biopsy sampling procedure causes behavioral change? What about powerful noises as those used in seismic surveys? Well, these are some examples of questions that have been addressed from LBS studies conducted around the world (e.g., Tyack 1981; Smultea 1994, Mobley et al. 1988; Bejder et al. 1999; Frankel and Clark 2000; Williams et al. 2002, 2006; Morete et al. 2003a, b, 2007a, b, 2008; Oviedo and Silva 2005; Coscarella et al. 2010, 2011; Araújo and Wang 2012; Sepúlveda et al. 2012; Dunlop et al. 2015, Danilewicz et al. 2016).

As shown, from a LBS it is possible to develop those "natural history" behavioral studies which form the structural knowledge of a species, as well as habitat use studies which attempt to describe, explain, and predict the distribution and abundance of organisms (Acevedo-Gutiérrez 2009), showing the way an animal uses physical and biological components in a habitat (Morrison 2002). Also, behavioral-response study (BRS), to determine the function and/or behavioral effects of conspecifcs, heterospecifics and anthropogenic stimuli (Dunlop et al. 2012). Likewise, studies of the impact of anthropogenic activities (as vessels traffic, dolphin/whalewatching, noise pollution caused by seismic activities, sonar signals) by means of short-term, punctual behavioral observation, and long-term behavioral and distribution changes.

Many studies around the world merge land-based observation with acoustics methods, as passive acoustic monitoring (PAM), in order to describe the animal's songs and sounds together with swimming speeds, direction and surface behavior (e.g., Frankel et al. 1995; Noad and Cato 2001, 2007). At Queensland coast, in Australia, there are sequences of LBS performing a combined visual and acoustic survey of humpback whales (*Megaptera novaeangliae*) while it migrates along to the coast. Observations of a whale or group of whale starts at one land-based station until it is within visual range and the next station continue observing the same group (Dunlop et al. 2015), making longer periods of focal follows (sampling method describe later in this chapter). The use of land-based platforms combined with other methods (e.g. acoustics, genetics) magnifies the scientific outcomes.

Fig. 4.1 (continued) (**h**) cetaceans at Ilha Vitoria, São Paulo (Brazil); (**i**) humpback whale at Cumuruxatiba Bahia (Brazil); (**j**) Bottlenose and Dusky dolphin at Valdes Peninsula (Argentina); (**k**) Humpback whale at Serra Grande, Bahia (Brazil); (**l**) Guiana dolphin at Caravelas Bahia (Brazil); (**m**) cetaceans at Mejillones Bay (Chile); (**n**) Bottlenose dolphin at Veracruz (Mexico); (**o**) Spotted dolphin at San Juan del Sur Nicaragua) (**p**) roof of a house at Praia do Forte Bahia (Brazil); (**q**) Humpback whale at Cerro la Mesa (Peru); (**r**) Bryde's Whale at Laje de Santos, São Paulo (Brazil)

4.2.1 Land-Base Station Aspects and Limitations

Studies from land-station, frequently suffer from limited geographic coverage and represent a small part of the species distribution, so it is necessary to treat the collected data with caution. Does the sampled area show what is observed in the entire distribution of the animals? What inferences can be derived from the data? Normally, because of the elevation of the land-based station, it is almost impossible to recognize individuals, except if the animal shows very conspicuous and uncommon marks. For this reason, it is important to note the possibility of the same individual being sampled more than once. Besides, animals that are farther away are less likely to be detected than those closer to shore, causing a skewed view of the actual distribution (Arranz et al. 2014). All these aspects and limitations should be taken into account. Scientists deal with these issues in a variety of ways, from time and spacing of sampling (e.g., Baker et al. 1982; Smultea 1994; Frankel and Clark 2002), up to the development of statistical models that allow the use of such data (Bart et al. 1998; Quinn and Keough 2002; Zuur et al. 2007, 2010).

4.2.2 Land-Base Station Spot and Tips

Not only the relatively close presence of the target species is relevant prior to defining the land-station spot; it is important to get some information about the locale. If the chosen place is inside a protected area, perhaps, special permits are necessary; if it is in a public area, gather information about the security of the place; also verify the possible occurrence of dangerous and poison animals. Likewise, it is pertinent to verify if the team presence won't jeopardized other animals' activities and behaviors, for example, nesting birds. Thus, there should be precautionary measures and the team should be aware of procedures to leave the area in case of any emergency.

Normally LBS are in open areas, without edifications, or constructions. Some kind of umbrella or tent, to protect from straight sunlight is advised. As environmental conditions may change suddenly, it is important to have some infrastructure to protect the team and the equipment when weather gets rough, as waterproof boxes, cases and proper clothes for protection against winds and cold.

The higher the land station the broader the field of view thus the study area. Although, it should not be too elevated spoiling the capacity of observations. If more than one observer will collect the data, or if observers will alternate functions within the study period, they should receive the same training by the same person to avoid inter-observers variations in data collection.

4.2.3 Land-Based Station Sampling Methods

Land-based station, as a platform to study animals without interference of observers on their behavior, enables the application of diverse methods. In face of the characteristic of the species studied, according to the research questions and the number of observers available, an adequate sampling method must be chosen. Among some authors (Altmann 1974; Martin and Bateson 1993; Bart et al. 1998; Mann 1999, 2000) there are differences in naming the methods, some divergences in definitions, but all of them discuss important considerations about the sampling methods, arising questions, restraints and problems about bias, over and underestimations, randomness, dependency, pseudo-replications, etc. It is crucial to understand all methods, assumptions and statistical consequences before defining the research protocol. A summary of the most frequent and common sampling methods used in LBS studies available in the literature is presented. However, for the complete information and assessment of the methods, search the above-mentioned references. Before starting, it is important to decide on systematic rules: which subjects to watch and when, how the behaviors will be recorded (Martin and Bateson 1993) and for how long (Mann 1999).

"*Ad libitum sampling*": there are no systematic constraints when the observer simply notes down whatever is visible and what seems relevant at the time (Martin and Bateson 1993). Altmann (1974) describes it as "field notes". This method will be clearly biased towards more conspicuous behaviors and data are not appropriate to estimate behaviors rates. It may be used during preliminary observations and pilot studies while creating the ethogram and delineating research questions (Mann 1999).

"*Focal sampling*" or "*focal-follow*": one individual or two (or some other focal unit within an adequate size for reliable data collection) are followed. The focal unit must have specific characteristics, or must behave in a way, which permit its identification and track. The unit will be observed for a specific amount of time, until it leaves the boundaries of the study area or until viewing conditions from land station gets rough due to glare, haze, or higher sea state. There should be rules in the observation protocol for deciding what have to be done when the focal unit is out of sight or "disappear" (Martin and Bateson 1993). Longer focal sampling observations are encouraged in order to avoid rare behaviors underestimation (Mann 1999), although observers fatigue should be taken into account in this context (Senigaglia 2011). The "*focal continuous sampling*" is when all behaviors performed by the focal unit are recorded with the exactly time of occurrence and duration, enabling to reproduce all that was observed (Martin and Bateson 1993). Continuous data are the richest source of information on social behavior, including information on details, sequences, rates and durations of behavior for individual animals (Mann 1999). This method is very demanding for observers and reliability of the data can be compromised when animals are active, or the focal unit is too large (Mann 1999). Altmann (1974) recommends this method for just one, or a maximum of two individuals.

"*Focal-group sampling*": a group (more than one individual) is followed; however, as already mention above, to observe and record all behaviors of all members of a group is very challenging leading to biased data. The "*focal-group continuous sampling*" is practically impossible without the use of video recording (Mann 1999).

"I*nstantaneous or point sampling*": the focal unit is sampled periodically on determined fixed moments, i.e. the behavior is recorded only if it occurs at the exactly moment of the sample point. As aquatic mammals dive between surface periods, they might be unavailable during some data collection points. The "*one-zero or timing sampling*" is when the focal unit is sampled periodically on fixed intervals and the observer records whether or not a behavior had occurred (Martin and Bateson 1993). Both, instantaneous sampling and one-zero sampling, do not give true behavior frequencies nor durations (e.g., Altmann 1974; Mann 1999), so their use must be specific to certain research goals where this kind of information is not necessary.

"*Scan sampling*": each individual, is observed for a short period and their behavior is recorded, but not in a continuous manner. There is no predetermined time for a scan, it can be conducted at regular intervals, or as quick as possible (Mann 1999). It may last few seconds to several minutes, depending on the size of the study area and the amount of individuals present, because in practice, the observation, classification, and recording of a behavior state takes time.

"*Survey*": individuals or groups are observed for a period of time, the number of animals, behavior, and location are recorded, providing information about demographic, reproductive and ecological factors (Mann 1999), mean and range of group sizes patterns (Mann 2000). It might serve as a kind of census. In order to properly characterize the groups, a survey can lasts longer because of the diving, swimming and sometimes cryptic behavior of target animals.

"*Behaviour sampling*": a whole group of subjects is under observation and just the occurrence of some behaviors is registered. It is basically used to record rare and significant behaviors (Martin and Bateson 1993).

It is often possible to use more than one sampling method in a study (Altmann 1974), actually, that is practical and advisable (Mann 2000). Also, adaptation and adequacies of methods must be made to fit research goals in the context of the target species. However, investigators should always minimize the biases inherent in different methods (Mann 2000), should recognize the limitations of their sample, methods and statistical inferences (Bart et al. 1998). If methods used to observe animals remain the same and sightability remains reasonably constant, comparisons among years should be valid (Salden 1988).

4.2.4 Preliminary Observations, Pilot Study, Protocol

A period of informal observation with quantitative recordings of behavior, aimed to understand and describe both, the species and the behavior intended to be measured, should preced an effective data collection. The preliminary observation provides the raw material for formulating hypothesis and also for choosing the most suitable recording method (Martin and Bateson 1993). The development of a protocol, describing all daily activities and procedures, the way equipment should be prepared, how to observe animals, the manner to register the data, how to summarize and tabulate data, etc., benefits and organize the research. Thus, a pilot study is recommended in order to build the ethogram, to test the observational protocol and to verify if the methods are adequate to reach the desired goals.

4.2.5 Ethogram

A comprehensive repertoire of the target species behavior is the first step in any behavioral study. It is crucial to extensively observe the animals in order to create a complete ethogram, which must be mutually exclusive and cumulative inclusive, i.e., each described behavior cannot include any other, nor give chance for misunderstanding among them, comprising all animal's behavioral events and behavioral states (as described by Altmann 1974: *behavioral events* are brief, short durations behaviors like breach, pectoral slap, lobtail, are measured in frequency, number of occurrence per unit of time, per example, the dolphin breached ten times per minute; *behavioral states* are behaviors of longer duration, like foraging, resting, are measured in units of time, the length of time which the state lasts, per example, the whale was resting for 25 min).

A list of well-defined behaviors, although challenging to develop, is essential. Each behavior must be clearly and specifically described (Mann 2000) allowing that different observers would interpret them the same way (Bart et al. 1998). Note that the meaning or function of a behavior event cannot be part of the ethogram, because the same behavior, the same movement, may occur in a variety of contexts and perhaps serve many different functions (Mann 2000). Better to use a very detailed ethogram at the field while collecting data, and at the moment of the analyses, if pertinent and/or necessary, cluster behaviors creating behavioral categories.

4.2.6 Land-Based Station Equipment

Another great advantage of land-based station studies is to be more cost-effective than other methods. Basically, the needed equipment is the observational device: binoculars, or spotting scope, and in studies where precise positioning is required, theodolite or total station. Theodolite and total stations a priori are relatively expensive devices, however, if they have proper maintenance and cleaning, if their measurement systems are checked and balanced annually by a professional, they can easily be used for decades, attenuating the initial investment. Also, as theodolite/ total station serves other purposes, they are often available for rent or to lend at some academic departments (e.g., Geology, Geography, Cartography).

Besides the viewing device, depending on the objective of the study, some other equipment's shall be used, as a VHF radio for communication purposes (i.e. with nearby vessels for example), a wind speedmeter used to monitor environmental conditions, photographic camera, a voice recorder or even a video camera to film animals. Some LBS studies use computers running dedicated software (described later in this chapter), and even in places where there is no energy supply, solar panel and batteries allow to overcome this limitation (Fig. 4.2).

4.2.7 Spotting Scope and Binoculars

A spotting scope is a small portable telescope with an erect image, optimized for the observation on land. Observer use just one eye and it has to be mounted on a tripod because of its magnification, usually ranging from 20× to 60×. Some have a zoom eyepiece to give a range of magnifications. However, more than 60× can lead to poorer image brightness, image shake and narrower field of view which difficult finding and following animals in movement.

Fig. 4.2 An example of a land-based station equipped with theodolite, binoculars, waterproof case, computer, solar panel and batteries to study humpback whales in the Abrolhos Archipelago, Bahia, Brazil

Binoculars are telescope mounted parallel pointing in the same direction, allowing the observer to use both eyes, providing more visual comfort. Normally the binoculars size permits to be held by both hands, and some may be also mounted on a tripod. All binoculars have numbers, such as 7×50 and 10×42. The first number refers to their magnification and the second number refers to objective lens diameter. The higher the magnification, smaller is the field of view. The bigger objective lens diameter the brighter image however, it makes the binocular heavier. Depending on the characteristic of the land-based station and the target species, different specifications for lens magnification, objective lens diameter and the use of a tripod may be applied.

Unlike spotting scope binoculars permit a three-dimensional image with an impression of depth of field, and although with less magnificent power than spotting scope, binocular allows to easily find animals, even in movement, what is especially necessary for observation of aquatic mammals with longer dive periods and fast displacement. Besides, binoculars are not expensive and are easy to carry and transport.

Some binoculars are provided with internal compass, vertical and horizontal reticles that enable the observer to have an approximate position of the target animal. To this end, although not mandatory, it is preferable to use the binoculars mounted on a tripod, which should be always at the exactly same known location. The binoculars should always be leveled (normally tripods present leveling bubbles). The height of the ground up to the eyepiece must be exactly measured and summed to the height of the land station in relation to the sea level, in order to obtain the approximate position of the target animals (see theodolite/total station section for measurements of cliff elevation and other factors that must be taken into to

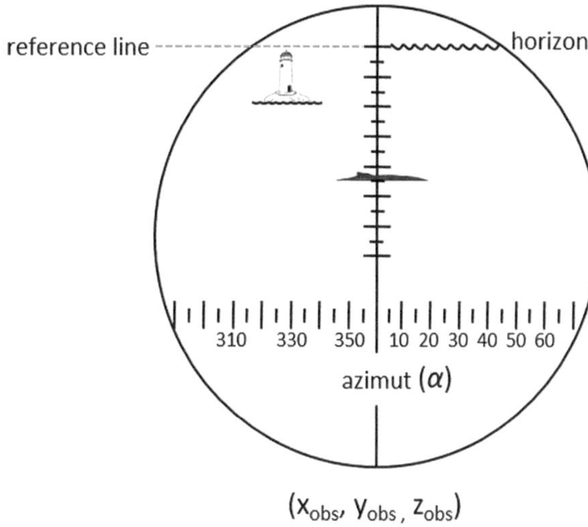

Fig. 4.3 Binocular field of view, showing vertical reticles, reference line: landmarks (as a lighthouse) or the horizon and azimuth from the internal compass. Xobs, Yobs, Z obs refers to the position of the observer (longitude, latitude and height) (Credit: Samuel Turgeon)

account). To get the position of the target, the top reticle of the binoculars should be at the horizon line. If there is no horizon line within the field of view, if the study area is within a bay, for example, the top reticle may be positioned on any fixed landmark, as a rock, a lighthouse, a tower, a building. In any case, every-day the top reticle must be set the same way. It is advisable to have other features to ensure the binoculars is always settled properly, minimizing positioning errors. When the target animal is in sight, the observer must read the closer reticle and its azimuth measured with the compass (remember that compass measures magnetic north, and there are local and time variation between magnetic and true north) (Fig. 4.3). Knowing the distance the animal is from the land station and the azimuth provided by the internal compass, the animal approximate position is obtained (for an overview about binocular sampling measurements check Buckland et al. 1993).

The binoculars aren't as accurate as a theodolite/total station, but it is possible to reach an optimal position estimate over medium distances with a precision good enough for most habitat use studies. Binoculars specification varies depending on brand or manufacturer; it is possible to find the scale value of one reticle and units at user's manual. When choosing binoculars and spotting scope, give preference to those that are water and fog-proof, as field weather conditions may be rough to equipment.

4.2.8 Theodolite and Total Station

In the seventies, with the awake of environmental consciousness, some young scientists started to observe whales and dolphins in their natural habitat. Roger and Katy Payne, Bernd and Melany Würsig (Fig. 4.4) were the pioneers utilizing theodolite to observe aquatic mammals at Valdes Peninsula. The theodolite "has

Fig. 4.4 Melany Würsig using the theodolite at the land-based station at Valdes Peninsula, in the 1970s (Credit: Bernd Würsig). Old models theodolites would produce an upside-down and inverted image, and measurement of the angles was just like a manual caliper rule (Credit: Maria Emilia Morete)

become the tool of choice when precise records of coastal movements of cetaceans were required, a way to observe cetaceans from afar, eliminating reliance on seagoing vessels, which can be disruptive to animals or restrictive to research budgets" (Samuels and Tyack 2000).

The total station and the theodolite are topographical equipments with monocular vision, normally with a magnification of 30×. The total station is a more complete device; its main advantage for animals tracking is the internal capacity of data storage. Both equipment's' enable the positioning of a target indirectly. They provide two angles, the vertical, which is the angle between the theodolite eye-piece and the target, and the horizontal, which is the angulation marking the azimuth, both in degrees (°), minutes (') and seconds ("). Theodolite and total station do not have an internal compass, it is necessary to mechanically input the right azimuth before collecting data, so the horizontal angle must be set at some fixed and known landmark in relation to the theodolite location, every time it is set up (Fig. 4.5). The theodolite/total station should always be on level (checked by levelling bubbles in the equipment) in order to give the correct vertical angle. With these two angles measured, plus the exact height of the station in relation to the sea level, considering the tide variation and Earth curvature, applying basic trigonometric calculus, the position of the target animal is obtained (Fig. 4.5). This conversion enables to faithfully reproduce the animals' movement and/or distribution (Fig. 4.6) and to extract desired movement parameters (e.g., speed, bearing, and turning angle) and measures of interest (e.g., distance among animals, distance between animal and vessel, between animal and sound source or hydrophone).

Data collection should be accomplished by a team of minimum three people: the principal theodolite observer, the binocular observer and the note taker. The binocular observer acts as a backup observer to assist the principal theodolite/total station observer. As theodolite/total station has a greater magnificent power, with a narrow

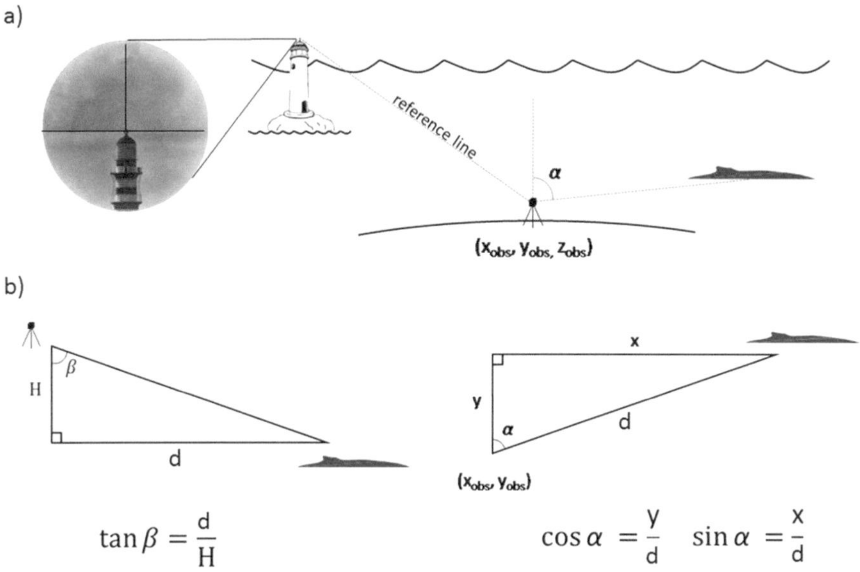

Fig. 4.5 (**a**) Fixed and known landmark should be used to reference the theodolite/total station horizontal angle (azimuth), in this example, the top of a lighthouse (Credit: Marina Leite Marques); (**b**) Trigonometric calculus (projected coordinate system) allow the position of the target, using the vertical (β) and horizontal (α) angles (measured by the theodolite/total station), plus the exact height (H) of the eye-piece above sea level (taking into to account tide variation). Xobs, Yobs, Zobs refers to the position of the observer (longitude, latitude and height), while x and y represent the distance (in these directions) between the observer and the target animal, while d is the closest distance (flight distance) between the observer and the target animal (Credit: Samuel Turgeon)

field of view, the binocular observer is essential as she/he can easier keep track of the animals, since binoculars have a larger field of view.

Basically, the higher the land-station, the more precise the positioning. However, caution must be taken, because there are many sources of haziness. It is essential to use a precise method in the establishment of the land-station height above sea level. It is indicated the use of a DGPS (Differential Global Positioning System), but a theodolite/total station itself allows calculating the station altitude (see details in Bailey and Lusseau 2004). The theodolite tripod has to be always placed at the exactly same spot, in order to avoid daily differences in positioning. Whenever theodolite/total station is mounted on a fixed tripod, the height from the ground up to the eyepiece must be registered, and it should be added to the land station height. Wursig and collaborators et al. (1991) present a table showing the impact of errors associated with incorrect measurements of cliff height, exemplifying for different land-station heights (Table 4.1). It is advised to estimate the error of the land-based station by comparing positions collected using a GPS and those obtained by positioning the boat along a track, though GPS positioning might also have small scale errors depending on cloud coverage, the number of satellites available, etc.

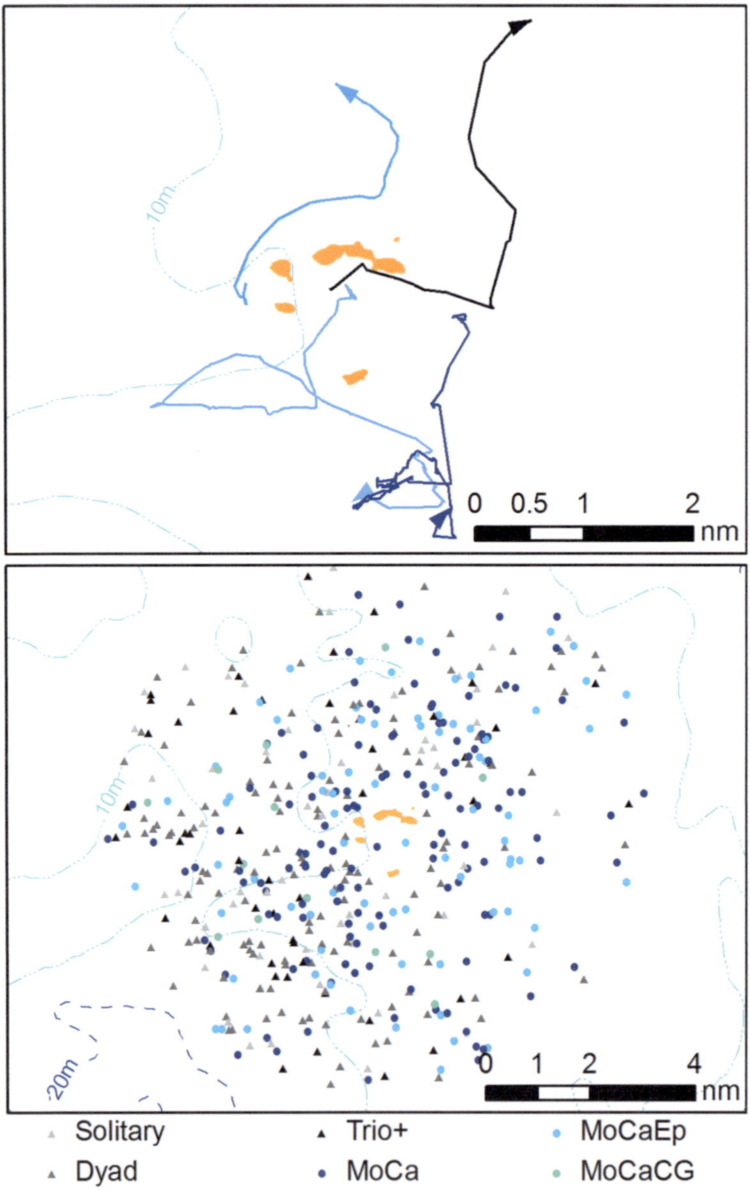

Fig. 4.6 An example of theodolite/total stations positioning data collected from Abrolhos Archipelago land-based station. The study area encompasses a 5 nm radius. (**a**) Examples of mother and calf humpback whale (*Megaptera novaeangliae*) focal follows showing their displacement around Abrolhos Archipelago (Modified from Bisi 2006). (**b**) Sightings of humpback whale groups: Solitary, Dyad (two adult whales), trio+ (three or more adult whales), MoCa (mother and calf), MoCaEp (mother, calf and one escort), MoCaCG (mother and calf within a competitive group) recorded during morning scans in 1998 season (IBJ unpublished data). The five islands of Abrolhos Archipelago are shown in *orange*

Table 4.1 Errors associated with incorrect measurements of cliff height

	Distance error (m)			
	True distance to position on the water			
Actual cliff height	Error in height	500 m	2500 m	5000 m
15 m	100 cm high	+34	+173	+388
	10 cm high	+4	+17	+39
	10 cm low	−3	−17	−38
	100 cm low	−30	−172	−379
30 m	100 cm high	+17	+85	+179
	10 cm high	+2	+8	+18
	10 cm low	−2	−9	−17
	100 cm low	−17	−85	−177
45 m	100 cm high	+12	+56	+117
	10 cm high	+2	+5	+12
	10 cm low	−1	−6	−11
	100 cm low	−11	−56	−116
100 m	100 cm high	+5	+25	+51
	10 cm high	+1	+2	+5
	10 cm low	0	−3	−5
	100 cm low	−5	−25	−51

Extracted from Würsig et al. 1991, on page 83
Note: Errors are not symmetrical above and below the actual cliff height because of the differential effects of curvature of the Earth at different heights and rounding of errors. (Errors of cliff height are usually due to an incorrect measurement of tidal height. Errors of positioning animals can also occur due to heat haze or swell moving the animals up and down. These non cliff height errors cannot easily be corrected, and if heat haze or swells are prodigious, theodolite tracking with one theodolite is not advised. Triangulating from two positions on shore is usually still possible)

Imprecise positioning can also be caused by the distinct visual acuity among different observers and by wind and swell (sea state). Morete (2007) demonstrated that after repeated readings of the same target the maximum variation of positioning inter and intra-observers were 20″. For example, from a LBS 37 m high, for a target distant 540 m, 20″ cause a variation of 0.8 m. However, with greater distances, a 20″ variation begins to portray a bigger positioning error, at a range of 3.5 km the error is about ±33 m, at 5 km the error is about 70 m (Fig. 4.7). Thus it is important the maintenance of the same observer within the same observation (survey, scan or focal follow). Restriction of data collection to lower winds (maximum 17 knots) and Beaufort sea state condition (maximum 4) when the study requires very precise positioning, as in studies of whalewatching interaction, in which distance between boat and target animals must be accurate. Also, there is some decrease of precision with increasing target's distance from the observer, because the further away, the same vertical angle amplitude, correspond to a larger area.

Old theodolite models had a maximum internal precision of 20″ being another source of imprecision. Whenever possible, it is advisable to use modern models of theodolite/total station, which has a lower intrinsic error, with 5″ precision. All these sources of imprecisions will depend on the height of the land-station, if it is

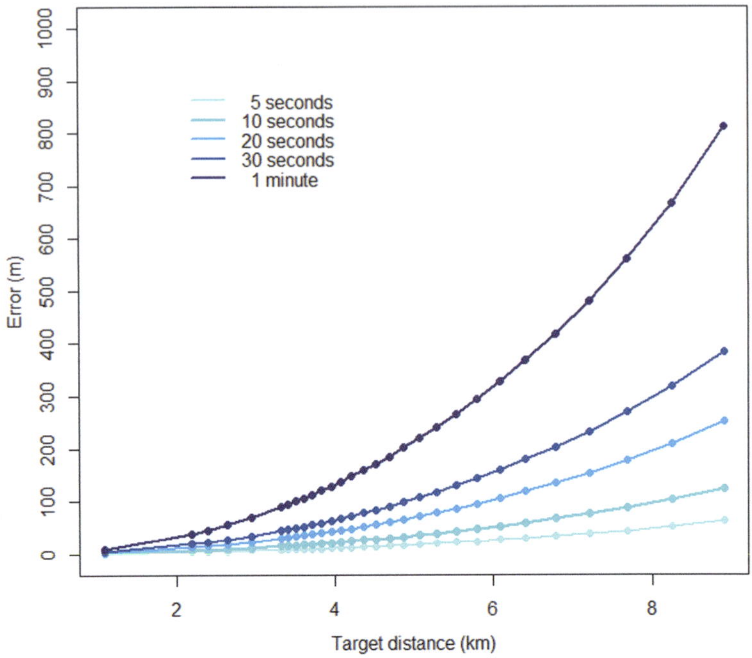

Fig. 4.7 Theoretical calculus (for a land-based station 37 m high) of positionings' errors generated by imprecise vertical readings (5 s to 1 min) of a target at different distances from the land-based station (Modified from Morete 2007)

relatively high, or the study area relatively small, the impact of the errors are minimal and might be disregarded. All errors and imprecision values can be calculated before starting any land-based studies, by using hypothetical angles values, comparing positions with variations on land-station height. This assessment is crucial in order to define the study area range and to have the knowledge of the amount of uncertainty the data may contain.

The conversion of theodolites' angles into geographic coordinates considering the height of the theodolite relative to sea level and tidal variations can simply be produced in an Excell spreadsheet (see Giralt 2012) however there are some computer software's specially developed for land-station studies. Some have continuous updates, evolving to newer, improved versions even changing names, like *Cyclops* to *VADAR* software (developed by Kniest 2004 and used by many land station studies e.g., Dunlop et al. 2015), and software *Mysticetus* (Steckler and Donlan 2016), used by e.g., MacKay et al. (2016). Some were developed for specific studies and are no longer available as *Pithagoras* (Gailey and Ortega-Ortiz 2000) used by e.g., Groch (2005), and *Aardvark* (Mills 1996), used by e.g., Frankel and Clark (1998); Morete et al. (2003a, 2007a). Basically, these programs allow recording data into computers in real time storing all behavioral information, translating the digital readings of the theodolite in geographic coordinates, correcting for all inputted parameters (e.g. equipment height above sea level tide, Earth curvature). Most software map animals and their displacement. Some merge technologies with GPS, acoustics devices, internet communication, mapping, and even cloud storage. Nowadays, with the ease of creating computer programs and advanced technological tools, software's may be ephemeral and become obsolete quickly, whilst novel and more elaborate ones arise.

It is possible to register data in a paper spreadsheet at the field and to entry the data afterwards into the dedicated computers programs. However, besides being a great source of writing errors and mistakes, it is necessary enormous effort and long periods to type and prepare data. So it is strongly advised to register data straight into the computer at the field.

4.3 Land-Based Studies in Latin American Countries

In this section, an overview of land-based studies of aquatic mammals conducted in Latin-American (LA) countries is presented. In order to provide a wide review, a web survey using the platform Survey Monkey (https://surveymonkey.com) was applied. The survey was released through the internet: by e-mail to certain researchers groups, through Aquatic Mammals' discussion list (MARMAM) and Aquatic Mammals Societies websites (SOLAMAC). Although the goal was to gather information about every LBS in Latin America, it is not possible to ensure that all have been comprised here.

Latin American sums up to 48 countries and territories. According to the review, land-based studies have been developed in at least 11 of them: Argentina, Brazil, Chile, Colombia, Ecuador, French Guiana, Mexico, Nicaragua, Peru, Uruguay and Venezuela (Fig. 4.8). Twenty-seven species of aquatic mammals were studied from land, six Mysticetes, 18 Odontocetes, two Pinnipeds and a Sirenian (Table 4.2).

It is usually thought that land-based stations are just for marine cetaceans observations, however it is interesting to note that the Araguaian River dolphin (*Inia araguaiaensis*) at Tocantins River (Araújo and Wang 2012[1]) and Amazon River dolphin (*Inia geoffrensis*) at Guamá River (Santos et al. 2012) habitat use and behavioral studies were developed from land-based stations. The West Indian manatee (*Trichechus manatus*) is also studied from land station at different locations at Brazilian north and north-east coast (Box 4.1). Besides, there are two Pinnipeds species studied from land-based station: South American sea lion (*Otaria flavescens*) at Cobquecura breeding colony at Central Coast of Chile (Sepúlveda et al. 2012), Isla Chañaral, North Central Chile (Pavez et al. 2011) and at Isla de Lobos, Uruguay (Franco-Trecu et al. 2015) where South American fur seal (*Arctocephalus australis*) (Box 4.2) are also studied (e.g. Franco-Trecu et al. 2010, 2014, 2016a).

[1] Note, at the time the study was conducted and published in 2012 there was no taxonomical differences between *Inia geoffrensis* and *Inia araguaiaensis*. This differentiation was accepted in 2014 (Hrbek et al. 2014).

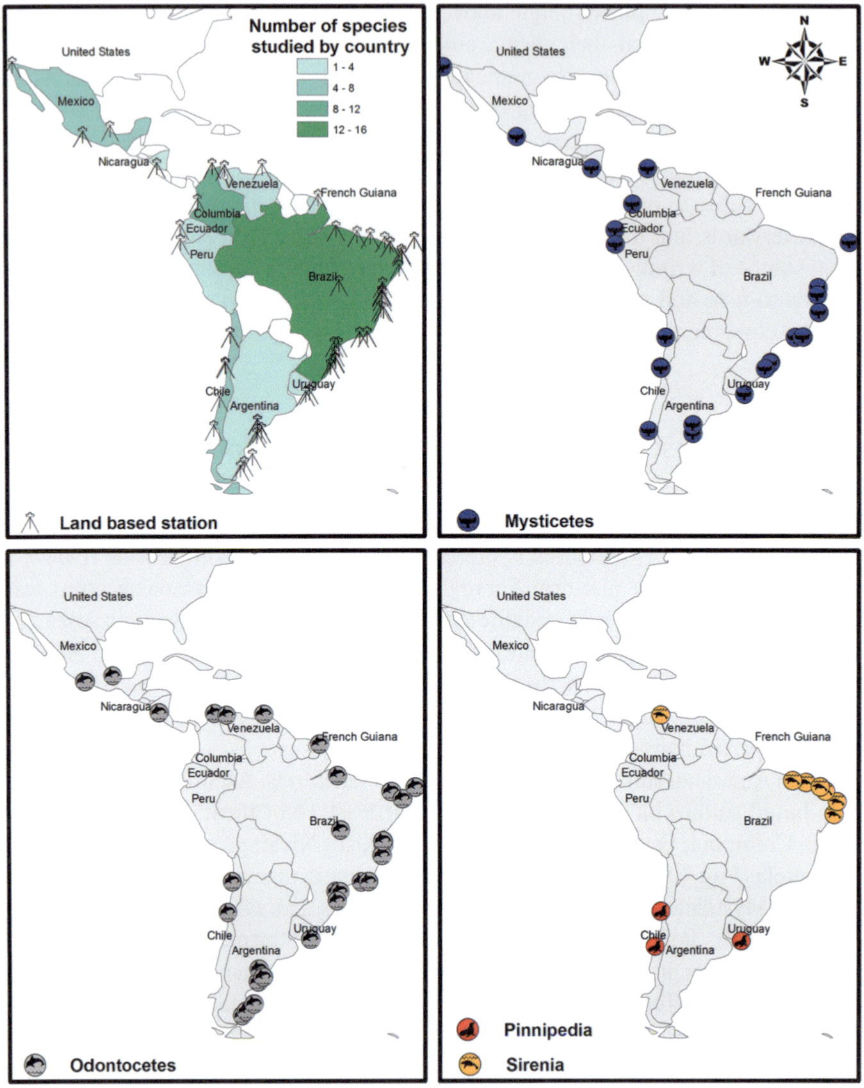

Fig. 4.8 Number of species studied from land-based stations in each Latin America country and an overview of their location classified by: Mysticetes, Odontocetes, Pinnipeds and Sirenian

There are some land stations that seems to be well structured and established, generating much research in a long-term period. For example, the land-based station of Abrolhos Archipelago (Brazilian Humpback Whale – Institute Instituto Baleia Jubarte) Bahia, Brazil, initiated in 1997 and although with gaps of few years, continues developing studies (Box 4.3). The land-based station at Isla de Lobos, Uruguay, utilizes complementary methods and techniques (e.g., biopsy samples, physiological samples) generating important ecological information about the target species, one of which is highlighted in the Box 4.2 (e.g., Franco-Trecu et al. 2015, 2016a, b).

In general, binoculars were the viewing device most used among these studies, and frequently for Odontocetes, while theodolite/total station was mostly used for mysticets studies. Basically, studies used focal follows and scans as sampling method. All these land stations resulted in several studies as peer-review articles, reports, PhD, Master and undergraduate thesis. Most of them were the description of species behavior, habitat use, evaluation of the impact of anthropogenic activities, density estimates and evaluation of the impact of climate change. There was also, a study combining land-based station observations with acoustics methods. Many Latin America land-based studies enabled conservation actions, implementation and improvement of coastal management (e.g. Heckel et al. 2001, Sheidat et al. 2004, Morete et al. 2007a, Silva and Silva Jr 2009, Pavez et al. 2011, Lundquist et al. 2013, Avila et al. 2015).

Information gathered from the review and web survey about the LBS studies conducted in LA are presented in Fig. 4.8 and Table 4.2. Below, a summary of the main characteristics, results and achievements of the land-based studies conducted in LA, published in peer-review articles, is provided by country.

Table 4.2 List of species studied from land-based stations in Latin America countries and respectively sources

	Species	Country	Sources
Mysticeti	*Balaenoptera edeni*	Brazil[a]	Siciliano et al. (2003)
		Colombia	Pardo and Palacios (2006)
	Balaenoptera musculus	Chile[b]	Cabrera et al. (2005) and Vernazzani et al. (2012)
	Balaenoptera physalus	Chile[c]	–
	Eschrichtius robustus	Mexico	Heckel et al. (2001)
	Eubalaena australis	Argentina	Payne (1986), Vermeulen et al. (2012), and Lundquist et al. (2013)
		Brazil[d]	Siciliano et al. (2003), Groch (2005), Danielski (2008), Mesquita (2011), Bottini (2014), Seyboth et al. (2015), and Danilewicz et al. (2016);
		Uruguay[e]	Costa et al. (2005)
	Megaptera novaeangliae	Brazil[f]	Siciliano et al. (2003), Petta et al. (2004), Bisi (2006), Morete (2007), Morete et al. (2003a, b, 2007a, b, 2008), and Abras (2015)
		Chile[g]	–
		Colombia	Ávila et al. (2015)
		Ecuador	Sheidat et al. (2004)
		Mexico[h]	–
		Nicaragua[i]	–
		Peru[j]	–

(continued)

Table 4.2 (continued)

	Species	Country	Sources
Odontoceti	*Cephalorhynchus commersonii*	Argentina	Failla et al. (2004), Iñíguez et al. (2007), Coscarella et al. (2010), and Castro et al. (2013)
	Delphinus capensis	Brazil	Siciliano et al. (2003)
	Delphinus delphis	Brazil[k]	–
		Chile[l]	–
	Globicephala macrorhynchus	Colombia	Pardo and Palacios (2006)
	Grampus griseus	Chile[m]	–
	Inia araguaiaensis	Brazil	Araújo and Wang (2012)
	Inia geoffrensis	Brazil	Santos et al. (2012)
	Lagenorhynchus obscurus	Argentina	Würsig and Würsig (1980)
		Chile[n]	–
	Orcinus orca	Brazil	Siciliano et al. (2003)
		Uruguay[o]	–
	Phocoena spinipinnis	Chile[p]	–
	Sotalia fluviatilis	Brazil	Santos et al. (2012)
	Sotalia guianensis	Brazil[q]	Santos et al. (2001, 2010), Favaro (2004), Domit (2006), Tosi (2007), Guilherme-Silveira and Silva (2009), Costa et al. (2012), and Izidoro and LePendu (2012)
		Colombia	Pardo and Palacios (2006)
		Venezuela[r] French	–
		Guiana[s]	–
	Stenella attenuata	Brazil[t]	–
		Colombia	Pardo and Palacios (2006)
		Mexico[u]	–
		Nicaragua[v]	–
	Stenella coeruleoalba	Colombia	Pardo and Palacios (2006)
	Stenella frontalis	Colombia	Pardo and Palacios (2006)
	Stenella longirostris	Brazil[w]	Silva-Junior (2005) and Silva and Silva Jr. (2009)
	Steno bredanensis	Mexico[x]	–
	Tursiops truncatus	Argentina	Würsig and Würsig (1978), Coscarella et al. (2011), and Vermeulen et al. (2015)
		Brazil[y]	Siciliano et al. (2003)
		Chile[z]	–
		Colombia	Pardo and Palacios (2006)
		Mexico[a2]	Morales-Rincon et al. (2016)
		Nicaragua[b2]	–
		Venezuela	Oviedo and Silva (2005)

(continued)

Table 4.2 (continued)

	Species	Country	Sources
Pinnipedia	*Arctocephalus australis*	Uruguay	Franco-Trecu et al. (2010, 2014, 2016b) and Pavés et al. (2016)
	Otaria flavescens	Chile[c2]	Pavez et al. (2011) and Sepúlveda et al. (2012)
		Uruguay	Franco-Trecu et al. (2015)
Sirenia	*Trichechus manatus*	Brazil[d2]	Alvite et al. (2002), Luna et al. (2011), and Normande et al. (2014)
		Venezuela[e2]	–

Small letters by the side of countries names refer to web survey respondents listed below
Web survey respondents: [a] Rodrigo Tardin, Salvatore Siciliano; [b] Maritza Sepúlveda; [c] Paola Hernandez Alvear, Maritza Sepúlveda, Ana Maria Garcia Cegarra; [d] Rodrigo Tardin; [e] Rodrigo Pingaro; [f] Bruna Rafaela Passos Araújo, Maria Isabel C. Golçalves, Rita Carli, José Martins da Silva Júnior Salvatore Siciliano, Silvana Bottini; [g] Paola Hernandez Alvear; [h] Katherina Audley; [i] Joëlle De Weerdt; [j] Ana Maria Garcia Cegarra, [k] Salvatore Siciliano; [l] Ana Maria Garcia Cegarra; [m] Ana Maria Garcia Cegarra; [n] Ana Maria Garcia Cegarra; [o] Rodrigo Pingaro; [p] Ana Maria Garcia Cegarra; [q] Yvonick Le Pendu, Ana Carolina Oliveira de Meirelles; Marcos Rossi-Santos; [r] Amandine Bordin; [s] Héctor Barrios-Garrido; [t] Rita Carli, José Martins da Silva Júnior; [u] Katherina Audley; [v] Joëlle De Weerdt; [w] Rita Carli, José Martins da Silva Júnior; [x] Katherina Audley; [y] Rodrigo Tardin, André Silva Barreto; [z] Paola Hernandez Alvear; [a2] Katherina Audley, Eduardo Morteo; [b2] Joëlle De Weerdt; [c2] Maritza Sepúlveda; [d2] Ana Carolina Oliveira de Meirelles; [e2] Héctor Barrios-Garrido

Box 4.1: West Indian Manatee, Northeast Brazil

The West Indian manatee is considered "vulnerable" to extinction by IUCN (Deutsch et al. 2008) throughout its range. Historically, they were distributed along the coastal areas and estuaries from Bahia (Brazil) up to Florida (USA) (Marsh et al. 1986). The Brazilian population was severely hunted and today is considered extinct in its southern range. The species actual distribution is from Alagoas to Amapá States, but with some gaps within this range (e.g., Lima 1997; Luna et al. 2011). In 1980, the Brazilian Federal Government established the Manatee Project (Projeto Peixe-boi Marinho) aiming to evaluate the species situation (Luna et al. 2011). Habitat loss and degradation are the main threats to the species, besides interaction with fisheries leading to accidental and intentional captures still occur in the Northern Brazilian Coast (Luna et al. 2011; Meirelles 2008). Annually, a high number of new born calves get separated from their mothers and strand along the coast during the summer months. Animals are rescued and conducted to rehabilitation centers (Normande et al. 2014). Afterwards, most animals are reintroduced in Alagoas and Paraíba States. Besides been tracked by radio and /or satellite (Normande et al. 2015) they are monitored from a network of land-based stations distributed in different locations: Paripueira (Alagoas), Rio Tinto (Paraiba), Icapuí (Ceará), Cajueiro da Praia (Piaui), Humberto Campos and São José de

(continued)

Ribamar (Maranhão), some of which are within Marine Protected Areas. The land-based monitoring program was established in 1987 as part of the species recovery plan (Alvite et al. 2002). From land stations observers characterize frequency of occurrence, density and abundance (Alvite et al. 2002). Behavior and interactions with anthropogenic activities are also observed serving in the establishment of new protected areas and supporting the development of management plans (Information gathered through the web survey: João Carlos Gomes Borges – Fundação Mamíferos Aquáticos; Ana Carolina Meirelles – Aquasis) (Fig. 4.9).

Fig. 4.9 Land-based station in Barra do Mamanguape (Credits: Collection of the Fundação Mamíferos Aquáticos); West Indian manatee and rehabilitation center on the Northeast of Brazil (Credits: Collection of AQUASIS)

Box 4.2: South American fur seal at Isla de Lobos, Uruguay

At Isla de Lobos, Uruguay, a long-term land-based study of the South American fur seal enables an in depth description of several aspects of the species behavior. The land station is located at the largest South American fur seal rockery and data collection takes place during the austral summer months (Fig. 4.10), which corresponds to the species reproduction season. Researches conduct scan or focal samplings (depending on study goals) using 10 × 50 binoculars from an observation platform allowing observations of individuals at short distances without apparent disturbance. This land-based research allowed the quantification of the occurrence of allo-sucking behavior (females feeding alien offspring) (Franco-Trecu et al. 2010), the determination of their mating system and quantification of male breeding success and reproductive tactics (using behavioral data and paternity assignment) (Franco-Trecu et al. 2014). Also the documentation of the parturition process, in the largest breeding colony of the species, allowed a phylogenetic comparison with other seal species (Franco-Trecu et al. 2016b). From the land-based station a putative hybrid female was identified based on phenotype traits that were later confirmed by genetic analysis (Franco-Trecu et al. 2016a). Moreover, the comparison of breeding seasons of the three geographic forms (Peruvian, South

(continued)

Chilean and Atlantic) provided support that these populations have distinct evolutionary histories (Pavés et al. 2016). The research conducted at Isla de Lobos is an excellent example of how to combine land-based station with other methods (Fig. 4.10) in order to enhance the knowledge of the target species ecology.

Fig. 4.10 Land-based station combined with genetics studies of South American fur seal at Isla de Lobos, Uruguay: Observer using binoculars; animals fighting; biopsy sampling procedures. Credits: Valentina Franco-Trecu

4.3.1 Argentina

Three species of odontocetes and the Southern Right whale were studied from land station in Argentina. One of the pioneer studies from land-based station was with Bottlenose dolphins (*Tursiops truncatus*) in the Golfo San Jose (Würsig and Würsig 1978), today a Marine Protected Area in northern Patagonia. The study describes general movement patterns, aerial and social behavior, and contributed to the knowledge of the population's ecology. About 30 years later, the sighting rates of the species at Golfo San Jose were re-assessed using aerial, boat and land-based surveys (Coscarella et al. 2011). The authors confirmed an abundance decline in the area that could be explained by increased natural and human-induced mortality and disturbance, resources depletion and environmental shift. Bottlenose dolphin was also studied from land-based stations in Bahía San Antonio (Vermeulen et al. 2015), using binoculars and a mix of scan and focal follows. Observations from land station were combined with boat-based observations allowing the characterization of the species seasonal behavior in the area, which is probably a result of seasonal variations in prey availability (Vermeulen et al. 2015).

The Commersons' dolphin (*Cephalorhynchus commersonii*) was studied by Coscarella and collaborators (2010) through scans conducted from a land-based station located 21 m above sea level (asl), at Bahía Enganõ (near the mouth of the Chubut River). The daily activity patterns and variation in relative abundance was verified. Seasonal changes in the frequency of behaviors were recorded and group size and abundance were related to sea surface temperature, suggesting a preference for lower temperatures (Coscarella et al. 2010). The species was also studied from another LBS (11 m asl) at Punta Quilla, Santa Cruz Estuary (Castro et al. 2013). The area was systematically scanned every 30 min during the austral summer in 2004 and 2005. This study was conducted in a pristine coastal area and allowed to identify a strong influence of the tide on Commerson's dolphin behavior, providing tools for management and conservation purposes (Castro, et al. 2013).

The Dusky dolphin (*Lagenorhynchus obscurus*) was also studied by Würsig and Würsig (1980) at Golfo San Jose using a theodolite. The study provided baseline information on the species behavior and aspects of its habitat use.

The Southern Right whale (*Eubalaena australis*) was studied from different LBS. One of the first studies from land in LA was designed to investigate this species in Valdes Peninsula in 1970's (Payne 1986). Together with aerial surveys, it was possible to define the whales' preferred depth during the breeding season (5 m bathymetric contour). The effect of swim-with-dolphin tourism was studied by Lundquist et al. (2013) from two cliffs within the El Doradillo Municipal Protected Area, an area where boat traffic is forbidden. From a land station 20 m asl equipped with theodolite authors followed a method of before/during/after comparison (e.g., Martin and Bateson 1993), with behavior and movement of whales before an interaction with boat and swimmers serving as control data for behavior and movement during and after the interaction. Based on gathered results, the authors recommended not permit any activity of swim-with-whale in Chubut Province until further investigations are done. The study of Vermeulen and collaborators (2012) at Baia San Antonio, was conducted from a cliff 20 m asl equipped with binoculars and spotting scope. Instead of using movement parameter as a proxy for whale disturbance, the behavioral data were analyzed as a series of time-discrete Markov chains. Whales reduced significantly their social interactions when confronted with human short-range presence. Whales did not appear to respond (e.g., avoid or approach the boat actively) until the boat entered a 500 m zone from the animal, which was defined as the *"impact-zone"* (quotes from the author). Maybe the use of a more precise positioning device would result in a more refined impact-zone.

4.3.2 Brazil

Land-based studies of 14 different species have been carried out in Brazil: ten Odontocetes, three Mysticetes and the West Indian manatee (Box 4.1). The Guiana dolphin (*Sotalia guianensis*) was studied from land-based station in several places along its distribution range. At Cananéia Estuary, São Paulo State, land-based stations were used to obtain photo-identifications of the animals allowing to characterize site fidelity (Santos et al. 2001) and influence of time of the day on their behavior (Santos et al. 2010). At Pontal Bay, Ilhéus, Bahia State, scan sampling was applied from four land station throughout the year of 2006, aiming to verify the influence of tide and water depth on dolphins' activity and spatial distribution. Another study conducted at Ilhéus (Izidoro and Le Pendu 2012) aimed to characterize the species play behavior. "Ad libitum" observation and scans were carried out to establish the ethogram and frequency of play behavior. At Canavieiras, Bahia State, from a land-based station located in a pier (5.5 m asl) near the mouth of Pardo River. The study compared ethnoecological knowledge of local artisanal fishermen, to behavioral data collected using binoculars and focal group sampling (Costa et al. 2012). A novel leap behavior called pirouette (one complete rotation around its own axis) first observed by a local fisherman was confirmed by researchers during land-based observations. The study emphasizes the importance of traditional knowledge in complementing scientific data. Authors mentioned researches safety as an important aspect of the land-based station location. Up North, at Enseada do Curral, Rio Grande do Norte State, the influence of time of the day and tide on the species behavior was assessed (Guilherme-Silveira and Silva 2009) and highlighted the need for regulating dolphinwatching activities that are increasing locally and might threat the population.

An ethnoecological study of the *Sotalia* genus and the Amazon River dolphin was conducted from land-based station located at Universidade Federal do Para campus in Belém, Para State, Northern Brazil, at the mouth of Igarapé do Tucunduba (Santos et al. 2012). The study used binoculars and applied scan sampling method allowing characterizing both species groups size and habitat use, and highlighted the need to conduct further studies to understand the species ecology in the area. The Araguaian River dolphin at Tocantins River was studied combining different methods (Araújo and Wang 2012[1]). The species' behavior and habitat use in the proximity of dams were described, as well as the species feeding activity and unusual behaviors such as object carrying.

A study conducted from a LBS at Fernando de Noronha Archipelago in Brazil investigated circadian and seasonal fluctuations in the daytime activities of the Spinner dolphin (*Stenella longirostris*) over a 4 years period (Silva and Silva Jr. 2009). The species is a major tourist attraction in the Archipelago, and although boat excursions were already regulated in the area (number of boats, speed and proximity of dolphins gathered results supported a time limitation to ongoing excursions (i.e. avoiding excursion from 7 to 11 am).

The Southern Right Whale was studied from land-based stations in Santa Catarina (e.g. Seyboth et al. 2015) and Rio Grande do Sul (Danilewicz et al. 2016). In order to investigate the relationship between southern right whale distribution off Santa Catarina State and environmental and temporal variables, observations from 14 land-based stations (20–90 m asl) were carried out between 2002 and 2008. Observation usually took place simultaneously in some land-based stations. Preferred calving areas were identified, supporting measures to enhance coastal management (Seyboth et al. 2015). In Torres (Rio Grande do Sul State), from a land-based station at Morro do Farol (45 m asl), 15 min scans with the naked eye followed by 5 min scans with 7 × 50 reticulated binoculars were conducted. The study characterized the area as part of the species breeding site along the Brazilian coast, as birth, nursing and mating were observed (Danilewicz et al. 2016).

Box 4.3: Humpback Whale at Abrolhos, Brazil

A remaining population of Humpback whales was rediscovered around the Abrolhos Archipelago in 1988, after the intensive hunting. This population was estimated in 1.379–1.887animals in 1995 (Kinas and Bethlem 1998), and the most recent estimate was of 16,410 (CV = 0.228, 95% CI = 10,563–25,495) animals in 2008 (Bartolotto et al. 2016).

The Abrolhos Archipelago is located 70 km off Caravelas city, southern Bahia State, Brazil. It is encompassed by the most important reef formation of the Southwestern Atlantic Ocean. The archipelago is within the first marine park created in Brazil (IBAMA/FUNATURA 1991).

The Abrolhos Archipelago land-based station project begun in 1997 with a pilot study that allowed the detailed description of the species ethogram. From 1998 to 2004, systematic scans and focal follows were conducted daily along the seasons. This middle-term study indexed temporal patterns of behaviors, group structure, seasonal and fluctuation of abundance, respiratory behavior, and calf behavior (Morete et al. 2003b, 2007b, 2008; Petta et al. 2004; Bisi 2006). The "tail-up" was first described, suggesting a culturally transmitted behaviour (Morete et al. 2003a). The finding that, around the Abrolhos Archipelago almost 50% of the groups contained calves, demonstrated the importance of the area for nursery (Morete et al. 2003b), and a study focused on the responses of the mother-calf pair to vessels was conducted, resulting in a recommendation to amend the existing Brazilian Legislation so that vessels maintained a 300 m caution zone, reducing the speed and avoid sudden changes around a mother with a calf (Morete et al. 2007a). Also, the 8-year data-set was used in a study about migration timing and environmental and climate change effects on the migration dynamics (Abras 2015). After a gap period, land-based systematic sampling was resumed in 2015, aiming to continue the legacy of knowledge achieved from this land-based station (Fig. 4.11).

Fig. 4.11 Humpback whales land-based station at Abrolhos Archipelago Bahia State, Brazil: observer using theodolite; humpback whale within land station field of view; note taker using computer running dedicated software (Credits: Daniela Abras e Maria E. Morete)

The Humpback whale is studied from LBS from different locations. A long-term systematic study is conducted from Abrolhos Archipelago (Bahia State) since 1997 (details are present in Box 4.3).

4.3.3 Chile

In Chile land-based stations were used to study Odontocetes, Mysticetes and Pinnipeds (Table 4.2, Figure 4.8). Despite its usual offshore distribution, Blue whales (*Balaenoptera musculus*) occur in close proximity to the coast in Chiloé. Cabrera et al. (2005) observed Balaenopterids and blue whales from land-based stations (38 and 58 m asl equipped with binoculars). The authors suggested that the implementation of a Marine Protected Area (MPA) in this Blue whale nursing and feeding ground of the Chiloé and Corcovado region should be a high priority measure for government authorities. Another blue whale study, combining aerial, boat and land-based surveys, aiming to monitor the species' occurrence in the area, was developed. From the land-based station located 104 m asl and equipped with 12 × 50 binoculars, trained observers recorded time, group size and estimated distance of whales from the coast (Vernazzani et al. 2012).

Land-based studies were conducted in two colonies of South American sea lion (*Otaria flavescens*): Isla Chañaral (Pavez et al. 2011), which is part of the Isla Chañaral Marine Reserve and the National Humboldt Penguin Reserve; and at Cobquecura breeding colony (Sepúlveda et al. 2012). The study at Isla Chañaral aimed to investigate the effects of tourist activities during the species breeding season. Sea lions response was negatively correlated to the distance at which boats approached the colony. Authors support limiting the distance of approach to 50 m in order to avoid disrupting animals' behavior (Pavez et al. 2011). The study at Cobquecura breeding colony was made from a distance of about 80 m using 10 × 15 binoculars and hand tally counters. Counts were conducted at 60 min. Fine-scale differences in haul-out behavior among age/sex classes were identified, as well as daily and seasonal differences in abundance. Authors pointed the importance of taken into consideration these aspects when planning surveys to estimate population abundance (Sepúlveda et al. 2012).

4.3.4 Colombia

As a first attempt to characterize cetacean occurrence on the central coast of the Colombian Caribbean, in the region of Santa Marta and the adjacent Tayrona National Natural Park, a combination of land-based and boat-based surveys was conducted. Land-based station observations were conducted at four places, all of them with heights greater than 20 m asl and with an effective radial field-of-view of about 2 km (Pardo and Palacios 2006). The occurrence of seven cetacean species was recorded, including first records for the area of short-finned pilot whale (*Globicephala macrorhynchus)* and a stranded Striped dolphin calf (*Stenella coeruleoalba*).

Humpback whales were studied at Bahía Málaga (Pacific coast) from a 20 m asl LBS using binoculars at the peak of the breeding season (Ávila et al. 2015). In the presence of whalewatching boats, whales reduced the number of blows per minute, increased swimming speed and moved more erratically increased breaching frequency, and decreased resting behavior. Authors highlight that despite the existence of local regulations it is important to focus on environmental education to promote compliance.

4.3.5 Ecuador

Humpback whales were studied at Isla de La Plata, Machalilla National Park, from a 90 m asl land-based station equipped with a theodolite (Sheidat et al. 2004). The research aimed to characterize the species movement patterns (swim speed and path directness) while in the absence and presence of boats. Based on gathered results it was recommended that boats limit their closest approach to 100 m, and that no more than two boats should be within 1 km of a target group during whalewatching trips. Although regulated, current guidelines need to be improved in order to enhance whales' protection and guarantee the sustainability of this important economic activity in Ecuador (Sheidat et al. 2004).

4.3.6 Mexico

Gray whale (*Eschrichtius robustus*) was studied during their migration in the vicinity of Todos Santos Bay from a land-based station in order to investigate the influence of whalewatching boats (Heckel et al. 2001). The study was conducted from a lighthouse (51.64 m asl) equipped with a theodolite during two consecutive boreal winter seasons. The goal was to verify if the current whalewatching regulation was well adapted to this area and to the whales' migratory behavior. In doing so, it highlights an important aspect of behavioral studies: the necessity of study target species' behavior in each different habitat instead of considering that the species behave in the same way no matter the habitat characteristics and the behavioral context (e.g., reproduction, feeding, migration). Heckel et al. (2001) recommended the avoidance of head-on approaches during whalewatching trips in this migratory distribution zone and highlighted the fact that whalewatching industries behave differently in each area, an aspect that must be taken into consideration.

4.3.7 Uruguay

The land-based studies conducted in Uruguay focus on Pinnipeds: South American sea lion and South American fur seal (research details in the Box 4.2) and Southern Right Whale. The Pinnipeds studies were conducted at the Marine Protected area Isla de Lobos. South American sea lion breeding success was quantified according to their breeding location within the colony (tide line vs. internal pool) and genetic samples. Two daily scan samplings were conducted using 10 × 50 binoculars to count the number of males and assign their spatial positions in a grid of 2 × 2 m. It was shown for the first time the coexistence of two distinct mating systems: males of tide line have a female-defence polygyny into floating territories, while males at the internal pools defended fixed territories establishing a resource-defence polygyny (Franco-Trecu et al. 2015).

Southern right whale presence was assessed with binoculars during simultaneous scans, from 9 land-based stations selected because of their appropriated height (from 10 m to 50 m asl) covering about 220 km occurrence area along the coast. The number of individuals, presence of calves and group size were recorded. The study suggested that the Uruguayan coast is not intensively used as calving area, when compared to other areas of occurrence as in Santa Catarina in Brazil and Valdes Peninsula in Argentina (Costa et al. 2005).

4.3.8 Venezuela

Combined land-based and boat-based surveys, aiming to establish the sighting frequency and relative abundance of Bottlenose dolphins, along the northeast coast of Margarita Island and Los Frailes Archipelago were developed (Oviedo and Silva 2005). The land-based observations were conducted from the Punta Ballenas lighthouse, and allowed to characterize it as the dominant species in the area.

4.4 Conclusions

Behavioral investigation of wild animals should form the baseline of ecological and conservation studies. Observing animals from Land-based stations in their natural habitat, year-round, during different seasons, in distinct activities, under diverse weather conditions is essential, in order to allow scientists to characterize the species, gathering information needed to the design and development of any further study project, for developing action plans for target species and management plans for protected areas. In fact, land-based station studies allowed the pioneers in the field to build baseline knowledge, for many species, but much more are still to be studied.

The knowledge of animals' natural behavior, habitat use, distribution and threats should be mandatory before the introduction of anthropogenic activities on their habitat. Aquatic mammals are suffering the consequences of human societies that are concentrated at the coastal zone, which is amongst the most modified areas on Earth (Bollmann 2010), with mega-cities, population concentration causing pollution, habitat degradation, overfishing and hunting. In fact, there are aquatic mammals species that have already been extinct, as the Sea Cow (*Hydrodamalis gigas*), Baiji Dolphin (*Lipotes vexillifer*), and others that are nearly extinct, like the Vaquita (*Phocoena sinus*) and the subspecies Maui's dolphin (*Cephalorhynchus hectori maui*), species that humans poorly knew. Meanwhile, there are new ones being recently discovered and described as Omura's Whale (Balenoptera omurai – Wada et al. 2003), as Araguaian dolphin (Hrbek et al. 2014). Thus, while we are discovering new species, others are being extinct, a fact already known for many taxa but surprisingly for marine mammals. How many species poorly known or even different ones, are out there? Some, perhaps along the coast, in places that could be easy and inexpensive to observe from a land-based station.

From our present taxonomical knowledge, in Latin America there are at least 67 species of aquatic mammals, amongst Mysticetes, Odontocetes, Pinnipeds, Sirenians and Mustelids inhabiting marine offshore areas, coastal regions, estuaries and rivers. From our survey, only 27 species are being studied from land-based station, in only 11 countries out of 48. Most parts of South Atlantic and South Pacific shoreline, as well as inner rivers systems are yet to be unveiled, lack knowledge and proper management. As LBS studies requires relatively low-budget to be implemented, it should be encouraged in all areas where aquatic mammals are present and observations from land are feasible, enabling diverse possibilities of research. Together with the important findings and conservation achievements of those LBS studies already in progress in Latin America, it would enhance our knowledge of aquatic mammals' ecology and of some anthropogenic impacts, providing valuable support to the development of effective conservation measures and to the proper management of their habitats.

Acknowledgments The authors would like to thanks to all respondents that participated of the web survey allowing to improve the coverage of the workpresented: Amandine Bordin, Ana Carolina Oliveira de Meirelles, Ana Maria Garcia Cegarra, André Silva Barreto, Bruna Rafaela Passos Araújo, Camila Morais, Claudia Brigagão Petta, Daniel Danilewicz, Eduardo Morteo, Flávio Lima, Héctor Barrios-Garrido, Isabel Cristina Avila, João Carlos Gomes Borges, Joëlle De Weerdt, José Martins da Silva Júnior, Karina Groch, Katherina Audley, Luiza Beirão Campos, Marcos Rossi-Santos, Maria Isabel C. Golçalves, Maritza Sepúlveda, Milene Novais, Paola Hernandez Alvear, Rita Carli, Rodrigo H. Tardin, Salvatore Siciliano, Silvana Bottini, Valentina Franco-Trecu and Yvonick Le Pendu. Ana Carolina Oliveira de Meirelles (Collection Aquasis), Ana Maria Garcia Cegarra, Bernd Würsig, Eduardo Morteo, Flavio Lima, João Carlos Gomes Borges (Collection Fundação Mamíferos Aquáticos), Joëlle De Weerdt, José Martins da Silva Júnior, Isabel Cristina Avila, Karina Groch, Melany Würsig, Samuel Turgeon, Tatiana Bisi, Valentina Franco-Trecu, provided images to illustrate the chapter. Julia Dombroski provided valuable comments. The authors would like to thanks David S. Janiger for his meaningful contribution to the study of aquatic mammals through his online library.

References

Abras DR (2015) Efeito de parâmetros ambientais na migração de baleias-jubarte (*Megaptera novaeangliae*) entre Mar de Scotia e Banco dos Abrolhos. Dissertation to the Universidade de São Paulo, São Paulo, 157 p

Acevedo-Gutiérrez A (2009) Habitat use. In: Perrin WF, Würsig B, Thewissen JGM (eds) Encyclopedia of marine mammals. Academic Press, San Diego, pp 524–529. 1355 p

Alcock J (1998) Animal behaviour – an evolutionary approach. Sinaur Associates, Sunderland. 640 p

Altmann J (1974) Observational study of behavior: sampling methods. Behaviour 49(3):227–267

Alvite CMC, Lima RP, Andrade JS (2002) Monitoramento do peixe-boi marinho (*Trichechus manatus*) em importantes áreas de ocorrência ao longo do litoral Nordeste do Brasil. In Libro de Resúmenes 10 RT y 4 Congreso SOLAMAC. Valdivia, Chile, p 59

Araújo CC, Wang JY (2012) Botos (*Inia geoffrensis*) in the upper reaches of the Tocantins River (Central Brazil) with observations of unusual behavior, including object carrying. Aquat Mamm 38(4):435–440

Arranz P, Borchers DL, de Soto NA, Johnson MP, Cox MJ (2014) A new method to study inshore whale cue distribution from land-based observations. Mar Mamm Sci 30:810–818

Avila IC, Correa LM, Parsons ECM (2015) Whale-watching activity in Bahía Málaga, on the Pacific coast of Colombia, and its effect on humpback whale (*Megaptera novaeangliae*) behavior. Tourism Mar Environ 11(1):19–32

Bailey H, Lusseau D (2004) Increasing the precision of theodolite tracking: new computation technique to calculate the altitude of land-based observation sites. Mar Mamm Sci 20:880–885

Baker SC, Herman LM, Bays BG, Stifel WF (1982) The impact of vessel traffic on the behaviour of humpback whales in the Southeast Alaska. Final report for the National Marine Mammal Laboratory (contract number 81-ABC-00114), Seattle, 39 p

Bart J, Fligner MA, Notz WI (1998) Sampling and statistical methods for behavioral ecologists. Cambridge University Press, Cambridge. 330 p

Bejder L, Dawson SM, Harraway JA (1999) Responses by Hector's dolphins to boats and swimmers in Porpoise Bay, New Zealand. Mar Mamm Sci 15:738–750

Bejder L, Samuels A, Whitehead H, Gales NJ, Mann J, Connor RC, Heithaus M, Watson-Capps J, Flaherty C, Krützen M (2006) Decline in relative abundance of bottlenose dolphins exposed to long-term disturbance. Conserv Biol 20:1791–1798

Bisi TL (2006) Comportamento de filhotes de Baleia Jubarte, *Megaptera novaeangliae*, na região ao redor do Arquipélago dos Abrolhos, Bahia (Brasil). Dissertation to the Universidade de São Paulo, São Paulo, 99 p

Bollmann MT (2010) World ocean review 2010: living with the oceans. Maribus, Hamburg. 232 p

Bortolotto GA, Danilewicz D, Andriolo A, Secchi ER, Zerbini AN (2016) Whale, whale, everywhere: increasing abundance of western south atlantic humpback whales (*Megaptera novaeangliae*) in their wintering grounds. PLoS One 11(10):e0164596. doi:10.1371/journal.pone.0164596

Bottini SS (2014) Distribuição espacial e padrões de movimento da Baleia-franca-austral (*Eubalaena australis*) em Torres, Rio Grande do Sul. Dissertation to the Universidade Federal do Rio Grande do Sul, Porto Alegre, 46 pp

Buckland ST, Anderson DR, Burnham KP, Laake JL (1993) Distance sampling: estimating abundance of biological populations. Chapman and Hall, New York/London, 446p

Cabrera E, Carlson C, Vernazzani BG (2005) Presence of blue whale (*Balaenoptera musculus*) in the northwestern coast of Chiloé Island, southern Chile. LAJAM 4(1):73–74

Castro RL, Dans SL, Coscarella MA, Crespo EA (2013) Living in an estuary: Commerson's dolphin (*Cephalorhynchus commersonii* (Lacépède, 1804)), habitat use and behavioural pattern at the Santa Cruz River, Patagonia, Argentina. Lat Am J Aquat Res 41(5):985–991

Corkeron PJ (1995) Humpback whales (*Megaptera novaeangliae*) in Hervey Bay, Queensland – behavior and responses to whale-watching vessels. Can J Zool 73(7):1290–1299

Coscarella MA, Pedraza SN, Crespo EA (2010) Behavior and seasonal variation in the relative abundance of Commerson's dolphin (*Cephalorhynchus commersonii*) in northern Patagonia, Argentina. J Ethol 28:463–470

Coscarella MA, Dans SL, Degrati M, Garaffo GV, Crespo EA (2011) Bottlenose dolphins at the southern extreme of the south-western Atlantic: local population decline? J Mar Biol Assoc UK 92(8):1843–1849

Costa P, Praderi R, Piedra M, Franco-Fraguas P (2005) Sightings of southern right whales, *Eubalaena australis*, off Uruguay. Lat Am J Aquat Mamm 4(2):157–161

Costa MEB, Le Pendu Y, Neto EMC (2012) Behaviour of *Sotalia guianensis* (van Bénéden, 1864) (Cetacea, Delphinidae) and ethnoecological knowledge of artisanal fishermen from Canavieiras, Bahia, Brazil. J Ethnobiol Ethnomed 8:18

Danielski ML (2008) Comportamentos de mães e filhotes de baleias-franca-austrais, *Eubalaena australis* (Desmoulins, 1822), em Santa Catarina, Brasil. Dissertation to the Universidade Federal de Juiz de Fora, Juiz de Fora, 114 pp

Danilewicz D, Moreno IB, Tavares M, Sucunza F (2016) Southern right whales (*Eubalaena australis*) off Torres, Brazil: group characteristics, movements, and insights into the role of the Brazilian-Uruguayan wintering ground. Mammalia, ISSN (Online) 1864–1547

Deutsch CJ, Self-Sullivan C, Mignucci-Giannoni A (2008). *Trichechus manatus*. The IUCN Red list of threatened species. http://dx.doi.org/10.2305/IUCN.UK.2008.RLTS.T22103A9356917. Downloaded on 04 November 2016.

Domit C (2006) Comportamento de pesca do boto-cinza, *Sotalia guianensis* (van Bénéden, 1864). Dissertation to the Universidade Federal do Paraná, Curitiba, 96 pp

Dunlop RA, Cato DH, Noad MJ (2012) Behavioral response studies: problems with statistical power. In: Popper AN, Hawkins A (eds) The effects of noise on aquatic life. Science + Business Media, LLC, pp 293–297. 695 p

Dunlop RA, Noad MJ, McCauley RD, Kniest E, Paton D, Cato DH (2015) The Behavioural response of humpback whales (*Megaptera novaeangliae*) to a 20 cubic inch air gun. Aquat Mamm 41(4):412–433

Ellis R (2009) Whaling, Aboriginal. In: Perrin WF, Würsig B, Thewissen JGM (eds) Encyclopedia of marine mammals. Academic Press, San Diego, pp 1227–1234. 1355 p

Failla M, Iñíguez MA, Fernandez-Juricic E, Tossenberger V (2004) Effect of vessel traffic on Commerson's dolphin (*Cephalorynchus commersonii*) in Bahia San Julian, Patagonia, Argentina. International Whaling Commission, Scientific Committee (IWC-SC) Report SC/56/WW7

Favaro EGP (2004) Utilização da Baía dos Golfinhos (Pipa, RN) pelo boto *Sotalia fluviatilis* (Cetacea: Delphinidae). Dissertation to the Universidade Federal de Pernambuco, Recife, 42 pp

Franco-Trecu V, Tassino B, Soutullo A (2010) Allo-suckling in the south American fur seal (*Arctocephalus australis*) at Isla de lobos, Uruguay: cost or benefit of living in group? Ethol Ecol Evol 22:143–150

Franco-Trecu V, Costa-Urrutia P, Schramm Y, Tassino B, Inchausti P (2014) Sex on the rocks: breeding tactics and reproductive success of the for south American fur seal males. Behav Ecol 25:1513–1523

Franco-Trecu V, Costa-Urrutia P, Schramm Y, Tassino B, Inchausti P (2015) Tide line versus internal pools: mating system and breeding success of south American sea lion males. Behav Ecol Sociobiol 69(12):1985–1996

Franco-Trecu V, Abud C, Feijoo M, Kloetzer G, Casacuberta M, Costa-Urrutia P (2016a) Sex beyond species: the first genetically analyzed case of intergeneric fertile hybridization in pinnipeds. Evol Dev 18(2):127–136

Franco-Trecu V, García-Olazábal MD, Tassino B, Acevedo J (2016b) Parturition process in an amphibian mammal species: new evidences in south American fur seal (*Arctocephalus australis*). Anim Biol 66:21–29

Frankel AS, Clark CW (1998) Results of low-frequency playback of M- sequence noise to humpback whales, *Megaptera novaeangliae*, in Hawaii. Can J Zool 76:521–535

Frankel AS, Clark CW (2000) Behavioral responses of humpback whales (*Megaptera novaeangliae*) to full-scale ATOC signals. J Acoust Soc Am 108:1930–1937
Frankel AS, Clark CW (2002) Atoc and other factors affecting the distribution and abundance of humpback whales (*Megaptera novaeangliae*) off the north shore of Kauai. Mar Mamm Sci 18(3):644–662
Frankel AS, Clark CW, Herman LM, Gabriele CM (1995) Spatial distribution, habitat utilization, and social interactions of humpback whales, *Megaptera novaeangliae*, off Hawaii, determined using acoustic and visual techniques. Can J Zool 73:1134–1146
Gailey G, Ortega-Ortiz J (2000) User's manual Pythagoras version 1.0: theodolite cetacean tracking. Texas A&M University, Galveston, 71 p
Gales N, Brennan A, Baker R (2003) Ethics and marine mammals research. In: Gales N, Hindel M, Kirkwood R (eds) Marine mammals: fisheries, tourism and management. CSIRO Publishing, Melbourne, pp 321–329. 460 p
Giralt O (2012) Tracking and calculations manual. Pacific Whale Foundation. https://drive.google.com/open?id=0B4KztXUThWN4V2dTM3hpMlk1TGM
Groch KR (2005) Biologia populacional e ecologia comportamental da baleia franca austral, *Eubalaena australis* (Desmoulins, 1822), Cetacea, Mysticeti, no litoral sul do Brasil. PhD Thesis to the Universidade Federal do Rio Grande do Sul, Porto Alegre, 168 p [In Portuguese and English]
Guilherme-Silveira FR, Silva FJL (2009) Diurnal and tidal pattern influencing the behaviour of *Sotalia guianensis* on the north-eastern coast of Brazil. Mar Biodiversity Rec 2. doi:10.1017/S1755267209000906
Heckel G, Reilly SB, Sumich JL, Espejel I (2001) The influence of whalewatching on the behaviour of migrating gray whales (*Eschrichtius robustus*) in Todos Santos Bay and surrounding waters, Baja California, Mexico. J Cetacean Res Manag 3(3):227–237
Hrbek T, da Silva VMF, Dutra N, Gravena W, Martin AR, Farias IP (2014) A new species of river dolphin from Brazil or: how little do we know our biodiversity. PLoS One 9(1):e0083623
IBAMA/FUNATURA (1991) Plano de manejo: Parque Nacional. Marinho dos Abrolhos, Brasília, 124 p
Iñíguez MA, Tossenberger VP (2007) Commerson's dolphins (*Cephalorhynchus commersonii*) off Ría Deseado, Patagonia, Argentina. Aquat Mamm 33(3):276–285
Izidoro FB, Le Pendu Y (2012) Estuarine dolphins (*Sotalia guianensis*, Cetacea, Delphinidae) play at Porto de Ilhéus harbor, Bahia, Brazil. Pan-Am J Aquat Sci 7(1):27–36
Kinas PG, Bethlem CP (1998) Empirical Bayes abundance estimation of a closed population using mark recapture data, with pplication to humpback whales, *Megaptera novaeangliae*, in Abrolhos, Brazil. Rep Int Whaling Commission 48:447–450
Kniest E (2004) Cyclopes. University of New Castle, Callaghan
Lima RP (1997) Peixe-boi marinho (*Trichechus manatus*): Distribuição, status de conservação e aspectos tradicionais ao longo do litoral nordeste do Brasil. Dissertation to the Universidade Federal de Pernambuco, Recife, 81 p
Luna FO, De Andrade MCM, Reis ML (eds) (2011) Plano de Ação Nacional para a Conservação dos Sirênios: Peixe-boi-da-Amazônia *Trichechus inunguis* e Peixe-boi-marinho *Trichechus manatus*. Instituto Chico Mendes de Conservação da Biodiversidade, Brasília, 81 p
Lundquist D, Sironi M, Würsig B, Rowntree V, Martino J, Lundquist L (2013) Response of southern right whales to simulated swim-with-whale tourism at Peninsula Valdés. Mar Mamm Sci 29(2):E24–E45
MacKay MM, Würsig B, Bacon CE, Selwyn JD (2016) North Atlantic humpback whale (*Megaptera novaeangliae*) hotspots defined by bathymetric features off western Puerto Rico. Can J Zool 94:517–527
Mann J (1999) Behavioral sampling methods for cetaceans: a review and critique. Mar Mamm Sci 15:102–122

Mann J (2000) Unraveling the dynamics of social life: long-term studies and observational methods. In: Mann J, Connor RC, Tyack PL, Whitehead H (eds) Cetacean societies: field studies of dolphins and whales. The University of Chicago Press, Chicago, pp 45–64. 448 p

Marsh H, O'shea TJ, Best RC (1986) Research on Sirenians. Ambio 15(3):177–180

Martin P, Bateson P (1993) Measuring behaviour: an introductory guide. Cambridge University Press, Cambridge, 222 p

Meirelles ACO (2008) Mortality of the Antillean manatee, *Trichechus manatus manatus*, in Ceará state, north-eastern Brazil. J Mar Biol Assoc UK 88(6):1133–1137

Mesquita CC (2011) Ocorrência, distribuição e composição de grupos de *Eubalaena australis* (Desmoulins, 1822), no litoral centro sul do estado de Santa Catarina e sua relação com fatores ambientais na temporada reprodutiva de 2010. Monograph to Universidade Federal Fluminense, Rio de Janeiro, 58 pp

Mills H (1996) Aardvark. Cornell University, Ithaca

Mobley JR, Herman LM, Frankel AS (1988) Responses of wintering humpback whales (*Megaptera novaeangliae*) to playback of recordings of winter and summer vocalizations and of synthetic sound. Behav Ecol Sociobiol 23:211–223

Morales-Rincon N, Morteo E, Martínez-Serrano I, Delfín-Alfonso CA (2016) Preliminary study of the behavior of dolphins (*Tursiops truncatus*) in interaction with artisanal fishing on the central coast of Veracruz, Mexico. XXXV International Reunion for the Study of Marine Mammals. Mexican Society of Marine Mammalogy (SOMEMMA). May 2–6th, La Paz, Baja California Sur, México

Morete ME (2007) Caracterização temporal da estrutura de grupos e do comportamento de baleias-jubarte (*Megaptera novaeangliae*) na área de reprodução da região do Arquipélago dos Abrolhos (Bahia Brasil). PhD Thesis to the Universidade De São Paulo, São Paulo, 147 p

Morete ME, Freitas AC, Engel MH, Pace RM III, Clapham PJ (2003a) A novel behavior observed in humpback whales on wintering grounds at Abrolhos Bank (Brazil). Mar Mamm Sci 19:694–707

Morete ME, Pace RM III, Martins CCA, Freitas AC, Engel MH (2003b) Indexing seasonal abundance of humpback whales around Arquipélago dos Abrolhos, Bahia, Brazil. LAJAM 2:21–28

Morete ME, Bisi TL, Rosso S (2007a) Mother and calf humpback whale responses to vessels around the Abrolhos archipelago, Bahia. Brazil J Cetacean Res Manag 9(3):241–248

Morete ME, Bisi TL, Rosso S (2007b) Temporal pattern of humpback whale (*Megaptera novaeangliae*) group structure around Abrolhos archipelago breeding region, Bahia, Brazil. J Mar Biol Assoc UK 87:87–92

Morete ME, Pace RM III, Rosso S (2008) Fluctuating abundance of humpback whales (*Megaptera novaeangliae*) in a calving ground off coastal Brazil. J Mar Biol Assoc UK 88:1229–1235

Morrison ML (2002) Wildlife restoration: techniques for habitat analysis and animal monitoring. Society for ecological restoration. Island Press, Washington, 209 p

National Research Council (NRC) (2005) Marine mammal population and ocean noise: determining when noise causes biologically significant effects. National Academy Press, Washington, 96 p

Noad MJ, Cato DH (2001) A combined acoustic and visual survey of humpback whales off southeast Queensland. Mem Queensl 47(2):507–523

Noad MJ, Cato DH (2007) Swimming speeds of singing and non-singing humpback whales during migration. Mar Mamm Sci 23:481–495

Normande IC, Luna FDO, Malhado ACM, Borges JCG, Junior PCV, Attademo FLN, Ladle RJ (2014) Eighteen years of Antillean manatee *Trichechus manatus manatus* releases in Brazil: lessons learnt. Oryx 49(02):338–344

Normande IC, Malhado ACM, Reid J, Viana PC, Savaget PVS, Correia RA, Ladle RJ (2015) Post-release monitoring of Antillean manatees: an assessment of the Brazilian rehabilitation and release programme. Anim Conserv 19(3):235–246

Oviedo L, Silva N (2005) Sighting frequency and relative abundance of bottlenose dolphins (*Tursiops truncatus*) along the northeast coast of Margarita Island and los Frailes archipelago, Venezuela. Rev Biol Trop 53(3–4):595–600

Pardo MA, Palacios DM (2006) Cetacean ocurrence in the Santa Marta region, Colombian Caribbean, 2004–2005. LAJAM 5(2):129–134

Pavés HJ, Schlatter RP, Franco-Trecu V, Páez E, Sielfeld W, Araos V, Cappozzo HL (2016) Breeding season of the south American fur seal (*Arctocephalus australis*, Otariidae: Carnivora): new data for establishing independent evolutionary histories? Rev Biol Mar Oceanogr 51(2):241–253

Pavez G, Muñoz L, Inostroza P, Sepúlveda M (2011) Behavioral response of south American sea lion *Otaria flavescens* to tourist disturbance during the breeding season. Rev Biol Mar Oceanogr 46(2):135–140

Payne R (1986) Long term behavioral studies of the southern right whale (*Eubalaena australis*). Report of the Inter Whal Comm 10:161–167

Petta CB, Andriolo A, Morete ME, Martins CCA (2004) Comportamento respiratório das baleias jubarte (*Megaptera novaeangliae*) em seu sítio reprodutivo no Banco dos Abrolhos, Bahia, Brazil. In: 11° Reunión de Trabajo de Especialistas en Mamíferos Acuáticos de América Del Sur – 5to Congreso de la Sociedad Latino Americana de Especialistas en Mamíferos Acuáticos, pp 155

Quinn GP, Keough MJ (2002) Experimental design and data analysis for biologists. Cambridge University Press, Cambridge, 537 p

Salden DR (1988) Humpback whale encounter rates offshore of Maui, Hawaii. J Wildlife Manage 52:301–304

Samuels A, Tyack PL (2000) Flukeprints: a history of Studing cetacean societies. In: Mann J, Connor RC, Tyack PL, Whitehead H (eds) Cetacean societies: field studies of dolphins and whales. The University of Chicago Press, Chicago, pp 9–44. 448 p

Santos MCO, Acuña LB, Rosso S (2001) Insights on site fidelity and calving intervals of the marine tucuxi (*Sotalia fluviatilis*) in south-eastern Brazil. J Mar Biol Ass UK 81:1049–1052

Santos UM, Alvarez MR, Schilling AC, Strenzel GMR, Le Pendu Y (2010) Spatial distribution and activities of the estuarine dolphin *Sotalia guianensis* (van Bénédén, 1864) (Cetacea, Delphinidae) in Pontal Bay, Ilhéus, Bahia, Brazil. Biota Neotrop 10(2):67–73

Santos GMA, Quaresma AC, Barata RR, Martins BML, Siciliano S, Silva JS Jr, Emin-Lima R (2012) Etho-ecological study of the Amazon River dolphin, *Inia geoffrensis* (Cetacea: Iniidade) and the dolphins of genus *Sotalia* (Cetacea: Delphinidae) in Guamá River, Amazonia. Mar Biodivers Rec 5:e23

Scheidat M, Castro C, Gonzalez J, Williams R (2004) Behavioural responses of humpback whales (*Megaptera novaeangliae*) to whalewatching boats near Isla de la Plata, Machalilla National Park, Ecuador. J Cetacean Res Manag 6(1):63–68

Senigaglia V (2011) Draft protocol to collect behavioural data from land (v. 1.0) In: The Large-scale Whalewatching Expermiment (LaWE) LaWEScience. Avaiable on https://sites.google.com/site/lawescience/protocols. Last Access on 15 Oct 2016

Sepúlveda M, Quiñones RA, Carrasco P, Pérez-Álvarez MJ (2012) Daily and seasonal variation in the haul-out behavior of the south American sea lion. Mamm Biol 77(4):288–292

Seyboth E, Groch KR, Secchi ER, Dalla Rosa L (2015) Habitat use by southern right whales, *Eubalaena australis* (Desmoulins, 1822), in their main northernmost calving area in the western South Atlantic. Mar Mamm Sci 31(4):1521–1537

Siciliano S, Hassel LB, Venturotti A, Magalhães FA (2003) Land based survey for whales and dolphins off Arraial do Cabo, Eastern coast of Rio de Janeiro, Brazil. Final Report to Yaqu Pacha and Cetacean Society International, 4 pp

Silva Jr JM (2005) Ecologia comportamental do golfinho-rotador (*Stenella longirostris*) em Fernando de Noronha. PhD Thesis to the Universidade Federal do Pernambuco, Recife, 121 pp

Silva FJL, Silva JM Jr (2009) Circadian and seasonal rhythms in the behavior of spinner dolphins (*Stenella longirostris*). Mar Mamm Sci 25(1):176–186

Smultea MA (1994) Segregation by humpback whale (*Megaptera novaeangliae*) cows with a calf in coastal habitat near the island of Hawaii. Can J Zool 72(5):805–811

Steckler D, Donlan P (2016) Mysticetus. Entiat River Technologies, Preston. Available www.mysticetus.com

Tosi CH (2007) Sincronismo comportamental em boto cinza (*Sotalia guianensis*). Dissertation to the Universidade Federal do Rio Grande do Norte, Natal, 113 pp

Tyack P (1981) Interactions between singing Hawaiian humpback whales and conspecifics nearby. Behav Ecol Sociobiol 8:105–116

Tyack PL, Connor RC, Mann J, Whitehead H (2000) Epilogue: the future of behavioral research on cetaceans in the wild. In: Mann J, Connor RC, Tyack PL, Whitehead H (eds) Cetacean societies: field studies of dolphins and whales. The University of Chicago Press, Chicago, pp 333–340. 448 p

Tyack P, Gordon J, Thompson D (2003) Controlled exposure experiments to determine the effects of noise on marine mammals. Mar Technol Soc J 37(4):41–53

Vermeulen E, Cammareri A, Holsbeek L (2012) Alteration of southern right whale (*Eubalaena australis*) behaviour by human-induced disturbance in Bahía San Antonio, Patagonia, Argentina. Aquat Mamm 38(1):56–64

Vermeulen E, Holsbeek L, Das K (2015) Diurnal and seasonal variation in the behaviour of Bottlenose dolphins (*Tursiops truncatus*) in Bahía San Antonio, Patagonia, Argentina. Aquat Mamm 41(3):272–283

Vernazzani BG, Brownell Jr. RL, Cabrera E, Carlson C, Sironi M (2012) Update on 2012 blue whale field season in Chile. International Whaling Commission, Scientific Committee (IWC-SC) Report SC/64/SH18

Wada S, Oishi M, Yamada TK (2003) A newly discovered species of living baleen whale. Nature 426:278–281

Williams R, Trites AW, Bain DE (2002) Behavioural responses of killer whales (*Orcinus orca*) to whale-watching boats: opportunistic observations and experimental approaches. J Zool 256(02):255–270

Williams R, Lusseau D, Hammond PS (2006) Estimating relative energetic costs of human disturbance to killer whales (*Orcinus orca*). Biol Conserv 133:301–311

Würsig B, Würsig M (1978) Behavior and ecology of the bottlenose dolphin, *Tursiops truncatus*, in the South Atlantic. Fish Bull 77(2):399–412

Würsig B, Würsig M (1980) Behavior and ecology of the Dusky dolphin, *Lagenorhynchus obscurus*, in the South Atlantic. Fish Bull 77(4):871–890

Würsig B, Cipriano F, Würsig M (1991) Dolphin movement and patterns. Information from radio and theodolite tracking studies. In: Pryor K, Norris KS (eds) Dolphin societies, discovery and puzzles. University of Califórnia Press, Oakland, pp 79–112. 397 p

Zuur AF, Leno EN, Smith GM (2007) Analysing ecological data. Springer Science & Business Media, New York

Zuur AF, Leno EN, Elphick CD (2010) A protocol for data exploration to avoid common statistical problems. Methods Ecol Evol 1(1):3–14

Chapter 5
Marine Mammal Bioacustics Using Towed Array Systems in the Western South Atlantic Ocean

Artur Andriolo, Franciele Rezende de Castro, Thiago Amorim, Gustavo Miranda, Juliana Di Tullio, Juliana Moron, Bruna Ribeiro, Gabriela Ramos, and Raíssa Rodrigues Mendes

Abstract Acoustic technologies have been applied in order to investigate and monitor underwater sound and have promoted achievements on the understanding of animal biology, behavior and ecology. Whales and dolphins produce sounds, which are unique, compared to other sounds in the marine environment. Passive acoustic surveys using a towed hydrophone array have become more accessible and widely used to explore patterns of occurrence, identifying critical habitats for several species of cetaceans and inferring about potential noise impacts over the populations.

A. Andriolo (✉)
Laboratório de Ecologia Comportamental e Bioacústica, Departamento de Zoologia, Instituto de Ciências Biológicas, Universidade Federal de Juiz de Fora, Juiz de Fora, MG, Brazil

Programa de Pós Graduação em Ecologia, Universidade Federal de Juiz de Fora, Juiz de Fora, MG, Brazil

Programa de Pós Graduação em Ciências Biológicas, Comportamento e Biologia Animal, Universidade Federal de Juiz de Fora, Juiz de Fora, MG, Brazil

Instituto Aqualie, Juiz de Fora, Juiz de Fora, MG, Brazil
e-mail: artur.andriolo@ufjf.edu.br

F.R. de Castro • T. Amorim
Programa de Pós Graduação em Ecologia, Universidade Federal de Juiz de Fora, Juiz de Fora, MG, Brazil

Laboratório de Ecologia Comportamental e Bioacústica, Departamento de Zoologia, Instituto de Ciências Biológicas, Universidade Federal de Juiz de Fora, Juiz de Fora, MG, Brazil

Instituto Aqualie, Juiz de Fora, Juiz de Fora, MG, Brazil

G. Miranda
Instituto Aqualie, Juiz de Fora, Juiz de Fora, MG, Brazil

J. Di Tullio
Instituto de Oceanografia, Universidade Federal de Rio Grande, FURG, Rio Grande, Brazil

M.R. Rossi-Santos, C.W. Finkl (eds.), *Advances in Marine Vertebrate Research in Latin America*, Coastal Research Library 22,
DOI 10.1007/978-3-319-56985-7_5

In this chapter we present characterization of acoustic signals produced by nine different cetacean species obtained form acoustic surveys. The species have species-specific qualities in their whistles and clicks. Acoustic methods can also offer population size estimates and identification of population structure.

5.1 Introduction

Acoustic technologies have been applied in order to investigate and monitor underwater sound and have promoted achievements on the understanding of animal biology, behavior and ecology. Due to physical characteristic the sound propagation is favored in the underwater environment and have contributed to evolution of acoustic features of marine life.

Whales and dolphins are top predators and are known to spend all their lives in the aquatic environment. As such, they can be considered sentinels of the ocean since they serve as indicators of the habitat health to which they are part of. Most importantly is that cetaceans produce sounds, which are unique, compared to other sounds in the marine environment.

Passive acoustic surveys have become more accessible and widely used to explore patterns of occurrence, identifying critical habitats (Risch et al. 2014) for several species of cetaceans and inferring about potential noise impacts over the populations (Rice et al. 2014; Pirotta et al. 2015). Acoustic methods can also offer population size estimates that are used to track large-scale displacement patterns (Mellinger and Barlow 2003) and long term population trends (Evans and Hammond 2004; Magera et al. 2013; Campbell et al. 2015).

The recent expansion in bioacoustics research was possible due to the development of available technologies on autonomous acoustic recorders (Passive Acoustic Monitoring – PAM), which can be fixed on the sea floor (e.g. Parks et al. 2007; Cerchio et al. 2001; Darling and Sousa-Lima 2005; Sousa-Lima et al. 2013) or

J. Moron
Laboratório de Ecologia Comportamental e Bioacústica, Departamento de Zoologia, Instituto de Ciências Biológicas, Universidade Federal de Juiz de Fora, Juiz de Fora, MG, Brazil

Instituto Aqualie, Juiz de Fora, Juiz de Fora, MG, Brazil

B. Ribeiro
Laboratório de Ecologia Comportamental e Bioacústica, Departamento de Zoologia, Instituto de Ciências Biológicas, Universidade Federal de Juiz de Fora, Juiz de Fora, MG, Brazil

Programa de Pós Graduação em Ciências Biológicas, Comportamento e Biologia Animal, Universidade Federal de Juiz de Fora, Juiz de Fora, MG, Brazil

Instituto Aqualie, Juiz de Fora, Juiz de Fora, MG, Brazil

G. Ramos • R.R. Mendes
Laboratório de Ecologia Comportamental e Bioacústica, Departamento de Zoologia, Instituto de Ciências Biológicas, Universidade Federal de Juiz de Fora, Juiz de Fora, MG, Brazil

which can be towed behind vessels (Jaquet and Whitehead 1999; Whitehead 2002; Watwood et al. 2006; Barlow et al. 2013), allowing to record sounds continuously.

Passive acoustic detection systems have been successfully used on a number of cetacean surveys (Gannier et al. 2002; Barlow and Taylor 2005; Leaper et al. 2003; Hastie et al. 2003; Lewis et al. 2007; Fais et al. 2016). Acoustic monitoring provides an opportunity to collect data in conditions unsuitable for visual observations such as darkness, poor visibility and high sea states (Evans and Hammond 2004; Mellinger et al. 2007; Ward et al. 2012). The use of simple towed hydrophones to monitor cetacean vocalizations enables quantifiable data to be collected at minimal cost (Whitehead 2002).

The implementation of acoustic survey associate to visual effort can maximize the results, knowing that a relative large amount of ship time would be lost due to bad weather or night fall, a passive acoustic towed-array device can increase search effort. Some researchers have demonstrated the advantages of acoustic towed arrays in comparison to visual survey methods specially to detect species that spend most of their time in deep divings, such as beaked whales and sperm whales (Jaquet and Whitehead 1999; Whitehead 2002; Watwood et al. 2006; Barlow et al. 2013; Yack et al. 2013; Schorr et al. 2014). On the other hand, this technique has disadvantages as the need to manage a heavy cable when the moving platform stops, loss of maneuverability of the moving platform, risk of damaging the propellers, requires an experienced crew and is a lengthy process to initiate the survey (Evans and Hammond 2004; Nielsen and Møhl 2006).

This chapter will present a general overview of the equipment which has being used to cetaceans acoustic survey in the Western South Atlantic Ocean. Also, will give a basic bioacoustic analysis introduction and will present a general description of acoustic parameters of different species registered. Recordings were obtained opportunistically from 2012 to 2015 during a survey mostly dedicated to visual monitoring of cetacean occurrence and distribution along the western South Atlantic continental shelf break. Research cruises were performed between 26°S and 38°S over the continental shelf break and slope. Acoustic tracklines comprised an average of 780 nm of effort per survey. Concurrent environmental and GPS data were logged automatically using WinCruz software. Visual positive identifications were associated to the acoustic recordings. The wave files were analyzed using partially automated detections tools complemented with visual and aural searched for species confirmation whenever possible.

5.2 Towed Array Equipment

The system consists of a three-elements hydrophone array (Fig. 5.1) on a cable towed 150–300 m behind the vessel) coupled to a recording system, configured to give a variable frequency response from 499 Hz (High Pass Filter) to 100,000 Hz (Fig. 5.2). This system is mostly used as an additional method to visual surveys commonly applied to marine mammals (Bittle and Duncan 2013), allowing real time or post data analysis. It has being used from different platforms and contexts

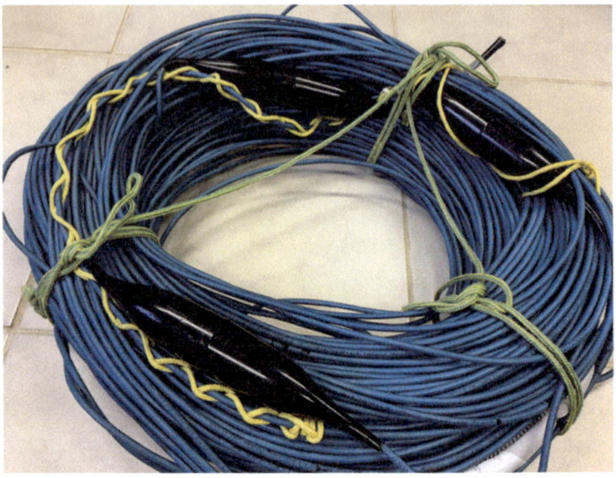

Fig. 5.1 Three-element hydrophone array in a 300 m cable. The *black* elements are the hydrophones covered by a polyurethane resin. (Reproduced with permission of Gustavo Miranda)

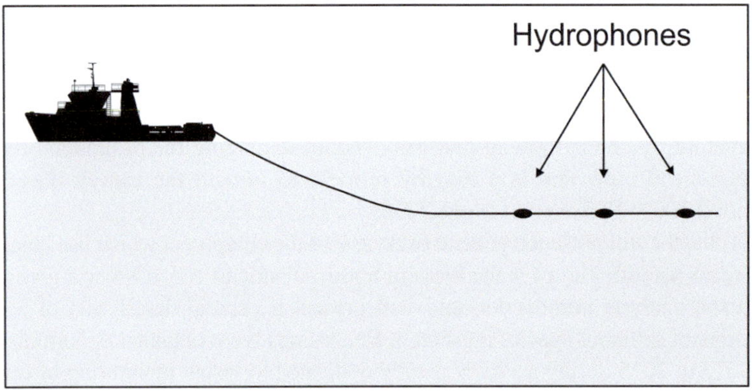

Fig. 5.2 Three-hydrophone array on a cable towed by a moving platform. (Reproduced with permission of Gustavo Miranda)

(e. g. Barlow and Taylor 2005; Lewis et al. 2007; Hastie et al. 2003; Leaper et al. 2003; Moretti et al. 2006; Swift et al. 2009).

Different types of equipment have been used over the years for different purposes. Their characteristics depend on the target species (Evans and Hammond 2004) and specific questions and analysis planned. At least two hydrophones in a linear array are necessary to apply the target motion analysis in order to define a spatial localization (Barlow and Taylor 2005; Lewis et al. 2007, Swift et al. 2009). Usually the recording system is composed by a fanless computer, a data acquisition board for frequencies higher than 192 kHz and an audio interface for frequencies below 192 kHz (Fig. 5.3). The use of a data acquisition board requires a high quality bandpass filter, given that the audio interface provides an anti-aliasing filter. The power supply of the system has to be considered with attention in order to reduce the electrical noise associated with the moving platform, but in some cases may be necessary an independent clean battery power (Rankin and Barlow 2010).

Fig. 5.3 Recording system composed by a fanless computer, a data acquisition board for frequencies higher than 192 kHz and an audio interface for frequencies below than 192 kHz. (Reproduced with permission of Franciele R. de Castro)

Regardless of the number of hydrophones in a linear array the ambiguity is present, because it is not possible to distinguish the side of the signal (Au and Hastings 2008). A greater number of hydrophones would enable the localization of the signal rather than a group of vocalizing individuals (Benda-Beckmann et al. 2013). However, this has been still discussed in the literature, especially considering the possible instability associated to towed systems.

In general the spacing between hydrophones considers the frequencies of vocalizations of the target species. The lower is the frequency, the greater will be the distance between hydrophones (Leaper et al. 1992, 2003; Benda-Beckmann et al. 2013; Gillespie 1997; Barlow and Taylor 1998, 2005; Lewis et al. 2007; Rankin and Barlow 2007; Swift et al. 2009; Gillespie et al. 2010). However, these systems are mostly indicated to species that produces signals above the noise frequency band produced by the moving platform and the flow noise of the cable, e.g. sperm whales (Barlow and Taylor 2005), beaked whales (Gillespie et al. 2010) and harbor porpoise (Sveegaard et al. 2011).

5.3 Basic Bioacoustics Data Analysis

For decades, scientists have been listening to the seas, looking forward to accurately record and understand the nature and purpose of whales and dolphins sound emissions. Marine mammals produce a variety of vocalizations, each of them suited to a particular behavior or situation. Bioacoustics is a cross-disciplinary science that is broadly concerned in understand: animal communication and associated behavior, sound production anatomy and neurophysiology, effects of human-made noise on animals, auditory capacities and auditory mechanisms of animals, effects of human-made noise on animals.

The primordial step in understanding acoustic behavior of a species is to describe the characteristics of a specific emission, with regard to what types of sounds make up the repertoire and what are the acoustical parameters features related to such sounds. Sound emissions by odontocetes can be classified into two broad categories of frequency-varying continuous tonal sounds referred to as whistles and pulsed sounds including broadband clicks including and burst sounds (Evans 1967). Depending on the main objectives of a research, bioacousticians rely on some technologies to make measurements in each broad category. Some examples are given below.

5.3.1 *Whistles*

The first method consists of visual inspection of spectrograms with a particular emphasis on establishing categories in the shape of the whistle contour. Despite of being very subjective, this method can be useful to determine if a particular contour is unique within the repertoire or whether is a minor variation to another contour. In order to specify a particular category of contour in which a whistle belongs to, we have been using six general broad categories based on the slope of the whistle fundamental frequency and the number of inflection points, defined as a point at which the slope of the contour reverses in direction:

1. Constant frequency: the frequency does not really remain constant over its duration, but has a minimum amount of frequency change (Fig. 5.4a).
2. Upsweep or ascending: the frequency is modulated with the instantaneous frequency increasing over duration and do not have any marked inflection points (Fig. 5.4b).
3. Downsweep or descending: – the frequency is modulated with the instantaneous frequency decreasing over duration and do not have any marked inflection points (Fig. 5.4c).
4. Concave or ascending-descending: the frequency modulated with the instantaneous frequency initially increasing with time, followed by an inflection point and an ending with the frequency decreasing over duration (Fig. 5.4d).
5. Convex or descending-ascending: the frequency is modulated with the instantaneous frequency initially decreasing with time, followed by an inflection point and an ending with the frequency increasing over duration (Fig. 5.4e).
6. Sinusoidal or multiple: the frequency is modulated with more than one repetition of a hill or a valley and the contour appearing somewhat like a sinusoidal signal with at least two inflection points (Fig. 5.4f).

A more quantitative method involves extract various parameters of the fundamental frequency directly from the spectrogram: beginning (starting) frequency, ending frequency, minimum frequency, maximum frequency, delta frequency, peak frequency, center frequency, duration, number of inflection points, number or presence of break in contour and number or presence of harmonics.

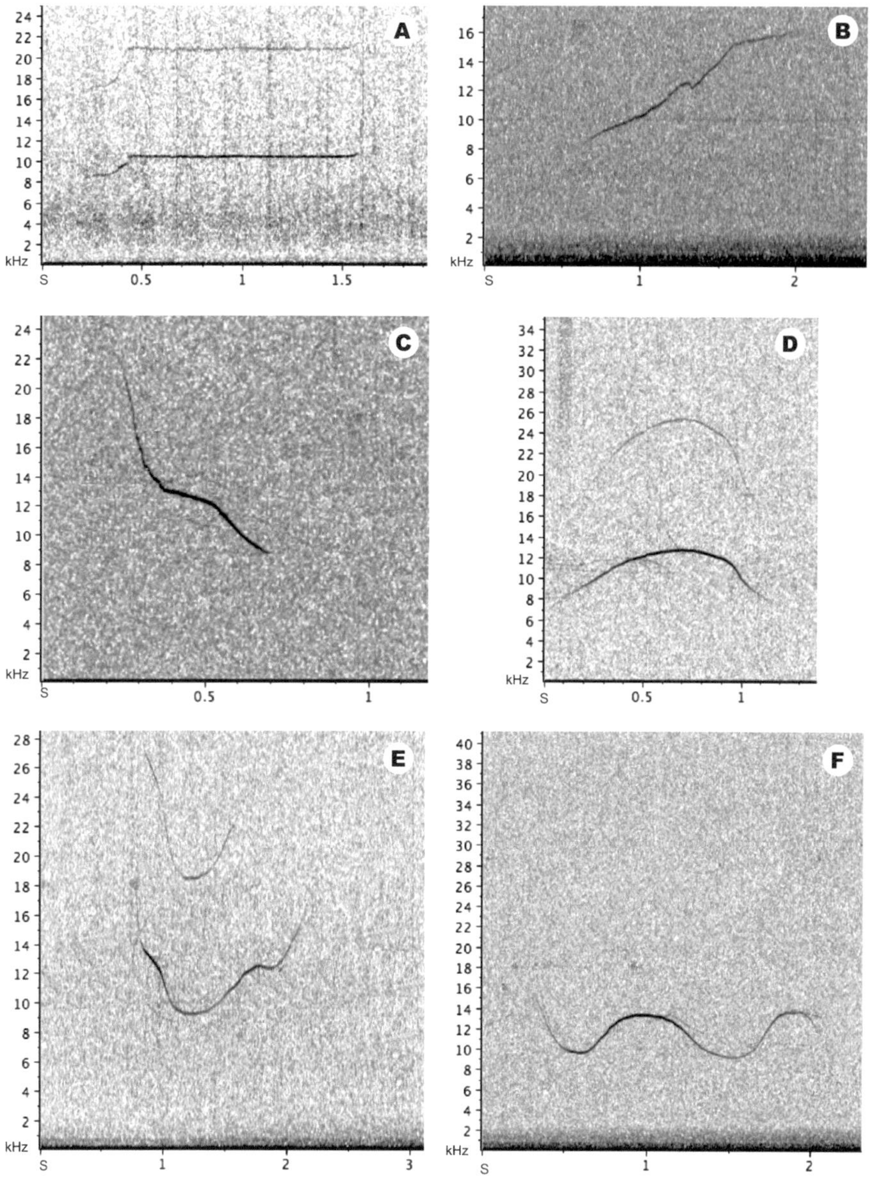

Fig. 5.4 Examples of spectrograms showing general categories of whistles contours. (**a**) Constant, (**b**) Upsweep or ascending, (**c**) Downsweep or descending, (**d**) Concave or ascending-descending, (**e**) Convex or descending-ascending and (**f**) Sinusoidal or multiple

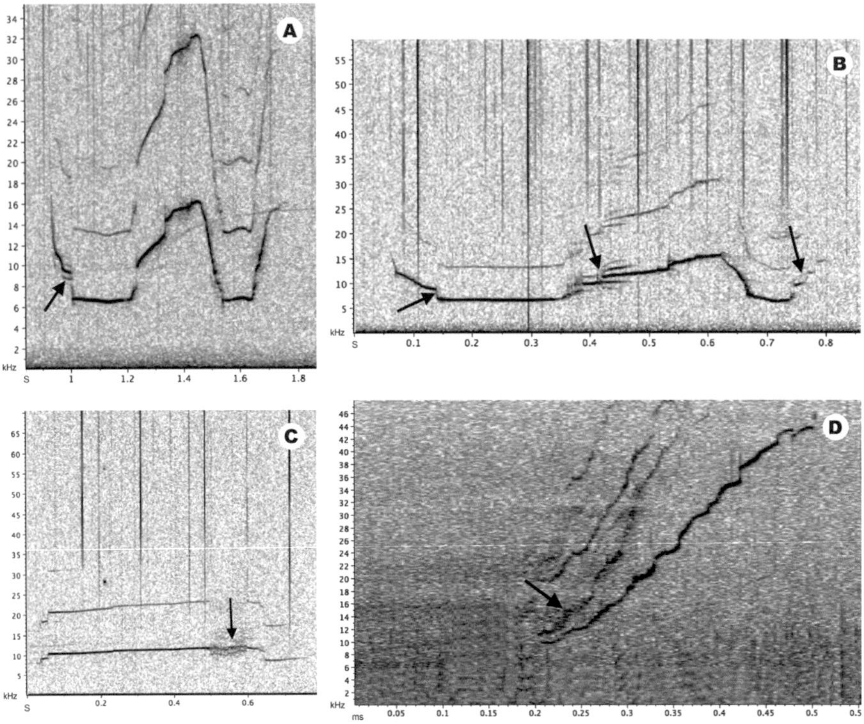

Fig. 5.5 Examples of non-linear phenomena. Arrows point to: (**a**) Subharmonics, (**b**) Frequency jump, (**c**) Deterministic chaos and (**d**) Biphonation

One set of relatively unexplored features, referred to as non-linear phenomena, can also be observed as a component of whistles emissions. This phenomenon includes subharmonics, frequencies jumps, biphonation and deterministic chaos. These complex features are produced by nonlinearities in the vocal production system, where rather simple neural commands to the system can result in highly complex and individually variable acoustic signals (Fitch et al. 2002). The spectrogram in Fig. 5.5 shows some non-linear acoustic phenomena.

5.3.2 Echolocation Clicks

As well as in whistles, we have been firstly categorizing echolocation into categories based on the visual inspection of spectrogram. The first category consists on clicks that can be aurally assigned to one vocalizing animal and does not show any other clicks belonging to a different train (Fig. 5.6A). The second category gathers clicks produced by more than one animal, and therefore it is possible to visualize many overlapped clicks (Fig. 5.6b). Again, depending on the main objective, both

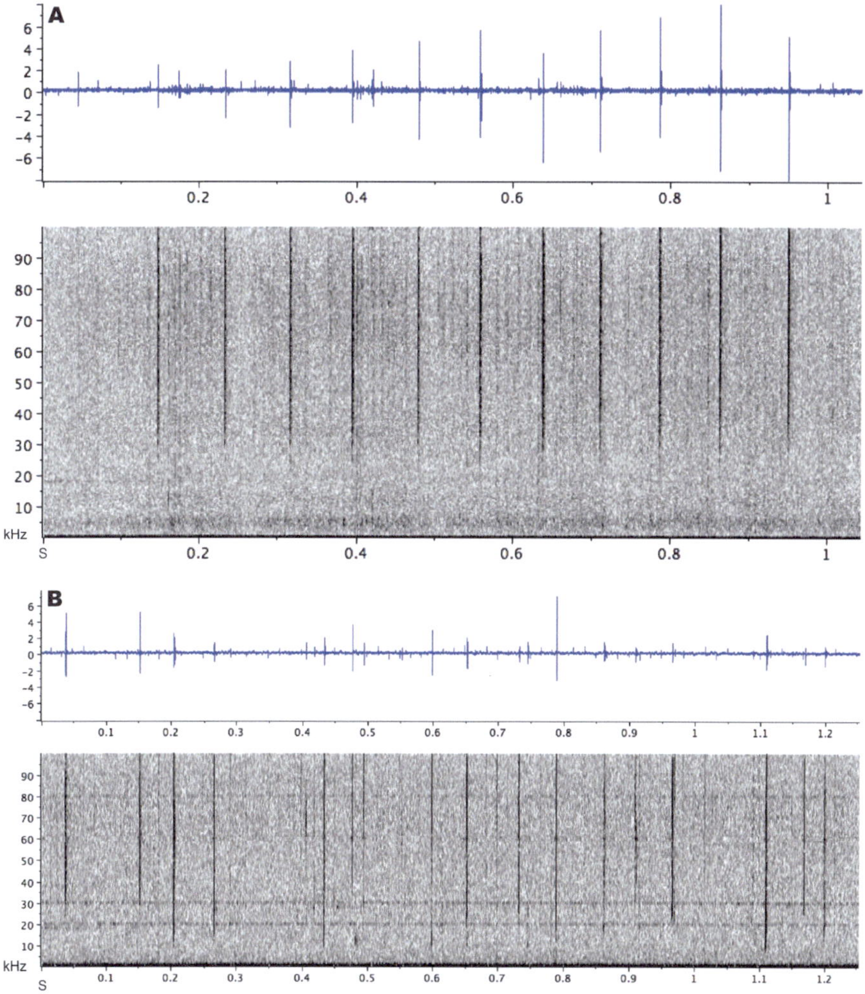

Fig. 5.6 Click trains. (**a**) Clicks can be aurally assigned to one vocalizing animal and do not show any other clicks belonging to a different train and (**b**) Overlapped clicks produced by more than one animal

types of trains can be used to investigate the general echolocation behavior. For example, the automatic measurement of inter-click interval (ICI) is greatly underestimated in overlapped clicks, but apart from that, other spectral and time parameters can be extracted using such clicks sequences without any inaccurate result.

For a general quantitative approach, some acoustical parameters can be automatically extracted: peak frequency, 3 dB bandwidth, 10 dB bandwidth, inter-click interval (ICI), sound pressure level (peak to peak), root mean square amplitude (rms) and number of clicks in train.

5.3.3 Burst Sounds

Burst pulse sounds are another major category of sound emissions by odontocetes. For dolphins and small whales, burst pulse sounds are characterized by a high repetition rate (greater than about 300 pulses per second) or low inter-pulse intervals (less than about 3 ms) (Au and Hastings 2008).

The great emphasis on studying whistles has led many to suppose that whistles are the primary mode of communication in odontocetes. It is likely that the reason for this relies on the wide band nature of burst pulse sounds compared to the frequency of the fundamental components of whistles, given that burst pulses can have frequencies components that extend beyond 100 kHz (Lammers and Au 1996). The Fig. 5.7 shows burst vocalizations with frequency components in the ultrasonic range. So far, our data set allows basic measurements as duration, peak frequency, center frequency and delta frequency. In addition, we have been observing the sequential and time patterns in which burst sounds appear in the spectrograms. Apparently, some burst sounds are emitted continuously to clicks trains (Fig. 5.7a) or they are discrete in time (Fig. 5.7b).

The basic description of sounds produced by marine mammals in South Atlantic Ocean, allows further investigations on geographical variation, signals classification systems, social structure, behavior and so many other science fields that reflect on acoustic communication. The next species sections in this chapter will present some acoustical parameters for whistles, clicks and burst sounds of delphinid species recorded in the South Atlantic continental shelf break.

5.4 Cetaceans Species Recorded and Identified in the Western South Atlantic

5.4.1 *Risso's Dolphin* (Grampus griseus)

Rissos dolphin (Fig. 5.8) is a small cetacean that can reach up to four meters long and is distributed in temperate and tropical waters worldwide (Jefferson et al. 2014). This species is distributed along the Brazilian continental shelfbreak and slope from northeast to south region (Bastida et al. 2007; Di Tullio et al. 2016).

The acoustic repertoire of the Risso's dolphin has not been fully detailed and examples of whistles and clicks are preseted in Fig. 5.9. Neves (2013) described the acoustic parameters of whistles and clicks in different regions (Australia, Azores, California, Egypt and Gran Canaria). Gannier et al. (2008) and Rendell et al. (1999) studied whistles in the waters of the western Mediterranean Sea and Azores, respectively. Philips et al. (2003) described the clicks emitted by the Risso's dolphin in captivity, and Madsen et al. (2004) studied click emissions of free-ranging animals in Sri Lanka. Corkeron and Van Parjis (2001), in addition, describing the acoustic parameters of pulsed sounds and whistles in Newcastle, Australia. Andriolo et al.

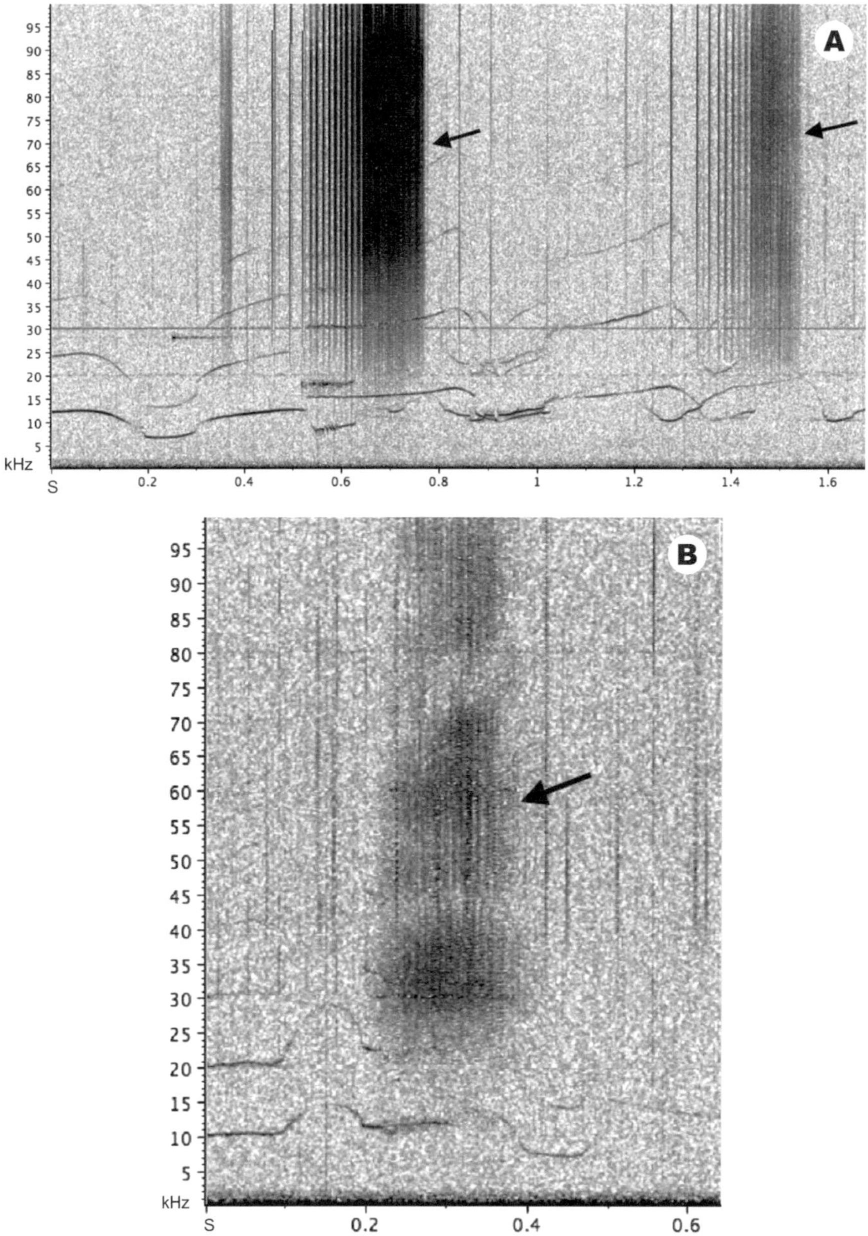

Fig. 5.7 Spectrogram of burst sounds. (**a**) Burst sounds emitted continuously to clicks trains and (**b**) Burst sounds emitted discrete in time

Fig. 5.8 Risso's dolphin photographed off the Brazilian coast (Reproduced with permission of Rodrigo Genoves)

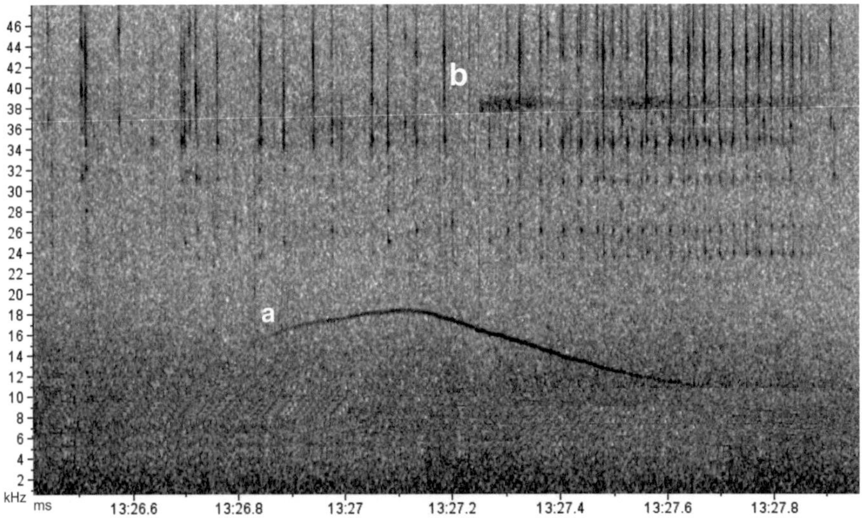

Fig. 5.9 Examples of Risso's dolphin whistles (**a**) and clicks (**b**)

(2013) conducted a study in Brazil describing the acoustic parameters of the tonal sounds of this dolphin.

*5.4.2 Rough-Toothed Dolphin (***Steno bredanesis***)*

The rough-toothed dolphin is a small delphinidae that can reach up to 2.85 ms long and is distributed in tropical and temperate warm waters worldwide (Bastida et al. 2007). In Brazilian waters this species can be found over the continental shelf shallow and deep waters from the northeast to south regions, whereas in Rio de Janeiro

Fig. 5.10 Rough-toothed dolphin photographed off the Brazilian coast (Reproduced with permission of Pedro Fruet)

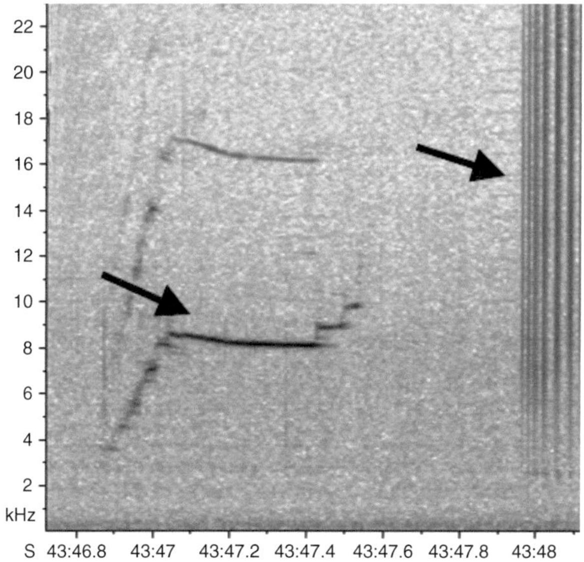

Fig. 5.11 *Steno bredanensis* acoustic signal presented in a spectrogram (Whistles and clicks, respectively)

state it is commonly sighted near the coast (Ott and Danilewicz 1996; Meirelles et al. 2009; Lodi and Hetzel 2012; Bastida et al. 2007) (Fig. 5.10).

As the majority of the delphinids species, the rough-toothed dolphin produces tonal (whistles) and pulsed sounds (echolocation clicks and burst pulses – broadband emissions) (Tyack and Clark 2000; Oswald et al. 2003, 2007; Rankin et al. 2008a, 2015; Hoelzel 2009; Baumann-Pickering et al. 2010; Corrêa 2012; Lima et al. 2012; Amorim et al. submitted) (Fig. 5.11). The acoustic signals produced by this species have been discussed in the literature through different approaches such as: acoustic repertoire characterization (Busnell and Dziedzic 1966; Watkins et al.

Fig. 5.12 Spinner dolphin photographed off the Brazilian coast (Reproduced with permission of Pedro Fruet)

1987; Oswald et al. 2003, 2007; Rankin et al. 2015), preliminary acoustic behavior description (Rankin et al. 2008a), including possible eavesdropping recording during synchronized swimming (Gotz et al. 2006), acoustic monitoring (Rankin et al. 2008b), and species identification (Oswald et al. 2003, 2007).

In Brazil, the rough-toothed dolphin acoustic signals have been recorded on the southeastern (Guanabara Bay, Rio de Janeiro state), northeastern (Abrolhos Bank) coast (Bezerra Filho 2012; Corrêa 2012; Lima et al. 2012) and, in a most recent effort, on the southern (Rio Grande do Sul state) coast (Amorim et al. submitted). Except for Bezerra Filho (2012), the others have described spectral and temporal whistles parameters for this species (Amorim et al. submitted; Corrêa 2012; Lima et al. 2012).

*5.4.3 Spinner Dolphin (***Stenella longirostris***) and Atlantic Spotted Dolphin (***Stenella frontalis***)*

Spinner dolphin is a small delphinidae, which can reach up to 2.30 m long and is considered to have a pantropical distribution. This species has been described to occur beyond the outer continental shelf in tropical waters of the Southwestern Atlantic Ocean and near oceanic islands (Bastida et al. 2007; Danilewicz et al. 2013; Amaral et al. 2015), however, a few sightings exist south of 31°S, in spring (Di Tullio et al. 2016) (Fig. 5.12).

The acoustic investigation of the whistles emitted by the spinner dolphin already presented some geographical variation (e.g. Bazúa-Durán and Au 2002; 2004; Moron et al. 2015; Bonato et al. 2015). However, so far the geographical isolation indicates no relation to whether spinner dolphin whistles. Thus is necessary to

Fig. 5.13 *Stenella frontalis* photographed off the Brazilian coast. (Reproduced with permission of Pedro Fruet)

investigate other factors handling those geographical differences (Moron et al. 2015) and even extend the comparison to pulsed signals. Therefore, more studies and partnerships between researchers studying other populations are important to be better explore and recognize the results found so far.

The Atlantic spotted dolphin is small, reaching approximately two meters long, and is restricted to the tropical and temperate warm waters of the Atlantic Ocean (Bastida et al. 2007). In Brazilian waters this species is found over the shallow continental shelf waters (20 m) and slope in southeastern and southern regiosn; however it appears to have a discontinuous distribution, with no records occurring between 6°S and 18°S (Moreno et al. 2005; Danilewicz et al. 2013). The Atlantic spotted dolphin is the only member of the genus *Stenella* that is commonly observed close to shore in Brazil (Moreno et al. 2005) (Fig. 5.13).

There are few published works about this species acoustic signals in the South Atlantic (Azevedo et al. 2010; Lima et al. 2016), being mostly studied in a few localities of the North Atlantic Ocean (e.g. Ding et al. 1995, Lammers et al. 2003, Herzing 2015). Previous acoustic studies were taken especially in the Bahama Islands, where a local population exists (e.g. Dudzinski 1996; Herzing 1996, 2015).

Both species presented several repeated contours, especially *S. frontalis*, indicating not only the existence of the so-called signature whistles (Moron and Andriolo 2015) but also indicating its importance. Nonlinear events were also identified, mostly on spotted dolphins' vocalizations (Moron and Andriolo 2016) in the form of different types of biphonation.

Overall, *S. longirostris* and *S. frontalis* emit complex whistles. Despite having many features in common, acoustic analysis is pointing on how to differentiate both species on the Western South Atlantic Ocean. Examples of both species acoustic behavior are presented in the Fig. 5.14.

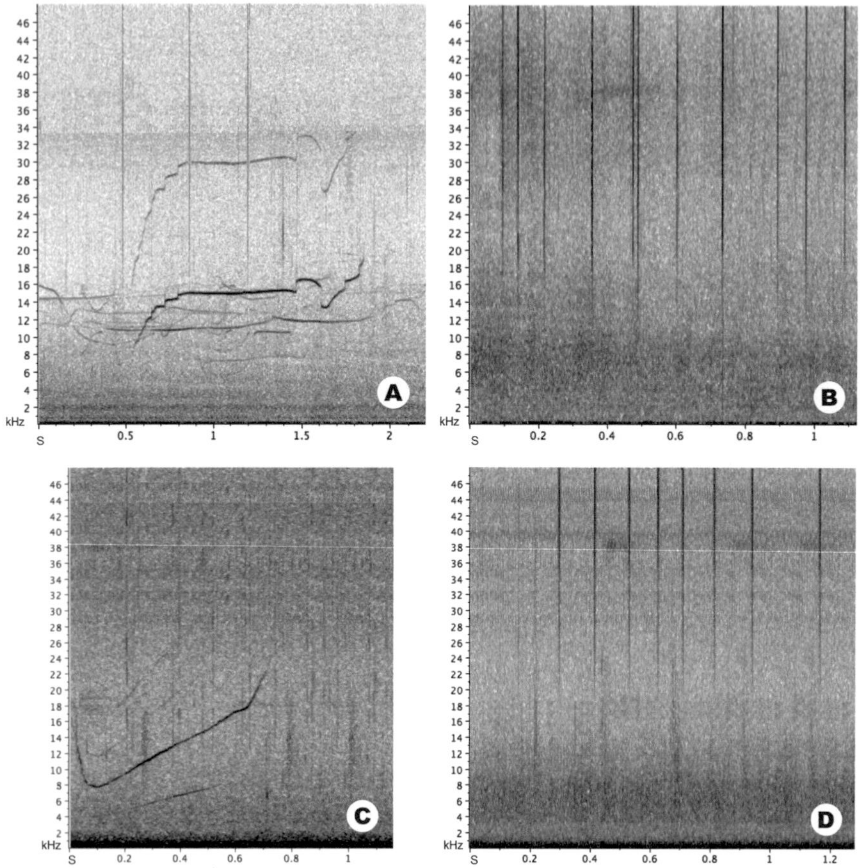

Fig. 5.14 Examples of spectrograms of spinner dolphins whistles (**a**) and (**b**) clicks. Examples of Atlantic spotted dolphins whistles (**c**) and clicks (**d**)

*5.4.4 Bottlenose Dolphin (***Tursiops truncatus***)*

The bottlenose dolphin are found in tropical and temperate waters worldwide and show a high variation in morphology according to geographical location, where adult lengths range between 2.5 and four meters approximately (Wells and Scott 2009). The distribution of bottlenose dolphins is spread along the Southwestern Atlantic at both nearshore (coastal ecotype) and offshore (oceanic ecotype) waters (Bastida et al. 2007). The coastal ecotype concentrates in areas near river discharge, estuaries and bays of Argentina, Uruguay and southern Brazil (Fruet et al. 2014). The oceanic ecotype, on the other hand, seems to be widely distributed in tropical and subtropical deep waters along the outer continental shelf and beyond and in association with oceanic islands (Bastida et al. 2007) (Fig. 5.15).

Fig. 5.15 *Tursiops truncates* photographed off the Brazilian coast. (Reproduced with permission of Rodrigo Genoves)

The bottlenose dolphin is probably one of the most studied cetacean species, therefore responsible for much of the knowledge about the dolphins sound emissions. Bottlenose dolphin acoustic repertoire consists of narrowband and modulated whistles; broadband clicks and pulsed calls (Tyack and Clark 2000). Efforts have been substantially focused in researches with whistles, mainly that named as signature whistles (e.g. Caldwell and Caldwell 1965; McCowan and Reiss 1995, 2001; Tyack 1997; Janik and Slater 1998; Sayigh et al. 2007; Harley 2008; Esch et al. 2009; Janik and Sayigh 2013; Kriesell et al. 2014).

So far, little is known about the acoustic of South Atlantic bottlenose dolphin populations. Azevedo et al. (2007) recorded bottlenose dolphins at Patos Lagoon estuary, southern Brazil. Lima et al. (2016) described and made a comparison among whistles emitted by four delphinid species, in the Rio de Janeiro State Coast, Brazil: *Sotalia guianensis*, *Stenella frontalis*, *Steno bredanensis* and *Tursiops truncatus* (Fig. 5.16).

*5.4.5 Short Beaked Common Dolphin (*Delphinus delphis*)*

The short beaked common dolphin is a small delphinidae, reaching up to approximately 2.60 m, and presents a warm-temperate distribution along the Atlantic and Pacific oceans (Bastida et al. 2007; Perrin 2009). Along the Brazilian water this species appears to have an isolated population in northern Brazil; however in the southeast this species is frequently sighted in shallower waters (18–70 m) and between southern Brazil and central Argentina this species is commonly found from the outer continental shelf to upper slope (100–1500 m) (Tavares et al. 2010; Cunha et al. 2015; Di Tullio et al. 2016) (Fig. 5.17).

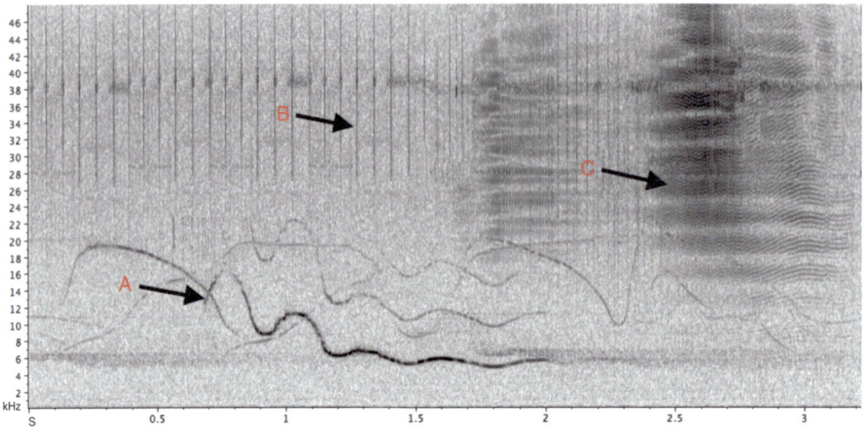

Fig. 5.16 Examples of spectrograms of (**a**) whistle, (**b**) clicks and (**c**) burst sound of bottlenose dolphin

Fig. 5.17 *Delphinus delphis* photographed off the Brazilian coast (Reproduced with permission of Pedro Fruet)

The common dolphin whistles are emitted usually between 3 and 23.51 kHz and duration between 0.01 and 2.1 s (Ansmann et al. 2007; Petrella et al. 2012). Whistles recorded in the western South Atlantic presented frequency range between 4.07 and 31.46 kHz and duration between 0.11 and 1.73 s (Fig. 5.18).

Acoustic studies with common dolphin whistles parameters are still rare and punctual. In Brazilian waters, this was the first effort to investigate common dolphin whistles repertoire. Moreover, in the current scenario of the taxonomic genus, the bioacoustics is a tool to help the classification of these dolphins and their phylogenetic relationships.

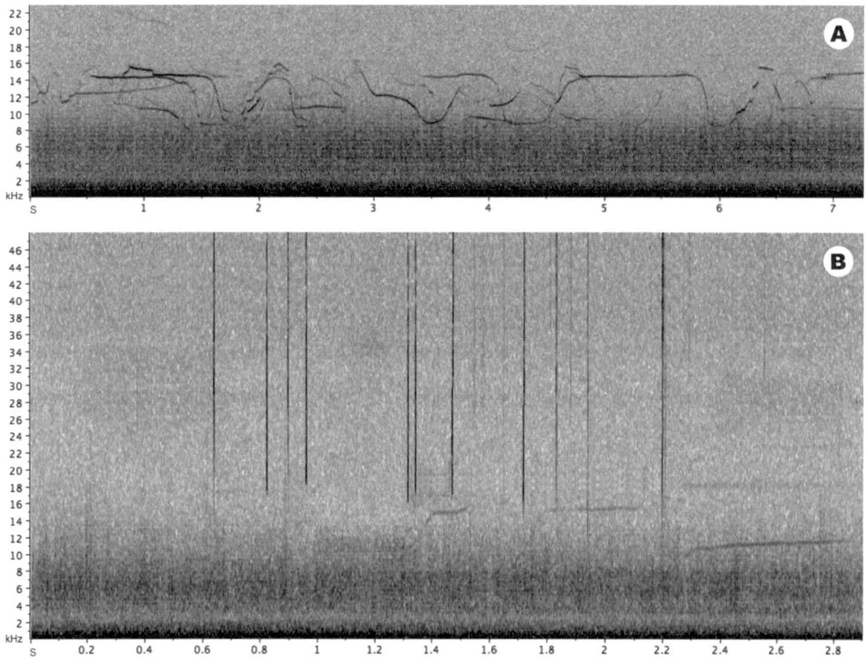

Fig. 5.18 Example of spectrogram of whistle (a) and clicks (b) of common dolphin

5.4.6 *Long Finned Pilot Whale* (Globicephala melas)

The long-finned pilot whale is a large delphinidae presenting a robust body and a slightly sexual dimorphism, which adult males are larger reaching up to six meters long (Olson 2009; Bastida et al. 2007). This species inhabits cold-temperate waters of the North Atlantic and southern Ocean and along Brazilian waters is sighted mainly south of 30°S and over depths between 500 and 1000 m (Pinedo et al. 2002; Zerbini et al. 2004; Bastida et al. 2007; Di Tullio et al. 2016) (Fig. 5.19).

The long finned pilot whales vocalizations have been still little described. The first studies reported whistles, pulsed calls (Busnel and Bziedzic 1966) and clicks (Busnel and Bziedzic 1966; Busnel et al., 1971) (Fig. 5.20) as part of their vocal repertoire. Taruski (1979) classified long finned pilot whales' whistles in seven broad categories, ranging from simple to complex types. Weilgart and Whitehead (1990) showed a strong relationship between the type and complexity of vocalization and the behavior displayed by animals.

Long finned pilot whale whistles recorded in the western South Atlantic presented frequency range between 2.77 and 7.50 kHz and duration between 0.23 and 0.58 s (Fig. 5.20).

Fig. 5.19 Long-finned pilot whale photographed off the Brazilian coast (Reproduced with permission of Rodrigo Genoves)

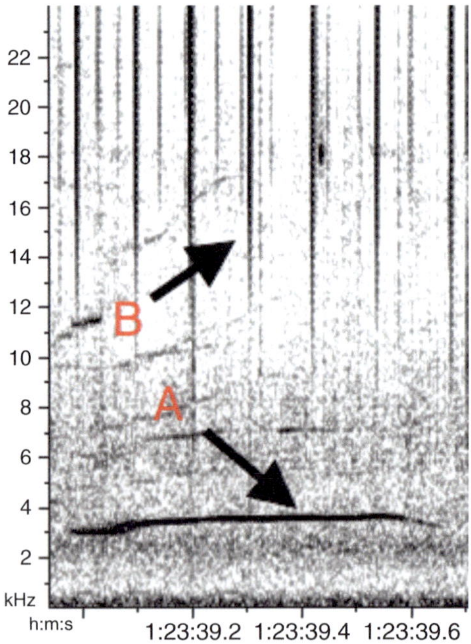

Fig. 5.20 Spectrogram of whistles (**a**) and clicks (**b**) of the long finned pilot whales

Fig. 5.21 Killer whale photographed in the western South Atlantic Ocean off the Brazilian coast (Reproduced with permission of Rodrigo Genoves)

*5.4.7 Killer Whale (**Orcinus orca**)*

Killer whales are widely distributed throughout all ocean basins, especially in areas of high ocean productivity (Ford 2009). This species presents a sexual dimorphism, which males develop larger total length (nine meters) and appendages than females (seven meters) (Bastida et al. 2007; Ford 2009). In Brazilian waters sightings of killer whale are more frequent in coastal waters of the southeast; however this species is known to frequently depredate the catch of longline fisheries near the shelf break and beyond in southern Brazil (Secchi and Vaske 1998; Dalla Rosa and Secchi 2007) (Fig. 5.21).

Killer whales have complex communication strategies (Filatova et al. 2012). Signals, such as whistles and pulsed calls, are probably culturally transmitted through vocal learning (Ford 1991; Deecke et al. 2002; Foote et al. 2006), and are highly stereotyped (Ford 1991; Riesch et al. 2006). When compared to pulsed calls, whistles are highly modulated signals and have lower sound pressure levels and higher fundamental frequencies (Ford 1989; Thomsen et al. 2001; Riesch et al. 2006). Discussions on the ecological role and possible ecological function of high frequency whistles (HFWs) produced by killer whales have recently emerged in the literature (Samarra et al. 2010, 2015; Simonis et al. 2012; Filatova et al. 2012; Trickey et al. 2014). This signal could be used for navigation (Simonis et al. 2012) or for private communication (Filatova et al. 2012).

Recent study on the western south Atlantic have described the acoustic parameters of killer whale whistles and highlighted the occurrence of high frequency whistles (Andriolo et al. 2015). The high frequency whistles (Fig. 5.22A) were highly stereotyped and were modulated mostly at ultrasonic frequencies. Compared to other contour types (Fig. 5.22B), the high frequency whistles are characterized by higher bandwidths, shorter durations, fewer harmonics, and higher sweep rates (Andriolo et al. 2015) (Table 5.1).

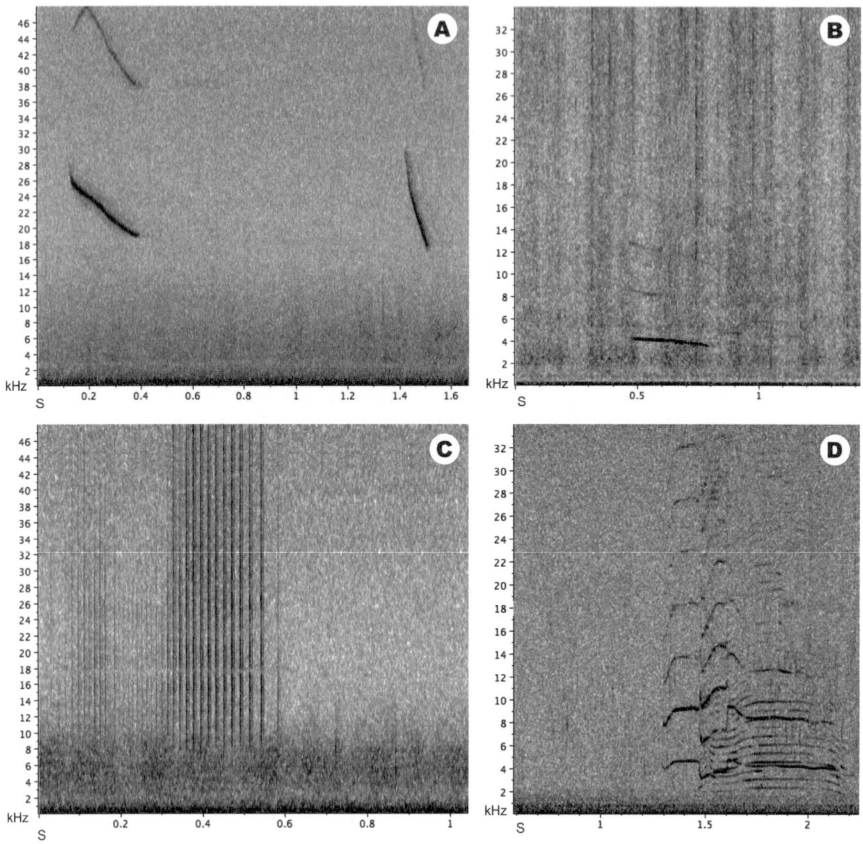

Fig. 5.22 Spectrograms of killer whale (A) high frequency whistle (HFW) (B) whistle contour, (C) clicks and (D) calls recorded in the western South Atlantic Ocean

5.5 Ecological Approaches of Bioacoustics Using Towed Array

5.5.1 Acoustic Discrimination Between Species

The extensive use of Passive Acoustic Monitoring for detecting and monitoring marine mammals has generated huge volumes of data (Mellinger and Barlow 2003). Species identification from acoustic recordings of marine mammal vocalizations can be challenging due to the high variability in many of the characteristics of sounds that can be easily measured or extracted from spectrograms, both within species and among species (e.g. Oswald et al. 2003, 2004; Roch et al. 2007; Soldevilla et al. 2010; Baumann-Pickering et al. 2015).

The analysis of cetacean sounds to the species level is an important step in processing long-term passive acoustic recordings made in a marine environment (Baumann-Pickering et al. 2010). Most of works on acoustic classification of odon-

Table 5.1 Spectral and temporal parameters of whistles and clicks for all species given as the median with the 10th and 90th percentile in parentheses. *MinF* minimum frequency, *MaxF* maximum frequency, *DeltaF* delta frequency, *PeakF* peak frequency, *CenterF* center frequency, *BeginF* beginning frequency, *EndF* ending frequency, Duration, *ICI* inter-click interval, *3dB bw* 3dB bandwidth, *10dB bw* 10dB bandwidth. Number of analysed vocalisations is given as N: number of whistles/number of clicks

	Whitles								Clicks		
Species	MinF (kHz)	MaxF (kHz)	DeltaF (kHz)	PeakF (kHz)	CenterF (kHz)	BeginF (kHz)	EndF (kHz)	Duration (ms)	ICI (ms)	3 dB bandwidth (kHz)	lOdB bandwidth (kHz)
Delphinus delphis	8.6	15.3	5.5	11.6	12.4	11.7	13.0	735.0	0.6	24.2	6.3
(N:397/202)	(5.6–12.9)	(10.7–20.2)	(1.8–10.3)	(8.3–17.4)	(8.4–17.9)	(7.5–17.1)	(6.8–17.0)	(303.0–1270.0)	(0.1–54.3)	(11.2–35.5)	(6.1–6.5)
Globicephala melas	3.6	4.8	1.1	4.3	4.2	4.0	4.5	390.0	30.8	28.1	12.3
(N:271/117)	(2.2–6.8)	(3.3–9.6)	(0.4–3.3)	(2.6–8.5)	(2.7–8.0)	(2.6–7.4)	(2.4–9.3)	(140.0–740.0)	(0.3–392.9)	(13.5–37.7)	(8.3–33.4)
Grampus griseus	7.1	15.4	7.7	11.1	11.0	10.7	9.7	690.0	0.1	25.7	9.6
(N:340/247)	(1.8–11.2)	(3.1–18.7)	(1.0–11.7)	(2.6–15.5)	(2.2–14.5)	(2.6–14.8)	(2.1–15.9)	(220.0–1390.0)	(0.08–0.2)	(15.8–38.7)	(7.2–15.3)
Orcinus orca	3.9	8.2	4.1	5.6	7.1	4.3	8.0	295.0	19.5	14.8	16.0
(N: 187/70)	(1.9–15.1)	(3.8–24.6)	(1.6–9.3)	(2.7–17.8)	(3.3–20.3)	(1.9–24.5)	(3.4–16.9)	(110.0–838.0)	(0.4–33.2)	(11.5–32.3)	(8.2–31.6)
Stenellafrontalis	9.6	16.9	7.7	15.4	14.5	9.9	16.8	320.0	37.0	31.1	12.9
(N: 892/98)	(5.4–11.9)	(15.8–19.6)	(4.5–12.9)	(12.2–17.1)	(13.3–15.5)	(5.7–12.7)	(15.4–19.4)	(212.8–499.5)	(3.5–63.3)	(18.7–38.6)	(9.4–30.0)
Stenella longirostris	9.9	15.9	4.6	13.8	13.5	11.8	13.7	690.0	86.4	26.2	19.4
(N:768/147)	(6.5–15.1)	(10.5–20.9)	(1.3–9.4)	(9.1–16.7)	(9.4–16.3)	(6.8–17.3)	(8.6–18.7)	(200.0–1416.0)	(16.1–166.8)	(20.4–34.2)	(10.9–37.5)
,SYe«0 *bredanensis*	5.8	8.1	2.2	7.4	6.9	6.3	7.6	409.0	8.9	32.0	19.0
(N:470/113)	(4.0–8.3)	(6.7–10.1)	(0.5–4.9)	(5.6–8.8)	(5.5–8.9)	(4.1–8.8)	(5.8–9.9)	(126.6–974.4)	(0.1–28.2)	(20.8–39.1)	(9.3–35.7)
Turstops truncatus	9.6	15.6	5.1	12.5	12.1	11.9	12.8	350.0	25.4	24.5	30.5
(N:1005/301)	(7.2–13.8)	(11.7–19.9)	(1.7–9.2)	(8.7–18.0)	(9.2–17.9)	(7.9–16.7)	(8.6–18.1)	(140.0–840.0)	(4.1–56.5)	(22.4–29.4)	(13.4–34.4)

tocetes have considered the acoustical emissions separately, examining only click trains (Verfub et al. 2007, 2013; Roch et al., 2011), or whistle sequences (Oswald et al. 2007). Lu et al. (2013) integrated features from both whistles and clicks for classification of bottlenose dolphin, spinner dolphin, melon-headed whale, short-beaked common dolphin and long-beaked common dolphin.

Proposing an integrative bioacoustics approach – combining whistles and clicks - we classified eight delphinid species: spinner (*Stenella longirostris*), Atlantic spotted (*Stenella frontalis*), rough-toothed (*Steno bredanensis*), Risso's (*Grampus griseus*), bottlenose (*Tursiops truncatus*), short-beaked common (*Delphinus delphis*) dolphins, killer (*Orcinus orca*) and long-finned pilot (*Globicephala melas*) whales from the Southwest Atlantic Ocean, at the Brazilian shelf break (Amorim et al. submitted). For that, we extract whistles acoustic parameters: maximum frequency, minimum frequency, frequency range, peak frequency, center frequency, beginning frequency, ending frequency and duration; and clicks parameters: inter-click interval, sound pressure level peak-to-peak (SPL), rms (root mean square) amplitude, 3 dB bandwidth and 10 dB bandwidth. The sequential classification analysis consisted of two methodological techniques, Discriminant Function Analysis (DFA) and Classification Tree Analysis.

The discriminant function showed that the overall whistle classification had the highest number of false classifications (40.7%, N = 475, Wilks' $\lambda = 0.18$), these numbers decrease when only clicks were analyzed (25.0%, N = 158, Wilks' $\lambda = 0.14$). The discrimination result improved with the combined analysis of whis-tles and clicks, given that the misclassification percentage was only of 5.8% (N = 30, Wilks' $\lambda = 0.01$). In classification tree analysis, the optimal classification whistle tree consisted of 28 splits and misclassification rates of 0.606. Click opti-mal tree consisted of 60 splits and misclassification of 0.260. When whistles and clicks were combined, optimal tree consisted of 90 splits and a false classification of 0.188.

In summary, the eight delphinid species showed species-specific qualities in their whistles and clicks. When taken individually, echolocation clicks presented greater efficiency in distinguishing species; this might be related to the behavioural context encoded in whistles and the relationship between echolocation signals features and animals' head morphology, which makes feasible to accurately determine a species from their clicks. Otherwise, analysing both signals in combination enhanced the correct classification scores. An integrati ve bioacoustics approach potentially presents a higher contribution in the classification process, once it considers the different signals produced by the species as part of a whole communication system employed in different ecological contexts.

5.5.2 *Sperm Whale Social Structure*

Sperm whales (*Physeter macrocephalus*) are the largest and long-lived odontocete, reaching between 8.3 and 20.5 m long and with lifespan of over 50 years old (Chivers 2009; Whitehead 2009). There is a marked sexual dimorphism in body

Fig. 5.23 Sperm whale photographed off the Brazilian coast (Reproduced with permission of Luciano Dalla Rosa)

length between adults; females reach reproductive maturity about nine meters long while males at approximately 16 m long. Although they are considered cosmopolitan, the distribution of females and young individuals are restricted to tropical and temperate deep waters and differs from adult male, which extends their distribution to polar waters (Whitehead 2009). Few records of sperm whales stranding occur along the Brazilian coast, however sightings in deep waters are more frequent, especially in southern Brazil at depths over 1000 m (Di Tullio et al. 2016) (Fig. 5.23).

Hinde in 1976 stated that social structure is the quality and nature of the interactions among individuals, and the relationships between them. Considering that, to study animal's societies, we must investigate interactions and describe relationships in order to achieve a model of social organization. Some methods are used and complement each other in studies of social structure: photo identification, molecular techniques, observation of behavior, visual and photographic estimates of length and genetic sexing and, passive acoustics monitoring and vocalizations descriptions (Whitehead 2003).

The sperm whale's long life span allows the formation of long-term social bonds among individuals (Christal et al. 1998). Within sperm whale social units, individuals prefer associations and avoidances. Such preferences suggest that sperm whales have individualized relationships within units and, as a consequence, the animals may rely on strategies to recognize individuals in order to adjust their behavior (Whitehead 2003).

Within the types of clicks that sperm whales emit, a series of stereotyped patter of clicks called codas stand out. Codas have been associated to communication and identification of a specific unit membership, besides of being used for individual recognition and identification (Watkins and Schevill 1977; Whitehead and Weilgart 1991) and consequently reflects family relationships (Weilgart and Whitehead 1997).

Sperm whales social units in the Pacific were grouped into vocal clans based on their coda repertoires (Rendell and Whitehead 2003a). Besides that, within vocal clans, unit members share elements of their coda repertoire, and eventually engage in codas exchanges in order to share a signal code (Whitehead 2003; Rendell and Whitehead 2004). It is know that codas repertoire varies geographically (Weilgart and Whitehead 1997) and are learnt within matrilineal social groups. Rendell and Whitehead (2005) found that similarities among repertoires showed a negative correlation with increasing distance of clans.

Codas were mostly studied in the Island of Dominica (Schulz et al. 2008; Antunes et al. 2011; Schulz et al. 2011; Gero et al. 2015; Gero et al. 2016), in the Mediterranean (Borsani et al. 1997; Pavan et al. 2000), in the Caribean (Moore et al. 1993), in Azores (Oliveira et al. 2016), in the Galápagos (Weilgart and Whitehead 1993; Rendell and Whitehead 2003b, 2004, 2005), in the Pacific (Rendell and Whitehead 2003a, Marcoux et al. 2006) and in the North western Atlantic (Watkins and Schevill 1977). As a first effort in studying codas and social structure of sperm whales in South Atlantic (Fig. 5.24), we have been making recordings since 2011 in the South, Southeast of Brazil. In collaboration with researchers of University of St. Andrews, the vocal clans and its dialects will be investigated. So far, a total of 939 codas were marked up using a custom-written software for analyzing sperm whale sounds called Rainbow Click (see Gillespie 1997; Leaper et al. 2000).

5.5.3 Abundance Estimation of Sperm Whales

Information on distribution and density of most marine mammals have been primarily obtained from visual surveys (Ward et al. 2012; Marques et al. 2013). However, the traditional methods of visual monitoring have been associated with: (1) training of qualified staff, (2) an intensive work effort, (3) a high implementation costs (Evans and Hammond 2004; Ward et al. 2012; Marques et al. 2011, 2013). They can be undertaken merely during daylight and in good weather conditions and, as the individuals are available for visual observation only when at the surface, just part of the animals present can be detected (Mellinger and Barlow 2003; Mellinger et al. 2007; Vallarta-Hernandez 2009; Ward et al. 2012; Yack et al. 2013).

PAM has been increasingly used in study of marine mammals, especially because many of them are more acoustically than visually available (Barlow and Taylor 2005; Zimmer 2013; Yack et al. 2013). Whether simultaneously conducted, the abundance estimates obtained through these methods are possibly more reliable

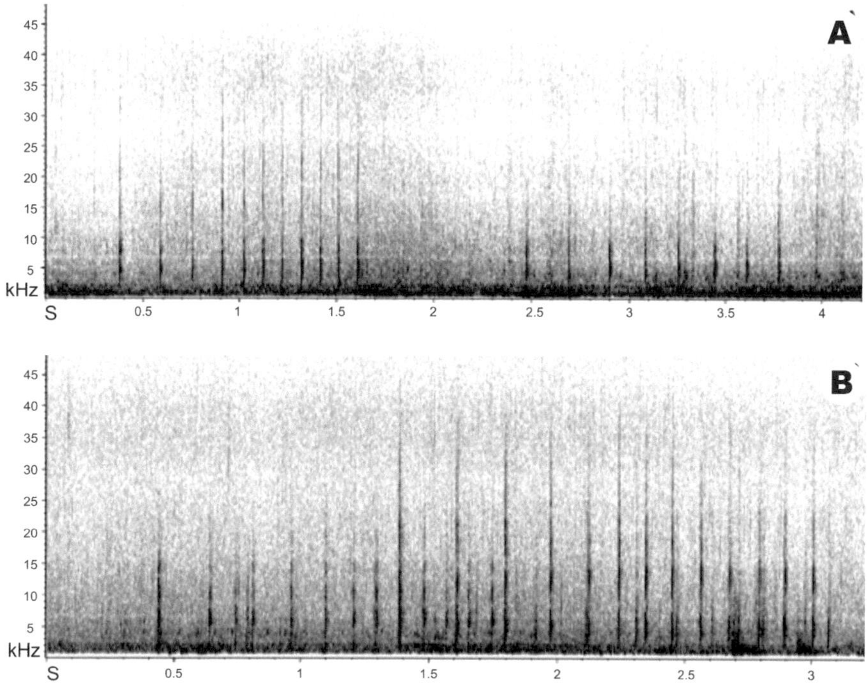

Fig. 5.24 Examples of codas from South/Southeast Brazilian continental shelf break. (**a**) Two different codas patterns and (**b**) Overlapped codas

(Whitehead 2002; Mellinger and Barlow 2003) as they tend to increase the probability of detecting individuals (Barlow and Taylor 1998, 2005; McDonald and Moore 2002; Whitehead 2003; Rankin et al. 2008c).

The long and deep dives performed by sperm whales can last approximately 1 h (Leatherwood et al. 1982; Barlow and Taylor 2005; Watwood et al. 2006). Thus, individuals spend short periods at the surface, reducing the probability of being visually detected (Whitehead 2003; Watwood et al. 2006; Ward et al. 2012). However, this species produces a variety of types of clicks (usual and slow clicks, creaks and coda), which are used in different contexts such as echolocation and communication (Zimmer et al. 2005).

According to Barlow and Taylor (2005), sperm whales are most likely to be acoustically detected among others cetacean's species. During their foraging dives, produce regular, audible and short duration clicks ('usual' clicks and creaks) with frequencies ranging from several hundred hertz to over 30 kHz (Watkins 1980; Gillespie 1997; Weilgart and Whitehead 1988; Madsen et al. 2002; Barlow and Taylor 2005). Several studies have been increasingly applied passive acoustic monitoring, some of which combined with visual surveys, trying to improve the methods of localization and density estimation of marine mammals, using sperm whales as a model (Gillespie 1997; Barlow and Taylor 1998, 2005; Gannier et al. 2002; Hastie et al. 2003; Leaper et al. 2003; Lewis et al. 2007; Swift et al. 2009; Whitehead 2009;

Ward et al. 2012; Von Benda-Beckmann et al. 2013; Tran et al. 2014; Fais et al. 2016). They have being carried out using a variety of methods and from different platforms. However, as the frequency of sperm whale clicks extends above the dominant range of the ship and water flow noise, it makes possible the use of towed passive acoustic system (Gillespie 1997; Barlow and Taylor 1998, 2005; Gannier et al. 2002; Hastie et al. 2003; Leaper et al. 2003; Lewis et al. 2007; Swift et al. 2009; Whitehead 2009; Fais et al. 2016). Once conducted opportunistically, it provides, as highlighted by Whitehead (2002), the coverage of large areas with relatively low cost resulting, additionally, in a more accurate abundance estimate for this species.

In Brazil, a recent effort has been conducted using this system, in association with a visual survey (Di Tullio et al. 2016), aiming to detect, localize and estimate the sperm whales population size in the south and southeast continental shelf break region through their acoustic signals. This study has being conducted as a partnership between the research groups of Federal Universities of Juiz de Fora and Rio Grande, as well as with the Centre for Research into Ecological & Environmental Modelling at the University of St. Andrews.

Acknowledgement We thank the crew of FURG's R/V "Atlântico Sul" for helping in all onboard activities. This study was funded by BG Group and Chevron Brasil Upstream Frade LTDA. We are grateful with the support of the Auset Company developing, building and providing assistance of the acoustical equipments. We also thanks Dr. Eduardo R. Secchi, Dr. Luciano Dalla Rosa, Dr. Alexandre Zerbini and all field team members. Scholarships and grants were supported by CNPq and CAPES.

References

Amaral KB, Alvares DJ, Heinzelmann L et al (2015) Ecological niche modeling of *Stenella* dolphins (Cetartiodactyla: Delphinidae) in the southwestern Atlantic Ocean. J Exp Mar Biol Ecol 472:166–179

Amorin TOS, Castro FR, Moron JR et al. (submitted) Acoustic classification of eight delphinid species in the Southwest Atlantic Ocean

Andriolo A, Reis SS, Amorim TOS et al (2013) Wide frequency range whistles of a Risso's dolphin (*Grampus griseus*) recorded in shelf waters of Southern Brazil. In: Anais XXIV International Bioacoustics Congress, Pirenópolis, Brazil, 8–13 Sept 2013

Andriolo A, Reis SS, Amorim TOS et al (2015) Killer whale (*Orcinus orca*) whistles from the western South Atlantic Ocean include high frequency signals. J Acoust Soc Am 138:1696–1701

Ansmann IC, Goold JC, Evans PGH et al (2007) Variation in the whistle characteristics of short-beaked common dolphins, *Delphinus delphis*, at two locations around the British Isles. J Mar Biol Assoc UK 87:19–26

Antunes R, Schulz T, Gero S et al (2011) Individually distinctive acoustic features in sperm whale codas. Anim Behav 81:723–730

Au WWL, Hastings MC (2008) Principles of marine bioacoustics. Springer, New York

Azevedo AF, Oliveira AM, Dalla Rosa L, Lailson-Brito J (2007) Characteristics of whistles from resident bottlenose dolphins (*Tursiops truncatus*) in southern Brazil. J Acoust Soc Am 121:2978–2983

Azevedo AF, Flach L, Bisi TL et al (2010) Whistles emitted by Atlantic spotted dolphins (*Stenella frontalis*) in southeastern Brazil. J Acoust Soc Am 127:2646–2651

Barlow J, Taylor BL (1998) Preliminary abundance of sperm whales in the Northeastern temperate Pacific estimated from a combined visual and acoustic survey. In: Donovan GP (ed) Forty-eighth report of the International Whaling Commission, Cambridge, 1998
Barlow J, Taylor B (2005) Estimates of sperm whale abundance in the northeastern temperate Pacific from a combined acoustic and visual survey. Mar Mam Sci 2:429–445
Barlow J, Tyack PL, Johnson M et al (2013) Trackline and point detection probabilities for acoustic surveys of Cuvier's and Blainville's beaked whales. J Acoust Soc Am 134:2486–2496
Bastida R, Rodríguez D, Secchi ER, Silva V (eds) (2007) Mamíferos Acuaticos de Sudamerica y Antartida. Vazquez Mazzini, Buenos Aires
Baumann-Pickering S, Wiggins SM, Hildebrand JA et al (2010) Discriminating features of echolocation clicks of melon-headed whales (*Peponocephala electra*), bottlenose dolphins (*Tursiops truncatus*), and Gray's spinner dolphins (*Stenella longirostris longirostris*). J Acoust Soc Am 128:2212–2224
Baumann-Pickering S, Simonis AE, Oleson EM et al (2015) False killer whale and short-finned pilot whale acoustic identification. Endanger Species Res 28:97–108
Bazúa-Dúran C, Au WWL (2002) The whistles of Hawaiian spinner dolphins. J Acoust Soc Am 112:3064–3072
Bazúa-Durán C, Au WWL (2004) Geographic variations in the whistles of spinner dolphins (*Stenella longirostris*) of the Main Hawaiian Islands. J Acoust Soc Am 116:3757–3769
Benda-Beckmann AMV, Beerens SP, Ijsselmuide SPV (2013) Effect of towed array stability on instantaneous localization of marine mammals. J Acoust Soc Am 134:2409–2417
Bezerra Filho GA (2012) Ecolocalização do golfinho-de-dente-rugoso (*Steno bredanensis*) no Parque Nacional Marinho dos Abrolhos. Monograph, Universidade Federal do Rio Grande do Norte
Bittle M, Duncan AJ (2013) A review of current marine mammal detection and classification algorithms for use in automated passive acoustic monitoring. In: Proceedings of Acoustics – 2013, Victor Harbour, Australia, 17–20 November 2013
Bonato M, Papale E, Pingitore G et al (2015) Whistle characteristics of the spinner dolphin population in the Comoros Archipelago. J Acoust Soc Am 138:3262–3271
Borsani JF, Pavan G, Gordon JCA, Notarbartolo di Sciara G (1997) Regional vocalizations of the Sperm Whale: Mediterranean codas. European Research on Cetaceans, Cambridge (UK) 10:78–81
Busnel R, Dziedzic A (1966) Acoustic signals of the pilot whale *Globicephala melaena* and of the porpoises *Delphinus delphis* and *Phocoena phocoena*. In: Norris KS (ed) Whales, dolphins and porpoises, 1st edn. University of California Press, Berkeley, pp 607–648
Busnel RG, Escudie B, Dziedzic A, Hellion A (1971) Structure des clics doubles d'echolocation du globicephale (Cetace odontocete). Comptes rendus del'Académie des sciences 272:2459–2461
Busnell RG, Dziedzic A (1966) Caracteristiques physiques de certains signaux acoustiques du *delphidide Steno bredanensis*, Lesson (Physical characteristics of certain acoustic signals of the delphinid *Steno bredanensis*, Lesson). Acad Sci Paris 268:143–146
Caldwell MC, Caldwell DK (1965) Individualized whistle contours in bottle-nosed dolphins (*Tursiops truncatus*). Nature 207:434–435
Campbell GS, Thomas L, Whitaker K et al (2015) Inter-annual and seasonal trends in cetacean distribution, density and abundance off southern California. Deep-Sea Res II 112:143–157
Cerchio S, Jacobsen JK, Norris TF (2001) Temporal and geographical variation in songs of humpback whales, *Megaptera novaeangliae*: synchronous change in Hawaiian and Mexican assemblages. Anim Behav 62:313–329
Chivers SJ (2009) Cetacean life history. In: Perrin WF, Würsig B, Thewissen JG (eds) Encyclopedia of marine mammals, 2nd edn. Academic, Amsterdam, pp 215–220
Christal J, Whitehead H, Lettevall E (1998) Sperm whale social units: variation and change. Can Jf Zool 76:1431–1440
Corkeron PJ, Van Parijs SM (2001) Vocalizations of eastern Australian Risso's dolphins (*Grampus griseus*). Can J Zool 79:160–164

Corrêa GC (2012) Repertório de assobios do golfinho de dentes-rugosos (*Steno bredanensis*), no Banco dos Abrolhos, Bahia. Monograph, Universidade Federal do Rio Grande do Norte
Cunha HA, de Castro RL, Secchi ER et al (2015) Molecular and morphological differentiation of common dolphins (*Delphinus* sp.) in the Southwestern Atlantic: testing the two species hypothesis in sympatry. PLoS One 10:e0140251
Dalla Rosa L, Secchi ER (2007) Killer whale (*Orcinus orca*) interactions with the tuna and swordfish longline fishery off southern and south-eastern Brazil: a comparison with shark interactions. J Mar Biol Ass 87:135–140
Danilewicz D, Ott P, Secchi E et al (2013) Occurrence of the Atlantic spotted dolphin, *Stenella frontalis*, in the southern Abrolhos Bank, Brazil. Mar Biodivers Rec 6:1–3
Darling JD, Sousa-Lima RS (2005) Songs indicate interaction between humpback whale (*Megaptera novaeangliae*) population in the Western and Eastern South Atlantic Ocean. Mar Mam Sci 21:557–566
Deecke VB, Slater PJB, Ford JKB (2002) Selective habituation shapes acoustic predator recognition in harbour seals. Nature 420:171–173
Di Tullio JC, Gandra TBR, Zerbini AN, Secchi ER (2016) Diversity and distribution patterns of cetaceans in the subtropical Southwestern Atlantic outer continental shelf and slope. PLoS One 11(5):e0155841. https://doi.org/10.1371/journal.pone.0155841
Ding W, Würsig B, Evans WE (1995) Comparison of whistles among seven odontocete species. In: Kastelein RA, Thomas JA, Nachtingall PE (eds) Sensory systems of aquatic mammals. The Spil Publishers, Woerden, pp 299–323
Dudzinski KM (1996) Communication and behavior in the Atlantic spotted dolphin (*Stenella frontalis*): relationships between vocal and behavioral activities. Dissertation, Texas A & M University
Esch HC, Sayigh LS, Wells RS (2009) Quantifying parameters of bottlenose dolphins signature whistles. Mar Mam Sci 25:976–986
Evans WE (1967) Vocalization among marine animals. In: Tavolga WN (ed) Marine bio-acoustics, 2nd edn. Pergamon, New York, pp 159–186
Evans PGH, Hammond PS (2004) Monitoring cetaceans in European waters. Mammal Rev 34:131–156
Fais A, Lewis TP, Zitterbart DP et al (2016) Abundance and distribution of sperm whales in the Canary Islands: can sperm whales in the Archipelago sustain the current level of ship-strike mortalities? PLoS One 11:e0150660
Filatova OA, Ford JKB, Matkin CO et al (2012) Ultrasonic whistles of killer whales (*Orcinus orca*) recorded in the North Pacific (L). J Acoust Soc Am 132:3618–3621
Fitch WT, Neubauer J, Herzel H (2002) Calls out of chaos: the adaptive significance of nonlinear phenomena in mammalian vocal production. Anim Behav 63:407–418
Foote AD, Griffin RM, Howitt D et al (2006) Killer whales are capable of vocal learning. Biol Lett 2:509–512
Ford JKB (1989) Acoustic behaviour of resident killer whales (*Orcinus orca*) off Vancouver Island, British Columbia. Can J Zool 67:727–745
Ford JKB (1991) Vocal traditions among resident killer whales (*Orcinus orca*) in coastal waters of British Columbia. Can J Zool 69:1454–1483
Ford JKB (2009) Killer whale - *Orcinus orca*. In: Perrin WF, Würsig B, Thewissen JG (eds) Encyclopedia of marine mammals, 2nd edn. Academic Press, Amsterdam, pp 650–657
Fruet PF, Secchi ER, Daura-Jorge F et al (2014) Remarkably low genetic diversity and hierarchical population structure in common bottlenose dolphins (*Tursiops truncatus*) from coastal waters of the Southwestern Atlantic Ocean. Conserv Genet 15:879–895
Gannier A, Drouot V, Goold JC (2002) Distribution and relative abundance of sperm whales in the Mediterranean Sea. Mar Ecol Prog Ser 243:281–293
Gannier A, Fuchs S, Oswald NJ (2008) Pelagic delphinids of the Mediterranean Sea have different whistles. New Trends Environ Monit Using Passive Syst 2008:1–4

Gero S, Whitehead H, Rendell L (2015) Individual, unit and vocal clan level identity cues in sperm whales codas. R Soc Open Sci 3:150372
Gero S, Bottcher A, Whitehead H, Madsen PT (2016) Socially segregated sympatric sperm whale clans in the Atlantic Ocean. R Soc Open Sci 3:160061
Gillespie D (1997) An acoustic survey for sperm whales in the Southern Ocean sanctuary conducted from the *RSV Aurora Australis*. In: Donovan GP (ed) Forty-seventh report of the International Whaling Commission, Cambridge, 1997
Gillespie D, Gordon J, Caillat M et al (2010) Detection of beaked whales using near surface towed hydrophones: prospects for survey and mitigation. J Acoust Soc Am 123:3774
Gotz T, Verfub UK, Schnitzler HU (2006) 'Eavesdropping' in wild rough-toothed dolphins (*Steno bredanensis*)? Biol Lett 2:5–7
Harley HE (2008) Whistle discrimination and categorization by the Atlantic bottlenose dolphin (*Tursiops truncatus*): a review of the signature whistle framework and a perceptual test. Behav Process 77:243–268
Hastie GD, Swift RJ, Gordon JC, Slesser G, Turrell WR (2003) Sperm whale distribution and seasonal density in the Faroe Shetland Channel. J Cetacean Res Manag 5:247–252
Herzing DL (1996) Vocalizations and associated underwater behavior of free-ranging Atlantic spotted dolphins, *Stenella frontalis* and bottlenose dolphins, *Tursiops truncatus*. Aquat Mamm 22:61–79
Herzing DL (2015) Synchronous and rhythmic vocalizations and correlated underwater behavior of free-ranging atlantic spotted dolphins (*Stenella frontalis*) and bottlenose dolphins (*Tursiops truncatus*) in the Bahamas. Anim Behav Cogn 2:14–29
Hinde RA (1976) Interactions, relationships and social structure. Man 11:1–17
Hoelzel AR (ed) (2009) Marine mammals biology: an evolutionary approach. Blackwell, Oxford
Janik VM, Slater PJB (1998) Context-specific use suggests that bottlenose dolphin signature whistles are cohesion calls. Anim Behav 56:829–838
Janik VM, Sayigh LS (2013) Communication in bottlenose dolphins: 50 years of signature whistle research. J Comp Physiol 199:479–489
Jaquet N, Whitehead H (1999) Movements, distribution and feeding success of sperm whales in the Pacific Ocean, over scales of days and tens of kilometers. Aquat Mamm 25:1–13
Jefferson TA, Weir CR, Anderson RC et al (2014) Global distribution of Risso's dolphin *Grampus griseus*: a review and critical evaluation. Mammal Rev 44:56–68
Kriesell HJ, Elwen SH, Nastasi A, Gridley T (2014) Identification and characteristics of signature whistles in wild bottlenose dolphins (*Tursiops truncatus*) from Namibia. PLoS One 9:e106317
Lammers MO, Au WWL (1996) Broadband recording of social acoustic signals of the Hawaiian Spinner and spotted dolphins. J Acoust Soc Am 100:2609
Lammers MO, Au WWL, Herzing HL (2003) The broadband social acoustic signaling behavior of spinner and spotted dolphins. J Acoust Soc Am 114:1629–1639
Leaper R, Chappell O, Gordon J (1992) The development of practical techniques for surveying sperm whale populations acoustically. In: Donovan GP (ed) Forty-second report of the International Whaling Commission, Cambridge, 1992
Leaper R, Gillespie D, Papastavrou V (2000) Results of passive acoustic surveys for odontocetes in the Southern Ocean. J Cetacean Res Manag 2:187–196
Leaper R, Gilliespie D, Gordon J, Matthews J (2003) Abundance assessment of sperm whales. Int Whal Comm SC/55/013
Leatherwood S, Goodrich K, Kinter AL, Truppo RM (1982) Respiration patterns and 'sightability' of whales. Rep Int Whal Comm 32:601–613
Lewis T, Gilliespie D, Lacey C et al (2007) Sperm whale abundance estimates from acoustic surveys of the Ionian Sea and Straits of Sicily in 2003. J Mar Biol Assoc UK 87:353–357
Lima IMS, Andrade LG, Carvalho RR et al (2012) Characteristics of whistles from rough-toothed dolphins (*Steno bredanensis*) in Rio de Janeiro coast, southeastern Brazil. J Acoust Soc Am 131:4173–4181

Lima IMS, Andrade LG, Bittencourt L et al (2016) Whistle comparison of four delphinid species in southeastern Brazil. J Acoust Soc Am 139:124–127
Lodi L, Hetzel B (2012) O golfinho-de-dentes-rugosos (*Steno bredanensis*) no Brasil. Rev Bioikos 12:29–45
Lu Y, Mellinger D, Klinck H (2013) Joint classification of whistles and echolocation clicks from odontocetes. In: Proceedings of meetings on acoustics, ICA, Montreal, 2–7 June 2013
Madsen PT, Wahlberg M, Møhl B (2002) Male sperm whale (*Physeter macrocephalus*) acoustics in a high-latitude habitat: implications for echolocation and communication. Behav Ecol Sociobiol 53:31–41
Madsen PT, Kerr I, Payne R (2004) Echolocation clicks of two free-ranging, oceanic delphinids with different food preferences: false killer whales *Pseudorca crassidens* and Risso's dolphins *Grampus griseus*. J Exp Biol 207:1811–1823
Magera AM, Mills Flemming JE, Kaschner K et al (2013) Recovery trends in marine mammal populations. PLoS One 8:e77908
Marcoux M, Whitehead H, Rendell L (2006) Coda vocalizations recorded in breeding areas are almost entirely produced by mature female sperm whales (*Physeter macrocephalus*). Can J Zool 84:609–614
Marques TA, Munger L, Thomas L et al (2011) Estimating North Pacific right whale *Eubalaena japonica* density using passive acoustic cue counting. Endanger Species Res 13:163–172
Marques TA, Thomas L, Martin SW et al (2013) Estimating animal population density using passive acoustics. Biol Rev 88:287–309
McCowan B, Reiss D (1995) Quantitative comparison of whistle repertoires from captive adult bottlenose dolphins (Delphinidae, *Tursiops truncatus*): a re-evaluation of signature whistle hypothesis. Ethology 100:194–209
McCowan B, Reiss D (2001) The fallacy of 'signature whistles' in bottlenose dolphins: a comparative perspective of 'signature information' in animal vocalizations. Anim Behav 62:1151–1162
McDonald MA, Moore SE (2002) Calls recorded from North Pacific right whales (*Eubalaena japonica*) in the eastern Bering Sea. J Cetacean Res Manag 4:261–266
Meirelles ACO, Monteiro-Neto C, Martins AMA et al (2009) Cetacean strandings on the coast of Ceará, north-eastern Brazil (1992–2005). J Mar Biol Assoc UK 89:1083–1090
Mellinger DK, Barlow J (2003) Future directions for marine mammal acoustic surveys: stock assessment and habitat use. In: Workshop held in La Jolla, CA, NOAA/PMEL, Seattle, 20–22 November 2002
Mellinger D, Stafford KM, Moore SE et al (2007) An overview of fixed passive acoustic observation methods for cetaceans. Oceanography 20:36–45
Moore KE, Watkins WA, Tyack PL (1993) Pattern similarity in shared codas from sperm whales (*Physeter catodon*). Mar Mam Sci 9:1–9
Moreno IB, Zerbini AN, Danilewicz D et al (2005) Distribution and habitat characteristics of dolphins of the genus *Stenella* (Cetacea: Delphinidae) in the southwest Atlantic Ocean. Mar Ecol Prog Ser 300:229–240
Moretti D, Dimarzio N, Morrissey R et al (2006) Estimating the density of Blainville's beaked whale (*Mesoplodon densirostris*) in the Tongue of the Ocean (TOTO) using passive acoustics. Oceans 2006:1–5
Moron JR, Andriolo A (2015) Preliminary evidence for signature and copied whistles among spinner dolphins in the Southwest Atlantic Ocean: Beacon purpose? J Acoust Soc Am 138:1904
Moron JR, Andriolo A (2016) *Stenella frontalis* screams record in the Southwest Atlantic Ocean. In: 1st listening for aquatic mammals in Latin America workshop, Natal, 21–23 June 2016
Moron JR, Amorim TOS, Sucunza F et al (2015) Spinner dolphin whistle in the Southwest Atlantic Ocean: is there a geographic variation? J Acoust Soc Am 138:2495–2498
Neves S (2013) Acoustic behaviour of Risso's dolphins, *Grampus griseus*, in the Canary Islands, Spain. Dissertation, University of St Andrews
Nielsen BK, Møhl B (2006) Hull-mounted hydrophones for passive acoustic detection and tracking of sperm whales (*Physeter macrocephalus*). Appl Acoust 67:1175–1186

Oliveira C, Wahlberg W, Silva MA et al (2016) Sperm whale codas may encode individuality as well as clan identity. J Acoust Soc Am 139:2860–2869
Olson P (2009) Pilot whales *Globicephala melas* and *G. macrorhynchus*. In: Perrin WF, Würsig B, Thewissen JG (eds) Encyclopedia of marine mammals, 2nd edn. Academic, Amsterdam, pp 847–852
Oswald JN, Barlow J, Norris TF (2003) Acoustic identification of nine delphinid species in the eastern tropical Pacific Ocean. Mar Mam Sci 19:20–37
Oswald JN, Rankin S, Barlow J (2004) The effect of recording and analysis bandwidth on acoustic identification of delphinid species. J Acoust Soc Am 116:3178–3185
Oswald JN, Rankin S, Barlow J, Lammers MO (2007) A tool for real-time acoustic species identification of delphinid whistles. J Acoust Soc Am 112:587–595
Ott PH, Danilewicz D (1996) Southward range extension of *Steno bredanensis* in the southwest Atlantic and new records of *Stenella coeruleoalba* for Brazilian waters. Aquat Mamm 22:185–189
Parks SE, Clark CW, Tyack PL (2007) Short- and long-term changes in right whale calling behavior: the potential effects of noise on acoustic communication. J Acoust Soc Am 122:3725–3731
Pavan G, Hayward TJ, Borsani JF et al (2000) Time patterns of sperm whale codas recorded in the Mediterranean Sea 1985-1996. J Acoust Soc Am 107:3487–3495
Perrin WF (2009) Common dolphins. In: Perrin WF, Würsig B, Thewissen JG (eds) Encyclopedia of marine mammals, 2nd edn. Academic, Amsterdam, pp 255–259
Petrella V, Martinez E, Anderson MG, Stockin KA (2012) Whistle characteristics of common dolphins (*Delphinus* sp.) in the Hauraki Gulf, New Zealand. Mar Mam Sci 28:479–496
Philips DJ, Nachtigall PE, Au WWL et al (2003) Echolocation in the Risso's dolphin, *Grampus griseus*. J Acoust Soc Am 113:605–616
Pinedo MC, Polacheck T, Barreto AS, Lammardo MP (2002) A note on vessel of opportunity sighting surveys for cetaceans in the shelf edge region off the southern coast of Brazil. J Cetacean Res Manag 4:323–329
Pirotta E, Merchant ND, Thompson PM et al (2015) Quantifying the effect of boat disturbance on bottlenose dolphin foraging activity. Biol Conserv 181:82–89
Rankin S, Barlow J (2007) Localization of a stationary sound source using a two-element towed hydrophone array In: Manuscript available from Shannon Rankin, SWFSC, 8604 La Jolla Shores Drive, La Jolla, CA 92937, 20 pages
Rankin S, Barlow J (2010) Acoustic recording system: a portable hardware system for shipboard passive acoustic monitoring of cetaceans using a towed hydrophone array. National Oceanic and Atmospheric Administration, La Jolla
Rankin S, Oswald JN, Barlow J (2008a) Acoustic behavior of dolphins in the Pacific Ocean: implications for using passive acoustic methods for population studies. Can Acoust 36:88–92
Rankin S, Barlow J, Oswald J, Balance L (2008b) Acoustic studies of marine mammals during seven years of combined visual and acoustic line-transect surveys for cetaceans in the Eastern and Central Pacific Ocean. NOAA Technical Memorandum, La Jolla
Rankin S, Barlow J, Oswald JN (2008c) An assessment of the accuracy and precision of a stationary sound source using a two-element towed hydrophone array. NOAA Technical Memorandum, La Jolla
Rankin S, Oswald JN, Simonis AE, Barlow J (2015) Vocalizations of the rough-toothed dolphin, *Steno bredanensis*, in the Pacific Ocean. Mar Mam Sci 31:1538–1548
Rendell L, Whitehead H (2003a) Comparing repertoires of sperm whales: a multiple methods approach. Bioacoustics 14:61–81
Rendell L, Whitehead H (2003b) Vocal clans in sperm whales (*Physeter macrocephalus*). Proc R Soc Lond B 270:225–232
Rendell L, Whitehead H (2004) Do sperm whales share coda vocalizations? Insights into coda usage from acoustic size measurement. Anim Behav 67:865–874
Rendell L, Whitehead H (2005) Spatial and temporal variation in sperm whales coda vocalizations: stable usage and local dialects. Anim Behav 70:191–198
Rendell LE, Matthews JN, Gill A et al (1999) Quantitative analysis of tonal calls from five odontocete species, examining interspecific and intraspecific variation. J Zool 249:403–410

Rice AN, Tielens JT, Estabrook BJ et al (2014) Variation of ocean acoustic environments along the Western North Atlantic Coast: a case study in context of the right whale migration route. Ecol Inform 21:89–99

Riesch R, Ford JKB, Thomsen F (2006) Stability and group specificity of stereotyped whistles in resident killer whales, *Orcinus orca*, off British Columbia. Anim Behav 71:79–91

Risch D, Castellote M, Clark C et al (2014) Seasonal migrations of North Atlantic minke whales: novel insights from large-scale passive acoustic monitoring networks. Mov Ecol 2:24

Roch MA, Soldevilla MS, Burtenshaw JC et al (2007) Gaussian mixture model classification of odontocetes in the Southern California Bight and the Gulf of California. J Acoust Soc Am 121:1737–1748

Roch MA, Klinck H, Baumann-Pickering S et al (2011) Classification of echolocation clicks from odontocetes in the Southern California Bight. J Acoust Soc Am 129:467–475

Samarra FIP, Deecke VB, Vinding K et al (2010) Killer whales (*Orcinus orca*) produce ultrasonic whistles. J Acoust Soc Am 128:EL205–EL210

Samarra FIP, Deecke VB, Simonis AE, Miller PJO (2015) Geographic variation in the time-frequency characteristics of high-frequency whistles produced by killer whales (*Orcinus orca*). Mar Mam Sci 31:688–706

Sayigh LS, Esch C, Wells RS, Janik VM (2007) Facts about signature whistles of bottlenose dolphins, *Tursiops truncatus*. Anim Behav 74:1631–1642

Schorr GS, Falcone EA, Moretti DJ, Andrews RD (2014) First long-term behavioral records from cuvier's beaked whales (*Ziphius cavirostris*) reveal record breaking dives. PLoS One 9:e92633

Schulz TM, Whitehead H, Gero S, Rendell L (2008) Overlapping and matching of codas in vocal interactions between sperm whales: insights into communication function. Anim Behav 76:1997–1988

Schulz T, Whitehead H, Gero S, Rendell L (2011) Individual vocal production in a sperm whale (*Physeter macrocephalus*) social unit. Mar Mam Sci 27:149–166

Secchi ER, Vaske JT (1998) Killer whale (*Orcinus orca*) sightings and depredation on tuna and swordfish longline catches in southern Brazil. Aquat Mamm 24:117–122

Simonis AE, Baumann-Pickering S, Oleson E et al (2012) High-frequency modulated signals of killer whales (*Orcinus orca*) in the North Pacific. J Acoust Soc Am 131:EL295–EL301

Soldevilla MS, Wiggins SM, Hildebrand JA (2010) Spatial and temporal patterns of Risso's dolphin echolocation in the Southern California Bight. J Acoust Soc Am 127:124–132

Sousa-Lima R, Norris TF, Oswald JN, Fernandes DP (2013) A review and inventory of fixed autonomous recorders for passive acoustic monitoring of marine mammals. Aquat Mamm 39:23–53

Sveegaard D, Teilmann J, Berggren P et al (2011) Acoustic surveys confirm the high-density areas of harbour porpoises found by satellite tracking. J Mar Sci 68:929–936

Swift RJ, Gillespie D, Vázquez JA et al (2009) Abundance of sperm whales (*Physeter macrocephalus*) estimated from acoustic data for Blocks 2, 3 and 4 (French and Spanish sectors). Cetacean Offshore Distribution and Abundance in the European Atlantic (CODA). Appendix IV. Available via DIALOG. http://biology.st-andrews.ac.uk/coda/. Accessed 1 Oct 2016

Taruski AG (1979) The whistle repertoire of the North Atlantic pilot whale (*Globicephala melaena*) and its relationship to behavior and environment. In: Winn HE, Olla BC (eds) Behavior of marine animals, 3rd edn. Plenum Press, New York, pp 345–368

Tavares M, Moreno IB, Siciliano S et al (2010) Biogeography of common dolphins (genus *Delphinus*) in the southwestern Atlantic Ocean. Mammal Rev 40:40–64

Thomsen F, Franck D, Ford JKB (2001) Characteristics of whistles from the acoustic repertoire of resident killer whales (*Orcinus orca*) off Vancouver Island, British Columbia. J Acoust Soc Am 109:1240–1246

Tran DD, Huang W, Bohn AC et al (2014) Using a coherent hydrophone array for observing sperm whale range, classification, and shallow-water dive profiles. J Acoust Soc Am 135:3352–3363

Trickey JS, Reyes MVR, Baumann-Pickering et al (2014) Acoustic encounters of killer and beaked whales during the 2014 SORP cruise. IWC Report, SC/65b/SM12re

Tyack PL (1997) Development and social functions of signature whistles in bottlenose dolphins *Tursiops truncatus*. Bioacoustics 8:21–46
Tyack PL, Clark CW (2000) Communication and acoustic behavior of dolphins and whales. In: Au WWL, Popper AN, Fay RR (eds) Hearing by whales and dolphins, 1st edn. Springer-Verlag, New York, pp 156–224
Vallarta-Hernández J (2009) The significance of passive acoustic array-configurations on sperm whale range estimation when using the hyperbolic algorithm. Thesis, Heriot-Watt University School of Engineering & Physical Sciences
Verfub UK, Honnef CG, Meding A et al (2007) Geographical and seasonal variation of harbour porpoise *Phocoena phocoena* presence in the German Baltic Sea revealed by passive acoustic monitoring. J Mar Biol Assoc UK 87:165–176
Verfub UK, Dähne M, Gallus A et al (2013) Determining the detection thresholds for harbor porpoise clicks of autonomous data loggers, the Timing Porpoise Detectors. J Acoust Soc Am 134:2462–2468
Ward JA, Thomas L, Jarvis S et al (2012) Passive acoustic density estimation of sperm whales in the Tongue of the Ocean, Bahamas. Mar Mam Sci 28:e444–e455
Watkins WA (1980) Acoustics and the behavior of sperm whales. In: Busnel RG, Fish JF (eds) Animal sonar systems. Plenum Publishing Corp, New York, pp 283–290
Watkins WA, Schevill WE (1977) Sperm whale codas. J Acoust Soc Am 62:1486–1490
Watkins WA, Tyack PL, Moore KE, Notarbartolo Di Sciara G (1987) Rough-toothed dolphin *Steno bredanensis* in the Mediterranean Sea. Mar Mam Sci 3:78–82
Watwood SL, Miller PJO, Johnson M et al (2006) Deep-diving foraging behaviour of sperm whales. J Anim Ecol 75:814–825
Weilgart LS, Whitehead H (1988) Distinctive vocalizations from mature male sperm whales (*Physeter macrocephalus*). Can J Zool 66:1931–1937
Weilgart LS, Whitehead H (1990) Vocalizations of the North Atlantic pilot whale (*Globicephala melas*) as related to behavioral contexts. Behav Ecol Sociobiol 26:399–402
Weilgart L, Whitehead H (1993) Coda communication by sperm whales (*Physeter macrocephalus*) of the Galápagos Islands. Can J Zool 71:744–752
Weilgart L, Whitehead H (1997) Group-specific dialects and geographical variation in coda repertoire in South Pacific sperm whales. Behav Ecol Sociobiol 40:277–285
Wells RS, Scott MD (2009) Common bottlenose dolphin - *Tursiops truncatus*. In: Perrin WF, Würsig B, Thewissen JG (eds) Encyclopedia of marine mammals, 2nd edn. Academic, Amsterdam, pp 249–255
Whitehead H, Weilgart L (1991) Patterns of visually observed behaviour and vocalizations in groups of female sperm whales. Behaviour 118:275–296
Whitehead H (2002) Estimates of the current global population size and historical trajectory for sperm whales. Mar Ecol Prog Ser 242:295–304
Whitehead H (2003) Sperm whales: social evolution in the Ocean. University of Chicago Press, Chicago
Whitehead H (2009) Sperm whales. In: Perrin WF, Würsig B, Thewissen JG (eds) Encyclopedia of marine mammals, 2nd edn. Academic, Amsterdam, pp 1091–1097
Yack TM, Barlow J, Calambokidis J et al (2013) Passive acoustic monitoring using a towed hydrophone array results in identification of a previously unknown beaked whale habitat. J Acoust Soc Am 134:2589–2595
Zerbini AN, Secchi ER, Bassoi M et al (2004) Distribuição e abundância relativa de cetáceos na Zona Econômica Exclu siva da região sudeste-sul do Brasil. In: Série de Documentos Revizee-Score Sul. Universidade de São Paulo, São Paulo
Zimmer WM (2013) Range estimation of cetaceans with compact volumetric arrays. J Acoust Soc Am 134:2610–2618
Zimmer WM, Tyack PL, Johnson MP, Madsen PT (2005) Three-dimensional beam pattern of regular sperm whale clicks confirms bent-horn hypothesis. J Acoust Soc Am 117:1473–1485

Chapter 6
Bioacoustic Techniques Applied to Odontocete Conservation and Management in Argentina

Mariana L. Melcón, M. Vanesa Reyes Reyes, and Miguel Iñíguez

Abstract Bioacoustics has become a tool of increasing importance to study cetaceans, due to its great potential to estimate population density, determine habitat use, and understand seasonal patterns, among other relevant ecological parameters. These applications convert passive acoustics into a non-invasive tool capable of gathering information for longer periods of time and in remote areas. After various efforts to conduct acoustic studies in the country, Argentina started to slowly grow in the field in the late 2000s. Since then, more reliable and systematic studies have been conducted on odontocetes as a baseline to better understand their biology, and to address potential effects of human activities for conservation purposes.

Given the advantages provided by this method, along with its relative low cost, we encourage other countries in Latin America to use this tool more often. Moreover, considering the most frequent difficulties and challenges we have reviewed from different research groups to carry out research on cetaceans using bioacoustics, we recommend establishing international collaborations and specialize human resources in the topic.

Keywords Bioacoustics • Odontocetes • Conservation • Management • Argentina

M.L. Melcón (✉)
Fundación Cethus, Gdor Luis Monteverde 3695 (B1636AEM),
Olivos, Provincia de Buenos Aires, Argentina
e-mail: marumelcon@gmail.com

M.V. Reyes Reyes • M. Iñíguez
Fundación Cethus, Gdor Luis Monteverde 3695 (B1636AEM),
Olivos, Provincia de Buenos Aires, Argentina

Whale and Dolphin Conservation, 7 Nelson Street, Plymouth, MA 02360, USA
e-mail: vanesa.reyes@cethus.org; miguel.iniguez@cethus.org

M.R. Rossi-Santos, C.W. Finkl (eds.), *Advances in Marine Vertebrate Research in Latin America*, Coastal Research Library 22,
DOI 10.1007/978-3-319-56985-7_6

6.1 History of Underwater Bioacoustics in Argentina

Argentina has a long history of attempting to develop acoustic studies on wild cetaceans by different groups of researchers. Despite decades of trials, the success of more systematic work on cetaceans using passive acoustics happened only recently (Fig. 6.1).

There were several projects since the 1970s which tried to incorporate sounds as a component of ongoing research. The first attempt, to our knowledge, was carried out by R. Payne and K. Payne, who recorded for the first time in Argentine waters the sounds of Southern right whales in Península Valdés, Chubut, for 12 h. They described the sound repertoire as hybrid calls (Payne and Payne 1971). In late June and early July 1971 as part of the R/V *Hero* expedition, W. Cummings recorded Southern right whales in Golfo San José, Argentina. He classified the sounds in their sample into five categories: belches, two kinds of "moans", miscellaneous sounds and pulses, being the most common the belch-like utterance with most energy below 500 Hz (Cummings et al. 1972). In 1972, he came back to Península Valdés and recorded 1750 sounds of Southern right whales. He mentioned that bellowings and moans of rising pitch were associated with sexual activity (Cummings et al. 1972); however, posterior studies by Clark in 1977 rejected Cummings' conclusions. After recording a total of 1571 h of whale sounds during 210 days in 1977, Clark characterized the acoustic repertoire of the Southern right whale (Clark 1982).

P. Tyack conducted acoustic studies on Bottlenose dolphins (*Tursiops truncatus*) from November 1974 to June 1975 in Golfo San José, Península Valdés. To record dolphins he stacked a hydrophone in a mudflat of the lower intertidal zone of the beach near the field station; if dolphins were far from the hydrophone, he used an inflatable boat and recorded the sounds from a portable hydrophone hung over the side. He categorized vocalizations and behavior of Bottlenose dolphins (Tyack 1976).

After a gap of 9 years without acoustic studies on cetaceans, B. Würsig and F. Cipriano recorded dusky dolphins (*Lagenorhynchus obscurus*) in Golfo San José, Península Valdés, displaying foraging behavior using a manual hydrophone deployed from an inflatable boat in January 1986, but no analyses were performed. Twelve years later, Würsig and colleagues returned and in December 1998, they recorded sounds from dusky dolphins groups resting, traveling and socializing while snorkeling, using a manually operated mobile video/acoustic system with stereo hydrophones with a 48 kHz sample rate (Vaughn-Hirshorn et al. 2012). They found that the inter-click interval (ICI) was a continuum and that dusky dolphin may use burst pulse sequences with a communication function.

M. Iñíguez made an attempt to record Commerson's dolphins (*Cephalorhynchus commersonii*) in Ría Deseado in 1986 and 1994 from a sailboat from Club Náutico "Capitán A. Oneto", using a portable hydrophone. This unsuccessful attempt was due to the lack of necessary specifications for the equipment to record Commerson's dolphins' high frequency clicks.

In 1990, 1991 and 1992, J. Ford, B. Ford and M. Iñíguez deployed a hydrophone (Fig. 6.2a) fixed to 200 m cable and to an audio cassette recorder, located at the

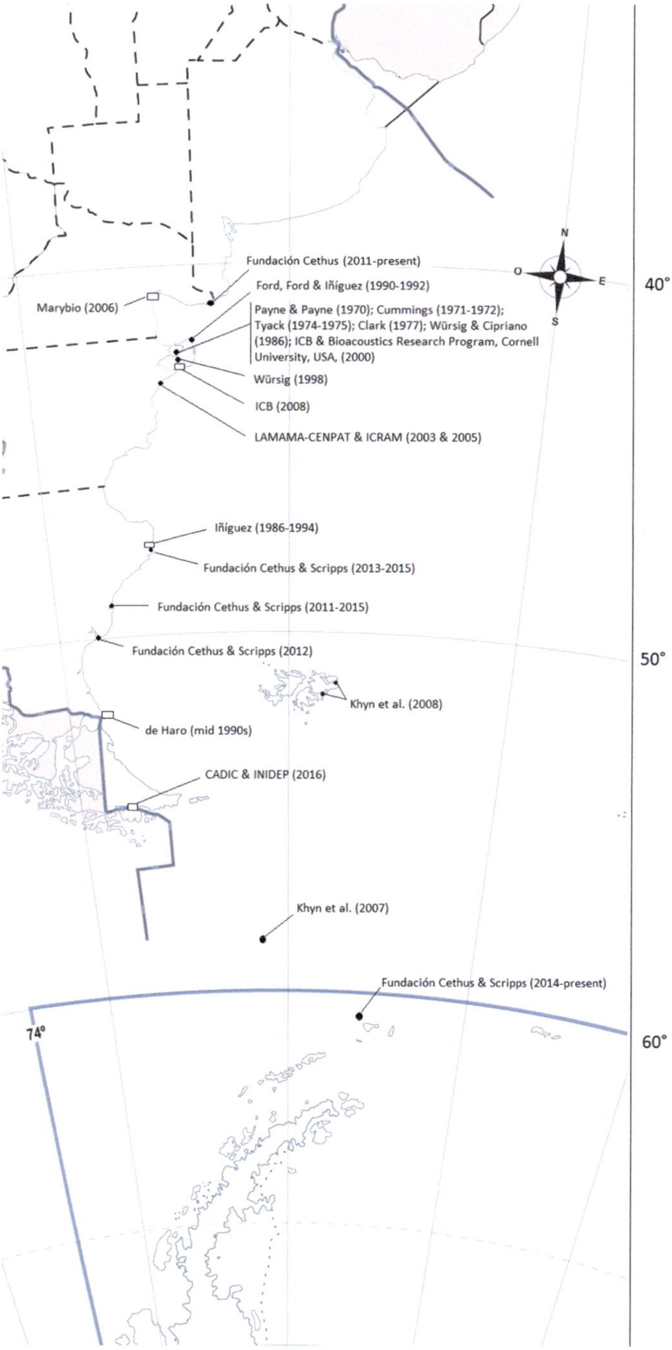

Fig. 6.1 Locations where successful (*black circles*) and failed (*rectangles*) attempts to collect acoustic recordings were made

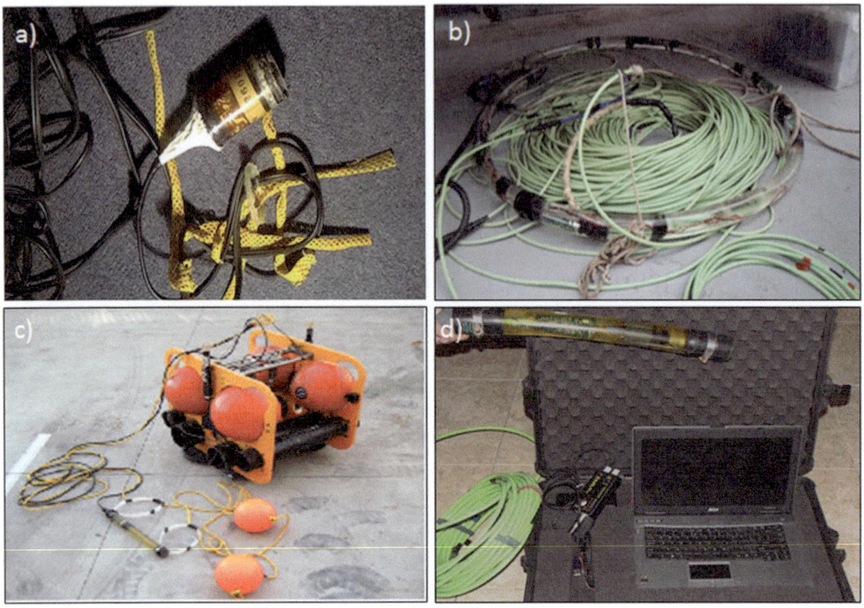

Fig. 6.2 Several devices to carry out sound recordings: (**a**) hydrophone used to collect sounds from killer whales in Punta Norte, Península Valdés by Ford, Ford & Iñíguez in the early 1990s; (**b**) towed hydrophone array; (**c**) high-frequency acoustic recording package (HARP); (**d**) Dipping broad-band hydrophone with built-in integrated preamplifier connected to an analog to digital converter and a PC to record the signals

Orcas channel in Punta Norte, Península Valdés, Chubut. They recorded the killer whales (*Orcinus orca*) that visit the area from February to May to hunt on South American sea lions (*Otaria flavescens*) and Southern elephant seal (*Mirounga leonina*). They found that these orcas were in complete silence previous to catch a prey but they vocalized once they succeeded (Iñíguez 1993). Unfortunately, the vocalization repertoires for the three orca's pod identified in this area were not described.

In the mid 1990s, a Fundación Cethus (Argentine NGO) research team led by C. de Haro, attempted to record Peale's dolphins (*Lagenorhynchus australis*) in Cabo Vírgenes, Santa Cruz, using a hydrophone deployed from an inflatable boat; however, due to bad weather conditions and facing logistic difficulties the project was discountinued.

More recently (in 2000), the Instituto de Conservación de Ballenas (ICB) research team in conjunction with the Bioacoustics Research Program, Cornell University, USA, conducted a brief study on the acoustic characterization of few bays in Golfo San José. The objective was to understand if Southern right whales will prefer those areas where they could hear their predators, such as killer whales, or improve their own communication, but the project was discontinued.

The Marine Mammals Lab (LAMAMA) of Centro Nacional Patagónico (CENPAT), working together with Istituto per la Ricerca Scientifica e Tecnologica

Applicata al Mare (ICRAM), Italy, recorded sounds of dusky dolphins and Commerson's dolphins to study the potential impact caused by tourist activity on dolphins between 2003 and 2005 in Northern Patagonia. Further studies and analyses by LAMAMA are still conducted.

In 2005, another Argentine non-governmental organization "Marybio", conducted studies on wild Franciscana in San Blas bay in collaboration with the University of Brussels, Belgium, and the Natural History Museum of Buenos Aires. In 2006, they also recorded *T. truncatus* in Bahía San Antonio, Río Negro, but the lack of human sources and the need to focus on higher-priority research lines made them postpon the project.

Kyhn and colleagues recorded Hourglass dolphins (*Lagenorhynchus cruciger*) in the Drake Passage with a four-element hydrophone array in 2007. This species produce narrow band high-frequency (NBHF) echolocation clicks and they concluded that offshore NBHF dolphins, such as hourglass dolphin, seem to produce clicks with higher source level than coastal dolphins do. A year later Kyhn recorded Peale's dolphins and Commerson's dolphins at Islas Malvinas (Falkland Islands) using a six-element hydrophone array. They obtained the first wide-band recordings of Peale's dolphins and showed that they produce NBHF clicks a well as Commerson dolphins, but with lower centroid frequency and higher source levels. They hypothesized that these differences may be a result of some degree of character displacements to allow species recognition in these two sympatric species.

The ICB planned to develop an acoustic study which included Southern right whales in Península Valdés in 2008, but they faced several problems to develop the study and it was cancelled.

By the end of 2010, Fundación Cethus began to develop research incorporating passive acoustics to study different species such as Franciscana dolphins (*Pontoporia blainvillei*) (Melcón et al. 2012a, 2016) and Bottlenose dolphins in the Río Negro Estuary, Commerson's dolphins in Bahía San Julián, Ría Santa Cruz and Ría Deseado (Santa Cruz) (Reyes Reyes et al. 2015, 2016a), Peale's dolphins in Ría Santa Cruz (Marino et al. in preparation). This was possible thanks to the collaboration of the Marine Physical Laboratory of Scripps Institution of Oceanography, which generously donated the necessary equipment for such research (Fig. 6.2d).

Since 2014 Fundación Cethus, in collaboration with Scripps Institution of Oceanography expanded the bioacoustics research line to the Antarctic, as part of the International Whaling Commission's Southern Ocean Research Partnership (IWC-SORP). In this long-term project we are using two different types of acoustic recording devices: a hydrophone array towed by an Argentine Coast Guard vessel (Fig. 6.2b), and the autonomous High-frequency Acoustic Recording Package (HARP) for broad-band recordings (Fig. 6.2c), which is deployed at 1000 m depth approximately. The main target species are beaked whales, killer whales and sperm whales, and blue whales (Trickey et al. 2015).

There were several attempts to record Burmeister's porpoise (*Phocoena spinipinnis*) at the Beagle Channel, Tierra del Fuego, using hydrophones made by the Laboratorio de Hidroacústica of Instituto Nacional de Investigacion y Desarrollo Pesquero (INIDEP) with Centro Austral de Investigaciones Científicas (CADIC).

For these purposes, two autonomous acoustic recording devices were deployed but unfortunately no sounds were recorded. Some challenges they have found in developing this project were bad weather conditions and difficult access to the study area, lack of human resources specialized in cetacean acoustics as well as counting on the proper equipment. However, they will explore the possibility of recording porpoises using a hydrophone made by the Laboratorio de Hidroacústica (INIDEP), which is more adequate to record the type of signals they aim to study.

Currently, the LAMAMA together with the Laboratorio de Hidroacústica, use active acoustic devices with eco-integrators to study presence, abundance and distribution of prey in presence of dolphins.

To summarize, historically in Argentina there have been many attempts to study wild cetaceans acoustically, but faced several difficulties which comprises mostly lack of trained researchers and difficulties to obtain the proper passive acoustic recording devices. Additionally, adverse weather conditions and accessibility to the area with existing resources may challenge the research even more. However, more recently several governmental and non-governmental institutions are turning the tide and began to explore and develop new research lines using acoustics, slowly learning to overcome some of the aforementioned obstacles.

6.2 Learning the Acoustic Repertoire of Cetaceans

Sound is a very efficient way for energy to propagate through water. This feature of aquatic environments together with a rapid attenuation of light along distance have been probably the main driving factors of the development of a complex sound production and reception systems that play a crucial role in marine organisms (Tyack and Clark 2000; Au and Hastings 2008).

Studying cetaceans in the wild is challenging because they spend most of the time submerged, out of the human vision range, and often inhabit remote areas with difficult access. Passive acoustics can overcome some limitations of the traditional visual methods to study cetaceans in the wild. For example, the time an animal spends underwater turns into an advantage using passive acoustics, since it improves the likelihood of recording its vocalizations. This is one of the reasons why some researchers are interested in learning the acoustic repertoire of different species of cetaceans, and even of different populations of cetaceans around the world. The acoustic repertoire comprises the different types of sounds that can be produced by animals within a species and/or population. Once the sounds produced by different species have been characterized, a combination of spectral and/or temporal features can generally be used to identify and discriminate among species.

Learning the acoustic repertoire of cetaceans consists of a dynamic process since the emitted sounds can change over spatio and/or temporal scales. These variations occur due to several factors: (1) Natural variations within the environmental features, e.g. coastal vs. oceanic habitats, or seasonalities. (2) Anthropogenic pressures (e.g. increase in ocean noise) can also lead animals to change sound types, patterns or features. (3) Intrinsic dynamics of the populations (e.g. body size, sexual selection) can also affect the acoustic repertoire produced by its individuals.

As we hinted earlier, characterization of the acoustic repertoire of animals at sea may be challenging, but it is necessary as a first step to conduct passive acoustic monitoring (PAM) of cetaceans, by means of which acoustic recordings can be classified and attributed to a particular species without the need of a visual match to verify the identity of the sound producing animal.

For some species, much of the information available comes from captive animals. While it may be the best approximation we can get to the species living in the wild, it is important to consider that the signals recorded and associated behavior displayed may not be representative of the behavior recorded of wild animals. For example, captive animals usually produce sounds with lower amplitude and frequency emphasis than animals in the field (Au 1993, Madsen and Wahlberg 2007).

In order to characterize the acoustic repertoire of a given species, we need to keep in mind what our goal is, but at the very least we need to be able to identify the species with a high degree of certainty; this is, either by using a hydrophone array that allows us to correlate the location of the animal with our observations, or we need to be in an area at a specific time in which there is only our species of interest. In some fortunate cases, however, sympatric species may produce very different sounds and we may be tempted to believe which sound is produced by which species. A typical example in our region would be studying the repertoire of Peale's and Dusky dolphins, two sympatric species that produce very different sounds: Peale's dolphins produce narrowband high-frequency (NBHF) clicks and it is believed that they do not whistle; while Dusky dolphins produce broadband (BB)[1] clicks and whistles are a major component of their acoustic repertoire. The risk may be that we associate all whistles to Dusky dolphins (because that's what we know) instead of trying to determine whether Peale's dolphins also produce whistles under certain circumstances. This assumption may hinder us from learning something new and promote spreading wrong information.

When it comes to the recordings themselves, a typical underwater recording system nowadays consists of a hydrophone, a preamplifier, and a band-pass filter connected to a digital recorder. The hydrophone is a transducer that converts sound pressure into voltage variations which is then amplified by the preamplifier. The band-pass filter reduces low frequency noise and filters out any signal component above the Nyquist frequency (half the sampling frequency of the digital recorder) which is the theoretical maximum frequency that can be digitized unambiguously. For example, if our digital recorder has a sampling frequency of 192 kHz, the recorded signals will be limited to frequencies below 96 kHz (Nyquist frequency), making this setup unsuitable to characterize properly the acoustic repertoire of species that produce sounds that exceed 96 kHz or to record NBHF clicks at all,

[1] NBHF refers to clicks produced by some species of odontocetes which have all their frequency content above 100 kHz, longer duration and a narrower frequency bandwidth than the short BB clicks produced by most delphinid species.

such as Franciscana dolphins. Finally, the analog voltage is digitized by the converter to create a discrete digital time series.

There is a variety of different devices that can be used to obtain sound recordings to study the acoustic repertoire of cetaceans: dipping hydrophones, towed hydrophone arrays, sonobuoys, bottom-moored autonomous recorders and acoustic tags.

Dipping Hydrophones are commonly submerged from a fixed platform or a vessel with the engine off. They can consist of a single hydrophone or of several hydrophones which allow characterization of source levels and directionality. The ideal situation when using this kind of devices for characterization is that researchers make the recordings when the species of interest is visually detected within a reasonable distance (a few meters up to a few kilometers depending on the species), in absence of any other odontocete species, and so the recorded sounds can be assigned to that species with a high level of confidence. Some advantages of this setup are that they are easily handled and can be used in shallow waters such as inshore bays or rivers. A disadvantage is the requirement of at least one person that should be at the site of deployment all the time while recording, which also limits the use of these gears in remote areas where navigation using small vessels is not suitable. Additionally, with the single hydrophone arrangement no accurate localization is possible, and source level estimates cannot be accurate.

Towed Hydrophone Arrays are usually towed from a moving vessel and used to localize vocalizing individuals through real-time calculations using the time difference of arrival of an acoustic signal at various hydrophones in the array. The key using this methodology is to obtain a visual match of the recorded signal to identify the species producing that particular sound. These systems have the advantage that they allow passive acoustic monitoring of species during the track line of large vessels. With this system localization is possible. Some cons are that they need to be towed a long distance far away from the moving vessel in order to avoid signals getting saturated with the vessel noise and so large vessels are required and they are not suitable for inshore or shallow waters.

Sonobuoys are designed as expendable devices to be dropped from a vessel and the signals detected by the hydrophone are transmitted by a radio transmitter to a receiver on the vessel. A specialized type of sonobuoys called DIFAR also allows to determine the direction of the sounds relative to the sonobuoys. The benefit of this is that using 3 or more of these instruments it is possible to determine the exact location of the sound source and then correlate the recorded signal to visual observations of cetacean species (mostly mysticetes, due to the limitations of sonobuoys to record high frequencies) at that location. This will allow us to associate which species produce certain sounds and also calculate the source levels of the signals for which the distance of the vocalizing animal to the hydrophone is required at a certain background noise level. Knowing the source level of the signals helps to optimize detection and classification routines frequently used for PAM, since sounds from different species may not only be distinguishable by their frequency content, bandwidth and durations but also by different source levels of the emitted sounds

(Madsen and Wahlberg 2007). A negative aspect of sonobuoys is that they are often used only one time and end up being trash in the ocean.

Autonomous Recorders there are different designs, but essentially all of them consist of a typical recording system described above, a set of batteries and hard drives to store the information, and a release system. These devices are deployed onto the seafloor and recovered after some time that could vary from several days to several months, depending on the device configuration and capabilities. Generally, when acoustic recordings are obtained from these autonomous recorders, there is no visual data to match the sounds to a particular species. In those cases, there are other sources of information that could provide some cues on species identification, e.g. the number of possible candidates can be reduced by studying which species are likely to inhabit the sampled area. Also, a preliminary and broad characterization of the signal, along with previous information on acoustic repertoire of some species could help to further reduce the candidates, e.g. for instance sperm whales can be excluded if the recorded signals are whistles or NBHF clicks; if recordings consist of upswept FM clicks any other taxa than beaked whales can be excluded. Later in this chapter we will explain with a real example how some additional information can help to have a good approximation of species identity. These systems are not usually used to study the acoustic repertoire of a species given the lack of visual correlation to confidently determine the species identity of the recorded sounds. Instead, they are suitable to study cetacean presence in a certain area, seasonal patterns and abundance estimates. In some case behavioral context is possible to be determined.

Acoustic Tags are attached on free-ranging animals and generally used to study the acoustic repertoire and correlate it with diving and movement patterns. They have sensors of sound, pressure and temperature, and a very high frequency (VHF) transmitter which stands for radio frequency electromagnetic waves to track the animal and recover the tag once it is detached from the animal (Johnson et al. 2009). An advantage of using tags is that usually the recorded sounds can be correlated with diving patterns of the species of interest. Also source properties of the signals can be described at an individual level. Among the cons, these tags are expensive and may result difficult to attach, especially in fast swimming, small odontocetes. Currently the deployment of the device is invasive and special permits are required to use them.

Equipment improvements over the last decades allowed scientists to get access to information that some years ago wasn't available because of the instrument limitations. For instance, the first studies on bioacoustics in Argentina were focused mostly on Southern right whales and killer whales, and the hydrophones and recorders used by researchers probably would not be able to record frequencies higher than 100 kHz. Thus, sounds from species like Commerson's dolphins, Franciscana dolphins, Peale's dolphins and porpoises which all produce NBHF clicks would not be recorded even if they were present and vocalizing. Thus, researchers aiming to conduct bioacoustics studies need to consider very carefully which technical requirements the equipment should meet in order to properly record and analyze the sounds of the species of interest. It is important to consider the sensitivity of the hydrophone (the sound pressure level required to generate 1 V), and frequency range which should cover the expected frequency range of the sounds of interest, ideally with a flat sensitivity over a broad range of frequencies, but if that is not the case, a proper calibration of the system will allow to correct the sensitivity as a function of frequency. The sampling frequency of the digital recorder which should be at least double than the maximum frequency of interest in the signal; and the dynamic range of the whole system (the range of signal's amplitudes that the system is able to process without being saturated or indistinguishable from the system noise floor). For example, if we are interested in recording sounds from a NBHF species we would need a digital recorder with a sampling frequency of at least 384 kHz which would allow us to obtain recordings up to 192 kHz, and a hydrophone able to record up to at least 160 kHz. However, if the main aim is to record the more intense low frequency clicks from sperm whales we don't need a digital recorded with such high sampling frequency but we may want to look for a system with a large dynamic range. This is because sperm whales' clicks are one of the loudest sounds in the oceans reaching up to 230 dB what can surpass the highest limit of the recording system and the result would be saturated signals.

Once we have the digital signal from the sounds recorded, typically there is such large amount of data obtained that the sound analysis can be very time consuming and computationally-intensive. There is a big variety of pieces of software such as Triton. Ishmael, Raven, Kaleidoskope, Avisoft SASLab, etc., and algorithms that can be used to visualize the signals of interest. For example, long-term spectral averages (LTSAs) can be calculated to avoid looking at each single audio file. LTSAs are long-term spectrogram with each time segment consisting of an average of hundreds of spectra, and the resulting plots allow visualization of large time series data sets and searching for and logging sounds of interest, like dolphin whistles or clicks for later analysis. Once the acoustic signals are noticeable in the LTSA, such audio files can be inspected more closely looking at each spectrum or time series.

After visually detecting and logging the acoustic events of interest we can use some of the pieces of software mentioned before or others to extract the spectral, temporal and energy properties of the signals. Alternatively, we may want to isolate each single signal/call for its characterization using custom-routines in MATLAB (Mathworks, Natick, MA) or R, to give some examples. Finally, once a large sample size of the acoustic repertoire of a species has been handled, usually researchers are interested in developing acoustic detectors, which are pieces of software that enable to detect and extract the target sounds in an automatic or semi-automatic manner. Hence, ideally acousticians will use one or more automatic or semi-automatic detectors to extract the signals of interest and characterize their spectral and temporal properties.

Several species of odontocetes can be found in Argentinean waters, all of which produce underwater sounds. Table 6.1 summarizes the available information on the acoustic repertoire of odontocete species distributed in Argentinean waters and around the western Antarctic Peninsula. There are several species for which there is no local data or the existing data is scarce such as Bottlenose dolphins (*Tursiops truncatus*), Common dolphins (*Delphinus delphis*), Killer whales, Dusky dolphins, Peale's dolphins, Burmeister's porpoise (*Phocoena spinipinnis*), spectacled

Table 6.1 Available data on the acoustic repertoire of species distributed along the waters off Argentina

Species	Type of sound	SL_{pp} (dB re 1 mPa p.–p.)	Peak frequency (kHz)	Bandwidth (kHz)	Duration (ms)	Source
Cephalorhynchus commersonii	NBHF clicks	165–190	119–139	16–31	0.05–0.14	Kyhn et al. (2010)
	NBHF clicks	NA	122–181	5–11	0.06–0.11	Reyes Reyes et al. (2015)
	BB clicks	NA	29–96	21–110	0.14–0.22	Reyes Reyes et al. (2016a)
	Whistles	NA	16	NA	13–27	Reyes Reyes et al. (2016a, b)
Pontoporia blainvillei	NBHF clicks	NA	139	19	0.21 ± 0.056	Melcón et al. (2012a)
Pontoporia blainvillei	Clicks/pulsed calls	NA	NA	123	NA	Melcón et al. (2016)
Orcinus orca	FM whistles	NA	14.5–23.3	0.9–12.1	20.0–157.0	Reyes Reyes et al. (2016b)
Antarctic beaked whale (*Hyperoodon planifrons*)	FM clicks	NA	26.6–32.6	4.1–9.4	0.39–0.78	Trickey et al. (2015)
Antarctic beaked whale (*Mesoplodon grayi*)	FM clicks	NA	36.0–37.1	5.1–6.9	0.61–0.70	Trickey et al. 2015
Lagenorhynchus australis	NBHF clicks	169–196	65–153	9–34	0.065–0.153	Kyhn et al. (2010)
	NBHF clicks	NA	125–189	6–11	0.02–0.08	Marino et al. (in preparation)
Lagenorhynchus obscurus	Burst pulses	NA	NA	NA	260 ± 430	Vaughn-Hirshorn et al. (2012)
Lagenorhynchus cruciger	NBHF clicks	190–203	122–131	5–11	0.08–0.18	Kyhn et al. (2009)

Note that out of the 28 species of odontocetes registered in Argentinean waters (Bastida and Rodríguez 2003), acoustic information exists for less than ten species

porpoise (*Phocoena dioptrica*), sperm whales (*Physeter macrocephalus*), long-finned pilot whales (*Globicephala melas*), and several species of beaked whales.

Along the coastline of Buenos Aires province down to the Rio Negro Estuary (RNE) in Rio Negro province, Franciscana dolphins can be identified through their stereotyped NBHF clicks (Melcón et al. 2012a). Recently, Melcón et al. (2016) also described sounds produced by the calves of this species, which have a much broader bandwidth than adults, allowing discrimination between calves and adults by means of PAM. Another species that can be found in the RNE is the Bottlenose dolphin which produces clicks and whistles with a lower frequency content than Franciscana dolphins and thus can be easily differentiated in the acoustic recordings. Killer whales can be found all along the waters of Argentinean shelf, including Patagonian shelf, and during February–April/ September–October can be frequently seen in the waters off Península Valdés in Chubut province. Here they use a unique technique to forage sea lions and elephant seals pups. However, the available acoustic information on this species is scarce and only recently the very first recordings on this species in the north part of the western Antarctic Peninsula were obtained (Table 6.1). During summer months, Dusky dolphins can be found in the waters of Golfo Nuevo. Moving to the south, Commerson's dolphins inhabit the waters of Playa Unión in Chubut province. This species is also distributed further south down to the Magellan Strait. Peale's dolphins and Commerson's dolphins are sympatric which means that can be found together in many areas of the Southern part of Argentina in Santa Cruz and Tierra del Fuego provinces. Both species produce NBHF clicks with very similar characteristics, making it challenging to unambiguously tell them apart by just conducting passive acoustic monitoring in those areas. A detailed characterization of the acoustic repertoire of both species along with the available classification and appropriate statistical tools may enable to accomplish this objective. Systematic studies of Commerson's dolphin's in Santa Cruz province in the last years showed that the acoustic repertoire of this species is broader than previously thought. BB clicks, whistles and variable NBHF clicks were described for the population of Bahía San Julián, Santa Cruz province (Reyes Reyes et al. 2015, 2016a).

At least five species of beaked whales are known to occur in the Southern Ocean waters, including Arnoux's beaked whales (*Berardius arnuxii*), Cuvier's beaked whales (*Ziphius cavirostris*), Gray's beaked whales (*Mesoplodon grayi*), Southern bottlenose whales (*Hyperoodon planifrons*), and strap-toothed whales (*Mesoplodon layardii*) (MacLeod et al. 2006). Given the elusive and deep-diving behavior of beaked whales, visual sightings are extremely difficult to obtain. However, the unique features of their upswept FM clicks facilitate the study of these cryptic odontocetes by means of their acoustic signals. Using baseline information on species distribution and the available knowledge on some sister species in the Northern hemisphere, we can infer which species are the most probable candidates producing certain sounds, as in the two examples shown in Table 6.1.

We have seen that the number of odontocetes that have been recorded at least once in Argentinean waters is large but acoustic data is only available for a few species. This means that there is a big potential for studying the acoustic repertoire of many odontocete species along all the coastal and platform waters of this country.

6.3 Understanding the Vocal Behavior

When it comes to studying the vocal behavior of marine mammals, we have a very important requirement: to have already learned the acoustic repertoire of the species we are studying. In addition to the vocal repertoire, we need to understand the behavioral context in which particular acoustic signals are being emitted. Given the difficulty of conducting experiments on odontocetes in the wild, when our intent is to understand their vocal behavior, we usually reach the level of correlation between sounds recorded and behavior displayed. While not the perfect experimental design, at least it allows us to advance in knowledge on these animals.

Alternatively, there have been studies in captivity that allow more controlled experimentation in odontocetes. However, the caveat is that the behavior observed in an aquarium is not always representative of what happens in nature. Additionally, animals may already have long-term effects of years in captivity, which may distort their displayed behavior and/or vocalizations even more than what we would expect in the wild. Thus, while we may learn from these experiments, we must be very careful in the assumptions we make regarding behavioral responses in the wild.

What is the application of understanding the vocal behavior of cetaceans? Historically we have been relying on visual data to describe their behavior. And this has its own challenges: to observe the behavior displayed both above the surface and underwater, to gather information for a long period of time, night time data, or even adverse weather conditions. These limitations on the visual data speak for the potential need to use complementary data, and acoustic information is one way to overcome some of those challenges. For example, in 2016 Reyes and colleagues published an article in which they associate certain acoustic signals produced by Commerson's dolphins with their displayed behavior. Should they decide to implement some kind of passive acoustic monitoring study in the future for that particular population of Commerson's dolphins, they will not only collect information of presence/absence over time, but also gather behavioral information thanks to knowing in which context particular sounds are produced. This type of data can be extremely useful when it comes to better understanding habitat use for conservation management, for instance.

In order to understand the vocal behavior, we need to combine acoustic recordings with visual information to study the correlation between particular sounds and the behavior displayed. For this task a passive acoustic recording system is needed, such as those introduced earlier. However, the data analysis is less demanding. Since the repertoire has already been described, most times having them categorized and associated to the behavior displayed by the animals is all we need to conduct the adequate exploratory or statistical analysis, such as multiple correspondence analysis (Hawkins and Gartside 2009; Reyes Reyes et al. 2016a).

6.4 Assessing Impact

Temporary or chronic exposure to noise can have an impact in cetaceans. That impact can range from changes in the behavior displayed to permanent hearing loss and potentially even death (e.g. Cox et al. 2006; Clark et al. 2009; Mooney et al. 2009; Mann et al. 2010). Passive acoustics can also be used to assess impact of noise on cetaceans. For this, two main approaches can be taken: experimental or opportunistic. The experimental consists of presenting a stimulus and measuring the response of cetaceans, such as has been done within the multidisciplinary project called SOCAL-BRS, in which researchers from several institutions try to assess impact of military sonar on different cetacean species in the South of California. Alternatively, one can choose an opportunistic approach, in which we measure the response of individuals using already existing alterations in the ambient noise levels. This last method is less invasive but also requires more time of acoustic data to make reliable comparisons.

One way of measuring impact consists of using a Bayesian approach (Melcón et al. 2012b). In this way, we can calculate the probability of a particular odontocete vocalizing given a particular anthropogenic noise and compare the probability of the same species vocalizing given no anthropogenic noise. In that particular paper, the authors studied blue whale foraging calls given sonar or no sonar. They found that with sonar animals are more silent than without sonar, probably implying that when blue whales hear sonar they stop foraging.

A similar approach can be taken to determine at which received level of sonar, for example, the probability of blue whales calling decreases to half. For this, we plot the probability of an animal vocalizing as a function of the received level of the noise, and we fit the data to a logit function (see Fig. 6.3).

Besides the change in behavior, ambient and anthropogenic noise can cause masking. This phenomenon can occur when there is an overlap of frequencies between the animal's acoustic signal and the noise, such that the noise "buries" partially the call of interest. To measure the effect of masking, one type of analysis that can be conducted consists of calculating the acoustic space reduction (Hermannsen et al. 2014; Clark et al. 2009). In other words, and based on the minimum signal-to-noise ratio required by the auditory system of the animal to evaluate the signal of interest, how much does the effective area get reduced with a certain level of background noise. This means that two individuals will have to be closer for communication to be effective with considerable background noise, or echolocation will be less effective at long distances.

While acoustics cannot address the whole spectrum of potential impacts that noise can have on marine mammals, there are certain measures that can be assessed using passive acoustics and can already tell us something about the effects of human activity in the water on marine life. Thus, while not overarching, we recommend the use of this method to learn more about the impact of noise pollution on cetaceans in Latin America, and in this way make the corresponding recommendations to advocate in an informed manner for the conservation of odontocetes in our regions.

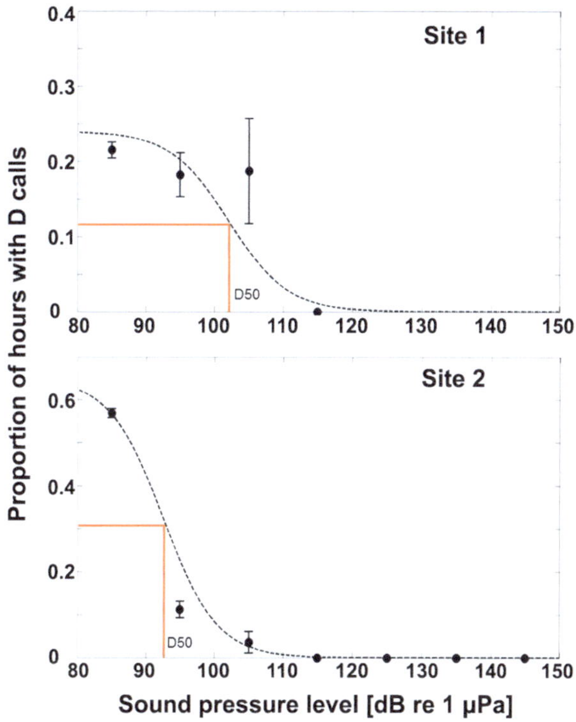

Fig. 6.3 Dose-response curve calculated for two different sites of California (figure not published). Represented are the probability of blue whale foraging calls (D calls) as a function of the received level of sonar at the whale's location. The louder the sonar, the lower the probability of blue whales emitting these sounds. Depicted in *orange* we can see the sound pressure level at which the probability of blue whales calling decreases to half in comparison to ambient noise only (D50), namely 103 dB for Site 1 and 92 dB for Site 2

6.5 Long-Term Studies

Long-term studies are hard to execute due to scarce resources, logistics, and sometimes additional difficulties to carry out a project continuously over time. However, we would like to stress out the importance of these studies: they can help us establish baseline knowledge on cetaceans in a particular area, understand habitat use, seasonal patterns, and others. By having a good baseline it allows us to detect changes over time, either caused by environmental alterations (gradual or abrupt), or by human activities. Conservation efforts often require these types of studies to detect trends in abundance, distribution, as well as changes in behavior. Other kinds of impact from cumulative and aggregate threats on a population or a species are also possible by conducting long-term studies, but complementary techniques may be needed, such as heavy metal analyses, histological studies from stranded animals, etc. In this section, we will go through a successful case study that we are currently developing in collaboration with other researchers and institutions.

Case Study: Antarctic Project

In 2009, a group of researchers from several countries conformed the Southern Ocean Researh Partnership (SORP), a collaborative consortium to develop and implement non-lethal methods for cetacean research in the Southern Ocean, which aims to maximize conservation efforts for cetaceans populations around Antarctica after many of them were almost depleted by commercial whaling from the end of nineteenth century and during the twentieth century. This program was endorsed by the International Whaling Commission becoming the IWC-SORP. Argentina, as one of the member countries of the SORP, has been participating in this program by conducting a yearly ship-based survey on-board an Argentinean Coastguard cutter, around the western Antarctic Peninsula during austral summer since 2014. The aim of this project is to determine year-round relative abundance, distribution, and seasonality of cetaceans and correlate them with oceanographic features and possible impact driven by climate change.

Line-transect surveys are conducted from the ship collecting simultaneous visual and acoustic data on cetacean presence. Acoustic recordings are collected using a linear array of hydrophones which is towed behind the vessel. Each element of the array is equipped with an omni-directional sensor with an approximately flat frequency response from 30 Hz to 200 kHz, and each sensor is connected to a custom-built preamplifier board and bandpass filter. One channel of the array records at 500 kHz sample rate and the others at 192 kHz which means that recordings are obtained up to 250 kHz and 96 kHz, respectively. Acoustic recordings are monitored in real time by an acoustician who is communicated by radio with the marine mammal observers located in the bridge. When sounds are recorded on several channels, the difference in the time of arrival at each sensor can be used to localize vocalizing animals and try to confirm the species producing that sound visually. The good thing of combining visual and acoustic monitoring is that there are some species that are more detectable through one method or the other, and in this way the detection probability overall increases. For example, given that the array is towed behind the vessel, the ship noise masks the low-frequency sounds from baleen whales preventing their detection. However, most toothed whales signals are in a frequency range that can be recorded by the array without being overlapped by ship noise, using the proper equipment. Besides, some deep diving species like sperm whales and beaked whales can be very hard to detect visually in the surface, but they can be easily recognized by their characteristic sounds.

One of the great advantages of using PAM to study cetaceans is that it gives us the possibility of collecting continuous data during day and night, all year around, in such a remote area as it is the Southern Ocean. Thus, in addition to the real time visual and acoustic monitoring, in February 2014, an acoustic bottom-mounted, autonomous recorder called HARP, was deployed near Elephant Island at a depth of approximately 1000 m in order to collect sounds for the whole year. This equipment samples at a rate of 200 kHz, obtaining acoustic signals from 10 Hz to 100 kHz, which covers all baleen whale calls and most of the odontocetes known to inhabit these waters. The HARP was successfully recovered and redeployed during a second SORP cruise in 2015, and again in a third cruise in 2016. Recovery, refurbishment and redeployed of the HARP is expected to be done every year during the shipboard surveys.

The data already collected and analyzed allowed us to described previously unknown echolocation signal types, likely produced by two Antarctic beaked whale species (Trickey et al. 2015), and also killer whale high-frequency whistles which had not been described before for an Antarctic killer whale ecotype (Reyes Reyes et al. 2016b). We were also able to estimate seasonal fin whale acoustic activity around Elephant Island, through the calculation of an acoustic index measured as the signal to noise ratio at 21 Hz (Baumann-Pickering et al. 2015).

This long-term project (planned for ten consecutive years) involves researchers from several countries in Latin America as well as researchers from the Scripps Institution of Oceanography, University of California where all the acoustic equipment was built.

6.6 Final Recommendations

The information shown in the present chapter shows a clear shortage in acoustic information of cetaceans in Argentina. While with time we have succeeded in implementing some systematic research in the country, a lot remains to be done.

Within the challenges observed, we note the need of specialized researchers in the topic. Many have tried to develop research without understanding the basic requirements of an equipment for a particular study, or without knowing how to analyze the data. Therefore, we recommend to either take courses on bioacoustics and/ or to seek for advice from experts to use the limited resources in an efficient way.

Another way of overcoming the problem of not knowing which type of equipment to use is to look for available literature proceeding from other areas in which the target species was studied. That may give us valuable information on potential recording devices as well as experimental design.

We have found that establishing international collaborations can be extremely useful when it comes to developing a new research line. Not only did they already go through the experience of starting up, but may help with equipment, tips and training of local researchers.

Multidisciplinary work usually enriches the research. Interacting with biologists, engineers, physicists, vets and technicians, among others, can help us cover more aspects within the same research project.

Finally, we would like to stress out the importance of collaboration. Particularly in Latin America, where certain resources are scarce, it is crucial to help each other to complement our strengths, seek for advice, and help each other.

Acknowledgements We would like to thank Pablo Bordino, Guillermo Caille, Claudio Campagna, Luis Capozzo, Iris Caceres, Silvana Dans, Cristian de Haro, Natalia Dellabianca, Diego Rodriguez, Mariano Sironi, Peter Thomas, Peter Tyack, Els Vermeulen and Bernd Wursig for their kindness to provide us with information on other groups doing research on bioacoustics, or facilitating contacts to conduct a more complete survey on professionals working on passive bioacoustics underwater.

References

Au WWL, Hastings MC (2008) Principles of marine bioacoustics. Springer, New York

Au WWL, Hastings, MC. (2008) Principles of marine bioacoustics. Springer, New York

Bastida R, Rodriguez D (2003) Mamíferos marinos de Patagonia y Antártida. Vazquez Mazzini, Buenos Aires

Baumann-Pickering S, Sirovic A, Trickey JS, Hildebrand JA, Reyes Reyes MV, Melcón ML, Iñíguez MA (2015) Cetacean presence near Elephant Island, Antarctica, based on passive acoustic monitoring. Paper SC/66a/SH18 presented to the IWC Scientific Committee, June 2015

Clark CW (1982) The acoustic repertoire of the southern right whale, a quantitative analysis. Anim Behav 30:1060–1071

Clark CW, Ellison WT, Southall BL, Hatch L, Van Parijs SM, Frankel A, Ponirakis D (2009) Acoustic masking in marine ecosystems: intuitions, analysis, and implication. Mar Ecol Prog Ser 395:201–222

Cox TM, Ragen TJ, Read AJ, Vos E, Baird RW, Balcomb K et al (2006) Understanding the impacts of anthropogenic sound on beaked whales. J Cetacean Res Manag 7(3):177–187

Cummings WC, Fish JF, Thompson PO (1972) Sound production and other behavior of southern right whales (*Eubalaena glacialis*). Trans San Diego Soc Nat Hist 17(1):1–13

Hawkins ER, Gartside DF (2009) Patterns of whistles emitted by wild Indo-Pacific bottlenose dolphins (*Tursiops aduncus*) during a provisioning program. Aquat Mamm 35:171–186

Hermannsen L, Beedholm K, Tougaard J, Madsen PT (2014) High frequency components of ship noise in shallow water with a discussion of implications for harbor porpoises (*Phocoena phocoena*). J Acoust Soc Am 136(4):1640–1653

Iñíguez MA (1993) Orcas de la Patagonia Argentina. Ed. Propulsora Literaria, Buenos Aires, 88 pp

Johnson M, Aguilar de Soto N, Madsen P (2009) Studying the behaviour and sensory ecology of marine mammals using acoustic recording tags: a review. Mar Ecol Prog Ser 395:55–73

Kyhn LA, Tougaard J, Jensen F, Wahlberg M, Stone G, Yoshinaga A, Beedholm K, Madsen PT (2009) Feeding at a high pitch: source parameters of narrow band, high-frequency clicks from echolocating off-shore hourglass dolphins and coastal Hector's dolphins. J Acoust Soc Am 125(3):1783–1791

Kyhn LA, Jensen FH, Beedholm K, Tougaard J, Hansen M, Madsen PT (2010) Echolocation in sympatric Peale's dolphins (*Lagenorhynchus australis*) and Commerson's dolphins (*Cephalorhynchus commersonii*) producing narrow-band high-frequency clicks. J Exp Biol 213(11):1940–1949

MacLeod CD, Perrin WF, Pitman R, Barlow J, Balance L, D'amico A, Gerrodette T, Joyce G, Mullin KD, Palka DL, Waring GT (2006) Known and inferred distributions of beaked whale species (Cetacea: Ziphiidae). J Cetacean Res Manag 7:271–286

Madsen PT, Wahlberg M (2007) Recording and quantification of ultrasonic echolocation clicks from free ranging toothed whales. Deep-Sea Res I 54(8):1421–1444. doi:10.1016/j.dsr.2007.04.020

Mann D, Hill-Cook M, Manire C, Greenhow D, Montie E, Powell J et al (2010) Hearing loss in stranded odontocete dolphins and whales. PLoS One 5(11):e13824. doi:10.1371/journal.pone.0013824

Marino A, Reyes Reyes MV, Thayre BJ, Melcón ML. Acoustic recordings of NBHF clicks from free-ranging Peale's dolphins (*Lagenorhynchus australis*) in the Patagonian shelf. In preparation

Melcón M, Failla M, Iñíguez M (2012a) Echocolocation behavior of Franciscana dolphins (*Pontoporia blainvillei*) in the wild. J Acoust Soc Am 131(6):EL448–EL453

Melcón ML, Cummins AJ, Kerosky SM, Roche LK, Wiggins SM, Hildebrand JA (2012b) Blue whales respond to anthropogenic noise. PLoS One 7(2):e32681

Melcón ML, Failla M, Iñíguez MA (2016) Towards understanding the ontogeny of echolocation in franciscana dolphins (*Pontoporia blainvillei*). Mar Mam Sci. doi:10.1111/mms.12336

Mooney TA, Nachtigall PE, Vlachos S (2009) Sonar-induced temporary hearing loss in dolphins. Biol Lett. doi:10.1098/rsbl.2009.0099

Payne RS, Payne K (1971) Underwater sounds of southern right whales. Zoologica 56:159–165

Reyes Reyes MV, Iñíguez MA, Hevia M, Hildebrand JA, Melcón ML (2015) Description and clustering of echolocation signals of Commerson's dolphins (*Cephalorhynchus commersonii*) in Bahía San Julián, Argentina. J Acoust Soc Am 138(4):2046–2053

Reyes Reyes MV, Tossenberger VP, Iñiguez MA, Hildebrand JA, Melcón ML (2016a) Communication sounds of Commerson's dolphins (*Cephalorhynchus commersonii*) and contextual use of vocalizations. Mar Mam Sci. doi:10.1111/mms.12321

Reyes Reyes MV, Baumann-Pickering S, Simonis AE, Trickey JS, Hildebrand JA, Melcón ML, Iñíguez MA (2016b). High-frequency modulated whistles of killer whales (*Orcinus orca*) in Antarctica. Abstract presented at the 1st LAMLA Worskhop, Brazil, July 2016

Trickey JS, Baumann-Pickering S, Hildebrand JA, Reyes Reyes MV, Melcón M, Iñíguez M (2015) Antarctic beaked whale echolocation signals near South Scotia Ridge. Mar Mamm Sci 31(3):1265–1274

Tyack P (1976). Patterns of vocalization in wild *Tursiops truncatus*. Senior thesis submitted to Harvard College, p 72

Tyack P, Clark C (2000) Communication and acoustic behavior of dolphins and whales. In: Au WW, Fay RR (eds) Hearing by whales and dolphins. Springer, New York, pp 156–224

Vaughn-Hirshorn RL, Hodge KB, Würsig B, Sappenfield RH, Lammers MO, Dudzinski KM (2012) Characterizing dusky dolphin sounds from Argentina and New Zealand. J Acoust Soc Am 132(1):498–506

Chapter 7
Novel Research Techniques Provide New Insights to the Sea Turtle Life Cycle

Maria Ângela Marcovaldi, Alexsandro S. Santos, Paulo H. Lara, and Milagros López-Mendilaharsu

Abstract The recent increase in the knowledge of the complex life cycle of sea turtles in the world has been possible by the use of novel technologies, such as satellite telemetry and molecular analyses. These technological advances have also provided the opportunity to assess future impacts of climate change. In this chapter, you will find a brief description of these techniques and an overview of the most relevant results obtained through scientific research of sea turtles in Brazil.

7.1 Introduction

Sea turtles have complex life histories that span many decades, habitats, and kilometers. Their *highly migratory behavior* makes them a *shared resource among many nations*. Similar to other species of marine mega vertebrates, sea turtles are an important conservation concern as the result of a range of past and ongoing threats. Direct exploitation (Marcovaldi and Marcovaldi 1999, Frazier 2003), incidental capture in fisheries (Domingo et al. 2006, Sales et al. 2008), habitat degradation (Santos et al. 2005, Tourinho et al. 2010) and fluctuating environmental conditions due to climate change (Hawkes et al. 2009) are among the main threats to sea turtle populations. Therefore, for an effective mitigation of direct threats, there is an acute need to understand how these animals use dynamic seascapes (Godley et al. 2008).

Among the seven species of sea turtles found around the world, five of them occur along the Brazilian coast (Marcovaldi and Marcovaldi 1999). All five species are threatened and included in the Brazilian Red List of Threatened Species (ICMBIO 2011), and in Appendix I in the CITES convention (CITES 2016). These species include the loggerhead sea turtle (*Caretta caretta*; Fig. 7.1a), hawksbill sea turtle (*Eretmochelys imbricata*; Fig. 7.1b), olive ridley sea turtle (*Lepidochelys olivacea*; Fig. 7.2a), leatherback sea turtle (*Dermochelys coriacea*; Fig. 7.2b) and

M.Â. Marcovaldi (✉) • A.S. Santos • P.H. Lara • M. López-Mendilaharsu
Fundação Pró Tamar, Rua Rubens Guelli, 134 sala 307, Salvador, Bahia, Brazil
e-mail: neca@tamar.org.br; alex@tamar.org.br; paulo.lara@tamar.org.br; milagros@tamar.org.br

M.R. Rossi-Santos, C.W. Finkl (eds.), *Advances in Marine Vertebrate Research in Latin America*, Coastal Research Library 22,
DOI 10.1007/978-3-319-56985-7_7

Fig. 7.1 Loggerhead sea turtle (**a**) and hawksbill sea turtle (**b**)

green sea turtle (*Chelonia mydas*; Fig. 7.2c). In 1980, the Sea Turtle Conservation Project—today known as TAMAR—was created to identify the species occurring in Brazil, their main breeding sites and threats to their survival (Marcovaldi et al. 2005). Prior to the creation of TAMAR, sea turtle populations in Brazil were heavily exploited through incidental take in fisheries, illegal poaching and hunting of turtle eggs and nesting females, resulting in a significant decline in turtle numbers. Exploitation for consumption is now regulated, but new threats have emerged and even increased, such as coastal development, intensified fishing, pollution, diseases and climate change (ICMBIO 2011). Hence, TAMAR adopted as its *modus operandi* an adaptive threats management framework to organize and orient priority research themes (da Silva et al. 2016).

Recent technological advances in genetics, telemetry, biochemical markers and stable isotope analysis are increasing our knowledge of the distribution and behavior of sea turtles. In the present chapter, we intend to review the most relevant information gained through research studies on (1) satellite telemetry, (2) conservation genetics and (3) climate change impacts.

Fig. 7.2 Olive ridley sea turtle (**a**) leatherback sea turtle (**b**) and green sea turtle (**c**)

7.2 Satellite Telemetry

Understanding the spatial behavior of marine turtles during their migrations and prolonged movement phases is central to elucidating key aspects of their complex life cycle and ecology. The development of satellite telemetry in the late 1980s allowed the first in-depth studies of animal movement patterns (Duron-Dufrenne 1987, Fancy et al. 1988). Since then, continuing improvements to the satellite network, transmitter miniaturization, and new deployment techniques have expanded the range of species that can be studied with satellite-tracking (Coyne and Godley 2005). In this way, satellite-tracking studies have become particularly prevalent in recent decades and has led to substantial progress in the field, as this technique allows the reconstruction of the actual routes followed by marked individuals (Godley et al. 2008). Adult females, for example, can migrate thousands of kilometers between breeding sites and feeding grounds, thus satellite telemetry provide the opportunity to map migratory routes and identify foraging and high-use areas.

7.2.1 Migratory Destinations and Foraging Areas of Adult Females

Initial studies using satellite telemetry in Brazil began in 2000–2001. Satellite transmitters were deployed on eight loggerhead-nesting females in Espírito Santo, and eight green turtles from foraging areas in Ceará, including juvenile and adult turtles. In the case of the loggerheads, it was not possible to determine any general pattern after they left the nesting beaches, as most of the transmissions were relatively short in duration (Lemke et al. 2003). On the other hand, green turtles remained in coastal areas; some exhibited high fidelity to their foraging areas while others displayed medium and long distance movements (Godley et al. 2003).

Between 2005 and 2006, 15 satellite transmitters (Fig. 7.3) were fitted to hawksbill sea turtles (*E. imbricata*) and 10 to loggerhead sea turtles (*C. caretta*) along the northern coast of the State of Bahia. After the inter-nesting period, all females migrated to distant feeding grounds up to 2000 km away from the beaches (Marcovaldi et al. 2010, 2012).

Fig. 7.3 Hawksbill sea turtle carrying a satellite transmitter at the nesting beach

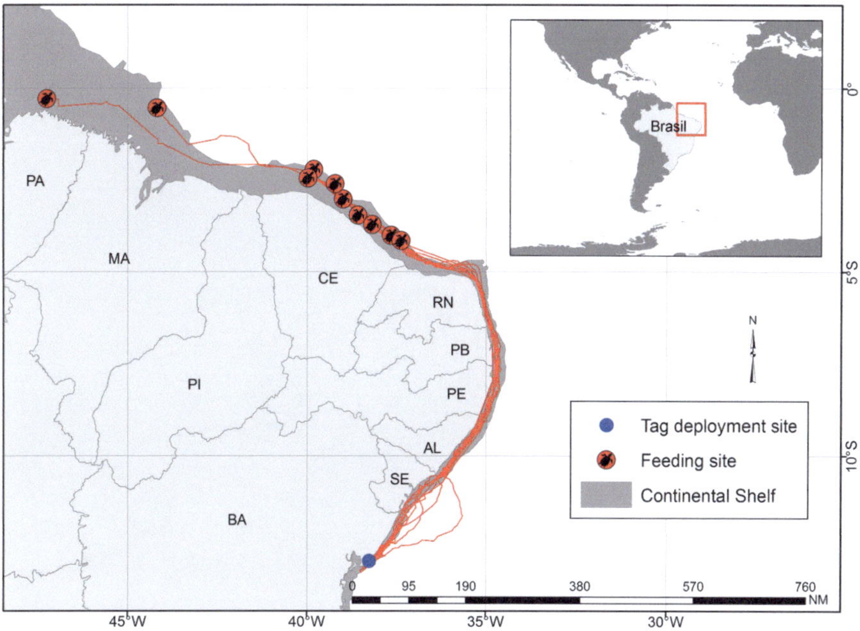

Fig. 7.4 Post-nesting migrations and feeding grounds of 10 loggerhead female turtles satellite-tracked from nesting beaches along the northern coast of Bahia State, Brazil

Study results revealed important aspects of the habitat use and identified a migratory corridor along the northeastern coast of Brazil. Main foraging areas of loggerheads turtles were located along the northern coast of Brazil, especially off the coast of Ceará (Fig. 7.4). Exceptionally long battery life of satellite transmitters allowed tracking nesting females through an entire remigration interval (i.e. round-trip migrations from foraging areas and breeding grounds). After a second post-nesting migration, five loggerhead turtles returned to the same foraging area indicating strong philopatry to specific foraging areas (Fig. 7.5). Concerning hawksbill sea turtles, some individuals migrated northwards and remained in feeding grounds between the States of Rio Grande do Norte and Pará, like the loggerheads, while others moved toward coral reef areas along the coast of Alagoas and southern Bahia (Fig. 7.6) (Marcovaldi et al. 2012).

In the State of Sergipe, 10 olive ridley female turtles (*L. olivacea*) were fitted with satellite transmitters. During the breeding season, the females remained along the coast of Sergipe and their locations overlapped with the area of operation of the shrimp trawl fleet, a major threat to this population. After leaving the nesting area, most of them migrated along the continental shelf towards feeding grounds along the northern, north-eastern and south-eastern coasts of Brazil, while two females migrated to oceanic waters near the equator (Fig. 7.7) (Da Silva et al. 2011).

Finally, four female leatherback sea turtles (*D. coriacea*) were tracked by satellite telemetry. Three turtles were tagged on nesting beaches in Espírito Santo state,

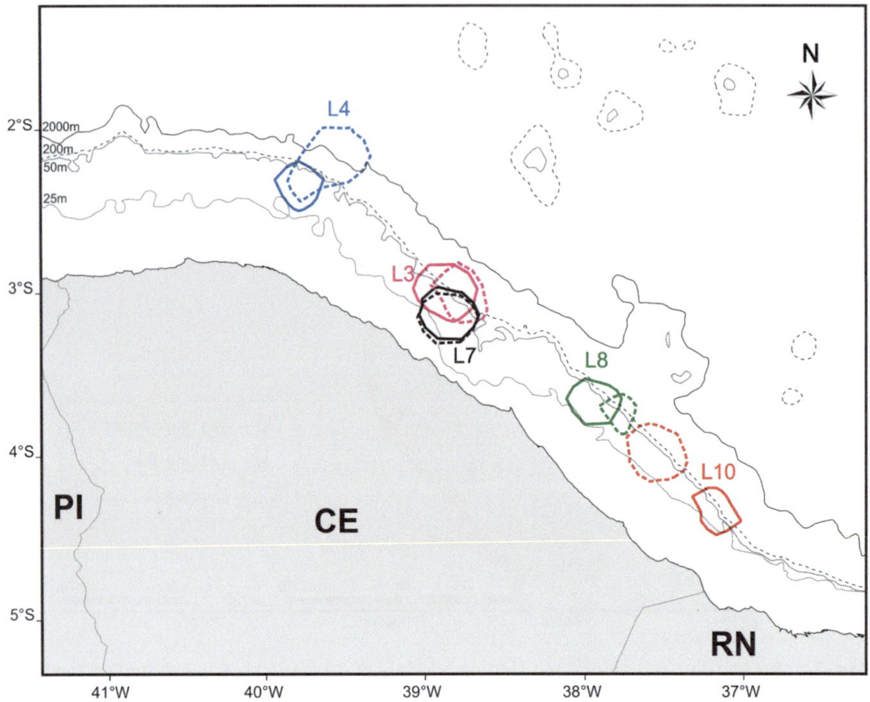

Fig. 7.5 Kernel-estimated home-range utilization 50% distributions of feeding areas of the five tracked loggerhead turtles following their first (*solid perimeter*) and second (*broken perimeter*) post-nesting migration. State abbreviations—*PI* Piauí, *CE* Ceará, *RN* Rio Grande do Norte (Data from Marcovaldi et al. 2010)

and one was incidentally capture in a driftnet off the coast of São Paulo State. Tracking in the breeding area revealed that between successive egg-laying trips, the turtles spread out and move up to 160 km off the coastline. The post-nesting migrations helped to identify foraging grounds between the South of Brazil and the Rio de la Plata estuary (Fig. 7.8) (Almeida et al. 2011). The overlap between the high-use areas of these animals and large-scale fishing zones was highlights as a major threat to these turtles.

7.2.1.1 Conservation Implications

Tracking data has helped us to identify migratory corridors and high-use areas of sea turtles in Brazilian waters. In addition, these findings were essential to better understand the impacts of human activities- such as fishing- across their habitats to improve coastal management. It is recognized that satellite telemetry can be a useful tool in determining the effectiveness of current marine protected areas (MPA) boundaries and where additional protected areas should be considered (Witt et al. 2008,

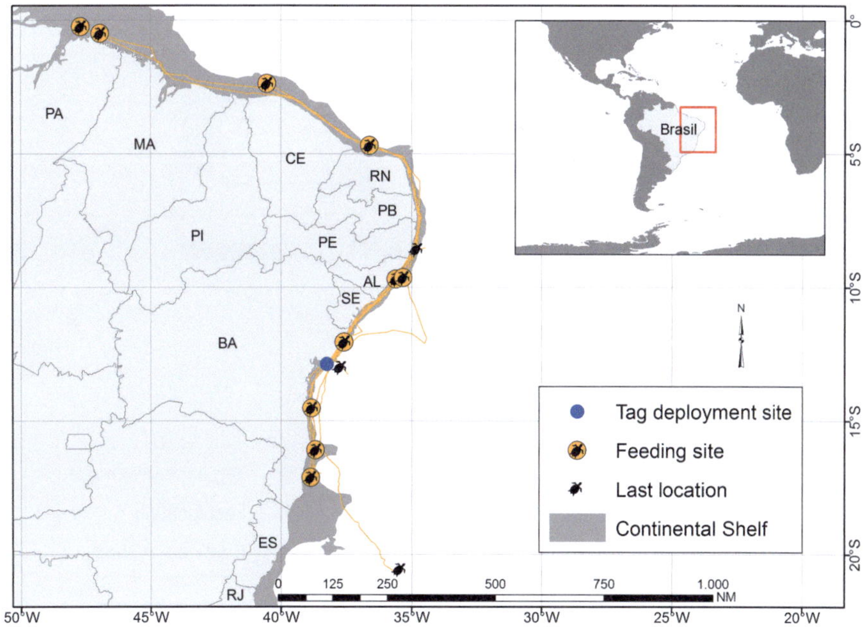

Fig. 7.6 Post-nesting migrations and feeding grounds of 15 hawksbill female turtles satellite-tracked from nesting beaches along the northern coast of Bahia State, Brazil

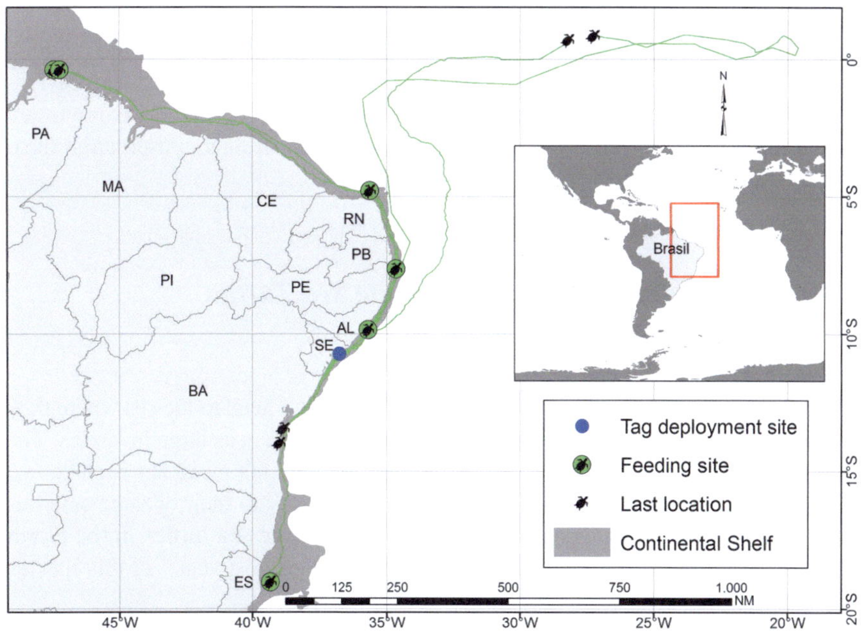

Fig. 7.7 Post-nesting migrations and feeding grounds of 10 olive ridley female turtles satellite-tracked from nesting beaches in Sergipe State, Brazil

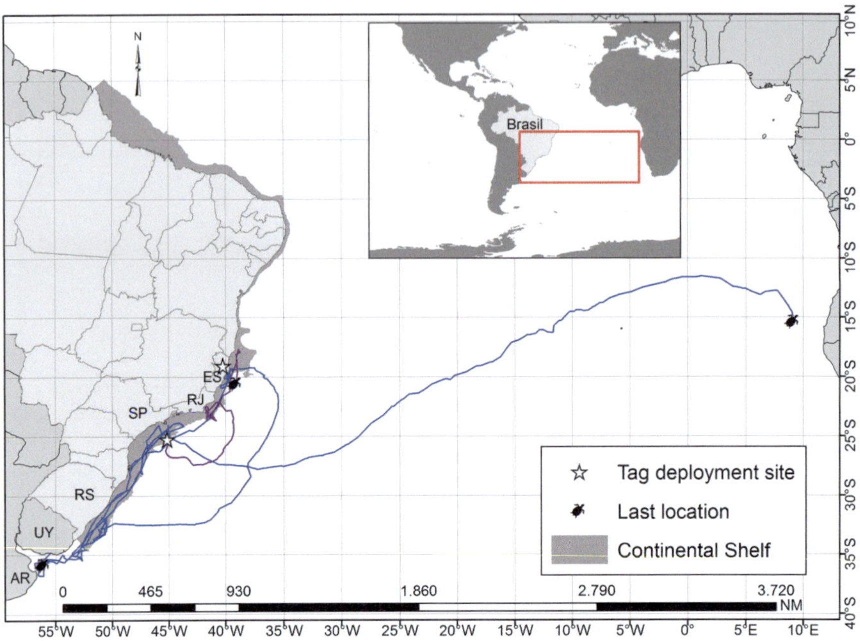

Fig. 7.8 Migration routes and high-use areas of two leatherback female turtles tracked from nesting beaches in Espírito Santo (*blue line*), and a female incidentally captured in a drift-net off the São Paulo coast, Brazil (*purple line*).

Shillinger et al. 2010). A factor of great importance is that the breeding and foraging grounds of most turtles tracked are located within Brazilian territorial waters. This provides a good starting point to ensure the long-term protection of this important portion of sea turtle's life cycle, as many political and national conservation measures are already established.

7.2.2 First Satellite Tracks of Neonate Sea Turtles in the South Atlantic

The "lost years" refers to the time after turtles' hatch and head to the sea where they remain for many years before returning to near-shore waters as large juveniles. The time period is often referred to as the "lost years" because not much has been known about where the young turtles go and how they interact with their oceanic environment. Satellite tracking devices attached to oceanic-stage sea turtles in the North Atlantic have provided new information on the so-called 'lost years' of this species (Mansfield et al. 2014).

Currently, TAMAR in collaboration with NOAA and the University of Central Florida are working to understand the migratory behavior and dispersal patterns of oceanic-stage loggerheads in the South Atlantic Ocean. For this, nineteen neonate

Fig. 7.9 Juvenile loggerhead turtle tracked using a solar-powered satellite tag

loggerhead sea turtles were satellite tagged using modified small-scale solar-powered satellite tags. Turtles were lab-reared until reaching a minimum size (>10 cm) to ensure the tags were <3–5% of the turtles' weight (Fig. 7.9). Then they were released at different times throughout the nesting season to coincide with changes in the current pattern along the coast of Praia do Forte (e.g. southern direction in early/mid season and northern direction in late season). Oceanographic drifters were released along with the turtles to collect concurrent information on current patterns, and to examine the probability that turtles were passively drifting with the prevailing ocean currents. Except for two turtles that transmitted for less than 5 days, the transmission duration of the tags varied from 1 to 4 months (n = 17). Throughout the tracking period the turtles traveled between 500 and 5000 km (e.g. as far as the island of Barbados in the Caribbean). Initially, all turtles followed a general circulation pattern observed with initial drifter trajectories following large-scale current patterns. While some of the turtles traveled along the continental slope (north or south depending on the prevailing current pattern), others also moved to oceanic waters. As a result young turtles in the South Atlantic appear to be influenced by seasonal changes in current regimes—possibly impacting the connectivity of Brazilian turtles in the western Hemisphere (ISTS).

7.3 Conservation Genetics

Maintaining genetic diversity is an important component of conservation. Conservation genetics aims to assess genetic status of populations and propose measures for maintaining genetic diversity to prevent risks that could affect

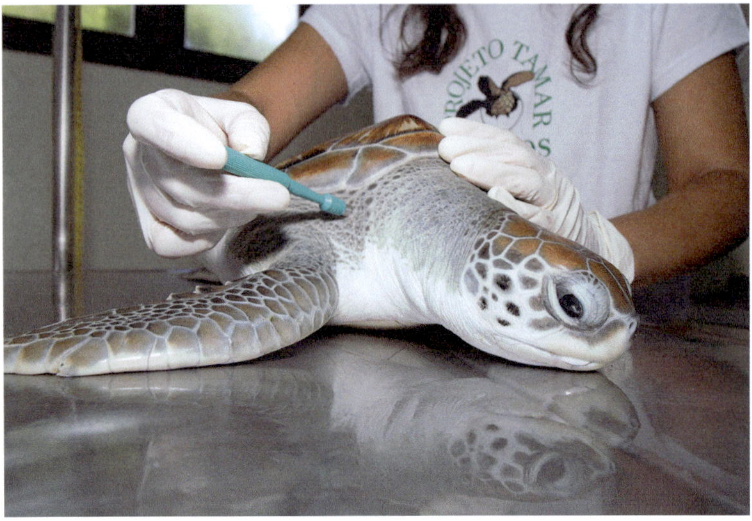

Fig. 7.10 Tissue sample collection from a green turtle for molecular genetic analyses

persistence of populations. However, in conservation, the application of molecular markers can also contribute to the understanding of evolutionary history, demography and ecology of endangered species (Vilaça et al. 2016).

Molecular genetic studies (Fig. 7.10) have improved our knowledge on sea turtle life history, in aspects such as sea turtle phylogeography, gene flow, dispersal, feeding groups, migratory behavior, mating systems, sex ratios of breeding populations, reproduction biology, hybridization, conservation and management (Vilaça et al. 2016).

Because sea turtles occupy broad geographic ranges including nesting and foraging areas utilized by adults, and in some cases geographically distinct ontogenetic habitats (Musik and Limpus 1997), defining the scale of management units (MUs) is challenging (Hamann et al. 2010). Furthermore, sea turtles exhibit complex population structure often influenced by sex-biased gene flow among nesting stocks, and varying degrees of overlap during post-hatchling migrations, in developmental habitats and on adult foraging grounds (Bowen and Karl 2007).

Thus, different molecular analyses can be applied to determine genetic stock structure for different demographic segments of a population (Wallace et al. 2010). Maternally inherited mitochondrial DNA (mtDNA) is useful for resolving nest site fidelity and homing behavior (Bowen et al. 2004). In addition to defining nesting populations, this genetic marker is also useful for resolving maternal origin of both males and females at various life stages and feeding habitats (Bowen and Karl 2007). Nesting females typically demonstrate philopatry to nesting areas. However, males do not restrict mating efforts to their ancestral breeding area, and apparently copulate with females from other regional nesting populations (FitzSimmons et al. 1997, Roberts et al. 2004). Because nuclear DNA (nDNA) reflects contributions of both males and females, analyses of nDNA markers (e.g. microsatellites) can resolve breeding or reproductive stocks that encompass multiple mtDNA-defined nesting stocks (Bowen and Karl 2007, Carreras et al. 2007).

Since the genetic differentiation of rookeries, by use of mtDNA, detailed studies using mixed stock analysis (MSA) - a method borrowed from fisheries research, to determine natal origins of foraging grounds- have aimed to explain the processes that generate the composition of turtles at mixed foraging areas (Jensen et al. 2013).

Genetic tools have also proven useful in answering questions about the reproductive behavior of sea turtles. Since the first study that documented multiple paternity in loggerhead turtles (Harry and Briscoe 1988), researchers have used dozens of microsatellite loci to understand mating systems in all species of sea turtles, and even so, questions remain regarding sperm storage and possible fitness benefits of different mating strategies (Phillips et al. 2013). Furthermore, a combination of mtDNA and microsatellite data has been used to document several cases of hybridization in sea turtles (Vilaça et al. 2012). Molecular genetic studies of sea turtles exist within broader genetic fields that are constantly evolving. New tools and techniques in this field promise to overcome some of the limitations of past studies (Jensen et al. 2013).

In the present section, we review key genetic studies for sea turtles populations in the Southwest Atlantic, including information on genetic composition of rookeries and foraging aggregates. For this purpose, we organized some of the most relevant results by species, as well as hybridization cases.

7.3.1 Loggerhead, Caretta caretta

Reis et al. (2009) assessed the genetic composition of Brazilian loggerhead rookeries and foraging aggregates based on mtDNA control region. The authors analyzed 329 samples from Brazilian rookeries (Rio de Janeiro, Espírito Santo, Bahia and Sergipe) and an oceanic foraging ground (Elevação do Rio Grande—ERG). Four distinct loggerhead haplotypes (380 bp fragment) were observed among the 204 turtles sampled from Brazilian rookeries: CC-A4 (86.3%), CC-A24 (6.4%), CC-A25 (0.5%), and CCxLO (6.8%) (Fig. 7.11). Results showed that endemic haplotypes (CC-A4, CC-A24, CC-A25), found only in Brazilian rookeries, create a unique Brazilian haplotype profile and suggest the existence of two genetic stocks: the northern stock (Sergipe and Bahia rookeries), and the southern stock (Espírito Santo and Rio de Janeiro rookeries). The CCxLO haplotype, only found in Sergipe, was attributed to specimens considered hybrids because they have the typical olive ridley mtDNA haplotype, but the external morphology of loggerheads or a mixture between loggerhead and olive ridley turtles.

When considering longer mtDNA fragments, CC-A4 haplotype was subdivided into three variants (Shamblin et al. 2014). Therefore, based on these results, Shamblin et al. (2014) proposed the recognition of three stocks in Brazil: Northern coast (Sergipe and Bahia), Espírito Santo, and Rio de Janeiro. In light of the subdivision of CC-A4 obtained with the longer control region fragments, an analysis with

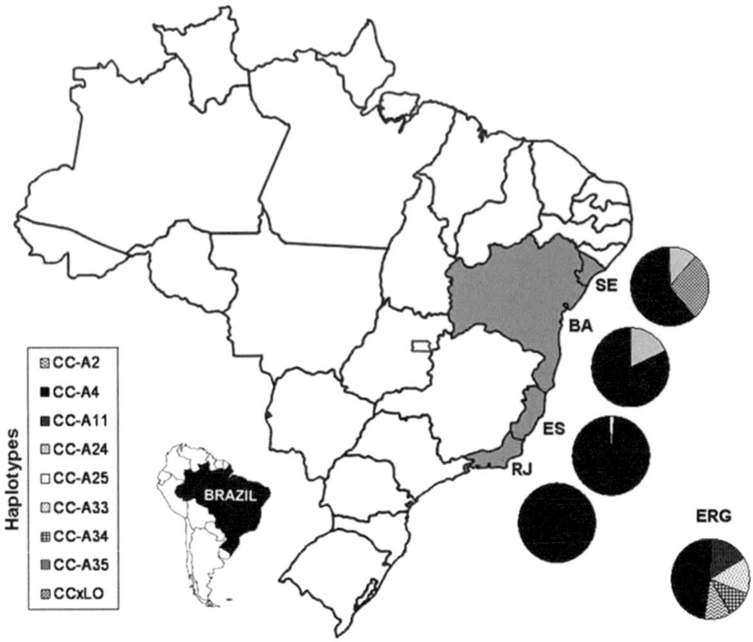

Fig. 7.11 Surveyed locations on the Brazilian coast and loggerhead turtle mtDNA haplotype frequencies for rookeries of Rio de Janeiro (*RJ*), Espírito Santo (*ES*), Bahia (*BA*) and Sergipe (*SE*), and for the foraging aggregation of Elevação do Rio Grande (*ERG*). Colors in the circles represent the different mtDNA haplotypes (Data from Reis et al. 2009)

larger sample sizes from the Brazilian rookeries is necessary to better resolve the number of MUs and their boundaries.

Mixed Stock Analysis (MSA) indicated a major contribution to the ERG foraging aggregation from Brazil (mean 59.5%). However, the ERG aggregation is a mixed stock with haplotypes from worldwide rookeries such as northwestern Atlantic, Mediterranean, Western Australia, Queensland, and New Caledonia (Fig. 7.12). However, the MSA results must be interpreted with care, since this analysis assumes that all source rookeries are known, but African and Indo-Pacific rookeries have been very poorly surveyed (Shamblin et al. 2014).

7.3.2 *Green Turtle,* Chelonia mydas

The genetic structure of nesting populations of green turtles nesting at the three known rookeries in Brazil were assessed: Trindade (n = 99), Rocas Atoll (n = 37), and Fernando de Noronha (n = 16) (Fig. 7.13). This study, based on short mtDNA control-region haplotypes suggested genetic structuring of green turtle populations, with haplotype frequencies significantly different between Trindade and the other

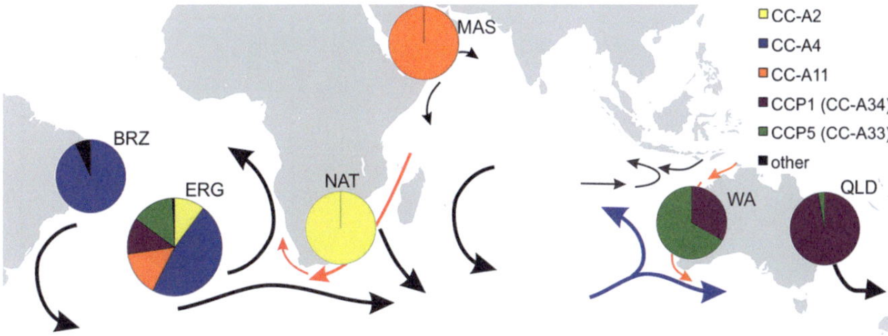

Fig. 7.12 Loggerhead turtle haplotype distribution for an oceanic foraging aggregation and major Indian Ocean rookeries. Control region haplotype (380 base pair) frequencies for the oceanic juvenile foraging aggregation from the Elevação do Rio Grande (*ERG*) and adjacent ridge and slope of the continental shelf in the South Atlantic Ocean and the regional management units (RMUs) in the South Atlantic, Indian, and South Pacific Ocean basins. Colors within the circles represent the different loggerhead haplotypes. *BRZ* is combined Brazilian rookeries; *NAT* is Natal, South Africa (Shamblin et al. 2014); *MAS* is Masirah Island, Oman (Shamblin et al. 2014); WA is Western Australia (FitzSimmons et al. 1996), *QLD* is Queensland, Australia rookeries (Boyle et al. 2009). The *arrows* indicate directionality of major surface currents: *orange* represents the Leeuwin Current; *blue* indicates the Western Australia Current, and *red*, the Agulhas Current (Data compiled by Shamblin et al. 2014)

two rookeries, and no significant difference between Rocas Atoll and Fernando de Noronha (Bjorndal et al. 2006).

Later, the stock structure of the Brazilian rookeries was reexamined using the mitochondrial short tandem repeat (mtSTR) marker and expanded control-region sequences. Results revealed significant structure between Rocas Atoll and Fernando de Noronha separated by only 150 km, indicating that these rookeries are discrete populations with respect to female natal homing (Shamblin et al. 2015).

Although feeding grounds (FGs) in Brazil were differentiated among them, intra-regional structure was less pronounced. Ascension was the primary natal source of the study FGs, with Surinam and Trindade as secondary sources (Naro-Maciel et al. 2012).

7.3.3 *Hawksbill,* Eretmochelys imbricata

Hawksbill turtles from the two main rookeries in Brazil (Bahia and Pipa) are inserted within the Southwest Atlantic Regional Management Unit (RMU) (Wallace et al. 2010). The two rookeries are related, even though they belong to different genetic stocks. By using sequences of the mtDNA control region and autosomal genes Vilaça et al. (2013) revealed striking differences between the two Brazilian nesting sites studied.

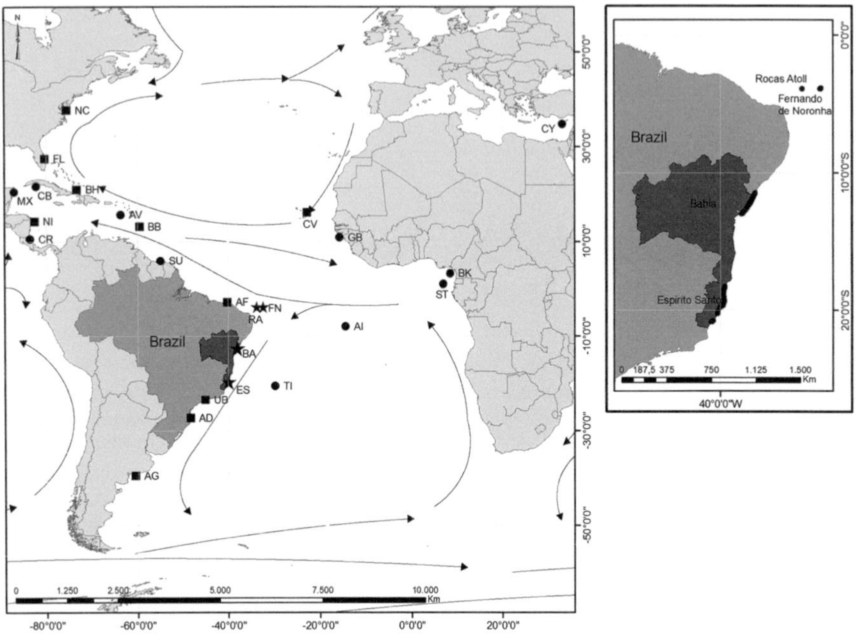

Fig. 7.13 Map of the Rocas Atoll (*RA*), Fernando de Noronha (*FN*), Bahia (*BA*), and Espirito Santo (*ES*) study sites (*symbolized by stars*) with respect to general oceanic circulation patterns shown as arrows, and other *C. mydas* groups previously subject to genetic analysis. Abbreviations for other FGs (*symbolized by squares*) are as follows: Almofala (*AF*); Argentina (*AG*); Arvoredo (*AD*); Bahamas (*BH*); Barbados (*BB*); Cape Verde (*CV*); Florida (*FL*); Nicaragua (*NI*); North Carolina (*NC*); Rocas Atoll (*RA*); and Ubatuba (*UB*). Rookeries that were assessed (*symbolized by circles*) were: Hutchinson Island, Florida, USA (*FL*); Quintana Roo, Mexico (*MX*); Aves Island, Venezuela (*AV*); Matapica, Surinam (*SU*); Lara Bay, Cyprus (*CY*); Cuba (*CB*); Tortuguero, Costa Rica (*CR*); Ascension Island, UK (*AI*); Poilão, Guinea Bissau (*GB*); Bioko Island, Equatorial Guinea (*BK*); Sao Tome (*ST*); Trindade Island, Brazil (*TI*); and Rocas Atoll, Brazil (*RA*). The inset map shows the sample collection sites (Adapted from Naro-Maciel et al. 2012)

Differences among Brazilian feeding aggregation sites and others around the world were also found (Vilaça et al. 2013, Proietti et al. 2014). Results from genetic analysis estimates showed that the studied Brazilian feeding aggregations are mostly composed of animals originating from the domestic rookeries Bahia and Pipa, but some contributions from African and Caribbean rookeries were also observed (Fig. 7.14).

7.3.4 *Leatherback,* Dermochelys coriacea

Comparing haplotypes distributions among nesting and foraging grounds, Vargas et al. (2008) revealed a major contribution to a Brazilian pelagic aggregate (Fig. 7.15) from Atlantic nesting populations (mean 96.1%) and the existence of six individuals

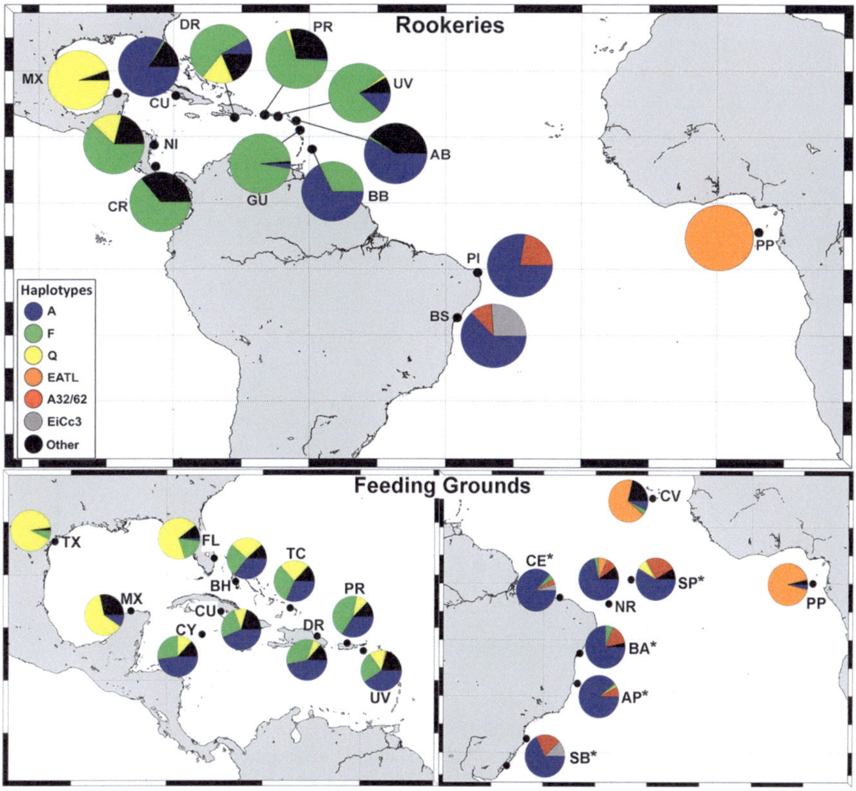

Fig. 7.14 Main haplotype frequencies of genetically described populations in the Atlantic. *study areas described in this work. Rookeries: *BS* Bahia/Sergipe, *PI* Pipa, *PP* Principe, *BB* Barbados, *GU* Guadeloupe, *AB* Antigua & Barbuda, *UV* U.S. Virgin Islands, *PR* Puerto Rico, *DR* Dominican Republic, *CR* Costa Rica, *NI* Nicaragua, *CU* Cuba, *MX* Mexico. Feeding grounds: *SP* São Pedro and São Paulo, *NR* Noronha/Rocas, CE Ceará coast, *BA* Bahia coast, *AP* Abrolhos Park, *SB* South Brazil, *PP* Principe, *CV* Cape Verde, *UV* U.S. Virgin Islands, *PR* Puerto Rico, *DR* Dominican Republic, *TC* Turks & Caicos, *BH* Bahamas, *FL* Florida, *CY* Cayman Islands, *MX* Mexico, *TX* Texas (Adapted from Proietti et al. 2014)

bearing an orphaned haplotype (Dc_A2) for leatherbacks. This haplotype was considered an orphaned haplotype because no match with nesting population were found at that moment. In a later comprehensive genetic analysis of worldwide stock structure, it was ascertained that the origins of the orphaned haplotype came from Gabon and/ or Western Africa (Dutton et al. 2013).

Later, Vargas et al. (2013) reanalyzed samples from a previous study and estimated the contribution of source rookeries to a Brazilian feeding ground, using population genetic measures and mixed stock analysis MSA, including two previously genetic undescribed rookeries: Gabon and Guadalupe/Martinique. Results of the MSA indicated major contribution to the Brazilian pelagic aggregate from Gabon (78%), the largest population of leatherbacks in the world.

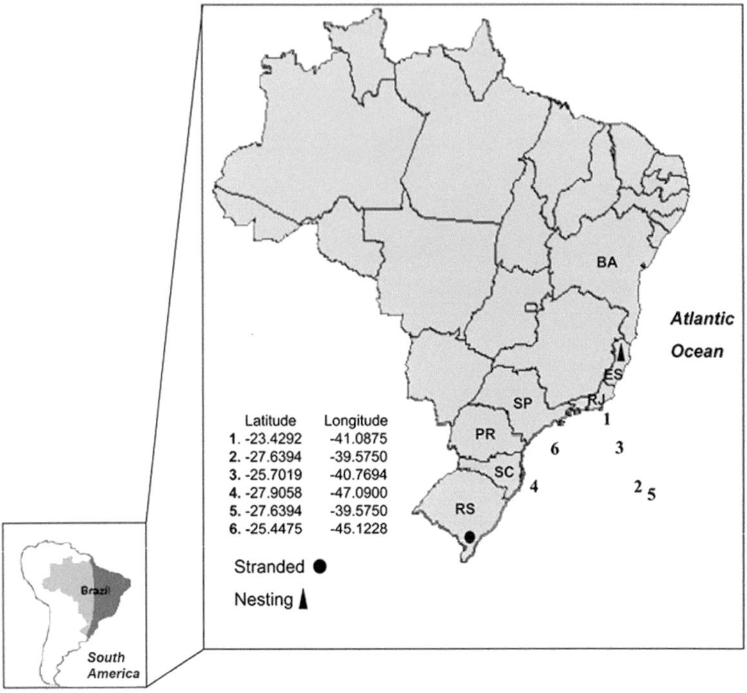

Fig. 7.15 Approximate location of sampling sites along the Brazilian coast (*RS*—Rio Grande do Sul State, *SC*—Santa Catarina State, *PR*—Paraná State, *SP*—São Paulo State, *RJ*—Rio de Janeiro State, *ES*—Espírito Santo State, and *BA*—Bahia State) and geographic coordinates for six sampling sites of pelagic (stranded/incidentally caught) animals (Adapted from Vargas et al. 2008)

7.3.5 *Olive Ridley,* Lepidochelys olivacea

The olive ridley sea turtle is widespread on tropical and subtropical oceans and is considered the most abundant sea turtle. In the Atlantic Ocean, this species has limited distribution and nesting has been reported mainly in Guinea Bissau, Suriname, French Guiana and northeastern Brazil. Hahn et al. (2012) characterized the genetic diversity and population structure of the olive ridley nesting populations in the Brazilian coast based on 92 samples sequenced for the mtDNA control region and 67 samples genotyped for 15 microsatellite loci. Although three mtDNA haplotypes were found, two previously unknown but very rare, the Brazilian nesting population presented one of the lowest mtDNA diversity known for the species. mtDNA data indicated a population expansion following a population decline in the past while microsatellite data suggested a scenario of demographic stability, supporting the scenario of colonization of Atlantic Ocean via a founder effect. Since this data could not find significant genetic differences between the studied olive ridleys nesting areas in the Brazilian coast, conservation strategies should consider the Brazilian olive ridleys as a single interbreeding population. (Hahn et al. 2012).

7.3.6 Hybrids

Vilaça et al. (2012) analysed nuclear markers to investigate the pattern of hybridization involving three species of sea turtles: hawksbill (*E. imbricata*), loggerhead (*C. caretta*) and olive ridley (*L. olivacea*). Results indicated that most of the individuals in the crossings olive ridley x hawksbill and olive ridley x loggerhead are F1 hybrids, whereas loggerhead x hawksbill crossings present F1 and backcrosses with both parental species. It is a remarkable find, because most of the nesting sites surveyed worldwide, including some in northern Brazil, presents no hybrids, and rare Caribbean sites present no more than 2% of hybrids. Thus, a detailed understanding of the hybridization process is needed to evaluate natural or anthropogenic causes of this regional phenomenon, which could be an important factor affecting the conservation of the Brazilian population. Of special interest is the high occurrence of loggerhead x hawksbill hybrids (42%), which display loggerhead mtDNA haplotypes but are characterized morphologically as hawksbills. The occurrence of several nesting individuals with identical mtDNA from another species may also suggest a long history of introgression between species producing likely F2 or further generation hybrids (Vilaça et al. 2012).

7.4 Climate Change Impacts

Sea turtles will be affected by changes to multiple climatic processes (e.g., increased air and sea temperature, sea level rise, precipitation, and storm activity) at all life stages and at different temporal and geographical scales (Fuentes et al. 2011, Hawkes et al. 2009). In addition, sea turtles will be affected indirectly by increased atmospheric CO_2 concentrations and declines in ocean pH. Arguably, the more detectable impacts of climate change to marine turtles will occur during their terrestrial reproductive phase (egg laying, egg incubation, and hatchling success) because there are clear, and relatively straightforward, effects of increased temperature, sea level rise, and cyclonic activity on nesting sites and reproductive output (Fuentes and Abbs 2010, Fuentes et al. 2009).

Successful incubation of sea turtle eggs occurs within a narrow thermal range of 25–33 °C. Incubation above the thermal threshold will result in hatchlings with higher morphological abnormalities and lower hatching success. Additionally, all species of sea turtle have temperature-dependent sex determination (TSD) (Mrosovsky and Yntema 1980). Warmer temperatures, above the pivotal temperature—where a 1:1 sex ratio is produced—yield more females while temperatures below the pivotal temperature shift the ratio towards more males. Consequently, projected increases in temperature are expected to result in highly female-biased populations, reduced hatching success and alteration of incubation duration (Fuentes et al. 2011). Potential consequences include a reduction in effective population size, the inability to find mates leading to reduced fecundity or female infertility, and

Fig. 7.16 Biologist collecting information on sand temperature in the field with a temperature datalogger

under more extreme climate projections, the potential for production of single sex cohorts. These changes are likely to reduce the viability of turtle populations (Fuentes et al. 2012). Consequently, knowledge of the primary sex ratio of populations with TSD is key for providing a baseline to inform management strategies and accurately predict how future climate changes may affect turtle populations. However, there is a lack of robust data on offspring sex ratio at appropriate temporal and spatial scales to inform management decisions. Today, with the advance of technology, electronic devices for temperature measurement (i.e. dataloggers; Fig. 7.16) allow monitoring temperature variations of beaches and nests. The use of such equipment in research studies makes possible the understanding of the potential impacts of climate change on populations of sea turtles and their habitats in Brazil.

Studies developed in the 1990s determined the pivotal temperature—the one that produces the sex ratio 1: 1—for loggerhead (Marcovaldi et al. 1997) and hawksbill for Brazilian populations (Godfrey et al. 1999). The conversion of this temperature to the pivotal incubation duration, which is the amount of days that the eggs incubated at constant pivotal temperature take to hatch, provide a method to estimate hatchling sex ratios (Marcovaldi et al. 1997), relatively easy to perform during fieldwork, creating the first baseline for these studies in Brazil.

Recent studies using long temporal coverages allowed us to estimate natural sex ratios to provide a baseline in advance of climate change and to assess management practices to inform future management decisions. Therefore, one of our first goals was to determine the historical variability of incubation duration and subsequent sex ratio produced at loggerhead and hawksbill nesting sites (Marcovaldi et al. 2014, 2016).

7.4.1 Hawksbill Nesting Grounds

We estimated the primary sex ratio of hawksbill hatchlings from incubation duration of 5514 in situ nests from 10 nesting beaches from two regions in Brazil over the last 27 years. A strong female bias was estimated in all beaches, with 96% and 89% average female sex ratios produced in Bahia (BA) and Rio Grande do Norte (RN). Both inter-annual (BA, 88–99%; RN, 75–96% female) and inter-beach (BA, 92–97%; RN, 81–92% female) variability in mean offspring sex ratio was observed (Figs. 7.17 and 7.18).

7.4.2 Loggerhead Nesting Grounds

The primary sex ratio of loggerhead hatchlings was estimated from incubation duration of 27,697 in situ nests from 21 nesting beaches used by two subpopulations of loggerhead turtles in Brazil over the last 25 years. A strong female bias (94%) was observed in all the areas used by the northern loggerhead stock, Sergipe (SE) and Bahia (BA), whereas a more balanced sex ratio (53% female) was estimated at the regions used by the southern loggerhead stock, Espírito Santo (ES) and Rio de Janeiro (RJ). Both inter-annual (SE: 83–99%, BA: 79.–98%, ES: 33–81%, RJ: 18–81%) and inter-beach variabilities (SE: 91–98%, BA: 89–96%, ES: 47–69%, RJ: 28–55%) in mean female offspring were observed (Fig. 7.19).

7.4.3 Conclusions

These findings provide evidence of bias in the production of female offspring in Brazil and significantly identify the beaches and months where males are produced. Identify and enhance the protection of male-producing beaches has been highlighted as high priority to mitigate potential impacts from projected increases in temperature on marine turtles (Hawkes et al. 2007, Fuentes et al. 2012, Marcovaldi et al. 2014).

This outcome has prompted us to explore the thermal profile of nesting areas used by loggerheads and hawksbills. For this, we deployed 60 data-loggers in 2013 along the whole extent of nesting areas and environments used by these species. Thermal profile will be monitored over the course of several years and will be used to estimate the current sex ratio being produced at each location. This data will be coupled with climate models to predict future sex ratio and thus help prioritize management in light of global warming.

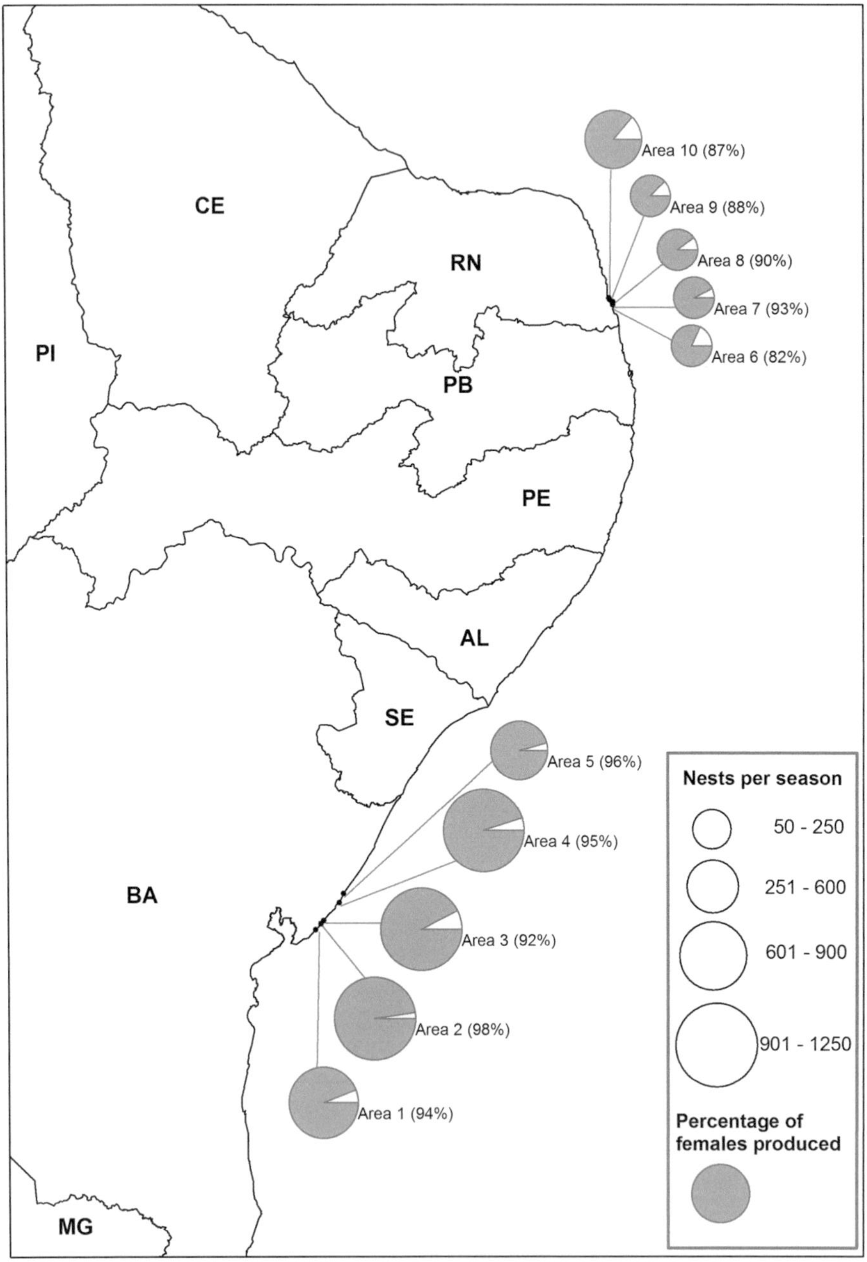

Fig. 7.17 Study sites with estimated proportion of females (*grey*) and male (*white*) hawksbill hatchlings in each area. Pie diameter is scaled according to magnitude of clutches laid each year in each area (Data from Marcovaldi et al. 2014).

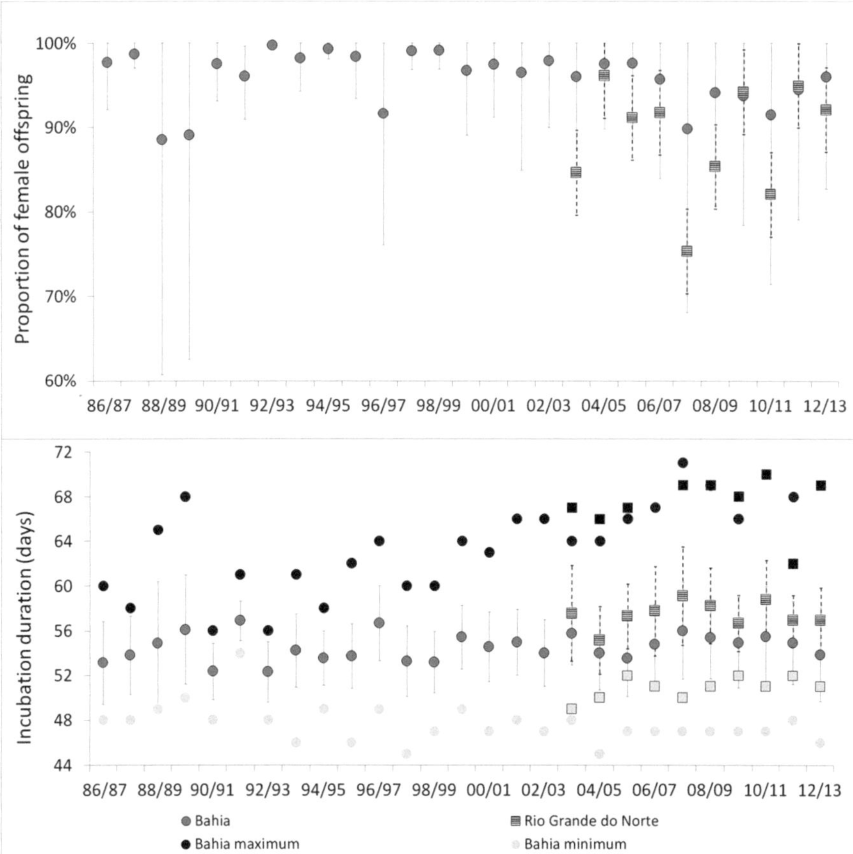

Fig. 7.18 Inter-annual (**a**) mean (SD), incubation duration of hawksbill clutches, and (**b**) mean percentage of females produced (SD) at Bahia and Rio Grande do Norte (Data from Marcovaldi et al. 2014).

7.5 Future Research and Perspectives for Sea Turtle Conservation

At the very beginning in the late 1970s, TAMAR faced a big challenge: obtain information on sea turtles occurrence along the 8.000 km of the Brazilian coastline, a mission with continental dimension. What was apparently a very basic gave us a starting point to explore a whole new universe. Collection of data, village by village, and observations in the field during the initial assessment, from 1980 to 1982, resulted in the first map of the sea turtle spatial and temporal distribution in our country and the main threats for these species (Marcovaldi and Marcovaldi 1999). As so many other sea turtle projects from other countries at that time, the key questions to be responded were related to the land part of the sea turtle life cycle.

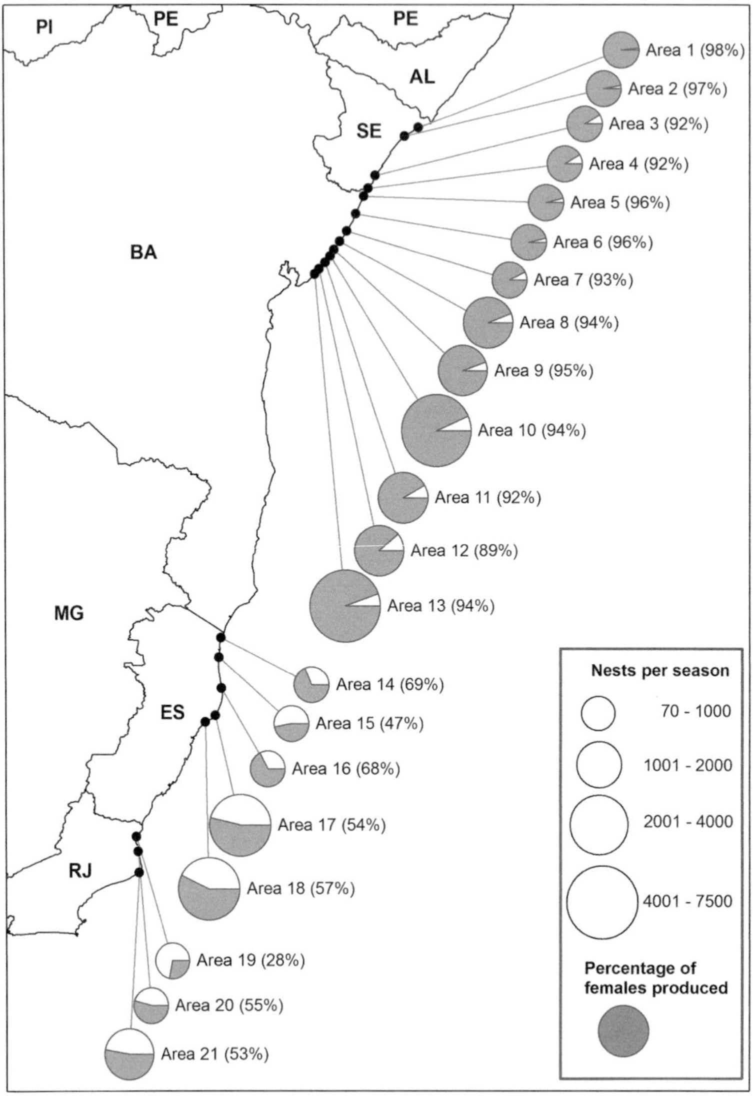

Fig. 7.19 Study sites with estimated proportion of females (grey) and male (white) loggerhead hatchlings in each area. Pie diameter is scaled according to magnitude of clutches laid each year in each area. Data from Marcovaldi et al. (2016)

The first tagging program started in Brazil in 1981 at the Biological Reserve of Atol das Rocas (Marcovaldi and Marcovaldi 1999). The tagging program gave us a little bit more knowledge to plan and increase our activities. Tags returns from long distance areas, sometimes, other nations or even continents were sent back to us, showing how much these individuals were able to travel. Standard flipping tags are still used throughout the world, to identify individual turtles, estimate demographic parameters, through mark recapture studies, and track turtle movements (Bjorndal 2011).

As time passed by, the use of new technologies enabled *many advances in our understanding* of *sea turtle biology.* The ability to collect fine-scale spatio-temporal location data on wide ranging species such as sea turtles is a great example. So far hundreds of satellite transmitters have been attached on sea turtles around the world which has allowed us to know fascinating facts about navigation, and the three-dimensional use of the ocean environment. Advances in tagging technology, such as tag miniaturization, have also enabled attachment of smaller tags to young animals providing important insights about early life stages. Molecular analyses turned to be an essential tool mainly to identify the natal origin of individuals and define management units for conservation. Hence, the integration of genetic data and telemetry have provided significant information on the distribution and migratory patterns of sea turtles highlighting the importance of international cooperation to protect these species.

Climate change threatens to alter the world in unpredictable ways, and globally migrating marine turtles are no exception. Potential consequences of global warming include a reduction in effective population size, the inability to find mates leading to reduced fecundity or female infertility, and under more extreme climate projections, the potential for production of single sex cohorts (Hawkes et al. 2009). These changes are likely to reduce the viability of turtle populations. Thus, it is crucial to understand both the male/female ratio necessary to sustain populations and the turtles' capacity to adapt to increases in temperature.

Even after decades of research, there are still many unanswered questions about sea turtles' natural history. These species are difficult to study due to their complex lives and elusive behaviors. For instance, it is difficult to find the animals in their early life stage since they span vast oceanic areas and encounters with males are very rare because they never leave the water.

The scientific community understanding of sea turtle ecology and the conservation strategies necessary for turtles' survival has progressed substantially over the years. Advances in *technology* are gradually revealing the secret lives of *sea turtles, especially during* cryptic *life* stages. However, the continued use of simple methods of research such as beach surveys and constant informal oral communication are still considered among the most valuable techniques for successfully studying sea turtles.

References

Almeida AP, Eckert SA, Bruno SC, Scalfoni JT, Giffoni B, López-Mendilaharsu M, Thomé JCA (2011) Satellite-tracked movements of female *Dermochelys coriacea* from southeastern Brazil. Endanger Species Res 15:77–86. doi:10.3354/esr00359

Bjorndal KA (2011) Introduction to the 2011 reissue of the book *So Excellent a Fishe* by Archie Carr. University Press of Florida, Gainesville, pp XI–XXI

Bjorndal KA, Bolten AB, Moreira L, Belini C, Marcovaldi MA (2006) Population structure and diversity of Brazilian green turtle rookeries based on mitochondrial DNA sequences. Chelonian Conserv Biol 5:262–268. doi:10.2744/1071-8443(2006)5[262:PSADOB]2.0.CO;2

Bowen BW, Karl SA (2007) Population genetics and phylogeography of sea turtles. Mol Ecol 16:4886–4907. doi:10.1111/j.1365-294X.2007.03542.x

Bowen BW, Bass AL, Chow SM, Bostrom M, Bjorndal KA, Bolten AB, Okuyama T, Bolker BM, Epperly S, Lacasella E, Shaver D, Dodd M, Hopkins-Murphy SR, Musick JA, Swingle M, Rankin-Baransky K, Teas W, Witzell WN, Dutton PH (2004) Natal homing in juvenile loggerhead turtles (*Caretta caretta*). Mol Ecol 13:3797–3808. doi:10.1111/j.1365-294X.2004.02356.x

Carreras C, Pascual M, Cardona L, Aguilar A, Margaritoulis D, Rees A, Turkozan O, Levy Y, Gasith A, Aureggi M, Khalil M (2007) The genetic structure of the loggerhead sea turtle (*Caretta caretta*) in the Mediterranean as revealed by nuclear and mitochondrial DNA and its conservation implications. Conserv Genet 8:761–775. doi:10.1007/s10592-006-9224-8

CITES (2016) Convention on international trade in endangered species of wild fauna and flora. Available at https://www.Cites.Org/eng/disc/what.Php - accessed 2016/10/14

Coyne MS, Godley BJ (2005) Satellite Tracking and Analysis Tool (STAT): an integrated system for archiving, analyzing and mapping animal tracking data. Mar Ecol Prog Ser 301:1–7

Da Silva ACCD, Dos Santos EAP, Lira F, Weber MI, Batista JAF, Serafini TZ, Castilhos JC (2011) Satellite-tracking reveals multiple foraging strategies and threats for olive ridley turtles in Brazil. Mar Ecol Prog Ser 443:237–247. doi:10.3354/meps09427

da Silva VRF, Mitraud SF, Ferraz MLCP, Lima EHSM, Melo MTD, Santos AJB, Marcovaldi MAG (2016) Adaptive threat management framework: integrating people and turtles. Environ Dev Sustain 18(6):1541–1558. doi:10.1007/s10668-015-9716-0

Domingo A, Bugoni L, Prosdocimi L, Miller P, Laporta M, Monteiro DS, Estrades A, Albareda D (2006) The impact generated by fisheries on sea turtles in the Southwestern Atlantic. WWF Progama Marino para Latino América y el Caribe, San José, Costa Rica, p 71

Duron-Dufrenne M (1987) Premier suivi par satellite en Atlantique d'une tortue Luth *Dermochelys coriacea*. Comptes rendus de l'Académie des sciences. Série 3, Sciences de la vie 304(15):399–402

Dutton PH, Roden SE, Stewart KR, LaCasella E, Tiwari M, Formia A, Thomé JC, Livingstone SR, Eckert S, Chacon-Chaverri D, Rivalan P, Allman P (2013) Population stock structure of leatherback turtles (*Dermochelys coriacea*) in the Atlantic revealed using mtDNA and microsatellite markers. Conserv Genet 14:625–636. doi:10.1007/s10592-013-0456-0

Fancy SG, Pank LF, Douglas DC, Curby CH, Garner GW (1988). Satellite telemetry: a new tool for wildlife research and management. US Department of Interior, Fish and Wildlife Service Resource Publication 172, 54p

FitzSimmons NN, Moritz C, Limpus CJ, Miller JD, Parmenter CJ, Prince R (1996) Comparative genetic structure of green, loggerhead, and flatback populations in Australia based on variable mtDNA and nDNA regions. In: Bowen BW, Witzell WN (eds) Proceedings of the international symposium on sea turtle conservation genetics. Miami, Florida: NOAA Tech Memo NMFS-SEFSC-396, pp 25–32

FitzSimmons NN, Limpus CJ, Norman JA, Goldizen AR, Miller JD, Moritz C (1997) Philopatry of male marine turtles inferred from mitochondrial DNA markers. Proc Natl Acad Sci U S A 94:8912–8917

Frazier J (2003) Prehistoric and ancient historic interactions between humans and marine turtles. In: Lutz PL, Musick JA, Wyneken J (eds) The biology of sea turtles, vol II. CRC Press, Boca Raton, pp 1–38

Fuentes MMPB, Abbs D (2010) Effects of projected changes in tropical cyclone frequency on sea turtles. Mar Ecol Prog Ser 412:283–292. doi:10.3354/meps08678

Fuentes M, Limpus CJ, Hamann M, Dawson J (2009) Potential impacts of projected sea-level rise on sea turtle rookeries. Aquat Conserv 20:132–139. doi:10.1002/aqc.1088

Fuentes MMPB, Limpus CJ, Hamann M (2011) Vulnerability of sea turtle nesting grounds to climate change. Glob Chang Biol 17:140–153. doi:10.1111/j.1365-2486.2010.02192.x

Fuentes MMPB, Fish MR, Maynard JA (2012) Management strategies to mitigate the impacts of climate change on sea turtle's terrestrial reproductive phase. Mitig Adapt Strat Glob Change 17:51–63. doi:10.1007/s11027-011-9308-8

Godfrey MH, D'Amato AF, Marcovaldi MA, Mrosovsky N (1999) Pivotal temperature and predicted sex ratios for hatchling hawksbill turtles from Brazil. Can J Zool 77:1465–1473. doi:10.1139/z99-117

Godley BJ, Lima EHSM, Åkesson S, Broderick AC, Glen F, Godfrey MH, Luschi P, Hays GC (2003) Movement patterns of green turtles in Brazilian coastal waters described by satellite tracking and flipper tagging. Mar Ecol Prog Ser 253:279–288. doi:10.3354/meps253279

Godley BJ, Blumenthal JM, Broderick AC, Coyne MS, Godfrey MH, Hawkes LA, Witt MJ (2008) Satellite tracking of sea turtles: where have we been and where do we go next? Endanger Species Res 4:3–22. doi:10.3354/esr00060

Hahn AT, Fernandes LB, Grazziotin FG, Castilhos JC, Bonatto SL (2012) Low mtDNA Diversity of the brazilian nesting populations of olive ridley turtles (*Lepidochelys olivacea*). In: Belskis LM. Frick M, Panagopoulou A, Rees AF, Williams K (eds) Proceedings of the twenty-nine annual Symposium on sea turtle biology and conservation. NOAA Tech Memo NMFS-SEFSC-630, p 80

Hamann M, Godfrey MH, Seminoff JA, Arthur K, Barata PCR, Bjorndal KA, Bolten AB, Broderick AC, Campbell LM, Carreras C, Casale P, Chaloupka M, Chan SKF, Coyne MS, Crowder LB, Diez CE, Dutton PH, Epperly SP, FitzSimmons NN, Formia A, Girondot M, Hays GC, Cheng IJ, Kaska Y, Lewison R, Mortimer JA, Nichols WJ, Reina RD, Shanker K, Spotila JR, Tomás J, Wallace BP, Work TM, Zbinden J, Godley BJ (2010) Global research priorities for sea turtles: informing management and conservation in the 21st century. Endanger Species Res 11:245–269. doi:10.3354/esr00279

Harry JL, Briscoe DA (1988) Multiple paternity in the loggerhead turtle (*Caretta caretta*). J Hered 79(2):96–99

Hawkes LA, Broderick AC, Godfrey MH, Godley BJ (2007) Investigating the potential impacts of climate change on a marine turtle population. Glob Chang Biol 13:923–932. doi:10.1111/j.1365-2486.2007.01320.x

Hawkes LA, Broderick AC, Godfrey MH, Godley BJ (2009) Climate change and marine turtles. Endanger Species Res 7:137–154. doi:10.3354/esr00198

ICMBio (2011) Plano de ação nacional para conservação das tartarugas marinhas. Available: http://www.iacseaturtle.org/docs/planes/Brazil.pdf. Accessed 22 out 2016

Jensen MP, FitzSimmons NN, Dutton PH (2013) Molecular genetics of sea turtles. In: Wyneken J (ed) Biology of sea turtles, vol III. CRC Press, Boca Raton, p 155

Lemke D, Frazier JG, Thomé JCA, Almeida, AP, Scalfoni, J (2003) Satellite telemetry of loggerheads in Brazil. In: Pilcher NJ (eds) Proceedings of the 23rd annual symposium on sea turtle biology and conservation, 17–21 March 2003, Kuala Lumpur, NOAA Tech Memo 536, pp 230–233

Marcovaldi MA, Marcovaldi GG (1999) Marine turtles of Brazil: the history and structure of Projeto TAMAR-IBAMA. Biol Conserv 91: 35–41. doi:http://dx.doi.org/10.1016/S0006-3207(99)00043-9

Marcovaldi MA, Godfrey MH, Mrosovsky N (1997) Estimating sex ratios of loggerhead turtles in Brazil from pivotal incubation durations. Can J Zool 75:755–770. doi:10.1139/z97-097

Marcovaldi MA, Patiri VJA, Thomé JC (2005) Projeto TAMAR-IBAMA: twenty five years protecting Brazilian sea turtles through a community-based conservation programme. MAST 3(2) and 4(1):39–62

Marcovaldi MA, Lopez GG, Soares LS, Lima EHSM, Thomé JCA, Almeida AP (2010) Satellite-tracking of female loggerhead turtles highlights fidelity behavior in northeastern Brazil. Endanger Species Res 12:263–272. doi:10.3354/esr00308

Marcovaldi MA, Lopez GG, Soares LS, López-Mendilaharsu M (2012) Satellite tracking of hawksbill turtles *Eretmochelys imbricata* nesting in northern Bahia, Brazil: turtle movements and foraging destinations. Endanger Species Res 17:123–132. doi:10.3354/esr00421

Marcovaldi MA, Santos AJB, Santos AS, Soares LS, Lopez GG, Godfrey MH, López-Mendilaharsu M, Fuentes MMPB (2014) Spatio-temporal variation in the incubation duration and sex ratio of hawksbill hatchlings: implication for future management. J Therm Biol 44:70–77. doi:10.1016/j.jtherbio.2014.06.010

Marcovaldi MA, López-Mendilaharsu M, Santos AS, Lopez GG, Godfrey MH, Tognin F, Baptistotte C, Thomé JC, Dias ACC, de Castilhos JC, Fuentes MMPB (2016) Identification of loggerhead male producing beaches in the south Atlantic: Implications for conservation. J Exp Mar Biol Ecol 477. Elsevier B.V.:14–22. doi:10.1016/j.jembe.2016.01.001

Mrosovsky N, Yntema CL (1980) Temperature dependence of sexual differentiation in sea turtles: implications for conservation practices. Biol Conserv 18:271–280. doi:http://dx.doi.org/10.1016/0006-3207(80)90003-8

Musik JA, Limpus CJ (1997) Habitat utilization and migration in Juvenile Sea turtles. In: Lutz P, Musick J (eds) The biology of sea turtles. CRC Press, Boca Raton

Naro-Maciel E, Bondioli ACV, Martin M, Almeida AP, Baptistotte C, Bellini C, Marcovaldi MA, Santos AJB, Amato G (2012) The interplay of homing and dispersal in green turtles: a focus on the southwestern atlantic. J Hered 103:792–805. doi:10.1093/jhered/ess068

Phillips KP, Jorgensen TH, Jolliffe KG, Jolliffe SM, Henwood J, Richardson DS (2013) Reconstructing paternal genotypes to infer patterns of sperm storage and sexual selection in the hawksbill turtle. Mol Ecol 22:2301–2312. doi:10.1111/mec.12235

Proietti MC, Reisser J, Marins LF, Rodriguez-Zarate C, Marcovaldi MA, Monteiro DS, Pattiaratchi C, Secchi ER (2014) Genetic structure and natal origins of immature hawksbill turtles (*Eretmochelys imbricata*) in Brazilian waters. PLoS One 9(2):e88746. doi:10.1371/journal.pone.0088746

Reis EC, Soares LS, Vargas SM, Santos FR, Young RJ, Bjorndal KA, Bolten AB, Lôbo-Hajdu G (2009) Genetic composition, population structure and phylogeography of the loggerhead sea turtle: colonization hypothesis for the Brazilian rookeries. Conserv Genet 11:1467–1477. doi:10.1007/s10592-009-9975-0

Roberts MA, Schwartz TS, Karl S (2004) Global population genetic structure and male-mediated gene flow in the green sea turtle (*Chelonia mydas*): analysis of microsatellite loci. Genetics 166:1857–1870. doi:10.1534/genetics.166.4.1857

Sales G, Giffoni BB, Barata PCR (2008) Incidental catch of sea turtles by the Brazilian pelagic longline fishery. J Mar Biolog Assoc UK 88:853–864. doi:10.1017/S0025315408000441

Santos IR, Friedrich AC, Barretto FP (2005) Overseas garbage pollution on beaches of Northeast Brazil. Mar Pollut Bull 50(7):783–786

Shamblin BM, Bolten AB, Abreu-Grobois FA, Bjorndal KA, Cardona L, Carreras C, Clusa M, Monzón-Argüello C, Nairn CJ, Nielsen JT, Nel R, Soares LS, Stewart KR, Vilaça ST, Türkozan O, Yilmaz C, Dutton PH (2014) Geographic patterns of genetic variation in a broadly distributed marine vertebrate: new insights into loggerhead turtle stock structure from expanded mitochondrial DNA sequences. PLoS One 9(1):e85956. doi:10.1371/journal.pone.0085956

Shamblin BM, Dutton PH, Bjorndal KA, Bolten AB, Naro-Maciel E, Santos AJB, Bellini C, Baptistotte C, Marcovaldi MA, Nairn CJ (2015) Deeper mitochondrial sequencing reveals cryptic diversity and structure in Brazilian green turtle rookeries. Chelonian Conserv Biol 4:167–172. doi:10.2744/CCB-1152.1

Shillinger GL, Swithenbank AM, Bograd SJ, Bailey H, Castelton MR, Wallace BP, Block B (2010) Identification of high-use internesting habitats for eastern Pacific leatherback turtles: role of the environment and implications for conservation. Endanger Species Res 10:215–232. doi:10.3354/esr00251

Tourinho PS, Ivar do Sul JA, Fillmann G (2010) Is marine debris ingestion still a problem for the coastal marine biota of southern Brazil? Mar Pollut Bull 60:396–401. doi:10.1016/j.marpolbul.2009.10.013

Vargas SM, Araújo FCF, Monteiro DS, Estima SC, Almeida AP, Soares LS, Santos FR (2008) Genetic diversity and origin of leatherback turtles (*Dermochelys coriacea*) from the Brazilian coast. J Hered 99:215–220. doi:10.1093/jhered/esm120

Vargas SM, Molfetti E, Vilaça ST, Monteiro D, Estima SC, Soares LS, Almeida AP, De Thiosy B, Naro-Maciel E, Santos FR (2013) Mixed stock analysis of leatherback turtles feeding in

Brazil: Records over four years. In: Tucker T, Belskis L, A. Panagopoulou A, Rees A, Frick M, K. Williams K, LeRoux R, Stewart K (eds) Proceedings of the thirty-third annual symposium on sea turtle biology and conservation. NOAA Tech. Memo. NMFS-SEFSC-645, p 246

Vilaça ST, Vargas SM, Lara-Ruiz P, Molfetti E, Reis EC, Lôbo-Hajdu G, Soares LS, Santos FR (2012) Nuclear markers reveal a complex introgression pattern among marine turtle species on the Brazilian coast. Mol Ecol 21:4300–4312. doi:10.1111/j.1365-294X.2012.05685.x

Vilaça ST, Lara-Ruiz P, Marcovaldi MA, Soares LS, Santos FR (2013) Population origin and historical demography in hawksbill (*Eretmochelys imbricata*) feeding and nesting aggregates from Brazil. J Exp Mar Biol Ecol 446:334–344. doi:10.1016/j.jembe.2013.06.004

Wallace BP, DiMatteo AD, Hurley BJ, Finkbeiner EM, Bolten AB, Chaloupka MY, Hutchinson BJ, Abreu-Grobois FA, Amorocho D, Bjorndal KA, Bourjea J, Bowen BW, Dueñas RB, Casale P, Choudhury BC, Costa A, Dutton PH, Fallabrino A, Girard A, Girondot M, Godfrey MH, Hamann M, López-Mendilaharsu M, Marcovaldi MA, Mortimer JA, Musick JA, Nel R, Pilcher NJ, Seminoff JA, Troëng S, Witherington B, Mast RB (2010) Regional management units for marine turtles: a novel framework for prioritizing conservation and research across multiple scales. PLoS One 5:1–11. doi:10.1371/journal.pone.0015465

Witt MJ, Broderick AC, Coyne M, Formia A, Ngouessono S, Parnell RJ, Sounguet GP, Godley BJ (2008) Satellite tracking highlights difficulties in the design of effective protected areas for leatherback turtles during the internesting period. Oryx 42:296–300

Chapter 8
Opportunistic Sightings of Dolphins (MAMMALIA, CETACEA, DELPHINIDAE) in the North Coast of Bahia State, Northeastern Brazil, Between 2001 and 2006

Mateus Souza de Carvalho, Marcos R. Rossi-Santos, Elitieri Santos-Neto, and Clarêncio G. Baracho

Abstract Using opportunistic platforms to study humpback whales (*Megaptera novaeangliae*) during their breeding season, from July to November, we analyzed the distribution of dolphin sightings in a poorly studied area, the north coast of Bahia State, Brazil. Between 2001 and 2006, more than 500 days of surveys were performed (2360 h of effort), in which 93 groups of dolphins were sighted from Itacaré (14° 53′S, 38° 15′W) to Subaúma (12° 30′S; 37° 63′W): 58 (62.4%) bottlenose dolphin, *Tursiops truncatus*; 21 (22.6%) estuarine dolphin, *Sotalia guianensis*; 7 (7.5%) -toothed dolphin, *Steno bredanensis*; 3 (3.2%) Clymene's Dolphin, *Stenella clymene*; 2 (2.1%) pygmy killer whale, *Feresa attenuata*; 1 (1.1%) melon-headed dolphin, *Peponocephala electra* and 1 (1.1%) Atlantic Spotted dolphin, *Stenella frontalis*. Group size and depth range were: 1–30 for *T. truncatus* (10–150 m); 1–35 for *S. guianensis* (4–60 m); 1–6 for *S. bredanensis* (27–45 m); 1–100 for *S. clymene* (42–121 m); 5–6 for *F. attenuata* (97–200 m); 2 for *S. frontalis* at 51 m and 20 for *P. electra* at 150 m. The more frequent sightings of *Tursiops truncatus* and *Sotalia guianensis* are probably related to their known distribution patterns in Brazilian coastal waters. The sightings reported to *S. guianensis* add new information in a previously unstudied area. The records of pelagic species such as *Stenella clymene*, *Feresa attenuata* and *Peponocephala electra*, can be related to the narrow continental shelf of the studied region. The sightings presented here confirm the occurrence of two additional species (*F. attenuata* and *S. frontalis*) in the Bahia state, and reveal the wide diversity of cetaceans in this area. This study is an initial step to document and collaborate with the knowledge about the dolphin species that occur in the shore of Bahia.

M.S. de Carvalho (✉) • M.R. Rossi-Santos • E. Santos-Neto • C.G. Baracho
Instituto Baleia Jubarte, Avenida do Farol, s/n, CP 92, Praia do Forte,
Mata de São João, Bahia, Brazil, 48280-000
e-mail: mateusscarvalho@gmail.com; http://www.baleiajubarte.org.br

M.R. Rossi-Santos, C.W. Finkl (eds.), *Advances in Marine Vertebrate Research in Latin America*, Coastal Research Library 22,
DOI 10.1007/978-3-319-56985-7_8

Keywords Dolphins • Opportunistic sightings • Bahia State • Brazil

8.1 Introduction

The distribution of the dolphins is clearly influenced by the marine environment, being frequently related to a environmental factor scale, including the sea surface temperature (Selzer and Payne 1988; Forney 2000; Baumgartner et al. 2001; Hamazaki 2002), salinity (Selzer and Payne 1988; Forney 2000), depth (Gowans and Whitehead 1995; Baumgartner 1997; Davis et al. 1998), and the bottom relief (Selzer and Payne 1988; Gowans and Whitehead 1995; Baumgartner 1997; Davis et al. 1998). However, the influence of these factors on the distribution of the animals seems to vary between regions and species (Hastie et al. 2005).

Information about dolphin sightings in all Southwestern Atlantic Ocean is scarce and primarily composed by punctual registers (eg. Zerbini et al. 2004; Baracho et al. 2005; Rossi-Santos et al. 2006a, b; Magalhães et al. 2007; Baracho et al. 2008, Carvalho and Rossi-Santos 2011).

The coast of the Bahia State, although it is the longest in Brazil, is still barely studied on its biodiversity and potential sources of natural resources. Despite the great variety of species and the existence of dolphin records by sightings and strandings (Lodi and Capistrano 1990; Hetzel and Lodi 1993; Batista et al. 2012), the information had been published individually and temporally distant, limiting results about the ecology and distribution of the respective species in the region.

There are few systematic cetacean research, which is centered mainly in the Guiana dolphin (*Sotalia guianensis*) (for revision of common name see Flores et al. 2008) along the coast, basically in estuaries and bays, mainly in Ilhéus coast (eg; Batista 2001; Santos 2001; Reis 2002; Batista et al. 2005; Rossi-Santos et al. 2007; Santos 2007; Assis 2008; Rossi-Santos and Reis 2008; Schiavetti et al. 2008; Reis et al. 2008; Izidoro 2009; Moraes 2010; Rossi-Santos et al. 2010; Santos 2010) and in the humpback whale (eg. Martins et al. 2001; Morete et al. 2003; Simões et al. 2005; Zerbini et al. 2006; Morete et al. 2007; Rossi-Santos et al. 2008; Andriolo et al. 2010; Wedekin et al. 2010; Andriolo et al. 2014).

This study aims to bring information about the occurrence and distribution of the dolphins in the north coast of the Bahia State, through opportunistic sightings, between 2001 and 2006.

8.2 Material and Methods

8.2.1 Study Area

The North Coast of the State of the Bahia occupies an area about 8.000 km^2, being formed by thirteen cities. Praia do Forte (main showed area) it is placed in the city of Mata de São João, and is located about 60 km from the capital Salvador. It possess a littoral band with approximately 14 km of extension, throughout the North Coast of the Bahia between the rivers Pojuca (12° 35′S, 38° 00′W) and Imbassaí (12° 30′S, 38° 00′W), limits south and north respectively. A narrow continental platform, with approximately 15 km of extension, it is the main oceanographic characteristic. Throughout the platform, its average of depth is of 40 m and the amplitude of tide varies between 0.1 and 2.6 m (DHN 1995).

The study area of the present work (Fig. 8.1) extends since Itacaré (14° 53′S; 38° 15′W) to Subaúma (12° 30′S; 37° 63′W), with a distance up to 15 nautical miles from the coast, until the 300 m isobath.

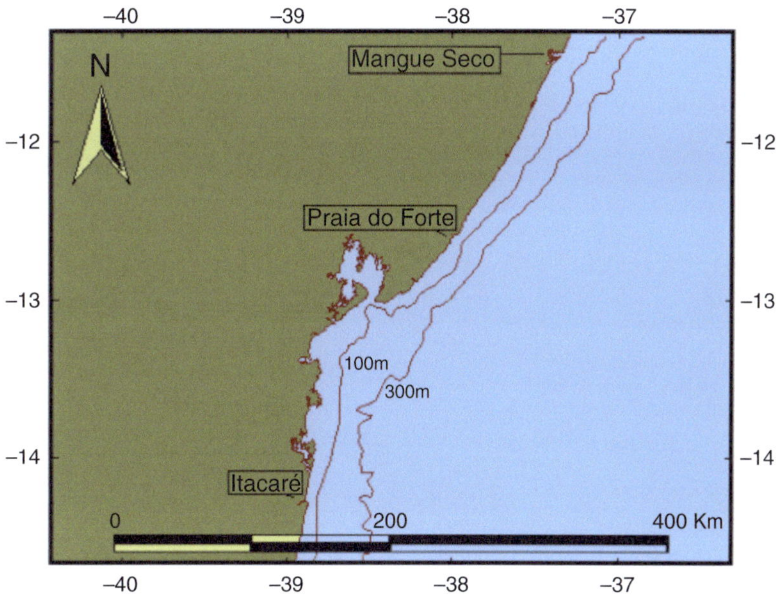

Fig. 8.1 Study area map, located in the Coast North of the Bahia State, Brazil. This place is characterized by a narrow continental shelf that embraces long sand beaches and coastal fringed coral reefs, making a great tropical scenario for the well established tourism industry

8.2.2 Field Procedures

Opportunistic sightings of dolphins occurred between July and October, during the years of 2001 and 2006, through the research cruises and the daily Whale Watching tours, in which were embarked staff technicians from the Instituto Baleia Jubarte/IBJ, in order to study the biology of humpback whales (*Megaptera novaeangliae*), in this breeding ground (see details in Rossi-Santos et al. 2008). During every year the research was conducted utilizing wood trawlers and schooners, with 13 m of length and equipped with a 120 HP center engine (Fig. 8.2).

For the record of the sailed routes and geographic position of sighted animals, a GPS Map 76 and a GPS Map 60c (Garmin®) was utilized, in the track function and way-point function, respectively. The dolphin species were photographed with cameras Nikon F5 and N 90s, with multiple 35 mm films, plus a recent digital Nikon D70, for posterior identification. The information about the number of individuals, composition of the groups, behaviors observed before and after approach, time of beginning and end of the sightings, depth, as well as the geographic coordinates and others, had been recorded in specific standard form.

Fig. 8.2 Opportunistic sightings of dolphins occurred between July and October, through the research cruises and the daily Whale Watching tours, in which a trained researcher went aboard, in order to study the biology of humpback whales (*Megaptera novaeangliae*) in this breeding ground (Photo: Marcos Rossi-Santos/IBJ)

8.2.3 Data Analyses

The data obtained with GPS had been transferred to a computer using the GPS Track Maker Professional Program (version 11.8). For posterior analysis and studies of the distribution of the dolphins in the showed area, the geographic coordinates of the sightings had been located in digital nautical letter using the program Arc View GIS (version 3.1), and the same program was used for elaboration of the distribution maps of the joined species.

8.3 Results

Between the period of July and November from 2001 to 2006, about 500 days of sampling effort were carried out, resulting in 93 groups of dolphins sighted in the north coast of Bahia: 58 (62.4%) bottlenose dolphin, *Tursiops truncatus*; 21 (22.6%) Guiana dolphin, *Sotalia guianensis*; 7 (7.5%) rough-toothed dolphin, *Steno bredanensis*; 3 (3.2%) clymene dolphin, *Stenella clymene*; 2 (2.1%) pygmy killer whale, *Feresa attenuata*; 1 (1.1%) melon-headed Whale, *Peponocephala electra* and 1 (1.1%) spotted dolphin, *Stenella frontalis* (Fig. 8.3).

Tursiops truncatus **(Montagu, 1821)**
The bottlenose dolphin, *Tursiops truncatus*, (Fig. 8.4) was the most representative species during all the period of study, occurring every year (during the period from 2001 to 2006). It was recorded a total of 274 individuals, divided in 58 groups,

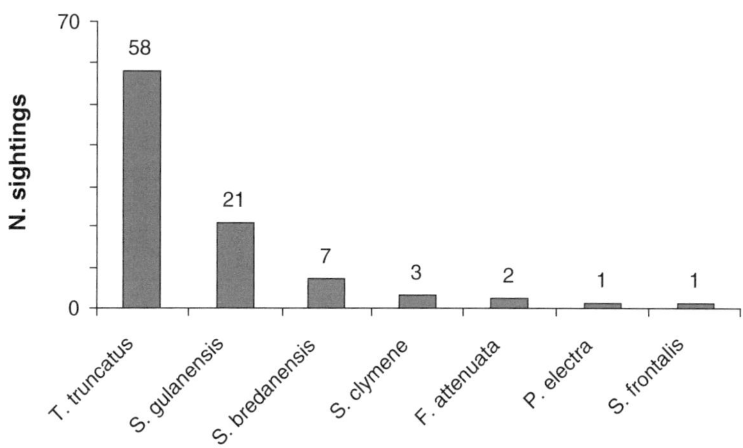

Fig. 8.3 Number of sightings of dolphin species registered in the North Coast of Bahia during the period of 2001 at 2006, affirming the bottlenose dolphin as the most frequent dolphin species during the whale-whatching surveys, followed by the Guiana dolphin, a common estuarine species along the Brazilian coast

Fig. 8.4 The bottlenose dolphin (*Tursiops truncatus*) the most sighted species in this study in the North Coast of Bahia during the period of 2001 to 2006 (Photo: Marcos Rossi-Santos/IBJ)

varying between 1 and 30 individuals for group, with 359 min of direct observation, a variation of 1–52 min. The species was registered in depths varying between 10 and 150 m. The sea conditions (Beaufort scales) varied from 0 to 4 and the intensity of the wind varied of 5–15 knots. The species was distributed throughout all area of study, from Itacaré to Praia do Forte, about 200 km distant. Most of the sightings occurred in front of Praia do Forte. The species was registered mostly near to the coast (10 m of depth), despite some sightings in deeper areas (150 m) (Fig. 8.5).

Sotalia guianensis **(P. J. Van Bénéden, 1864)**
The Guiana dolphin, *Sotalia guianensis*, (Fig. 8.6) was the second more representative species during the period of the study, occurring between the years of 2003 and 2006. A total of 115 individuals was recorded, divided into 21 groups, varying from 1 to 35 individuals per group. The time of direct observation of the species was of 344 min, ranging from 1 to 20 min. The species occurred between depths of 4 and 60 m. The sea conditions varied of 0–3 and the intensity of the wind varied of 3–12 knots. The species also was distributed throughout all area of study, however with intervals in some specific areas. Its presence always occurred next to the coast, entered of rivers and Baía de Todos os Santos (Fig. 8.7).

Steno bredanensis **(Lesson, 1828)**
For the rough-toothed dolphin, *Steno bredanensis,* (Fig. 8.8) 27 individuals were registered, divided in 7 groups, with a variation of 1–6 individuals for group size. It had 26 min of direct observation, varying of 1–5 min. It occurred between depths of 27 and 45 m. The sea conditions varied of 1–5 and the intensity of the wind varied of 9–16 knots. Only one of all the sightings did not occur in the north coast of Bahia,

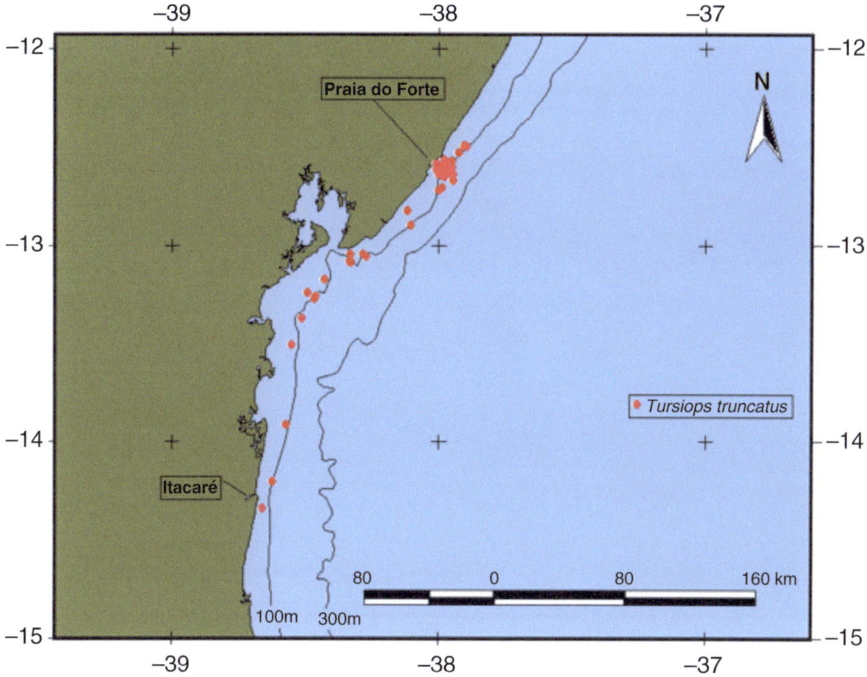

Fig. 8.5 The bottlenose dolphin (*Tursiops truncatus*) the most sighted species in this study in the North Coast of Bahia during the period of 2001 to 2006

Fig. 8.6 The Guiana dolphin (*Sotalia guianensis*) was the second most sighted species in this study in the North Coast of Bahia during the period of 2001 to 2006 (Photo: Marcos Rossi-Santos/IBJ)

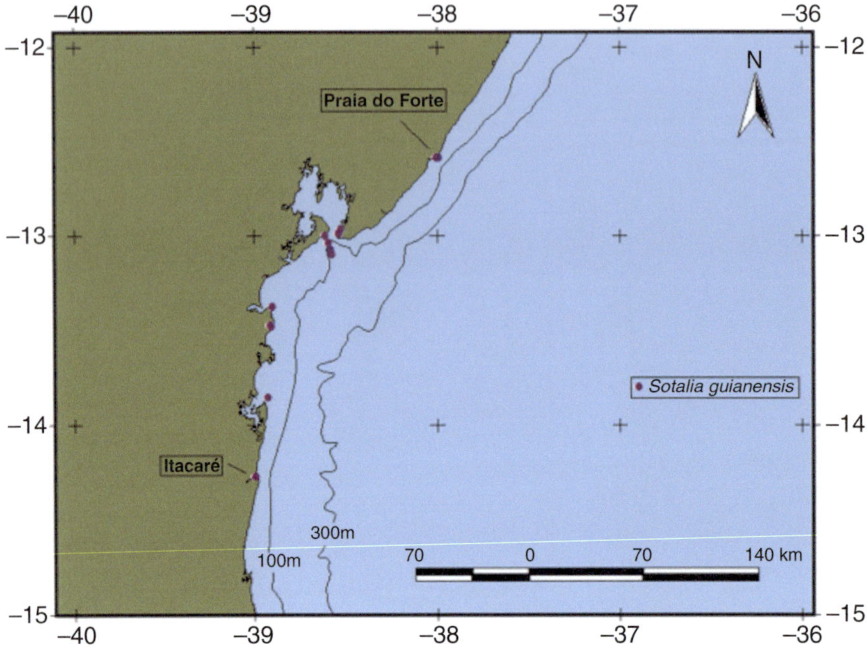

Fig. 8.7 The Guiana dolphin (*Sotalia guianensis*) was the second most sighted species in this study in the North Coast of Bahia during the period of 2001 to 2006

Fig. 8.8 The rough-toothed dolphin (*Steno bredanensis*) was the third common sighted species in this study in the North Coast of Bahia during the period of 2001–2006 (Photo: Marcos Rossi-Santos/IBJ)

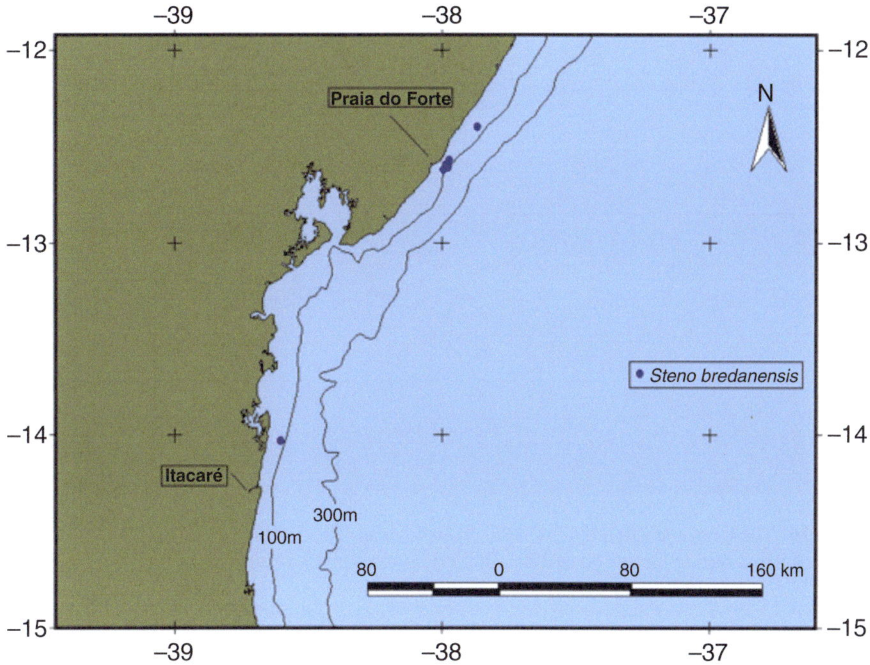

Fig. 8.9 The rough-toothed dolphin (*Steno bredanensis*) was the third common sighted species in this study in the North Coast of Bahia during the period of 2001–2006

but in the south limit of the study area, next to Barra Grande (14° 53′S; 38° 15′W). Five groups had been sighted in front of Praia do Forte (12° 96′S; 38° 13′W) and the other group in front of Imbassaí (12° 30′S; 38° 00′W). The species presented a patched instead continuous distribution throughout the coast, between latitudes 12° 30′S; 38° 00′W and 14° 53′S; 38° 15′W, limits north and south, respectively (Fig. 8.9).

Stenella clymene **(Gray, 1850)**
The Clymene's dolphin, *Stenella clymene* (Fig. 8.10), was only recorded in the years of 2001 and 2003. The total number was 113 individuals, divided between 3 groups, with a variation of 1–100 individuals per group. We performed 23 min of direct observation, varying between 3 and 20 min per group. The species occurred between depths of 42 and 121 m. The sea conditions varied of 2–4 and the intensity of the wind variation was 9–14 knots. The species had two sightings in front of Praia do Forte and one another near to Salvador (Fig. 8.11).

Feresa attenuata **(Gray, 1874)**
The pygmy killer whale, *Feresa attenuata*, was registered twice in the same day, during the year of 2006 (Fig. 8.12). The total number was of 11 individuals, divided in 2 groups, with 5 and 6 individuals respectively. There was a total of 30 min of

Fig. 8.10 The Clymene's dolphin (*Stenella clymene*) was also identified in this study in the North Coast of Bahia during the period of 2001–2006 (Photo: Daniel Lewis)

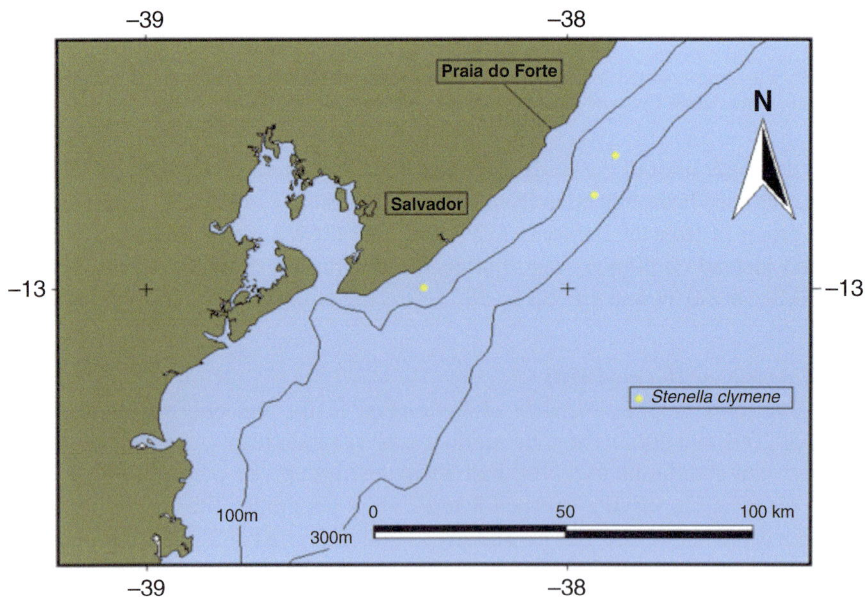

Fig. 8.11 The Clymene's dolphin (*Stenella clymene*) was only recorded in the years of 2001 and 2003. The total number was 113 individuals, divided between 3 groups, with a variation of 1–100 individuals per group, in this study in the North Coast of Bahia during the period of 2001 to 2006

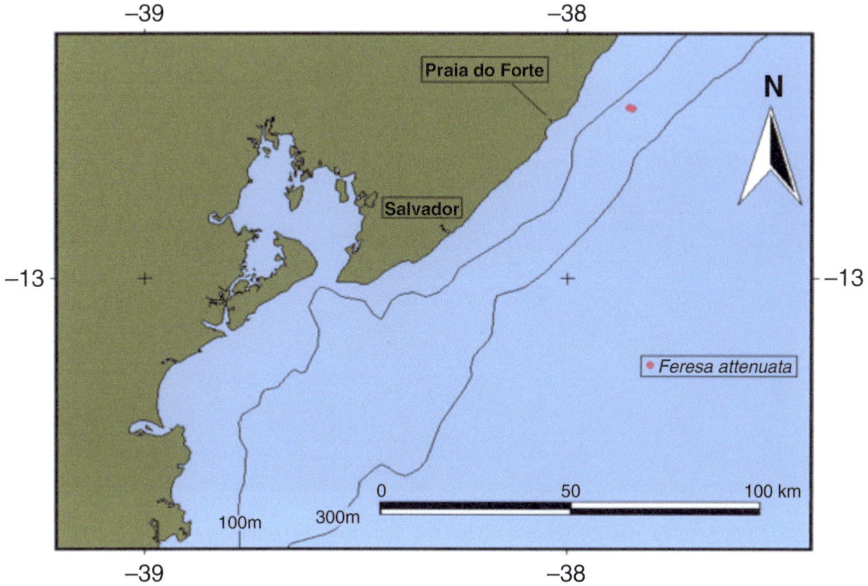

Fig. 8.12 Sightings of the pygmy killer whale, Feresa attenuata, in the north coast of Bahia State, Brazil, between 2001 and 2006

direct observation, being the first sighting of 10 min and the second of 20 min. The sightings occurred in the depths of 97 and 200 m. The sea conditions were presented in scale 1 and the wind intensity of 8 knots for both sightings. The two sightings had occurred near to Praia do Forte (Fig. 8.13).

Peponocephala electra **(Gray, 1846)**
This species, the melon-headed dolphin, *Peponocephala electra*, was also recorded only once, in 2004. The total number was of 20 individuals, observed in 1 group, with 20 min of direct observation. The sea conditions were presented in scale 1 with intensity of the wind was of 16 knots. Its single record occurred near to Praia do Forte (Fig. 8.14).

Stenella frontalis (*G. Cuvier*, **1829)**
In the present study, the Atlantic spotted dolphin, *Stenella frontalis*, (Fig. 8.15) was sighted just once, in the year of 2004, in the coordinates 12° 16′S and 38° 51′W. There were 02 individuals in total, observed for 5 min, bow riding the research boat. The local depth was of 51 m, with the sea conditions in scale 1 and intensity of the wind of 9 knots. The sighting occurred near to Salvador (Fig. 8.16).

The total number of individuals recorded in the study period (2001 to 2006) was 562 animals. *T. truncatus*, *S. guianensis* and *S. clymene* were the most representative species, with 274, 115 and 113 individuals recorded, respectively. The species *S. bredanensis*, *P. electra*, *F. attenuata* and *S.frontalis* were the least sighted with

Fig. 8.13 Pygmy killer whales (Feresa attenuata) showing characteristics as the rounded head (A, D), clearly demarcated dorsal cape and white tooth marks (B), Pointed dorsal fin (C) in an unusual group observed in Praia do Forte, Bahia State, Northeastern Brazil (Photos: Marcos Rossi-Santos/IBJ)

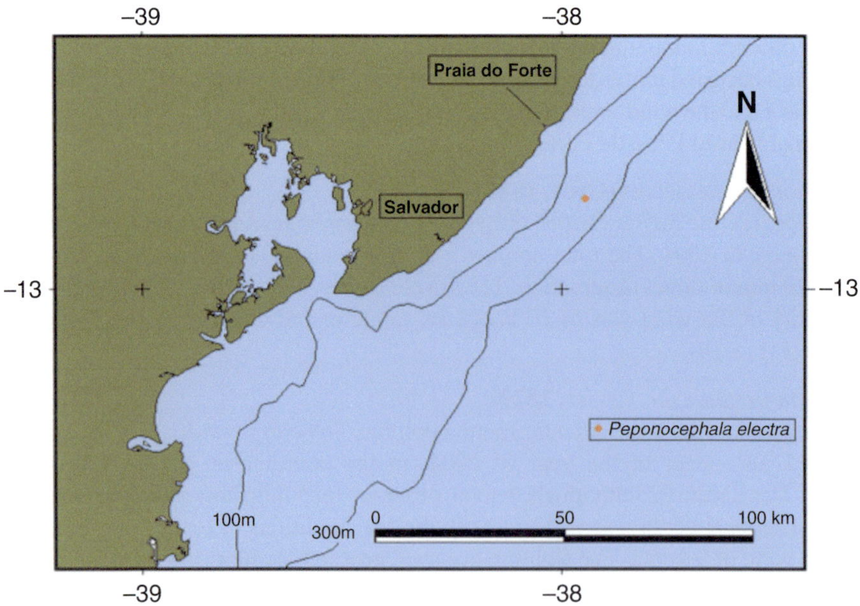

Fig. 8.14 Sightings of the melon-headed dolphin, *Peponocephala electra*, in the north coast of Bahia State, Brazil, between 2001 and 2006

Fig. 8.15 The Atlantic spotted dolphin, *Stenella frontalis*, in the north coast of Bahia State, Brazil, between 2001 and 2006 (Photo: Marcos Rossi-Santos)

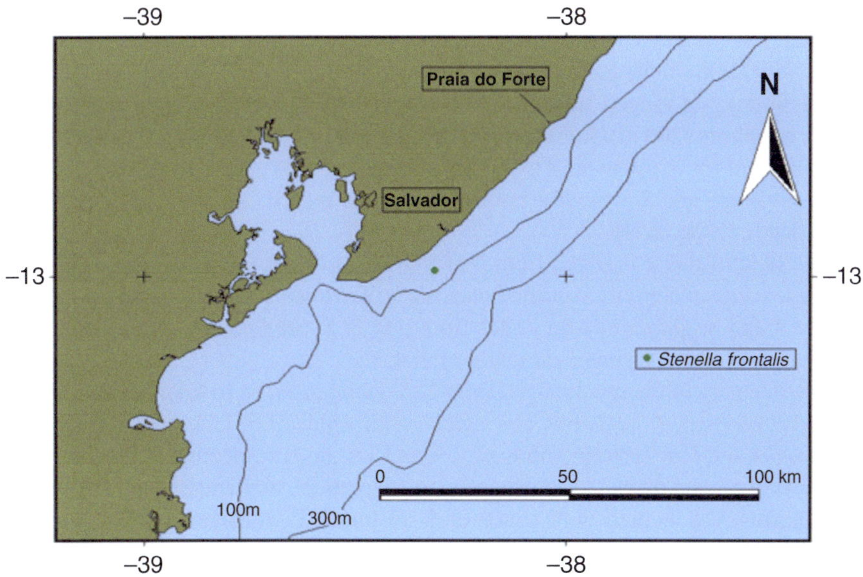

Fig. 8.16 Sightings of the Atlantic spotted dolphin, *Stenella frontalis*, in the north coast of Bahia State, Brazil, between 2001 and 2006.

Table 8.1 Total number of individuals, variation of individuals by group and years of occurrence of each species in the period from 2001 to 2006

Species	Total number of individuals	Δ Variation of individuals by group	Years of occurrence
Tursiops truncatus	274	01–30	2001–2006
Sotalia guianensis	115	01–35	2003–2006
Steno bredanensis	27	01–06	2002, 2004, 2005 and 2006
Stenella clymene	113	01–100	2001 and 2003
Feresa attenuata	11	05–06	2006
Peponocephala electra	20	20	2004
Stenella frontalis	02	02	2004

27, 20, 11 and 02 individuals, each. The species that had the largest variation in the number of individuals per group was *S. clymene*, 01 to 100 individuals. *S. guianensis* and *T. truncatus* were the species that had the most variation in the number of individuals per group after *S. clymene*, with 01 to 35 and 01 to 30 individuals, respectively. The only species that was recorded in all the years of the study was *T. truncatus*, followed by *S. guianensis* and *S.bredanensis* that were not present in 2 years of the study. *S. clymene* was present in only 2 years of study (2001 and 2003). *P. electra* and *S. frontalis* were only sighted in the year 2004, and *F. attenuata* in the year 2006 (Table 8.1).

Depth logging ranged from 4 to 200 m. The species recorded in greater depth was *F. attenuata* (200 m), followed by *T. truncatus* (150 m). The lowest depth record was recorded for *S. guianensis* (4 m), but the species was also sighted in 60 m deep sites. *P. electra* and *S. frontalis*, both with a single record, were sighted at 150 and 51 m depth, respectively.

The state of the sea ranged from 0 to 5 (Beaufort scale). *F. attenuata, P. electra* and *S. frontalis* were only recorded on scale 1. *T. truncatus* was recorded on scales of 0 to 4 and *S. guianensis* on scales from 0 to 3. *S. bredanensis* was recorded on scales of 1–5 and *S. clymene* on scales of 2–4.

The intensity of the wind ranged from 3 (*S. guianensis*) to 16 knots (*S. bredanensis* and *P. electra*). *T. truncatus* was sighted with winds of 5–15 knots. *F. attenuata, S. frontalis* and *P. Electra* in winds of 8, 9 and 16 knots, respectively. The intensity of the wind ranged from 3 (S. guianensis) to 16 knots (S. bredanensis and P. electra). T. truncatus was sighted with winds of 5–15 knots. F. attenuata, S. frontalis and P. Electra in winds of 8, 9 and 16 knots, respectively. All these data is shown on the Table 8.2.

Table 8.2 Depth variation (meters), sea state variation (Beaufort scale) and variation of wind intensity (nodes) between the years 2001 and 2006

Species	Δ Depth variation	Δ Sea state variation	Δ Variation of wind intensity
Tursiops truncatus	10–50	0–4	05–15
Sotalia guianensis	04–60	0–3	03–12
Steno bredanensis	27–45	1–5	09–16
Stenella clymene	42–121	2–4	09–14
Feresa attenuata	97–200	1	08
Peponocephala electra	51	1	09
Stenella frontalis	150	1	16

8.4 Discussion

In this work we found that tail up behavior varied in frequency according to the depth. It is probable that the narrow continental shelf is influencing this fact, once circa 60% of the sightings for humpback whales in this area are situated in waters of 35–55 m deep (Rossi-Santos et al. 2008).

Seven different species of dolphins had been sighted during the research conducted through the coast of the North shore of Bahia, between the years 2001 and 2006. Al these species are confirmed by literature (eg. Pinedo et al. 1992; Hetzel and Lodi 1993), except for the *Feresa attenuata* that had its first record of sighting for the Brazilian coast (Rossi-Santos et al. 2006b) and *Stenella frontalis* that had its first record of sighting for the coast of Bahia State. Species of both coastal and pelagic habits had been sighted during the study, fact that could be associated with oceanographic characteristic, such as the narrow continental platform, with about 7 nautical miles, in the study area (DHN 1995).

The bottlenose dolphin, *T. truncatus*, was the most frequent species during the period of the study. It is the most common dolphin species and more studied in all the oceans (e.g. Leatherwood and Reeves 1983; Bearzi 2005; Blasi and Boitani 2012). In Brazil, despite it is not the most studied species, the dolphin have been increasingly investigated (eg., Sampaio and Reis 1998; Simões-Lopes and Fábian 1999; Silva et al. 2002; Baracho et al. 2005; Rossi-Santos et al. 2006a; Baracho et al. 2008; Lodi et al. 2008, Carvalho and Rossi-Santos 2011; Tardin et al. 2013). Although the majority of the studies are about coastal populations, the species had being sighted in oceanic waters (Sampaio and Reis 1998; Cremer et al. 2002; Silva et al. 2002; Caon and Ott 2004; Baracho et al. 2005; Baracho et al. 2008; Carvalho and Rossi-Santos, 2011).

Group size varied between 1 and 30 individuals. *T. truncatus* is found generally in relatively small groups of 1–20 animals (Shane et al. 1986; Bastida et al. 2007; Carvalho and Rossi-Santos 2011; Lodi and Monteiro-Neto 2012). An notable exception is mentioned for *T. aduncus* in the South Africa, where the size of the group varied of 3–1000 and the average of animals for group was of 140 (Saayman and Tayler 1973).

In the present work, the species occurred in depths of 10–150 m. Studies related to the *T. truncatus* in Brazil, indicate the presence of the specie in areas of depths varying between 15 and 300 m (Skaf and Secchi 1994; Baracho et al. 2005; Baracho et al. 2008; Carvalho and Rossi-Santos 2011; Tardin et al. 2013), what does not corroborate with the data found in the present work.

The Guiana dolphin, *Sotalia guianensis*, was the second more representative species during the period of study. This species is so far the most studied dolphin species in Brazil, mainly because it inhabits inside of bays and estuaries along all the Brazilian coast (eg., de Oshima and de Santos 2016; Godoy 2011; Lima 2010; Santos et al. 2010; de Santos et al. 2010; Araújo et al. 2007; Rossi-Santos 2006; Spinola 2006; De Jesus 2004; Lodi 2003; Reis et al. 1993). Here we raise special concern to the fact of the location of the sightings presented in this work are inside an area considered as a gap of knowledge for the species (Hubner et al. in press). There are some small rivers in the north coast of Bahia that could contribute for occurrence of the species in this unstudied area. Rossi-Santos et al. (2006a, b), studying distribution and use of the habitat of the small cetaceans in the Abrolhos Bank, found that *Sotalia guianensis* was the most common species, followed by *Steno bredanensis* and *Tursiops truncatus.*

In the present work, group size varied from 1 to 35 individuals. In Baía do Pontal, Ilhéus, southern Bahia, the group size varied form 1–7 individuals (Santos 2010a). In another study in the same area Imamura (2015) found groups raging 2–10 individuals.

Along the Brazilian coast, groups of 2–20 individuals are the most sighted number observed in Sepetiba Bay, Rio de Janeiro State, despite groups up to 280 were recorded (Flach et al. 2008). For the Baía de Paraty, in this same State, group size varied from 1 to 100 individuals (Lodi 2003). For the Norte Bay, Santa Catarina State, and southern limit of distribution to the species, studies indicate an average group size of 29 individuals (Daura-Jorge et al. 2005) and maximum group size between 60 and 80 animals (Flores and Bazzalo 2004). In estuarine regions, such as Cananéia, São Paulo State and Caravelas, Bahia State, small group sizes (1–14 animals) are reported as the most frequent category for the Guiana dolphin (Geise et al. 1999; Rossi-Santos 2006; Godoy 2011).

Here, we reported the species occurring in depths of 4–60 m, and sightings of *S. guianensis* in waters deeper than 30 m are quite rare to the species (see Borobia et al. 1991). Innumerous works have reported for the species, its habitat use of shallow waters. For example, in Baía do Pontal, Ilhéus, southern Bahia, the highest frequency of sightings occurred in deeper areas. Nearly half of the records (49.3%) occurred in quadrants with 3–6.79 m depth (Santos et al. 2010).

In Caravelas, southern Bahia, another estuarine environment, *S. guianensis* occurred more frequently in shallow water areas (0–6 m) (Rossi-Santos 2006), corroborating this preference of the species habitat use. In Cananéia, it was observed Guiana dolphins in feeding behavior in depths varying between 1 and 15 m (Geise et al. 1999; Godoy 2011). However, in the same study, the animals, when in traveling behavior, utilized areas with approximately 10 m. In the Baía de Guanabara, Rio de Janeiro, it was also reported the species occurring in areas with depth of 10 m

(Azevedo et al. 2005) while in the Baía de Sepetiba, *S. guianensis* occurred in areas with depths varying between 3 and 30 m (Simão and Poletto 2002).

Our results allied to other opportunistic sightings of the species, during the summer months, support the affirmation that the estuarine dolphin occurs frequently in this poorly studied stretch of the coast.

These factors associated with the short extension of the continental platform in this stretch of the Bahia State, offer a mosaic of environments, where beyond coastal/ estuarine, widely reported as area of occurrence of the species, exists also the oceanic/ pelagic where we evidence that the specie also frequents.

The presents observations in relation to the distribution and depth differ from world-wide literature on the occurrence of the *Steno bredanensis*, rough-toothed dolphin, that usually record the species occurring in deep waters, out of the edge of the continental platform (Leatherwood and Reeves 1983; Miyasaki and Perrin 1994). On the other hand, *S. bredanensis* is registered occurring in coastal flat waters in Brazil (Lodi and Capistrano 1990; Ott and Danilewicz 1996; Flores and Ximenez 1997; Lodi and Hetzel 1998, 1999; Wedekin et al. 2004).

Considering that the continental platform of the state of Bahia, Brazil, is relatively narrow, we could say that *S. bredanensis* in the north coast of Bahia possibly have preferences for shallow and coastal waters. In the Bank of Abrolhos and throughout the coastal archipelago of the Biological Reserve of Arvoredo in the state of Santa Catarina, *S. bredanensis* also occurs in shallow waters, what probably it can be a standard for the coast of Brazil (Lodi and Hetzel 1998).

Group size varied could be characterized as small, from 1 to 6 individuals, when compared to records of more than 50 animals registered in tropical eastern Pacific (Leatherwood et al. 1988), five groups in Japanese waters involving of 23–53 animals, or even smaller groups of 10–20 that seem to be more usual (Miyasaki and Perrin 1994). Our study corroborates a previous revision about the species in Brazilian waters, that suggests group size of *S.bredanensis* between 1 and 10 animals (Lodi and Hetzel 1998).

In Brazil, the *Stenella clymene*, have been early registered for Santa Catarina and northeast region (Simões-Lopes et al. 1994; Zerbini et al. 2004).

The depth data of in this work contradict with the literature (eg. Moreno et al. 2005; Davis et al. 1998; Bastida et al. 2007), that suggests this species of dolphin as having preference for deep waters, generally of 1000–4500 m, beyond the continental shelf. According to Davis et al. (1998), sightings of *S. clymene* whitin of the continental shelf are not common.

Groups of *S. clymene* generally are numerous, however in the Brazilian coast it may be more variant, from about 8 to 300 individuals (Moreno et al. 2005). This species was also considered as the most frequent stranded dolphin species between 2000 and 2002 in Northeastern Brazil (Alvite et al. 2004), what can coincide with the period of sightings made for the research in the coast of Bahia.

There is no sighting record for the spotted dolphin, *Stenella frontalis*, in the coast of Bahia, first considered as an hiatus of distribution to the spotted dolphin (Moreno et al. 2005), therefore this work reports on the first register of sighting for the species in this area.

Different of the other species of the genus *Stenella*, the *S. frontalis* has preference for more coastal waters, habiting regions with depths varying between 20 and 961 m (Moreno et al. 2005) and is considered as the most common in the south and Southeast Brazil (Zerbini et al. 2004). This pattern of distribution of the *S. frontalis* is similar to one observed throughout the Southeastern coast of the United States and Gulf of Mexico, where the species occurs in regions of depths between 20 and 200 m (Jefferson and Schiro 1997; Würsig et al. 2000).

The pygmy killer whale, *Feresa attenuata*, is one of the less studied dolphin species in the world because its pelagic habits. The observed sightings contribute to the diversity of cetaceans in the study area. The fact of this species was observed near to the coast could be a result of oceanographic features, mainly the narrow continental shelf or even to prevent any type of noise in oceanic waters, such as ongoing seismic activity off Brazilian coast (Rossi-Santos et al. 2006b).

In general, *F. attenuata* is seen in small groups, with 50 individuals being more typical (Carwardine et al. 2002; Bastida et al. 2007), what it corroborates with the sightings of the work.

There are many records of strandings of *Peponocephala electra* for the northeast of Brazil, including a massive strand of at least 240 individuals, in Itacaré (Lodi et al. 1990). Sightings of the species are rare, with only one record reported until the present (Zerbini et al. 1999).

Is common to see groups of 100–500 individuals (Carwardine et al. 2002; Bastida et al. 2007). Our results bring information about smaller groups of *P. electra*. There is another record about groups of 14 individuals for the Southeast Caribbean (Watkins et al. 1997). Despite this similarity, in the Caribbean, the species was registered occurring in waters of 1200 m of depth (Watkins et al. 1997), what is different of the depth found in this work.

Anthropogenic factors as traffic of boats, activity of production of oil and gas and seismic exploration can also influence the distribution of the dolphins (eg., Richardson et al. 1995). It is believed that the first influence on the cetacean distribution is probably the aggregation of prey species (Davis et al. 1998), which is determinant in the occurrence, social organization, composition and group size of dolphins (Wells et al. 1980).

8.5 Conclusions

There is a considerable difference between the spatial distribution for the three more frequent species, where the *T. truncatus* if distributed for all coast, the *S. guianensis* showed preference for coastal waters, the *S. bredanensis* is limited to two places of the coast, Praia do Forte and Itacaré, making a great interval in the occurrence of the species inside of the study area.

Differences in the selection of habitat among the species can reflect distinct habits of foraging strategies, and also can contribute for reduction of the competition between species that occur in the same geographic area. In our work, according oceanographic features of the study area it seems that the depth and extension of the continental shelf are contributing to shape the odontocete distribution, what deserves further investigation.

References

Alvite CMC, Jesus AH, Meirelles ACO, Aguiar C, Silva CPN, Castro DF, Júnior ÉS, Luna FO, Lima F, Macedo GD, Sazima I, Borges JCG, Vergara-Parente JE, Júnior JMS, Nascimento LF, Dórea-Reis LW, Engel MH, Silva MB, Marcondes MCC, Medeiros PIAP, Maia-Nogueira R, Nunes SF, Godoy T, Lima RP (2004) Northeastern aquatic mammals strandings network, Brazil (REMANE): 2000 to 2002. 11a. Reunião de Trabalho de Especialistas em Mamíferos Aquáticos da América do Sul (RT). Equador, Quito

Andriolo A, Kinas PG, Engel MH, Martins CCA, Rufino AM (2010) Humpback whales within the Brazilian breeding ground: distribution and population size estimate. Endanger Species Res 11:233–243. doi:10.3354/ esr00282

Andriolo A, Zerbini AN, Moreira S, Pizzorno JL, Danilewicz D, Maia YG, Mamede N, Castro FR, Clapham P (2014) What do humpback whales *Megaptera novaeangliae* (Cetartiodactyla: Balaenopteridae) pairs do after tagging? Fortschr Zool 31(2):105–113

Araújo JP, Araújo ME, Souto A, Parente CL, Geise L (2007) The influence of seasonality, tide and time of activities on the behavior of Sotalia guianensis (Van Bénéden) (Cetacea, Delphinidae) in Pernambuco. Brazil Rev Bras Zool 24:1122–1130

Assis CV (2008) Comportamento, padrões de pesca e caracterização dos grupos de boto-cinza, Sotalia guianensis (van Benéden, 1864) (CETACEA: DELPHINIDAE) no Porto Comercial de Ilhéus, Bahia. Dissertação de mestrado, Universidade Estadual de Santa Cruz, Ilhéus, Bahia

Azevedo AF, Viana SC, Oliveira AM, Sluys MV (2005) Group characteristics of marine tucuxis (*Sotalia fuviatilis*) (Cetacea: Delphinidae) in Guanabara Bay, south-eastern Brazil. J Mar Biol Assoc UK 85:209–212

Baracho C, Cipolotti S, Marcovaldi E, Apolinário M, Silva MB (2005) Ocorrência do golfinho-nariz-de-garrafa (*Tursiops truncatus*) na reserva biológica do atol das rocas. Workshop "Ilhas Oceânicas Brasileiras – Da Pesquisa ao Manejo" Museu Nacional (UFRJ), Rio de Janeiro

Baracho C et al (2008) The occurrence of bottlenose dolphins (*Tursiops truncatus*) in the biological reserve of Atol das Rocas in north-eastern Brazil. J Mar Biol Assoc UK 2 – Biodiver Rec 1:1–4

Bastida R, Rodriguez D, Secchi E, Silva V (2007) Mamíferos Acuáticos de Sudamérica y Antártica, 1ª edn. Vázquez Mazzini Editores, Buenos Aires

Batista RLG (2001) Estudo dos padrões comportamentais do boto cinza *Sotalia fluviatilis* (Gervais, 1853) (Cetacea, Delphinidae) no litoral de Ilhéus-BA. Monografia de Bacharelado, Universidade Estadual de Santa Cruz, Ilhéus

Batista RLG, Bastos BL, Maia-Nogueira R, Reis MSS (2005) Rescue and release of two Estuarine Dolphins, *Sotalia fluviatilis* (Gervais, 1853), found confined in a natural pool of the Cachoeira River, Ilhéus, Southern Bahia. Brazil Aquat Mamm 31(4):434–437. http://dx.doi.org/10.1578/AM.31.4.2005.434

Batista RLG, Schiavetti A, Santos UA, Reis MSS (2012) Cetaceans registered on the coast of Ilhéus (Bahia), northeastern Brazil. Biota Neotrop 12(1): http://www.biotaneotropica.org.br/v12n1/en/abstract?article+bn00312012012

Baumgartner MF (1997) The distribution of Risso's dolphin (*Grampus griseus*) with respect to the physiography of the northern Gulf of Mexico. Mar Mamm Sci 13:614–638

Baumgartner MF, Mullin KD, May LN, Leming TD (2001) Cetacean habitats in the northern Gulf of Mexico. Fishery Bulletin US 99:219–239

Bearzi M (2005) Aspects of the ecology and behaviour of bottlenose dolphins (*Tursiops truncatus*) in Santa Monica Bay, California. J Cetacean Res Manag 7(1):75–83
Blasi MF, Boitani F (2012) Modelling fine-scale distribution of the bottlenose dolphin Tursiops truncatus using physiographic features on Filicudi (southern Thyrrenian Sea, Italy). Endanger Species Res 17:269–288. http://dx.doi.org/10.3354/esr00422
Borobia M, Siciliano S, Lodi L, Hoek W (1991) Distribution of South American dolphin *Sotalia fluviatilis*. Can J Zool 69:1025–1103
Caon G, Ott PH (2004) Populações oceânicas de golfinhonariz-de-garrafa (Tursiops truncatus): fotoidentificação e fidelidade no Arquipélago de São Pedro e São Paulo, nordeste do Brasil. In: Resúmenes 11ª Reunión de Trabajo de Especialistas en Mamíferos Acuáticos de America del Sur, Resúmenes 5° Congreso de La Sociedad Latinoamericana de Especialistas en Mamíferos Acuáticos; 2004; Quito. Quito, p 44
Carvalho MS, Rossi-Santos MR (2011) Sightings of the bottlenose dolphin (Tursiops truncatus) in Trindade island, South Atlantic Ocean. Mar Biodiver Rec 4:1–4. http://dx.doi.org/10.1017/S1755267211000029
Carwardine M, Hoyt E, Fordyce RE, Gill P (2002) Whales & Dolphins: the ultimate guide to marine mammals. Haper Collins Publishers, London
Cremer MJ, Barreto AS, Hardt FA, Tomello Jr AJ, Mounayer R (2002) Aspectos comportamentais da baleia-minke-anã (*Balaenoptera acuturostrata*) e do golfinho-nariz-de-garrafa (*Tursiops truncatus*) nas proximidades de uma plataforma de petróleo em águas oceânicas do sul do Brasil. In: Reunion de Trabajo de Especialistas em Mamíferos Acuáticos de América Del Sur, 10ª. Resumos. Valdivia. Universidad Austral do Chile, p 115
Daura-Jorge FG, Wedekin LL, Piacentini VQ, Simões-Lopes PC (2005) Seasonal and daily patterns of group size, cohesion and activity of estuarine dolphin, *Sotalia guianensis* (P.J. Van Bénéden) (Cetacea, Delphinidae), in southern Brazil. Revista Brasileira de Zoologia 22(4):1014–1021
Davis RW, Fargion GS, May N, Leming TD, Baumgartner M, Evans WE, Hansen LJ, Mullin K (1998) Physical habitat of cetaceans along the continental slope in the north, central and western Gulf of Mexico. Mar Mamm Sci 14:490–507
De Jesus AH (2004) Perfil comportamental do boto cinza (*Sotalia fluviatilis*) (Gervais, 1853) (Delphinidae: Cetacea) no litoral sul do Rio Grande do Norte, Brasil. Master thesis, Universidade Federal do Rio Grande do Norte, Brazil
de Oshima JEF, de Santos MCO (2016) Guiana dolphin home range analysis based on 11 years of photo-identification research in a tropical estuary. J Mammal 97(2):599–610
de Santos MCO, Oshima JEF, Pacífico ES, Silva E (2010) Group size and composition of Guiana dolphins (Sotalia guianensis) (Van Bénèden, 1864) in the Paranaguá Estuarine Complex, Brazil. Braz J Biol 70:111–120
DHN/Departamento de Hidrografia e Navegação (1995) Tábuas das marés. Marinha do Brasil. Diretoria de hidrografia e navegação, pp 90–92
Flach L, Flach P, Chiarello AG (2008) Aspects of behavioral ecology of *Sotalia guianensis* in Sepetiba Bay, Southeast Brazil. Mar Mamm Sci 24(3):503–515
Flores PAC, Bazzalo M (2004) Home range and movement patterns of the marine tucuxi, *Sotalia fluviatilis*, in Baía Norte, Southern Brazil. Lat Am J Aquat Mamm 3:37–52
Flores PAC, Ximenez A (1997) Observations on the rough-tootthed dolphin *Steno bredanensis* off Santa Catarina Island, southern Brazilian coast. Biotemas 10(1):71–79
Flores PAC, Bazzalo M, Caballero S, Santos MCO, Trujillo F, Azevedo AF, Cremer MJ, Simoes-Lopez PC, Callado LM, Flach L, Barrios H, Montiel MG, Rossi-Santos MR, Meirelles ACO, Bolaños J, Silva FJL (2008) Proposed English common name of *Sotalia guianensis* to the IWC list of recognised cetacean species. 3pp. In: Paper SC/60/SM16 presented to the IWC Scientific Committee
Forney KA (2000) Environmental models of cetacean abundance: reducing uncertainty in population trends. Conserv Biol 14:1271–1286
Geise L, Gomes N, Cerqueira R (1999) Behaviour, habitat use and population size of *Sotalia fluviatilis* (Gervais, 1853) (CETACEA, DELPHINIDAE) in the Cananéia estuary region, São Paulo, Brazil. Rev Bras Biol 59(2):183–194

Godoy DF (2011) Utilização de hábitat pelo boto-cinza, *Sotalia guianensis* (Cetacea, Delphinidae), na região do complexo estuarino Lagunar de Cananéia, São Paulo. Dissertação de mestrado, Universidade Federal de Juiz de Fora, Juiz de Fora – Minas Gerais

Gowans S, Whitehead H (1995) Distribution and habitat partitioning by small odontocetes in the Gully, a submarine canyon on the Scotian shelf. Can J Zool 73:1599–1608

Hamazaki T (2002) Spatiotemporal prediction models of cetacean habitats in the mid-western North Atlantic Ocean (from Cape Hatteras, North Carolina, U.S.A. to Nova Scotia, Canada). Mar Mamm Sci 18:920–939

Hastie GD, René JS, George S, Paul MT, Willian RT (2005) Environmental models for predicting oceanic dolphin habitat in the Northeast Atlantic. J Mar Sci 62:760–770

Hetzel B, Lodi L (1993) Baleias, Botos e Golfinhos: Guia de Identificação para o Brasil. Nova Fronteira, Rio de Janeiro

Hubner A, Pansard KCA, Gondim MA, Medeiros PIAP, Ananias SMA, Yamamoto ME (in press) Relatório do I Workshop do Nordeste de Pesquisa e Conservação de *Sotalia guianensis*. In: Hubner A, Medeiros PIAP, Silva FJL (eds) Boto-cinza *Sotalia guianensis*: Pesquisa e Conservação no Nordeste do Brasil. I Workshop do Nordeste de Pesquisa e Conservação de *Sotalia guianensis*, vol 1. UERN, Mossoró

Imamura MM (2015) Reações de *Sotalia guianensis* (Cetartiodactyla, Delphinidae) na presença de embarcações: efeitos sobre o comportamento de superfície. Dissertação de mestrado, Universidade Estadual de Santa Cruz, Ilhéus

Izidoro FB (2009) Identificação e Descrição das atividades de jogo em boto-cinza (Sotalia guianensis) no Porto de Ilhéus BA. Dissertação de mestrado, Universidade Estadual de Santa Cruz, Ilhéus

Jefferson TA, Schiro AJ (1997) Distribution of cetaceans in the offshore Gulf of Mexico. Mammal Rev 27:27–50

Leatherwood S, Reeves RR (1983) Whales and dolphins (The Sierra Club Handbook). Sierra Club, San Francisco. ISBN:0-87156-340-1

Leatherwood S, Reeves RR, Perrin WF, Evans WE (1988) Ballenas, delfines y marsopas del Pacifico nororiental y de las aguas articas adyacentes: una guía para su identificación. In: Comision Interamericana del Atun Tropical, Special Report No 6, La Jolla, California, pp178–183

Lima MS (2010) Uso de Habitat e Comportamento do Boto-cinza Sotalia guianensis em uma Zona Estuarina do Estado de Sergipe, Nordeste do Brasil. Dissertação de mestrado, Universidade Federal do Rio Grande do Norte, Natal – RN.

Lodi L (2003) Tamanho e Composição de grupo dos botos-cinza, *Sotalia guianensis* (van Bénéden, 1864) (Cetacea, Delphinidae), na Baía de Paraty, Rio de Janeiro, Brasil. Atlântica, Rio Grande 25(2):135–146

Lodi L, Capistrano L (1990) Capturas acidentais de pequenos cetáceos no litoral norte do Rio de Janeiro. Biotemas 3(1):47–65

Lodi L, Hetzel B (1998) O Golfinho-de-Dentes-Rugosos (*Steno bredanensis*) no Brasil. Bioikos 12(1):29–45

Lodi L, Hetzel B (1999) Rough-toothed dolphin, *Steno bredanensis*, feeding behaviors in Ilha Grande Bay, Brazil. Biociencias 7(1):29–42

Lodi L, Monteiro-Neto C (2012) Group size and composition of Tursiops truncatus (Cetacea: Delphinidae) in a coastal insular habitat off southeastern Brazil. Biotemas 25:157–164. http://dx.doi.org/10.5007/2175-7925.2012v25n2p157

Lodi L, Siciliano S, Capistrano L (1990) Mass stranding of Peponocephala electra (Cetacea, Globicephalinae) on Piracanga Beach, Bahia, Northeastern Brazil. Sci Rep Cetacea Res Tokyo 1(1):79–84

Lodi L, Wedekin LL, Rossi-Santos MR, Marcondes MC (2008) Deslocamentos do golfinho-nariz-de-garrafa (*Tursiops truncatus*) no estado do Rio de Janeiro, sudeste do Brasil. Biota Neotropica 8(4): http://www.biotaneotropica.org.br/v8n4/pt/abstract?article+bn00808042008. ISSN:1676-0603

Magalhães FA, Garri RG, Tosi CH, Siciliano S, Chellapa S, Silva JL (2007) Primeiro registro confirmado de *Feresa attenuata* (Delphinidae) para o litoral norte do Brasil. Biota Neotrop 7(2):313–318

Martins CCA, Morete ME, Engel MH, Freitas AC, Secchi ER, Kinas PG (2001) Aspects of habitat use patterns of humpback whales in the Abrolhos Bank, Brazil, breeding ground. Memoirs of the Queensland Museum 47:563–570

Miyasaki N, Perrin WP (1994) Rough-toothed dolphin, *Steno bredanensis* (Lesson, 1828). In: Ridgway SH, Harrison R (eds) Handbook of marine mammals, vol 5. Academic, London, pp 1–21

Moraes BC (2010) Aplicação da fotoidentificação com botos-cinza, Sotalia guianensis (van Bénédén, 1864), no litoral Ilheense, Bahia: Residência, padrões de associação estimativa populacional. Monografia de Graduação, Universidade Estadual de Santa Cruz, Bahia

Morete ME, Pace RM, Martins CCA, Freitas A, Engel MH (2003) Indexing seasonal abundance of humpback whales around Abrolhos arquipelago, Bahia, Brazil. Lat Am J Aquat Mamm 2:21–28. doi:10.5597/ lajam00027

Morete ME, Bisi TL, Rosso S (2007) Temporal pattern of humpback whale (*Megaptera novaeangliae*) group structure around Abrolhos Archipelago breeding region, Bahia, Brazil. J Mar Biol Assoc UK 87:87–92

Moreno IB, Zerbini AN, Danilewicz D, Santos MCO, Simões-Lopes PC, Lailson-Brito J Jr, Azevedo AF (2005) Distribution and habitat characteristics of dolphins of the genus *Stenella* (Cetacea: Delphinidae) in the southwest Atlantic Ocean. Mar Ecol Prog Ser 300:229–240

Ott PH, Danilewicz D (1996) Southward rang extension of *Steno bredanensis* in the southwest Atlantic and new records of *Stenella coeruleoalba* for Brazilian waters. Aquat Mamm 22(3):185–189

Pinedo MC, Rosas FC, Marmontel M (1992) Cetáceos e Pinípedes do Brasil. Uma revisão dos registros e guia para identificação das espécies. Manaus. UNEP/FUA, 213p

Reis MSS (2002) O boto *Sotalia fluviatilis* (Gervais, 1853) (Cetacea, Delphinidae) no litoral de Ilhéus-BA: Comportamento e interações com atividades pesqueiras. Dissertação de mestrado, Universidade Estadual de Santa Cruz, Ilhéus

Reis MSS, Reis LWD, Queiroz ELA (1993) Observações sobre comportamento de *Sotalia fluviatilis* (Gervais, 1853) (Cetacea, Delphinidae) na baía de Todos os Santos, Bahia, Brasil. XI Encontro Anual de Etologia. Novembro, 1993, Bauru – SP

Reis MSS, Bauer LM, Lopes PCC, Brito APD (2008) Comportamento de botos-cinza Sotalia guianensis (Cetacea: Delphinidae) na Baía de Todos os Santos, Bahia, Brasil: Uso do habitat, horário de atividades, distribuição e tamanho de grupo. Abstract 012 presented in the XIII Reunion de Trabajo de Especialistas em Mamíferos Acuáticos de America del Sur y 7° Congreso de la Sociedade Latino-Americana de Especialistas em Mamíferos Acuáticos, Montevideo, Uruguay, p 85

Richardson WJ, Greene CR Jr, Malme CI, Thomson DH (1995) Marine mammals and noise. Academic, San Diego. 576pp

Rossi-Santos MR (2006) Ecologia comportamental do boto cinza, *Sotalia guianensis* (Van Bénedén, 1874) (Cetacea:Delphinidae) na região extremo sul do estado da Bahia, Nordeste do Brasil. Tese de Mestrado, Universidade Federal do Paraná, Brazil

Rossi-Santos M, Reis MSS (2008) Relatório do II Workshop do Nordeste sobre pesquisa e conservação de Sotalia guianensis. Editora Editus, pp 11–39

Rossi-Santos MR, Wedekin LL, Sousa-Lima RS (2006a) Distribution and habitat use of small cetaceans off Abrolhos Bank, Eastern Brazil. LAJAM 5(1):23–28

Rossi-Santos M, Baracho C, Neto ES, Marcovaldi E (2006b) First sightings of the pygmy killer whale, *Feresa attenuata*, for the Brazilian coast. JMBA2 – Biodiver Rec

Rossi-Santos MR, Wedekin LL, Monteiro-Filho ELA (2007) Residence and site fidelity of Sotalia guianensis in the Caravelas River Estuary, eastern Brazil. J Mar Biol Assoc UK 87:207–212

Rossi-Santos MR, Neto ES, Baracho CG, Cipolotti SR, Marcovaldi E, Engel MH (2008) Occurrence and distribution of humpback whales (*Megaptera novaeangliae*) on the north coast of the State of Bahia, Brazil, 2000–2006. ICES J Mar Sci 65:667–673

Rossi-Santos MR, Wedekin LL, Monteiro-Filho ELA (2010) Habitat use of the Guiana dolphin, Sotalia guianensis (Cetacea, Delphinidae), in the Caravelas river estuary, eastern Brazil. Lat Am J Aquat Mamm 8(1–2):111–116

Saayman GS, Tayler CK (1973) Social organization of inshore dolphins (*Tursiops aduncus* and *Sousa*) in the Indian Ocean. J Mammal 54:993–996

Sampaio CLS, Reis MSS (1998) Registros de cetáceos na costa nordestina. Resumos da VIII Reunião de Especialistas em Mamíferos Aquáticos da América do Sul. II Congresso da Sociedade Latino-americana de Especialistas em Mamíferos Aquáticos – SOLAMAC. Olinda, 25–29 de outubro de 1998, p 187

Santos UA (2001) Movimentação do boto cinza *Sotalia fluviatilis* (Cetacea, Delphinidae), na baía do Pontal, Ilhéus, Bahia. Monografia de Bacharelado, Universidade Estadual de Santa Cruz, Ilhéus

Santos UA (2007) Uso do hábitat e atividades do Boto-cinza, Sotalia guianensis (Cetacea, Delphinidae), na Baía do Pontal, Ilhéus, Bahia. Dissertação de mestrado, Universidade Estadual de Santa Cruz, Ilhéus

Santos UA, Alvarez MR, Schilling AC, Strenzel GMR, Pendu YL (2010) Spatial distribution and activities ofthe estuarine dolphin Sotalia guianensis (van Bénédén, 1864) (Cetacea, Delphinidea) in Pontal Bay, Ilhéus, Bahia, Brazil. Biota Netrop 10(2):1–7

Santos MC de O, Oshima JEF, Pacífico ES, Silva E (2010a). Group size and composition of Guiana dolphins (Sotalia guianensis) (Van Bénèden, 1864) in the Paranaguá Estuarine Complex, Brazil. Braz J Biol, 70, p. 111–120

Santos UA, Alvarez MR, Schilling AC, Strenzel GMR, Pendu YL (2010) Spatial distribution and activities of the estuarine dolphin Sotalia guianensis (van Bénédén, 1864) (Cetacea, Delphinidea) in Pontal Bay, Ilhéus, Bahia, Brazil. Biota Netrop 10(2):1–7

Schiavetti A, Alarcon DT, Santos UA, Velozo RQ, Lê Pendu Y, Alvarez M RV, Assis CV, Izidoro FB, Santos MS, Batista RLG, Reis MSS (2008) Atuação do grupo de pesquisa em mamíferos aquáticos de Ilhéus, Bahia, Brasil. Abstract 203 presented in the XIII Reunión de Trabajo de Especialistas em Mamíferos Acuáticos de América del Sur y 7° Congreso de la Sociedade Latino-Americana de Especialistas em Mamíferos Acuáticos, Montevideo, Uruguay, p 201

Selzer LA, Payne PM (1988) The distribution of white-sided (*Lagenorhynchus acutus*) and common dolphins (*Delphinus delphis*) vs. environmental features of the continental shelf of the northeastern United States. Mar Mamm Sci 4:141–153

Shane SH, Wells RS, Würsig B (1986) Ecology, behavior and social organization of the bottlenose dolphin: a review. Mar Mamm Sci 2(1):34–63

Silva ED, Carneiro ADVN, Moreira SC, Fernandes T, de Alecastro PMR, Ramos RMA (2002) Ocorrência de cetáceos em áreas profundas da bacia de Santos, região sudeste do Brasil. In: Reunion de Trabajo de Trabajo de Especialistas em Mamíferos Acuáticos de América Del Sur, 10ª. *Resumos*. Valdivia. Universidad Austral do Chile. p 115

Simão SM, Poletto FR (2002) Áreas Preferenciais de pesca e dieta do ecotipo marinho do boto-cinza (*Sotalia fluviatilis*) na Baía de Sepetiba, Rio de Janeiro. Floresta e Ambiente 9(1):18–25

Simões DG, Macedo RHF, Engel MH (2005) Turismo de Observação de Cetáceos como Ferramenta no Estudo do Comportamento de Baleias Jubarte (Megaptera novae angliae). Rev Etol 7(1):3–14

Simões-Lopes PC, Fabián ME (1999) Residence patterns and site fidelity in bottlenose dolphins, *Tursiops truncatus* (Montagu) (Cetacea, Delphinidae) off Southern Brazil. Revista Brasileira de Zoologia, Curitiba 16(4):1017–1024

Simões-Lopes PC, Praderi R, Paula GS (1994) The clymene dolphin, *Stenella clymene* (Gray, 1846), in the southwestern south Atlantic Ocean. Mar Mamm Sci 10(2):213–217

Skaf MK, Secchi E (1994) Avistagens de cetáceos na travessia do Atlântico: Santos-Tenerife. Reunião de Trabalho de Especialistas em mamíferos Aquáticos da América do Sul. *Resumos*, 6 ª. Florianópolis, p 72

Spínola JL (2006) Atividade comportamental diurna do boto-cinza *Sotalia guianensis* (Van Bénédén, 1964) (Cetacea, Delphinidae), na Barra do Rio Paraguaçu, estado da Bahia. Master's thesis, Universidade Federal do Paraná, Paraná, Brazil, 79p

Tardin RH, Simão SM, Alves MAS (2013) Distribution of Tursiops truncatus in Southeastern Brazil: a modeling approach for summer sampling. Natureza & Conservação 11(1):65–74

Watkins WA, Daher MA, Samuels A, Gannon DP (1997) Observations of *Peponocephala electra*, the Melon-headed Whale, in the Southeastern Caribbean. Caribb J Sci 33(1–2):34–40

Wedekin LL, Freitas A, Engel MH, Sazima I (2004) Rough-toothed dolphins (*Steno bredanensis*) catch diskfishes while interacting with humpback whales (*Megaptera novaeangliae*) off the Abrolhos Bank breeding ground, Southwest Atlantic. Aquat Mamm 30(2):327–329

Wedekin LL, Neves MC, Marcondes MCC, Baracho C, Rossi-Santos MR, Engel MH, Simões-Lopes PC (2010) Site fidelity and movements of humpback whales (*Megaptera novaeangliae*) on the Brazilian breeding ground, southwestern Atlantic. Mar Mamm Sci 26:787–802. doi:10.1111/J.1748- 7692.2010.00387.X

Wells RS, Irvine AB, Scott MD (1980) The social ecology of inshore odontocetes. In: HERMAN LM (ed) Cetacean behavior : mechanisms and functions. Wiley, New York, pp 263–317p

Würsig B, Jefferson TA, Schmidly DJ (2000) The marine mammals of the Gulf of México. Texas A&M University Press, College Station

Zerbini AN; Da Rocha JM, Andriolo A, Siciliano S, Moreno IB, Lucena A (1999) Report of a sighting survey conducted on the former Brazilian whaling ground off the Northeastern coast of Brazil. Paper SC/51/O10 presented to the IWC Scientific Committee, 16pp

Zerbini AN, Secchi ER, Bassoi M, Dalla-Rosa L (2004) Distribuição e abundância relativa de cetáceos na Zona Econômica Exclusiva na Região Sudeste-Sul do Brasil, Série Documentos Revizee—Score Sul. Instituto Oceanográfico, Universidade de São Paulo, São Paulo

Zerbini AN, Andriolo A, Heide-Jørgensen MP, Pizzorno JL, Maia YG, Van Blaricom GR, DeMaster DP, Simões-Lopes PC, Moreira S, Bethlem C (2006) Satellite-monitored movements of humpback whales *Megaptera novaeangliae* in the Southwest Atlantic Ocean. Mar Ecol Prog Ser 313:295–304

Chapter 9
Using Chemical Elements to the Study of Trophic and Spatial Ecology in Marine Mammals of the Southwestern Atlantic Ocean

Elisa Seyboth, Silvina Botta, and Eduardo Secchi

Abstract The ecology of marine mammals has been largely benefited from the use of biogeochemical markers. In this chapter we compiled information about studies using these tracers, with the aim of highlighting their relevance to understand some aspects of the trophic and spatial ecology of cetaceans, pinnipeds and sirenians from southwestern Atlantic Ocean (SAO). Stable isotope analysis is currently the mostly applied technique for such studies and its use increased considerably in the last few years. It is based on the premise that stable isotope composition in the tissues of the consumers reflects the composition of their diet that is effectively assimilated. Depending on the element being analyzed, information about habitat use patterns, diet composition and trophic position of the organisms can be inferred. Trace element analysis is underexplored as ecological tracers for marine mammals. There is only one study that analyzed Ba/Ca ratios in teeth of marine and freshwater dolphins to describe their patterns of habitat use in Brazil. Despite the known limitations of these chemical tracers, their use have shed light into different aspects of marine mammals' ecology that would be difficult or impossible to obtain with alternative approaches.

Keywords Chemical tracer • Stable isotope • Ba/Ca • Cetacean • Pinnipeds • Sirenians

E. Seyboth (✉)
Programa de Pós Graduação em Oceanografia Biológica, Instituto de Oceanografia, Universidade Federal do Rio Grande, Rio Grande CEP 96201-900, Brazil

Laboratório de Ecologia e Conservação da Megafauna Marinha, Instituto de Oceanografia, Universidade Federal do Rio Grande, Rio Grande CEP 96201-900, Brazil
e-mail: elisaseyboth@gmail.com

S. Botta • E. Secchi
Laboratório de Ecologia e Conservação da Megafauna Marinha, Instituto de Oceanografia, Universidade Federal do Rio Grande, Rio Grande CEP 96201-900, Brazil
e-mail: silbotta@gmail.com; edu.secchi@furg.br

M.R. Rossi-Santos, C.W. Finkl (eds.), *Advances in Marine Vertebrate Research in Latin America*, Coastal Research Library 22,
DOI 10.1007/978-3-319-56985-7_9

9.1 Introduction

Biogeochemical markers such as stable isotopes and trace elements have been used to investigate aspects of trophic and spatial ecology of marine mammals. Their use has been increasing notably over the last decade. Together with traditional approaches these markers can enhance considerably the knowledge about the trophic ecology and patterns of habitat usage of the animals (Graham et al. 2010; Hobson et al. 2010) and hence, their conservation.

Stable isotope analysis (SIA) has been largely applied to study the ecology of marine megafauna, including foraging, diet, habitat use, movements, ecological partition within and between species and physiology (see reviews by Hobson 1999; Kelly 2000; Newsome et al. 2010). Isotope fractionation, also called trophic discrimination or trophic enrichment, is considered the difference in the isotopic composition between an animal's tissue and its diet (Martínez del Rio et al. 2009). It results from different factors, including multiple ecological, behavioral, and physiological processes of different ecosystem components, as light, temperature, and precipitation (Ben-David and Flaherty 2012). For example, the food preference (herbivore, omnivore or carnivore), the tissue being analyzed and the diet macronutrient abundance of stable isotopes of different elements (commonly carbon, nitrogen, oxygen, hydrogen and sulfur) have effects over the fractionation (Barnes et al. 2009). Stable isotopes of carbon ($^{13}C/^{12}C$) and nitrogen ($^{15}N/^{14}N$) are the main components of all biological material, thus they are the elements most used in SIA. For the application of SIA, it is assumed that the stable isotope composition of an organism reflects that of its assimilated diet (DeNiro and Epstein 1978, 1981; Peterson and Fry 1987), with some predictable difference found between the sources and the consumers (Peterson and Fry, 1987). These ratios are usually presented in the conventional delta notation (δ), which means they are in parts per-mil (‰) of the ratios found in international measurement standards (e.g. Vienna Pee Dee Belemnite for carbon and atmospheric N_2 for nitrogen) (Ben-David and Flaherty 2012).

Isotopic composition of carbon and nitrogen depends on the biogeochemical characteristics of the ecosystem, to the element fractionation, the composition and assimilation efficiency, and the animal thermoregulation system (ectothermic or endothermic) (Caut et al. 2009; Martínez del Rio et al. 2009; Robbins et al. 2010; Borrell et al. 2012; Martínez del Rio and Carleton 2012).

Nitrogen isotopic values of consumer tissues are generally ^{15}N-enriched relative to its diet (DeNiro and Epstein 1981; Peterson and Fry 1987). This increment is tissue-dependent (e.g. Lesage et al. 2001; Caut et al. 2009), but a general enrichment of 3–4‰ is widely considered in trophic studies (Post 2002). Due to this significant increase along the food web, the nitrogen isotope composition of a consumer is commonly used to estimate its trophic position (DeNiro and Epstein 1981; Minagawa and Wada 1984; Post 2002).

In the case of carbon nitrogen isotopes, although some enrichment occurs along food webs, it presents a lower modification in its composition and thus this element is mainly used to determine the basal sources of food webs (Kelly 2000).

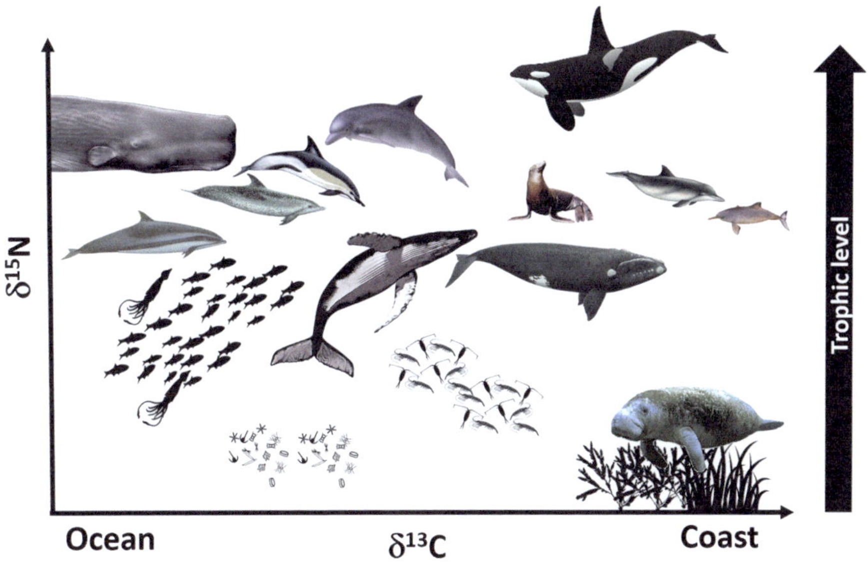

Fig. 9.1 Hypothetic $\delta^{13}C$ and $\delta^{15}N$ values of marine mammals in relation to their habitat use and trophic position. Species figures are not in scale

Furthermore, the values of marine primary producers vary depending on a number of factors including the photosynthetic route, taxonomic differences in isotopic fractionation among others (see review by Graham et al. 2010). Therefore, this variability determines latitudinal, inshore *vs.* offshore, marine *vs.* freshwater and benthic *vs.* pelagic $\delta^{13}C$ gradients (Hobson et al. 1994; Vander Zanden and Rasmussen 1999; Graham et al. 2010) (Fig. 9.1). These spatial gradients allow isotopic data to be used to study habitat use, movements among habitats, and migrations at regional to ocean basin scales (Newsome et al. 2010; McMahon et al. 2013).

Tissues differ in their turnover rates (i.e. the length of time taken to incorporate new compounds) and then they integrate information that represent different period of the animals lifetime. For example, blood is considered a tissue with a fast turnover rate, reflecting the assimilated diet of the last days of an animals' life, while for skin and muscle this period varies from weeks to months. In the case of inert tissues such as tooth, baleen, bones and vibrissae, it can represent the diet of several years or even the entire life of the animal (Koch 2007). Therefore, the choice of the tissue to be used is an important step in SIA and must be carefully made depending on the main aim of the study.

SIA has been widely used to determine the diet composition of predators, through the analysis of the most probable mixture of prey with different stable isotope composition that results in the value observed in consumer's tissues. One of the advantages of SIA over the traditional scat or stomach content analysis is that the former informs about the assimilated diet rather than on the observed prey ingested in the last meals. Moreover, scat and stomach content analysis is prone to biases due to the variation on digestion rates among different prey items.

Furthermore, stable isotope compositions in tissues of marine mammals are also used as an alternative way to investigate the spatial, trophic, and temporal dimensions of the ecological niche (i.e. n-dimensional hypervolume [Hutchinson 1957]) (e.g., Bearhop et al. 2004; Franco-Trecu et al. 2012). In this context, some dimensions of the ecological niche of a predator can be determined through $\delta^{15}N$ (trophic component) and $\delta^{13}C$ and $\delta^{15}N$ (space component) values (Bearhop et al. 2004; Newsome et al. 2007). This isotopic niche can be represented in a multivariate space forming a δ-space with isotopic values as axes and is related to the mixture of food items consumed by a predator. With the increase in the use of this methodology, tools have been developed to visualize and measure the isotopic niche as proposed by Layman et al. (2007). In this approach, the $\delta^{13}C$ and $\delta^{15}N$ values of organisms are the base for the construction of convex hulls, considered the extent of their trophic diversity. Then, statistical mixing models, also called stable isotope mixing models (SIMM), were criated, allowing researchers to deal with various food sources to the predators they were studying under the estimation of the proportional contribution of each source (Phillips 2012). They are mathematical equations that are used to explain the diet of a consumer taking into account the mixture of the isotope values of its assimilated diet (Phillips 2012). Examples of them are the Mix-SIR (Moore and Semmens 2008) and the Stable Isotope Analysis in R (SIAR) (Parnell et al. 2010). Later, Bayesian metrics were incorporeted to represent the isotope niche of an individual, using standard ellipses that represent a species isotope niche (e.g. Jackson et al. 2011; Phillips 2012). This tool can considerably improve determination of diet, especially when the source geometry of the mixing model is poor (e.g. Ward et al. 2010). One of the most used is the Stable Isotope Bayesian Ellipses in R (SIBER) (Jackson et al. 2011), a package developed for the free statistical software R (R Development Core Team). It has the great advantadge of allowing the use of relative small data set and the comparison of the niches among diferente sample size data sets. One of the limitations to which mixing models are subject is the overlap or great similarity that can occur among the food resources of a consumer; the more isotopic distinct they are, better the diet estimation is (Phillips 2012). With the advantages of providing information on the isotope variation within a population, and then on how its ecological niche varies, a novel approach was developed (Semmens et al. 2013). It is called Mix-SIAR, and it can indicate variations in the diet a population in relation to diferente aspects, such as sex, age and geographical location (Semmens et al. 2013).

Carcasses of marine mammals washed ashore or incidentally caught in fishing gear are valuable sources for sampling different tissues, at low cost, for biological and ecological studies through SIA. In different regions of southwestern Atlantic Ocean (SAO), research groups survey the coast and collect carcasses or samples from them. In addition, scientific biopsy sampling of animals from different environments and at different time frames also provide useful material for these studies. As a result, several single- and multi-species studies have been published in recent years, which represent an important contribution to the ecology of these marine mammals.

In this chapter we compilled peer-reviewed published studies that adopted SIA to investigate such ecological aspects of marine mammals in the SAO. First we explore several studies for which this methodology was used to describe trophic and/or spatial ecology of cetaceans, pinnipeds and sirenians (Table 9.1). Then we show results of the only application of Ba/Ca ratios to study the spatial ecology of marine mammals. Finally we conclude our review with some recommendations on future applications that will help encouraging research groups to take further advantage of these increasingly available chemical techniques. We also highlighted the importance of long-term sampling effort and scientific collections as source of valuable material for the use of SIA and trace elements to enhance the knowledge on important aspects regarding the spatial and trophic ecology of marine mammals.

9.2 Stable Isotopes

In this section, we present the main findings of researchers using the stable isotope technique, refered in the introduction of the chapter (Table 9.1), separated by the main groups of marine mammals.

9.2.1 Cetaceans

Cetaceans are highly mobile organisms with sometimes large home ranges. These characteristics together with their complete aquatic habitat, including remote oceanic regions, make the study of their biology and ecology a difficult task. Therefore, SIA has been largely applied to answer variety of research questions about the ecology of this taxon (see review by Newsome et al. 2010). Several research groups in Brazil, Uruguay and Argentina are increasingly applying these chemical tracers as tools for investigating different aspects of the ecology of several cetacean species from SIA. Figure 9.2 ilustrates, though schematically, the outcome of its application for describing the spatial ecology of some species inhabiting the region.

The habitat use and trophic level of several dolphin species in a coastal-oceanic gradient were analyzed by means of SIA in areas ranging from tropical (Bisi et al. 2013; Baptista et al. 2016) to sub-tropical (Botta et al. 2012) and sub-Antarctic (Riccialdelli et al. 2010). The main aim of these studies was to reveal patterns of habitat use and trophic positions of a variety of dolphin species within the coastal to oceanic marine food webs. Stable isotopic values in muscle of bottlenose dolphins (*Tursiops truncatus*), rough-toothed dolphins (*Steno bredanensis*), Guiana dolphins (*Sotalia guianensis*) (Bisi et al. 2013; Baptista et al. 2016), franciscanas (*Pontoporia blainvillei*) (Baptista et al. 2016), Atlantic spotted dolphin (*Stenella frontalis*), pantropical spotted dolphin (*S. attenuata*), spinner dolphin (*S. longirostris*), Risso's dolphin (*Grampus griseus*), Fraser's dolphin (*Lagenodelphis hosei*), common dolphins (*Delphinus* sp.) and false killer whales (*Pseudorca crassidens*) (Bisi et al.

Table 9.1 Major publications regarding stable isotope with marine mammals' tissues in southwestern Atlantic Ocean prior to September 2016

Source	Location	Species	Obs	Tissue(s)	Carbon	Nitrogen	*n* (individuals)	Main topic
Albernaz et al. (2016)	Brazil	*Arctocephalus australis*	Male	T	−12.6 ± 0.4	19.2 ± 0.8	10	TE
			Female		−12.5 ± 0.4	17.0 ± 0.7	12	
		Arctocephalus gazella	Male		−17.7 ± 2.0	11.1 ± 1.7	4	
			Female		−16.8 ± 1.2	10.9 ± 1.1	6	
		Arctocephalus tropicalis	Male		−14.8 ± 1.9	13.8 ± 0.5	6	
			Female		−14.4 ± 0.0	14.8 ± 2.0	2	
Baptista et al. (2016)	Brazil	*Pontoporia blainvillei*		M	−15.5 to −16.1	13.6 to 14.6	16	TE
		Sotalia guianensis			−15.1 to −16.2	14.1 to 15.6	28	
		Steno bredanensis			−15.8 to −17.6	15.2 to 17.8	9	
		Tursiops truncatus			−15.2 to −15.9	15.9 to 16.7	10	
Bisi et al. (2012)	Brazil	*Sotalia guianensis*	Guanabara Bay	M	−14.0 ± 0.7	14.2 ± 0.9	20	TE
			Sepetiba Bay		−14.5 ± 1.0	14.0 ± 0.6	44	
			Ilha Grande Bay		−16.6 ± 0.4	14.2 ± 0.6	10	

Bisi et al. (2013)	Brazil	*Delphinus delphis*		M	−14.5 ± 0.3	15.3 ± 0.9	2	H/TE
		Grampus griseu			−15.7	11.3	1	
		Lagenodelphis hosei			−16.5 ± 0.3	12.6 ± 1.0	10	
		Pseudorca crassidens			−14.5 ± 0.3	15.3 ± 0.9	2	
		Sotalia guianensis	Guanabara Bay		−13.8 ± 0.7	14.2 ± 0.9	26	
			Sepetiba Bay		−14.6 ± 0.9	14.1 ± 0.6	49	
			Ilha Grande Bay		−16.6 ± 0.4	14.2 ± 0.2	10	
			"Região dos Lagos"		−14.9 ± 0.6	13.7 ± 0.7	5	
		Stenella frontalis			−16.0 ± 0.4	13.5 ± 1.0	13	
		Stenella attenuata			−15.8 ± 0.2	11.4 ± 0.2	2	
		Stenella longirostris			−17.1	12.1	1	
		Steno bredanensis			−14.9 ± 0.5	14.4 ± 0.3	3	
		Tursiops truncatus			−15.5 ± 0.5	14.9 ± 1.7	7	
Botta et al. (2012)	Brazil	*Delphinus delphis*		T	−11.7 ± 0.32	16.6 ± 0.59	4	H/TE
		Lagenodelphis hosei			−12.8	14.9	1	
		Orcinus orca			−10.5 ± 0.23	17.7 ± 0.05	2	
		Pseudorca crassidens	Type A		−12.0 ± 1.65	19.2 ± 0.25	2	
			Type B		12.2 ± 0.20	11.2 ± 0.62	6	
		Stenella frontalis			−11.9 ± 0.36	16.6 ± 0.91	5	
		Steno bredanensis			−12.3 ± 1.18	16.7 ± 0.82	5	
		Tursiops sp.			−10.6 ± 0.51	17.6 ± 0.93	24	

(continued)

Table 9.1 (continued)

Source	Location	Species	Obs	Tissue(s)	Carbon	Nitrogen	*n* (individuals)	Main topic
de Castro et al. (2015)	Argentina	*Lagenorhynchus obscurus*	Golfo San Matías/Male	SK	−16.6 ± 0.303	18.4 ± 0.309	7	TE
			Golfo San Matías/Female		−16.6 ± 0.3	18.3 ± 0.391	8	
			Golfo San José/Male		−16.2 ± 0.243	18.8 ± 0.323	8	
			Golfo San José/Female		−16.1 ± 0.199	18.9 ± 0.306	7	
			Golfo Nuevo/Male		−15.8 ± 0.227	18.7 ± 0.324	8	
			Golfo Nuevo/Female		−15.9 ± 0.291	18.6 ± 0.403	6	
			Golfo San Jorge/male		−16.5 ± 0.313	17.9 ± 0.201	3	
			Golfo San Jorge/Female		−16.5 ± 0.303	17.9 ± 0.471	12	
de Castro et al. (2016)	Argentina	*Delphinus delphis*	Buenos Aires	SK	−16.3 ± 0.3	17.6 ± 0.4	20	TE
			Golfo San Matías offshore		−16.5 ± 0.5	18.1 ± 0.5	18	
			Golfo San Matías inshore		−16.2 ± 0.3	19 ± 0.5	24	
Ciotti et al. (2014)	Brazil	*Trichehcus manatus manatus*		T	−12.4 ± −2.5	9 ± 1.3	6	TE
				BO	−12.9 ± 3.4	8.5 ± 1.0	6	
Di Beneditto et al. (2011)	Brazil	*Pontoporia blainvillei*		M	−15.3 ± 0.1	11.8 ± 0.6	2	TE
		Sotalia guianensis		M	−17.4 ± 0.3	12.3 ± 0.6	2	

Di Beneditto et al. (2013)	Brazil	*Pontoporia blainvillei*		L	−15.2 ± 0.6	14.6 ± 1	16	TE
				M	−16 ± 0.5	13.8 ± 1	16	
Drago et al. (2009a)	Argentina	*Otaria flavescens*		BO	−12.2 ± 1	22 ± 1.2	96	TE
Drago et al. (2009b)	Argentina	*Otaria flavescens*	Male	BO	−13.6 to −10.7	19.1 ± 26.2	40	TE
			Female		−14.8 to −11.1	20.1 to 25.9	40	
Drago et al. (2010)	Argentina	*Otaria flavescens*		BL	−15.7 ± 0.3	20.9 ± 0.6	26	TE
				S	−15.8 ± 0.3	23.1 ± 0.6	26	
Drago et al. (2015)	Brazil and Uruguay	*Otaria flavescens*	Male	SK	−13.9 ± 0.2	21.9 ± 0.4	12	TE
			Female		−14.2 ± 0.2	19.8 ± 0.4	10	
			Male	BO	−12.4 ± 0.6	21.6 ± 0.5	6	
			Female		−12.4 ± 0.9	20.5 ± 1.4	10	
Franco-Trecu et al. (2012)	Uruguay	*Arctocephalus australis*		S	−17.90 ± 0.18	18.53 ± 0.59	26	TE
				SK	−15.80 ± 0.16	17.08 ± 1.18	10	
		Otaria flavescens		S	−16.19 ± 0.24	21.37 ± 0.35	23	
				SK	−13.68 ± 0.53	20.30 ± 0.74	24	
Franco-Trecu et al. (2013)	Uruguay	*Arctocephalus australis*		SK	Values not available		35	TE
		Otaria flavescens					10	

(continued)

Table 9.1 (continued)

Source	Location	Species	Obs	Tissue(s)	Carbon	Nitrogen	*n* (individuals)	Main topic
Franco-Trecu et al. (2014)	Uruguay	*Arctocephalus australis*	Male	W	−14.9 (−16.6 to −13.7)	17.6 (14.3 to 20.2)	3	TE
			Female		−15.3 (−16.6 to −14.1)	17.2 (14.1 to 19.1)	5	
		Otaria flavescens	Male		−13.2 (−14.8 to −12.3)	19.9 (19.1 to 21.3)	4	
			Female		−12.8 (−13.9 to −12.2)	20.0 (18.5 to 21.6)	5	
Fruet et al. (2015)	Brazil	*Tursiops truncatus*	Age 0	T	−13.45 ± 0.98	21.50 ± 1.91	4	TE
			Age 0.5		−11.45 ± 0.87	19.14 ± 1.49	3	
			Age 1		−11.40 ± 0.58	18.48 ± 1.20	8	
			Age 2		−11.22 ± 0.70	18.30 ± 0.65	14	
			Age 3		−10.99 ± 0.45	17.96 ± 0.87	6	
			Age 4		−11.13 ± 0.37	18.51 ± 0.57	7	
			Age 5		−11.32 ± 0.33	18.57 ± 1.08	5	

Hardt et al. (2013)	Brazil	*Pontoporia blainvillei*	Estuary	M	−15.86 ± 0.88	15.55 ± 1.16	16	TE
			Nearshore		−15.5 ± 1.10	15.16 ± 0.52	7	
		Sotalia guianensis	Estuary		−16.32 ± 1.57	15.64 ± 0.94		
Hückstädt et al. (2007)	Chile	*Otaria flavescens*		H/V	−12.48 ± 0.68	20.97 ± 0.77	27	TE
Kehrig et al. (2013)	Brazil	*Sotalia guianensis*		M	–	12.1 to 15.9	21	TE
Lewis et al. (2006)	Argentina	*Mirounga leonina*		V	−21.19 to −12.43	10.01 to 20.75		TE
Ott et al. (2016)	Brazil	*Tursiops truncatus*	São Paulo/Paraná State	T	−13.1	15.7	5	TE
			Rio Grande do Sul State		−10.6	17.8	26	
Riccialdelli et al. (2010)	Argentina	*Cephalorhynchus commersonii*		BO	−12.7 ± 1.3	17.2 ± 0.7	121	TE
		Grampus griseus	Ecotype GG1		−11.6 ± 0.5	19.5 ± 1.3	23	
			Ecotype GG2		−13.9 ± 0.9	15.5 ± 1.4	25	
		Lagenorhynchus australis			−10.8 ± 1.2	19.3 ± 1.6	39	
		Lagenorhynchus cruciger			−17.2 ± 1.0	10.2 ± 0.8	5	
		Lissodelphis peronii			−12.9 ± 0.9	15.3 ± 2.4	37	
		Phocoena dioptrica			−12.8 ± 0.4	13.1 ± 0.7	27	
		Phocoena spinipinnis			−17.9 ± 1.2	10.1 ± 0.9	87	
		Pseudorca crassidens			−13.0 ± 1.0	17.9 ± 0.7	7	
Riccialdelli et al. (2012)	Argentina	*Grampus griseus*		BO	−15.6 to −10.9	13.1 to 21.6	48	TE
Riccialdelli et al. (2013)	Argentina	*Cephalorhynchus commersonii commersonii*		BO	−13.3 ± 1.2	17.4 ± 1.0	220	TE

(continued)

Table 9.1 (continued)

Source	Location	Species	Obs	Tissue(s)	Carbon	Nitrogen	*n* (individuals)	Main topic
Riccialdelli & Goodall (2015)	Argentina	*Pseudorca crassidens*		BO	−13 ± 0.6	13.6 ± 1.6	29	TE
Rowntree et al. (2001)	Argentina	*Eubalaena australis*		BA	−20.5 to −26	5 to 10	7	H
Rowntree et al. (2008)	Argentina	*Eubalaena australis*		BA	−18.2 to −23.8 (means intervals)	–	5	TE
Saporiti et al. (2014)	Argentina	*Arctocephalus australis*	Modern	BO	−13.32 ± 0.54	20.59 ± 0.99	29	TE
			Pre-contact		−13.26 ± 0.80	18.45 ± 2.12	5	
		Otaria flavescens	Modern		−12.17 ± 0.66	22.30 ± 0.78	36	
			Pre-contact		−12.77 ± 0.91	21.52 ± 1.60	26	

Saporiti et al. (2016)	Argentina	*Arctocephalus australis*	Northern Patagonia/Males	BO	−13.4 ± 0.5	20.6 ± 1	28	TE
			Southern Patagonia/Males		−13.2 ± 0.7	19.0 ± 1.6	6	
	Brazil		Brazil/Males		−13.6 ± 0.8	20.6 ± 0.6	54	
			Brazil/Females		−13.6 ± 0.5	18.8 ± 1.0	5	
	Argentina	*Otaria flavescens*	Northern Patagonia/Males		−11.9 ± 0.6	22.2 ± 1.0	14	
			Northern Patagonia/Females		−12.3 ± 0.6	22.3 ± 0.5	11	
			Southern Patagonia/Males		−13.1 ± 1.1	20.7 ± 1.2	31	
			Southern Patagonia/Females		−12.3 ± 0.7	20.9 ± 1.0	10	
	Brazil		Brazil/Males		−11.9 ± 0.7	20.7 ± 0.5	15	
			Brazil/Females		−12.0 ± 0.3	20.4 ± 0.3	4	
Secchi et al. (2016)	Brazil	*Tursiops truncatus*	1970s	T	−11.2 ± 0.6	18.0 ± 1.0	5	TE
			1980s		−11.1 ± 0.6	18.2 ± 1.1	20	
			1990s		−11.0 ± 0.7	19.0 ± 0.8	9	
			2000s		−11.0 ± 0.5	17.9 ± 0.8	40	
			2010s		−11.2 ± 0.7	18.2 ± 0.7	13	
Tatsch et al. (2016)	Brazil	*Berardius arnuxi, Mesoplodon densiristris, Mmesoplodon hectori, Mesoplodon grayi, Mesoplodon layardii and Ziphius cavirostris*	Bulk	B	−18.0 ± 2.9	13.6 ± 1.4	13	M
			Lipid extracted		−16.8 ± 2.5	13.8 ± 1.4		
			Acidified		−17.8 ± 2.8	13.6 ± 1.3		
			Lipid extracted and acidified		−16.9 ± 2.6	13.3 ± 1.5		
Troina et al. (2016)	Brazil	*Pontoporia blainvillei*		T	−10.9 ± 0.7	19.1 ± 1.1	125	TE

(continued)

Table 9.1 (continued)

Source	Location	Species	Obs	Tissue(s)	Carbon	Nitrogen	*n* (individuals)	Main topic
Valenzuela et al. (2009)	Argentina	*Eubalaena australis*		SK	−20.8 ± 1.3	8.0 ± 1.9	131	H
Valenzuela et al. (2010)	Argentina	*Eubalaena australis*	Calves 2003	SK	−20.5 ± 1.4	8.7 ± 1.3	12	N
			Calves 2004		−21.6 ± 1.6	8.0 ± 2.1	20	
			Calves 2005		−20.3 ± 1.7	8.7 ± 1.9	10	
			Mothers		−21.2 ± 1.3	7.8 ± 2.0	42	
Vales et al. (2013)	Brazil	*Arctocephalus australis*		BO	−13.56 ± 0.79	20.58 ± 0.62	54	TE
Vales et al. (2015)	Argentina	*Arctocephalus australis*		BO	−13.7 ± 0.7	20.4 ± 1.4	61	TE
				V	−14.8 ± 0.8	18.1 ± 1.4	9	
Vighi et al. (2014)	Argentina and Brazil	*Eubalaena australis*	δO:29.54 ± 1.17 (*n* = 122)	BO	−20.39 ± 3.08 (*n* = 120)	9.31 ± 2.3 (n = 118)	~120	H
Zenteno et al. (2013)	Argentina and Brazil	*Otaria byronia* (= *Otaria flavescens*)	δO	BO	Values not available		112	H
Zenteno et al. (2014)	Brazil	*Otaria flavescens*	1986–1988/<9year	T	−12.6 ± 1.3	20.3 ± 0.6	8	TE
			1986–1988/≥9year		−11.5 ± 0.5	21.0 ± 0.7	8	
			1994–1998/<9year	BO	−11.7 ± 0.8	20.7 ± 0.5	6	
			1994–1998/≥9year		−11.9 ± 0.9	20.8 ± 0.5	12	
			2000–2009/<9year		−12.3 ± 0.7	20.8 ± 1.9	6	
			2000–2009/≥9year		−11.7 ± 0.7	20.4 ± 0.5	14	
Zenteno et al. (2015)	Argentina	*Otaria flavescens*		BO	−12.3 ± 0.8	21.2 ± 0.9	54	TE
				V	−13.1 ± 0.8	22.5 ± 1.5	8	DF

Location indicates where samples were collected. Tissues: *BO* = bone, *BA* = baleen, *BL* = blood, *H* = hair, *L* = liver, *M* = muscle, *S* = serum, *SK* = skin, *T* = teeth, *V* = vibrissae, *W* = whiskers. Main topics: *DF* = discrimination factor, *H* = habitat use, *M* = methodological, *N* = nutritional stress and *TE* = trophic ecology. Values are given in the format mean ± SD, range or mean (range) of the isotope values in accordance to the available information in the publication

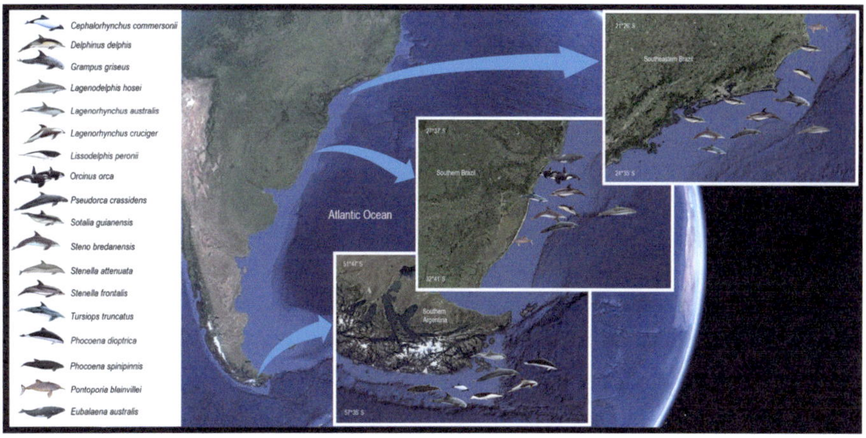

Fig. 9.2 Schematic outcome of the application of stable isotope analysis for describing the spatial ecology cetaceans inhabiting different ecosystems in the southwestern Atlantic Ocean

2013) have been reported. The stable isotope composition of these species reflected the Guiana dolphin and bottlenose dolphin coastal habitat (i.e. higher $\delta^{13}C$ values), and more offshore habitats for the other species, with the oceanic Fraser's and spinner dolphins showing the lowest $\delta^{13}C$ values. Furthermore, intra-specific different isotopic values from Guiana dolphins inhabiting the three different bays previously presented by Bisi et al. (2012) were compared to other species. The similarity of the $\delta^{13}C$ values between Guiana dolphins from Ilha Grande Bay and Atlantic spotted dolphins indicated that both feed in areas under the influence of the South Atlantic Central Water. Bottlenose dolphins and false killer whales presented the highest $\delta^{15}N$ values, which is in accordance to their high trophic position in comparison with the other species considered in this study.

In subtropical (Botta et al. 2012) and sub Antarctic waters (Riccialdelli et al. 2010), tooth and bone samples from stranded marine mammals collected along decades and deposited in two of the largest scientific collections of this group in Latin America (Laboratory of Ecology and Conservation of Marine Megafauna - Federal University of Rio Grande-FURG, Brazil and Acatushún Museum, Tierra del Fuego, Argentina) were analyzed to determine their C and N stable isotopic composition. These multispecies studies included several species of Delphinidae and Phocoenidae, such as bottlenose dolphins, rough-toothed dolphins, Atlantic spotted dolphins, common dolphins, Fraser's dolphins, killer whales (*Orcinus orca*) (Botta et al. 2012), false killer whales (Riccialdelli et al. 2010; Botta et al. 2012), Risso's dolphins, spectacled (*Phocoena dioptrica*) and Burmeister's (*P. spinipinnis*) porpoises, Commerson's dolphins (*Cephalorhynchus commersonii*), Pale's dolphins (*Lagenorhynchus australis*), hourglass dolphins (*L. cruciger*) and Southern right whale dolphins (*Lissodelphis peronii*) (Riccialdelli et al. 2010). Findings reported in Botta et al. (2012) showed that the coastal ecotype of bottlenose dolphins (−10.6‰) and killer whales (−10.5‰), had the highest $\delta^{13}C$ values, while species from deep pelagic regions, such as Fraser's dolphin (−12.8‰), showed the most ^{13}C-depleted $\delta^{13}C$ values. In the sub Antarctic region, Riccialdelli et al. (2010) also showed that isotopic carbon values clearly reflect the inshore-offshore decreasing gradient of $\delta^{13}C$ (France 1995; Clementz and Koch 2001). On the other hand, $\delta^{15}N$ values also showed this gradient and evidenced the relative trophic level of the species in their habitats. Bottlenose dolphins and killer whales, for example, had high trophic levels within coastal sub tropical species (Botta et al. 2012), while Pale's and hourglass dolphins occupied high trophic positions among the sub Antarctic dolphin species (Riccialdelli et al. 2010).

Stable isotopes were also used as evidence of putative ecotypes of some odontocetes in the SAO (Riccialdelli et al. 2010, 2012; Botta et al. 2012; Ott et al. 2016). Two putative ecological stocks of Risso's dolphins were recognized in sub Antarctic waters based on significant differences in $\delta^{13}C$ and $\delta^{15}N$ values from bone samples (Riccialdelli et al. 2010, 2012). Both groups were associated to the southern Argentina continental shelf break, one inhabiting northern areas and the other foraging in higher latitudes. In the SAO off southern Brazilian waters, Botta et al. (2012) found isotopic evidence for the existence of more than one putative ecotype of bottlenose dolphins and false killer whales.

In tropical SAO, two studies used stable isotopes as ecological tracers of the trophic structure of food webs that sustain coastal small cetaceans. Furthermore, both studies also analyzed trace-metals, mercury (Hg) (Bisi et al. 2012; Kehrig et al. 2013) and selenium (Se) (Kehrig et al. 2013) in order to understand their distribution throughout the food web. In southern Rio de Janeiro state, Bisi et al. (2012) used SIA and total Hg of several components of the biota, including the Guiana dolphins, to investigate the structure of food webs of Sepetiba, Guanabara and Ilha Grande bays. Individuals from Guanabara bay had the lowest $\delta^{15}N$ values, suggesting the contribution of atmospheric nitrogen fixation by cyanobacteria in the region. Considering the range of $\delta^{15}N$ found in Guiana dolphins from the three bays, results corroborate the stomach content analyses, indicating that the species feed on prey of different trophic levels. Contrary to what was observed in Guanabara bay, some prey showed higher $\delta^{15}N$ than dolphins in Sepetiba and Ilha Grande bays. These locations have high biodiversity, and dolphins there seem to be feeding upon a wider range of taxa, especially those of lower trophic levels, in comparison to the former. A similar pattern was observed in a coastal region under the influence of the Paraiba do Sul river. Stable nitrogen isotopes of Guiana dolphins, fish, cephalopods, crustaceans and mesoplankton revealed five trophic positions along this food web, with Guiana dolphins and the largehead hairtail (*Trichiurus lepturus*) occupying the top of the trophic web (Kehrig et al. 2013). Hg web magnification was higher than Se, being positively correlated with $\delta^{15}N$ values. These studies emphasized the useful application of stable isotopes as tracers for disentangling the complex trophic structure of tropical food webs. Furthermore, the combined use of these tracers provides important information about biomagnification processes that could affect top-predators.

Various researches were conducted with the main aim of revealing diet of single species and/or the trophic overlap between sympatric species. To achieve this, most studies used SIMM to estimate the likely diet composition. In northern Rio de Janeiro state, C and N isotopic values suggested that franciscanas and Guiana dolphins, two of the most common coastal dolphin species along southeastern and southern Brazil, have a considerable overlap in their isotopic niches, despite the larger niche areas of the latter (Di Beneditto et al. 2011; Di Beneditto & Monteiro 2015). Evidence of an overlap in diet between the two species, based on the similarity of $\delta^{13}C$ and $\delta^{15}N$ values, was also observed in the estuarine region of Babitonga bay, in southern Brazil (Hardt et al. 2013). This pattern contrasts to previous stomach content studies (Cremer et al. 2012) and might be influenced by the simple visual exploration of the isotopic values of prey and consumers in the δ-space in the study of Hardt et al. (2013). Thus further research on this topic is needed for a better understanding of the trophic interactions between these sympatric species in Babitonga Bay.

Regional, seasonal and gender-related differences in the diet of dusky dolphins were investigated in four gulfs of Peninsula Valdés, Argentina (de Castro et al. 2015). The contribution of potential prey, as inferred by SIMM, indicated an opportunistic feeding of dusky dolphins upon pelagic items. While information from stomach contents pointed to a predominant predation on Argentine anchovies (*Engraulis anchoita*) SIA indicated butterfish (*Stromateus brasiliensis*) as the main prey for dusky dolphins in this region. However, the potential of SIAR for discriminating among these items was limited due to their overlapped $\delta^{13}C$ and $\delta^{15}N$ values. In addition, SIA suggested that feeding preference varies among gulfs but not between warm and cold periods. Spatial isotope differences in the diet composition were also observed between short-beaked common dolphins from the continental shelf off Buenos Aires, where the Argentine hake (*Merluccius hubbsi*) was indicated as the main prey and the inshore waters of Golfo San Matías, northern Patagonia, where the Argentine anchovy appeared to be more important (de Castro et al. 2016). Compared to previous stomach content analysis, SIA indicated a more complex diet of this species, with differences in the habitat use and foraging strategies between groups.

Long-term and gender-related variation in the diet of the bottlenose dolphin in southern Brazil was determined through stomach content analysis and $\delta^{13}C$ and $\delta^{15}N$ values in tooth dentine of individuals that washed ashore in Rio Grande do Sul state coast from 1977 to 2012 (Secchi et al. 2016). A decrease in the contribution of the whitemouth croaker (*Micropogonias furnieri*) and the kingcroaker (*Menticirrhus* sp.) and an increased relative importance of the largehead hairtail were observed for the period after 1990. This was probably a consequence of the decrease in abundance of the former fish species due to intensified fishing activity in the region during recent decades. Despite a slightly higher importance of the mullet (*Mugil liza*) for females, it seems that males and females share the same resources. The results also show that bottlenose dolphins from this area are opportunistic feeders and prey upon the most available demersal and demersal–pelagic teleost fish.

Studies focusing on ontogenetic changes in habitat use and/or feeding habits, including those associated with the weaning process in cetaceans have been largely benefited from the application of SIA (Knoff et al. 2008; Newsome et al. 2009). In the SAO, the ontogenetic differences in diet and/or weaning ages of bottlenose dolphins (Fruet et al. 2015), franciscanas (Troina et al. 2016), false killer whales (Riccialdelli et al. 2015), Commerson's dolphins (Riccialdelli et al. 2013) and southern right whales (*Eubalaena australis* [Rowntree et al. 2008]) were reported. Stable isotopes have been successfully applied to estimate the weaning age in some cetaceans (e.g. killer whales [Newsome et al. 2009], beluga whales, *Delphinapterus leucas* [Matthews and Ferguson 2015]). The rational for this approach is based on the fact that offsprings nursing on their mother's milk will present higher $\delta^{15}N$ relative to the mother's signal because it is produced by catabolism of her own tissues. Likewise, they should also present lower $\delta^{13}C$ values due to the influence of high lipid content of milk and, consequently, ^{13}C-depleted $\delta^{13}C$ values (Hobson and Sease 1998; Newsome et al. 2009). In this context, the time of weaning can be defined as the age when a change in trend of $\delta^{13}C$ (i.e., a marked increase followed by a stabilizing or reduction phase) and $\delta^{15}N$ (i.e., a marked decrease followed by a stabilizing or increasing phase) is detected (Newsome et al. 2009). This approach was used based on isotopic values from tooth dentine of known-age bottlenose dolphins from the Patos Lagoon estuary and adjacent coastal areas. Trends in stable isotope values showed a clear change in the $\delta^{13}C$ and $\delta^{15}N$ profiles near age 2 indicating the most probable weaning age (Fruet et al. 2015). Following the same rationale, trends in C and N stable isotope values in tooth dentine of franciscana dolphins from southern Brazil, were positively ($\delta^{13}C$) and negatively ($\delta^{15}N$) related to age, a pattern attributed to lactation, also in this species (Troina et al. 2016). However, the authors highlighted the need of caution due to the fact that the amount of dentin deposited during the first year of life in this species' tooth is larger than that deposited later in life, then having a great contribution to the total tooth composition and leading to inaccurate estimations of the age at weaning. Bone collagen from Commerson's dolphins and false killer whales from sub Antarctic regions have been also analyzed to determine their stable isotopic composition (Riccialdelli et al. 2013, 2015). Although Commerson's dolphins from all age classes seem to use the same foraging areas in Tierra del Fuego, juveniles consume higher proportions of Falkand sprats (*Sprattus fueguensis*) than adults while the latter consume more cephalopods such as the Argentine shortfin squid (*Illex argentinus*) and the Patagonian squid (*Loligo gahi*) as evidenced by SIMM (Riccialdelli et al. 2013). Similarly, false killer whales from a mass stranding in the Strait of Magellan showed a trend of increased C and N isotopic values between young and older age classes (Riccialdelli et al. 2015). Both studies found relatively high $\delta^{15}N$ and low $\delta^{13}C$ values in calves, attributing this to their lactation period.

Although southern right whales have been studied since 1970s in Argentina and since 1986 in Brazil, much about its spatial and trophic ecology remain unknown. SIA has contributed to the understanding of some aspects of the species' distribution and indicated the existence of different feeding grounds of southern right whales. Baleen plates from adult and sub-adult right whales from Península Valdés

(PV) were sampled every 2 cm from base to tip in order to get chronologically ordered $\delta^{15}N$ and $\delta^{13}C$ values (Rowntree et al. 2001, 2008). These baleen plates showed oscillation in $\delta^{13}C$ values that corresponded to annual cycles where whales alternate their feeding between the Patagonian continental shelf (higher $\delta^{13}C$ values) and higher latitudes areas (i.e. South Georgia, with lower $\delta^{13}C$ values) (Rowntree et al. 2008). Furthermore, differences in the range of the annual oscillations among whales were attributed to an individual variation in the feeding grounds of specimens from PV. The combination of the stable isotope signatures and molecular analysis of biopsy samples from PV revealed that southern right whales have a maternally inherited site fidelity to summer foraging grounds, as isotopically similar whales shared the same haplotypes (Valenzuela et al. 2009).

Stable isotope compositions of C, N and O of bone samples of southern right whales evidenced a significant difference in the $\delta^{13}C$ and $\delta^{18}O$ between PV and southern Brazil, the two main breeding sites of the species in the SAO. In addition, lack of significant variation in $\delta^{15}N$ suggests similar trophic position of their prey (Vighi et al. 2014). Results were in accordance with the isotope baselines of the breeding areas used by the species, suggesting that substantially feeding occur also in these areas.

9.2.2 *Pinnipeds*

The understanding on several aspects of the ecology of South American sea lion (*Otaria flavescens*), fur seals (*Arctocephalus* spp.) and the Southern elephant seal (*Mirounga leonina*) along their distribution in SAO have also been enhanced through the use of SIA.

To investigate the influence of sealing and industrial fishing in the diet of sea lions in northern Patagonia, Argentina, skull samples collected from 1940 and 2007 were used in the study of Drago et al. (2009a). They observed a decrease in the $\delta^{13}C$ for both males and females for the period from 1940 to 1970, indicating a decrease in the consumption of benthic coastal preys. Then, a significant increase in $\delta^{13}C$ was detected, which was related to an increase in the predation upon pelagic prey. The period during which a coastal feeding predominated was related to the hunting of the species. This led to a marked population decrease, and probably the remaining animals concentrated in the coastal area (Drago et al. 2009a). In another study with the same species and in the same area, Drago et al. (2010) noticed that pups suckling from females that feed on the pelagic offshore preys seem to present faster growth rates than those that feed on the benthic coastal prey. Although preying in the coastal region might be more convenient because it can reduce the intervals between suckling bouts (e.g. Costa 2008), preys in this region have lower energy densities than pelagic ones (Eder and Lewis 2005). The difference in the foraging habitat among females of this population was related to performance, which, in turn, depends on the individual's age, body size, and/or foraging skills (Drago et al. 2010). The overlap between the diet of male and female sea lions was also investigated trough SIA of skin and bone samples from individuals collected in Brazil and Uruguay (Drago et al. 2015). The species seems to perform a wide variety of feeding strategies throughout the year, though feeding is reduced during the pre-breeding. Also, during this period there seems to be no overlap in the isotopic niche between sexes. The dietary shift is probably a strategy for maximizing energy intake to prepare for a period of high-energy demand (Costa 2008). The difference between sexes in the pre-breeding period can be attributed to the higher forage capacity of males compared to females. The larger body size and related physiological attributes of males, such as energy conservancy, allow them to have easier access to larger and more energetically preys in deeper waters than females (Page et al. 2005).

Assuming that the diet of females' otariids can be inferred by the diet of their pups, Drago et al. (2010) and Franco-Trecu et al. (2012) used serum and skin samples of pups and concluded that there is no trophic overlaps between females of South American fur seal (*Arctocephalus australis*) and South American sea lion. The higher $\delta^{13}C$ and $\delta^{15}N$ values for sea lion pups indicate that the lactating females of this species consumed nearshore resources, while fur seals probably foraged further offshore and included anchovy and squids in their diet. As the isotopic values are related to the tissues' turnover rates (Tieszen et al. 1983), authors related the serum isotope composition to the prepartum period of females' diet and the skin, to the postpartum period. Both species showed a shift between these periods, but for different reasons. For fur seals, it was attributed to lower variation in trophic levels of the preys, whereas for sea lions it was related to a decrease in the diversity of foraging areas. The authors have also raised the possibility of some females foraging at different latitudes to explain the large variation in their $\delta^{13}C$ values. These movements could benefit the species by minimizing intraspecific competition, mainly during the reproductive period.

With the aim of understanding the reasons for the different population trends of these sympatric species in Uruguay, Franco-Trecu et al. (2014) assessed their forage ecology through SIA of individuals' whiskers. A minimum overlap between the isotopic niche both at individual and population levels was observed. At the intraspecific level, in spite of using different foraging areas, a large overlap between sexes was observed for both species. Nevertheless, the niche was wider and the individuals were considered more generalists in fur seals than in sea lions individuals, which presented a considerable specialization. The narrower niche and the high degree of individual specialization, among others, were considered by the authors possible explanations for the population declining of the latter species.

The overlap between the diet of these two species was also evaluated in areas to the south (northern and southern Argentina) and to the north (southern Brazil) of the La Plata river estuary (Saporiti et al. 2016). In this study only adults were considered. The diet of female sea lion overlapped with larger male sea lions, but not with male fur seals. Morphological similarity was significantly correlated with the isotope niche represented by the ellipses generated by $\delta^{13}C$ and $\delta^{15}N$ of each group in the La Plata river estuary.

The long-term diet of these species was assessed by the isotope composition of individuals' bones (Vales et al. 2013; Zenteno et al. 2014). The tested hypothesis was that the diet of the males of these species changed in the last years as a response of the high fishing effort in their foraging habitats. No change was observed in the $\delta^{13}C$ and $\delta^{15}N$ values of fur seals collected from 1994 to 2011 and of sea lions from 1986 to 2009 in Brazil. Vales et al. (2013) also pointed small pelagic fish and squid as important prey for fur seals and Zenteno et al. (2014) observed that sea lions had a mixed diet of pelagic and demersal fish along the study period.

On a larger time scale, the diet of sea lions was also reconstructed since the late Holocene from stable isotope composition of bones remains collected from archeological sites in Patagonia (Zenteno et al. 2015). The main finds were that the trophic levels of sea lion increased during this long time period and that the modern diet of the species includes more benthic prey than in the past. The results also indicated that the main changes in the species' diet occurred in the twentieth century, after the considerable removal of individuals by sealing, leading them to use other food sources and then expanding their trophic niche.

Saporiti et al. (2014) conducted a study to evaluate the influence of anthropogenic activities in the structure of coastal marine food webs in the Argentinian Patagonia. It was expected that these activities, such as sealing and fishing, shortened the food webs by the preferential removal of top predators. However, the results showed that the modern food webs in the region are longer and less overlapping than those from ancient times. The authors attributed this to the great impact of sealing rather than to the fishing activity. It means that the seals significantly decreased in number and, although the fishes were affected by fishery, the size of the fishes preyed by the seals did not change significantly over time. Consequently, the intraspecific competition in South American fur seal diminished while they continue to have access to large-sized fish prey.

Ontogenetic changes in the diet of males South American fur seals were observed in the Argentinian Patagonia (Vales et al. 2015). Besides the expected higher $\delta^{15}N$ and lower $\delta^{13}C$ values found in pups, this study also found differences in the diet of subsequent age classes. Juveniles seemed to feed more on pelagic and lower trophic level prey than both sub-adults and adults, for which the Argentine shortfin squid was considered the main prey item. These findings corroborated previous studies that used traditional methods for diet analysis and classified *A. australis* in Patagonia as a trophic generalist, despite its preference for pelagic prey. Also, it pointed to the importance of the lobster krill (*Munida gregaria*), in the diet of adults, which was only perceived through SIA method. In addition, as no regular oscillation in the isotope values was detected in the vibrissae of the animals, the authors considered them as no regularly migratory, although displacements of relatively small magnitude could not be ruled out (Vales et al. 2015).

Intra- and interspecific differences in spatial and trophic ecology of three species of the genus *Arctocephalus* (*A. australis*, *A. tropicalis* and *A. gazella*, South American fur seal, Subantarctic fur seal and Antarctic fur seal, respectivelly) were investigated through the analysis of $\delta^{13}C$ and $\delta^{15}N$ values in the dentine of each annual growth layer group (Albernaz et al. 2016). Ontogenetic variation was observed in the diet of the three species. The results also indicated that the intra- and interspecific competition is minimized due to the spatial and trophic segregation in their foraging.

The understanding about the feeding ecology of the southern elephant seal (*Mirounga leonina*) has also improved through the use of SIA. The $\delta^{13}C$ and $\delta^{15}N$ composition of the vibrissae of individuals from Peninsula Valdés, Argentina, and one male from Malvinas/Falkland Islands, revealed that males presented a great between- groups variability in isotopic composition (up to 6‰ $\delta^{13}C$ and 7‰ $\delta^{15}N$), indicating considerable differences in their main prey item and in their trophic level (Lewis et al. 2006). Males where considered as specialist foragers with but variable preferences between groups. The authors attributed this variation to intra and interspecific competition near the breeding site. Females, on the other hand, presented a consistent diet (feeding behavior), with a considerable contribution of invertebrates.

Although $\delta^{18}O$ are much less used than $\delta^{13}C$ and $\delta^{15}N$ in ecological studies, it was employed to investigate the relationship of sea surface temperature (SST) changes in the past and with the sealing of South American fur seal (Saporiti et al. 2013). The oxygen-isotope values used was not from this marine mammal species, but from the bivalves *Aulocomya atra atra* and *Mytilus edulis*. The shell of these organisms assimilates variations in the oxygen-isotope ratio of the water during their biomineralization, and then their $\delta^{18}O$ can be used as a proxy to SST values from the past (Dettman et al. 1999). No relationship was found between fur seals' densities and the $\delta^{18}O$ values in the bivalves' band shells. Then, the fluctuations in the number of the seals were attributed to variations in the primary productivity during the second half of the Holocene, because it is related to the availability and quality of food to the fur seals.

Assuming that the bone apatite also registers $\delta^{18}O$ of the animals' diet, it was used to evaluate the overlap of the areas used by South American sea lion (*Otaria byronia = flavescens*), from southern Brazil and from Patagonia and Tierra del Fuego, in Argentina (Zenteno et al. 2013). The $\delta^{18}O$ values differed significantly from one region to another, with Patagonia individuals presenting the higher ones. This result was expected, once tissues tend to be enriched in ^{18}O as temperature declines (Kim and O'Neil 1997) and reveal that they have restricted dispersal movements between the three areas. In addition, it was also found was that the variability of the $\delta^{18}O$ values of juveniles of both sexes was wider than for adults, suggesting that they disperse during this earlier life stage.

9.2.3 *Sirenia*

The application of SIA for ecological studies in sirenians from the southwestern South Atlantic, is restricted to only one study on Antillean manatee (*Trichechus manatus manatus*) (Ciotti et al. 2014). Values of $\delta^{13}C$ and $\delta^{15}N$ were obtained from samples of bones and teeth at different positions of the manatee's mandible. One interesting characteristic of manatee dentition is the continuous tooth replacement, in which the new tooth emerges in the distal part and moves towards the frontal part of the arcade, allowing for a chronological analysis of the isotopic composition.

Variations in the isotopic values of teeth suggest temporal differences in the contributions of macroalgae and seagrasses during the life of some individuals. In addition, interindividual variations were related to differences in the use of the feeding area. Individuals with lower variations in the isotope values were assumed to have higher site fidelity, while those with more variable values, were associated to more mobile animals.

9.3 Trace Elements

This method is based on the connection between the concentrations of chemical elements in tissues of an organism to the environment where it lives. The most common elements used for the application of this technique are barium (Ba) and strontium (Sr). Considering that their bioavailability are, respectively, negatively and positively correlated to the salinity of aquatic environments (Jacquet et al. 2005), they can be used as biochemical tracers to track animals habitat use (e.g. Elsdon and Gillanders 2005; Albuquerque et al. 2012).

In this case, the use of inert tissues, as tooth and bones, are very useful, because they have marks that can be related to the age of the animal, and then to a period of life in which it used habitats with different biogeochemical characteristics. To our knowledge, among marine mammals, just cetaceans have already been studied using this technique. Botta et al. (2015) used Ba/Ca and Sr/Ca ratios in teeth from dolphins stranded or incidentally killed in fishing gear in Brazil and Uruguay. Two of the species (the Amazon River dolphin, *Inia geoffrensis*, and tucuxi, *Sotalia fluviatilis*) were typical of freshwater environments and other two (franciscana and Guiana dolphin), from marine coastal regions subject to different degrees of freshwater discharge. Ba/Ca results well represented the salinity gradient and this ratio was considered useful as natural marker for habitat use in marine mammals.

9.4 Final Considerations

Some advantages of the SIA method can be recognized considering the studies presented here and the contribution of them to different species knowledge and conservation. Compared to the traditional method of gastro-intestinal content, the main advantages of stable isotope to be considered are: (i) the resolution of the pathways in the food web if the pathways are well distinguished by $\delta^{13}C$ values; (ii) the possibility of indicate the nutritional roles of prey items, as stable isotopes are a result of the diet absorved and not consumed by the predator; (iii) the estimation of the trophic level of the organisms, if the basal isotope composition of the food web is known (e.g. Polunin and Pinnegar 2002); and (iv) the possibility of following ontogenetic changes and individual consistency and specialization through longitudinal analysis of chronologically ordered structures such as teeth and vibrissae.

A specific comparison of the advantage of using the SIMMsto the traditional methods was made for two species of marine mammals, the South American fur seal (*Arctocephalus australis*) and the South American sea lion (*Otaria flavescens*), in Uruguay (Franco-Trecu et al. 2013). The SIMMs can include (SIMMs-IP) or not (SIMMs-UP) some informative priors to the analysis. Both were compared to the scat analysis of these species. The two SIMMs methods resulted in different diet composition. Based on the SIMM-UP, the diet of the fur seals was dominated by pelagic fishes and no clear dominance was detected for sea lions. The SIMM-IP indicated a dominance of other fish species, including the striped weakfish (*Cynoscion guatucupa*), in accordance with the scat analysis, highlighting the importance of squids for the diet of both otariid species.

Despite the great advantages of methodologies considered here, they have limitations and care should be take when interpreting their results and estimating the diet composition of a species. When compared to the gut content analsysis, the main disadvantages are: (i) classification of broad cathegories of preys (not at the smaller or species level); (ii) limited possibility of detect short-term differences/changes in the diet of large predators (it will depended on the tissue turnover rate) and (iii) restricted capacity of detecting spatial changes in the diet of the organisms (considerable changes in $\delta^{13}C$ among areas are necessary to occour) (Polunin and Pinnegar 2002).

These limitations should be taken into consideration for the application of SIA and for the interpretation of its results. Despite them, considerable progress have been achieved through the use of this tool. In this chapter we presented several examples of how the understanding on the ecology of marine mammals in the SAO have been improved by SIA and trace elements methods. Althoug the advances in this field are fast, such as in the use of compound-specific SIA (e.g. McMahon et al. 2013), we hope this chapter highlights the importance of these tools and encourage researchers to use them as well as other chemical tracers methods, such as fatty acids, for a better understanding of the spatial and trophic ecology of marine mammals.

References

Albernaz TL, Secchi ER, Oliveira LR et al (2016) Ontogenitic and gender-related variation in the isotopic niche within and between three species of fur seals (genus *Arctocephalus*). Hydrobiologia. doi:10.1007/s10750-016-2950-0

Albuquerque CQ, Miekeley N, Muelbert JH et al (2012) Estuarine dependency in a marine fish evaluated with otolith chemistry. Mar Biol 159:2229–2239

Baptista G, Kehrig HA, Di Beneditto APM et al (2016) Mercury, selenium and stable isotopes in four small cetaceans from the Southeastern Brazilian coast: influence of feeding strategy. Environ Pollut 218:1298–1307

Barnes C, Jennings S, Barry JT (2009) Environmental correlates of large-scale spatial variation in the $\delta^{13}C$ of marine mammals. Estuar Coast Shelf Sci 81:368–374

Bearhop S, Adams CE, Waldron S et al (2004) Determining trophic niche width: a novel approach using stable isotope analysis. J Anim Ecol 73:1007–1012

Ben-David M, Flaherty EA (2012) Stable isotope in mammalian research: a beginner's guide. J Mammal 93:312–328
Bisi TL, Lepoint G, Azevedo AF et al (2012) Trophic relationships and mercury biomagnification in Brazilian tropical coastal food webs. Ecol Indic 18:291–302
Bisi TL, Dorneles PR, Lailson-Brito J et al (2013) Trophic relationships and habitat preferences of Delphinids from the southeastern Brazilian coast determined by carbon and nitrogen stable isotope composition. PLoS One 8:e82205
Borrell A, Abad-Oliva E, Gómez-Campos E et al (2012) Discrimination factors of stable isotope in fin whales tissues and application to diet assessment in cetaceans. Rapid Commun Mass Spectrom 26:1596–1602
Botta S, Hohn AA, Macko SA et al (2012) Isotopic variation in delphinids from the subtropical western South Atlantic. J Mar Biol Assoc UK 92:1689–1698
Botta S, Albuquerque C, Hohn AA et al (2015) Ba/Ca ratios in teeth reveal habitat use patterns of dolphins. MEPS 521:249–263
Caut S, Angulo E, Courchamp F (2009) Variation in discrimination factors (Δ^{15}N and Δ^{13}C): the effect of diet isotopic values and applications for diet reconstruction. J Appl Ecol 46:443–453
Ciotti LL, Luna FO, Secchi ER (2014) Intra- and interindividual variation in δ^{13}C and δ^{15}N composition in the Antillen manatee *Trichehcus manatus manatus* from northeastern Brazil. Mar Mamm Sci 30:1248–1247
Clementz MT, Koch PL (2001) Differentiating aquatic mammal habitat and foraging ecology with stable isotopes in tooth enamel. Oecologia 129:461–472
Costa DP (2008) A conceptual model of the variation in parental attendance in response to environmental fluctuation: foraging energetics of lactating sea lions and fur seals. Aquat Conserv 17:S44–S52
Cremer MJ, Pinheiro PC, Simões-Lopes PC (2012) Prey consumed by Guina dolphin *Sotalia guianensis* (Cetacea, Delphinidae) and franciscana dolphin *Pontoporia blainvillei* (Cetacea, Pontoporiidae) in an estuarine environment in southern Brazil. Iheringia, Sér Zool 102:131–137
de Castro RL, Saporiti F, Vales DG et al (2015) Feeding ecology of dusky dolphins *Lagenorhynchus Obscurus*: evidence from stable isotopes. J Mammal 97:310–320
de Castro RL, Saporiti F, Vales DG et al (2016) What are you eating? A stable isotope insight into the trophic ecology of short-beaked common dolphins in the southwestern Atlantic Ocean. Mamm Biol 81:571–578
DeNiro MJ, Epstein S (1978) Influence of diet on the distribution of carbon isotopes in animals. Geochim Cosmochim Acta 42:495–506
DeNiro MJ, Epstein S (1981) Influence of diet on the distribution of nitrogen isotopes in animals. Geochim Cosmochim Acta 45:341–351
Dettman DL, Reische AK, Lohmann KC (1999) Controls on the stable isotope composition of seasonal growth bands in aragonitic fresh-water bivalves (Unionidae). Geochim Cosmochim Acta 63:1049–1057
Di Beneditto APM, Souza CMM, Kehrig HA et al (2011) Use of multiple tools to assess the feeding preference of coastal dolphins. Mar Biol 158:2209–2217
Di Beneditto APM, Rezende CE, Camargo PB et al (2013) Trophic niche comparison between two predators in northern Rio de Janeiro state, Brazil: a stable isotopes approach. Biota Neotrop 13:000–000
Di Beneditto AP, Monteiro LR (2015) Isotopic niche of two coastal dolphins in a tropical marine area: specific and age class comparisons. J Mar Biol Assoc UK 96:853–858
Drago M, Crespo EA, Aguilar A et al (2009a) Historic diet change of the south American sea lion in Patagonia as revealed by isotopic analysis. MEPS 384:274–286
Drago M, Cardona L, Crespo EA et al (2009b) Ontogenic dietary changes in south American sea lions. J Zool 279:251–261
Drago M, Cardona L, Aguilar A et al (2010) Diet of lactating south American sea lions, as inferred from stable isotopes, influences pup growth. Mar Mamm Sci 26:309–323
Drago M, Franco-Trecu V, Zenteno L et al (2015) Sexual foraging segregation in south American sea lions increases during the pre-breeding period in the Río de la Plata plume. MEPS 52:261–272

Eder EB, Lewis MN (2005) Proximate composition and energetic value of demersal and pelagic prey species from the SW Atlantic Ocean. MEPS 291:43–52
Elsdon TS, Gillanders BM (2005) Alternative life-history patterns of estuarine fish: barium in otoliths elucidates freshwater residency. Can J Zool 62:1143–1152
France RL (1995) Carbon-13 enrichment in benthic compared to planktonic algae: foodweb implications. MEPS 124:307–312
Franco-Trecu V, Aurioles-Gamboa D, Arim M et al (2012) Prepartum and postpartum trophic segregation between sympatrically breeding female *Arctocephalus australis* and *Otaria flavescens*. J Mammal 93:514–521
Franco-Trecu V, Drago M, Riet-Sapriza F et al (2013) Bias in diet determination: incorporating traditional methods in Bayesian mixing models. PLoS One 8:e80019
Franco-Trecu V, Aurioles-Gamboa D, Inchausti P (2014) Individual trophic specialisation and niche segregation explain the contrasting population trends of two sympatric otariids. Mar Biol 161:609–618
Fruet PF, Genoves RC, Möller LM et al (2015) Using mark-recapture and stranding data to estimate reproductive traits in female bottlenose dolphins (*Tursiops truncatus*) of the southwestern Atlantic Ocean. Mar Biol 162:661–673
Graham BS, Koch PL, Newsome SD et al (2010) Using isoscapes to trace the movements and foraging behavior of top predators in oceanic ecosystems. In: West J et al (eds) Isoscapes: understanding movement, pattern and process on Earth through isotope mapping. Springer-Verlag, New York, pp 299–318
Hardt FAS, Cremer MJ, Tonello AJ Jr et al (2013) Use of carbon and nitrogen stable isotopes to study the feeding ecology of small coastal cetacean populations in southern Brazil. Biota Neotrop 13:90–98
Hobson KA (1999) Tracing origins and migration of wildlife using stable isotopes: a review. Oecologia 120:314–326
Hobson KA, Sease JL (1998) Stable isotope analysis of tooth annuli reveal temporal dietary records: an example using Steller sea lions. Mar Mamm Sci 14:116–129
Hobson KA, Piatt JF, Pitocchelli J (1994) Using stable isotopes to determine seabird trophic relationships. J Anim Ecol 63:786–798
Hobson KA, Barnett-Johnson R, Cerling T (2010) Using isoscapes to track animal migration. In: West J et al (eds) Isoscapes: understanding movement, pattern and process on Earth through isotope mapping. Springer, New York, pp 273–298
Hückstädt LA, Rojas CP, Antezana T (2007) Stable isotope analysis reveals pelagic foraging by the Southern sea lion in central Chile. J Exp Mar Biol Ecol 347:123–133
Hutchinson GE (1957) Concluding remarks. Cold Spring Harb Symp Quant Biol 22:415–427
Jackson AL, Parnell AC, Inger R et al (2011) Comparing isotopic niche widths among and within communities: SIBER – stable isotope Bayesian ellipses in R. J Anim Ecol 80:595–602
Jacquet SHM, Dehairs F, Cardinal D et al (2005) Barium distribution across the Southern Ocean frontal system in the Crozet-Kerguelen Basin. Mar Chem 95:149–162
Kehrig HA, Seixas TG, Malm O et al (2013) Mercury and selenium biomagnification in a Brazilian coastal food web using nitrogen stable isotope analysis: a case study in an area under the influence of the Paraiba do Sul River plume. Mar Pollut Bull 75:283–290
Kelly JF (2000) Stable isotopes of carbon and nitrogen in the study of avian and mammalian trophic ecology. Can J Zool 78:1–27
Kim ST, O'Neil JR (1997) Equilibrium and nonequilibrium oxygen isotope effects in synthetic carbonates. Geochim Cosmochim Acta 61:3461–3475
Knoff A, Hohn AA, Macko SA (2008) Ontogenetic diet changes in bottlenose dolphins (*Tursiops truncatus*) reflected through stable isotopes. Mar Mamm Sci 24:128–137
Koch PL (2007) Isotopic study of the biology of modern and fossil vertebrates. In: Michener R, Lajtha K (eds) Stable isotopes in ecology and environmental science, 2nd edn. Blackwell Publishing, Boston, pp 99–154
Layman CA, Arrington DA, Montana CG et al (2007) Can stable isotope ratios provide for community-wide measures of trophic structure? Ecology 88:42–48

Lesage V, Hammill MO, Kovacs KM (2001) Marine mammals and the community structure of the Estuary and Gulf of St Lawrence, Canada: evidence from stable isotope analysis. Mar Ecol Prog Ser 210:203–221

Lewis R, O'Connell T, Lewis M et al (2006) Sex-specific foraging strategies and resource partitioning in the southern elephant seal (*Mirounga leonina*). Proc R Soc Lond B Biol Sci 273:2901–2907

Martínez del Rio C, Carleton SA (2012) How fast and how faithful – the dynamics of isotopic incorporation into animal tissues. J Mammal 93:353–359

Martínez del Rio C, Wolf N, Carleton SA et al (2009) Isotopic ecology ten years after a call for more laboratory experiments. Biol Rev 84:91–111

Matthews CJD, Ferguson SH (2015) Weaning age variation in beluga whales (*Delphinapterus leucas*). J Mammal 96:425–437

McMahon KW, Hamady LL, Thorrold (2013) Ocean ecogeochemistry: a review. Oceanogr Mar Biol Annu Rev 51:327–374

Minagawa M, Wada E (1984) Stepwise enrichment of ^{15}N along food chains: further evidence and relation between δ^{15}N and animal age. Geochim Cosmochim Ac 48:1135–1140

Moore JW, Semmens BX (2008) Incorporating uncertainty and prior information into stable isotope mixing models. Ecol Lett 11:470–480

Newsome SD, del Rio CM, Bearhop S et al (2007) A niche for isotopic ecology. Front Ecol Environ 5:429–436

Newsome SD, Etnier MA, Monson DH et al (2009) Retrospective characterization of ontogenetic shifts in killer whale diets via δ^{13}C and δ^{15}N analysis of teeth. Mar Ecol Prog Ser 374:229–242

Newsome S, Clementz MT, Koch PL (2010) Using stable isotope biogeochemistry to study marine mammal ecology. Mar Mamm Sci 26:509–572

Ott PH, Barreto AS, Siciliano S et al (2016) Report of the working group on taxonomy and stock identity of bottlenose dolphins in the Southwest Atlantic Ocean. LAJAM 11:16–28

Page B, McKenzie J, Goldsworthy SD (2005) Dietary resource partitioning among sympatric New Zealand and Australian fur seals. Mar Ecol Prog Ser 293:283–302

Parnell AC, Inger R, Bearhop S et al (2010) Source partitioning using stable isotopes: coping with too much variation. PLoS One 5:e9672

Peterson BJ, Fry B (1987) Stable isotopes in ecosystem studies. Annu Rev Ecol Syst 18:293–320

Phillips DL (2012) Converting isotope values to diet composition: the use of mixing models. J Mammal 93:342–352

Polunin N, Pinnegar J (2002) Trophic ecology and the structure of marine food webs. In: Hart P, Reynolds J (eds) Handbook of fish and fisheries. Blackwell Science, Oxford, pp 301–320

Post DM (2002) Using stable isotopes to estimate trophic position: models, methods, and assumptions. Ecology 83:703–718

Riccialdelli L, Goodall RNP (2015) Intra-specific trophic variation in false killer whales (*Pseudorca crassidens*) from the southwestern South Atlantic Ocean through stable isotopes analysis. Mammal Biol 80:298–302

Riccialdelli L, Newsome SD, Fogel ML et al (2010) Isotopic assessment of prey and habitat preferences of a cetacean community in the southwestern South Atlantic Ocean. MEPS 418:235–248

Riccialdelli L, Newsome SD, Goodall RNP et al (2012) Insight into niche separation of Risso's dolphin (*Grampus griseus*) in the southwestern South Atlantic *via* δ^{13}C and δ^{15}N values. Mar Mamm Sci 28:E503–E515

Riccialdelli L, Newsome SD, Dellabianca NA et al (2013) Ontogenetic diet shift in Commerson's dolphin (*Cephalorhynchus commersonii commersonii*) off Tierra del Fuego. Polar Biol 36:617–627

Robbins CT, Felicetti LA, Florin ST (2010) The impact of protein quality on stable isotope ratio discrimination and assimilated diet estimation. Oecologia 162:571–579

Rowntree VJ, Payne RS, Schell DM (2001) Changing patterns of habitat use by southern right whales (*Eubalaena australis*) on their nursery ground at Península Valdés, Argentina, and their long-range movements. J Cetacean Res Manag (Special Issue) 2:133–143

Rowntree VJ, Valenzuela LO, Franco Fragas P et al (2008) Foraging behavior of southern right whales (*Eubalaena australis*) inferred from variations of carbon stable isotope ratios in their baleen. International Whaling Commission Document SC/60/BRG23

Saporiti F, Bala LO, Crespo EA et al (2013) Changing patterns of marine resource exploitation by hunter-gatherers throughout the late Holocene of Argentina are uncorrelated to sea surface temperature. Quat Int 299:108–115

Saporiti F, Bearhop S, Silva L et al (2014) Longer and less overlapping food webs in anthropogenically disturbed marine ecosystems: confirmations from the past. PLoS One 9:e103132

Saporiti BS, Vales DG et al (2016) Resource partioning among air-breathing marine predators: are body size and mouth diameter the major determinants? Mar Ecol 37:957–969

Secchi ER, Botta S, Wirgand MM et al (2016) Long-term and gender-related variation in the feeding ecology of common bottlenose dolphins inhabiting a subtropical estuary and the adjacent marine coast in western South Atlantic. Mar Biol Res. doi:10.1080/17451000.2016.1213398

Semmens B, Stock B, Ward E et al (2013) MixSIAR: A Bayesian stable isotope mixing model for characterizing intrapopulation niche variation. Presented at Ecological Society of America, Minneapolis, MN, 04–09 August 2013

Tatsch ACC, Secchi ER, Botta S (2016) Effects of acidification, lipid removal and mathematical normalization on carbon and nitrogen stable isotope compositions in beaked whale (Ziphiidae) bone. Rapid Commun Mass Spectrom 30(3):460–466

Tieszen LL, Boutton TW, Tesdahl KG et al (1983) Fractionation and turnover of stable carbon isotopes in animal tissues: implications for $\delta^{13}C$ analysis of diet. Oecologia 57:32–37

Troina G, Botta S, Secchi ER (2016) Ontogenetic and sexual characterization of the feeding habits of franciscanas, Pontoporia blainvillei, based on tooth dentin carbon and nitrogen stable isotopes. Mar Mamm Sci 32:1115–1137

Valenzuela LO, Sironi M, Rowntree VJ et al (2009) Isotopic and genetic evidence for culturally inherited site fidelity to feeding grounds in southern right whales (*Eubalaena australis*). Mol Ecol 18:782–791

Valenzuela LO, Sironi M, Rowntree VJ (2010) Interannual variation in the stable isotope differences between mothers and their calves in southern right whales (*Eubalaena australis*). Aquat Mamm 36:138–147

Vales DG, Saporiti F, Cardona L et al (2013) Intensive fishing has not forced dietary change in the South American fur seal *Arctophoca* (= *Arctocephalus*) *australis* off Río de la Plata and adjoining areas. Aquat Conserv Mar Freshw Ecosyst 24:745–759

Vales DG, Cardona L, García NA et al (2015) Ontogenetic dietary changes in male South American fur seals *Arctocephalus australis* in Patagonia. MEPS 525:245–260

Vander Zanden MJ, Rasmussen JB (1999) Variation in $\delta^{15}N$ and $\delta^{13}C$ trophic fractionation: implications for aquatic food web studies. Limnol Oceanogr 46:2061–2066

Vighi M, Borrell CEA et al (2014) Stable isotopes indicate population structuring in the Southwest Atlantic population of right whales (*Eubalaena australis*). PLoS One 9:e90489

Ward EJ, Semmens BX, Schindler DE (2010) Including source uncertainty and prior information in the analysis of stable isotope mixing models. Environ Sci Technol 44:4645–4650

Zenteno L, Crespo E, Goodall A et al (2013) Stable isotopes of oxygen reveal dispersal patterns of the South American sea lion in the southwestern Atlantic Ocean. J Zool 291:119–126

Zenteno L, Crespo E, Vales D et al (2014) Dietary consistency of male South American sea lions (*Otaria flavescens*) in southern Brazil during three decades inferred from stable isotope analysis. Mar Biol. doi:10.1007/s00227-014-2597-1

Zenteno L, Borella F, Otero JG (2015) Shifting niches of marine predators due to human exploitation: the diet of the South American sea lion (*Otaria flavescens*) since the late Holocene as a case study. Paleobiology 41:387–401

Part III
Conservation Ecology of Marine Vertebrate Research in Latin America

Chapter 10
Elasmobranchs Consumption in Brazil: Impacts and Consequences

Hugo Bornatowski, Raul R. Braga, and Rodrigo P. Barreto

Abstract Commercial fisheries struggle to apply regulatory and legal mechanisms that depend on reliable species-specific data, and the shark industry faces an even greater obstacle to transparency with sellers changing product names to overcome consumer resistance. Fraudulent representation or mislabeling of fish, including sharks and rays, has been recorded in some countries. In Brazil, for instance, sharks are sold as "cação" – a popular name attributed for any shark or ray species; however, according to consumer's knowledge of a large city of southern Brazil, more than 70% of them are often unaware that "cação" refers to sharks. Today, the Brazilian market has a high interest in encouraging people to eat "cação" meat, mainly because of their attractive prices. This raise a number of questions, mainly in respect to the knowledge of people/consumers, as what are they eating, and why the Brazilian meat market has grown so much in the last years.

10.1 Introduction

The Chondrichthyes, or cartilaginous fish (i.e sharks, rays, skates and chimaeras), are among the oldest taxa of vertebrates and have survived on planet earth for over 400 million years, including 4 mass extinction events (Camhi et al. 1998; Musick 1999). It can be considered an evolutionary successful group because it has diverse reproductive strategies (including parthenogenesis), ranging from planktivorous to top predators, occupying practically all aquatic niches (Priede et al. 2006; Snelson

H. Bornatowski (✉)
Centro de Estudos do Mar, Universidade Federal do Paraná, Curitiba, Brazil
e-mail: anequim.bio@gmail.com

R.R. Braga
LEC-Laboratório de Ecologia e Conservação/Universidade Federal do Paraná, Curitiba, Brazil

R.P. Barreto
Centro Nacional de Pesquisa e Conservação da Biodiversidade Marinha do Sudeste e Sul do Brasil (CEPSUL/ICMBio), Itajaí, Brazil

M.R. Rossi-Santos, C.W. Finkl (eds.), *Advances in Marine Vertebrate Research in Latin America*, Coastal Research Library 22,
DOI 10.1007/978-3-319-56985-7_10

et al. 2008). Sharks, skates, and rays composes the subclass Elasmobranchii (i.e. elasmobranchs) currently comprising a group of approximately 1150 species (Nelson 2006). Due to the development of a differentiated mandibular apparatus and an accurate sensorial system, elasmobranchs became predominantly predators, usually occupying higher or intermediate trophic levels (Camhi et al. 1998; Musick 1999). Therefore, they usually occur in relatively smaller abundances than species of lower trophic levels (Walker 1998; Stevens et al. 2000) though playing an important role in maintaining marine ecosystems healthy (e.g. control of genetic quality of populations, Holden 1974).

Elasmobranchs are long lived species characterized by low productivity rates which generally implies in small litters, slow growth rates and late onset of sexual maturity (Cortes et al. 2002; Garcia et al. 2008). These features are associated to high vulnerability and extinction risk, whereas populations have limited recovery capacity following overexploitation by fishing and even habitat losses and/or degradation (Smith et al. 1998; Stobutzki et al. 2002). A great diversity of elasmobranchs has been explored by modern fisheries, regularly caught as by-catch (i.e incidental) and more recently (>1990) as target, especially in longline, trawl and gill nets fisheries (Bonfil 1994; Baum et al. 2003; Barreto et al. 2016).

Population declines have been documented for several elasmobranchs species worldwide (Hutchings 2000; Baum et al. 2003; Ferretti et al. 2010; Clarke et al. 2011; Barreto et al. 2016). There are well-documented cases of elasmobranchs populations that collapsed such as *Lamna nasus* in the North Atlantic; *Galeorhinus galeus* in California and Australia; *Cetorhinus maximus* in England; *Squalus acanthias* in the North Sea and British Columbia; *Pristis pectinata* in Florida and Louisiana, and large coastal sharks off the west coast of the United States, such as *Carcharias taurus* and *Carcharhinus obscurus* (International Union for Conservation of Nature – IUCN). The IUCN reports that approximately 30% of elasmobranchs are threatened with extinction (Dulvy et al. 2014). Worryingly, 47% of extant species do not have minimal information to be evaluated under the IUCN classifying system (Dulvy et al. 2014; Mace et al. 2008).

10.2 Elasmobranch Fisheries in Brazil

South Atlantic fisheries, specifically assessed from Brazil during the late 1990s reached critical levels for spinny angel shark *Squatina guggenheim*, hidden angel shark *S. occulta*; Brazilian guitarfish *Rhinobatos horkelli*, scalloped hammerhead shark *Sphyrna lewini*, sand tiger shark *Carcharias taurus*, school shark *Galeorhinus galeus* and narrownose smooth-hound *Mustelus schmitti* (Vooren 1997; Lessa et al. 1999). More recently (between 2010 and 2012), Brazil reassessed elasmobranchs following the IUCN's system, resulting in 33% of species in threatened categories (VU = 19; EN = 8; CR = 28), and 36% for which the available information did not allow for any sort of categorization (i.e. Data Deficient-DD) (ICMBio 2016). This result overcomes the global rate of threatened species (approximately 25% – Dulvy et al. 2014).

The situation of large pelagic elasmobranchs currently is very worrying. According to Barreto (2015), which analyzed CPUE and demography of pelagic shark species from Brazil, most populations are currently depleted. Around 13 species have been caught by local and international longline fishing fleets since the 1950s, specifically: shortfin mako shark (*Isurus oxyrhinchus*), longfin mako shark (*I. paucus*), porbeagle shark (*Lamna nasus*), blue shark (*Prionace glauca*), oceanic whitetip shark (*Carcharhinus longimanus*), dusky shark (*C. obscurus*), night shark (*C. signatus*), crocodile shark (*Pseudocarcharias kamoharai*), common thresher shark (*Alopias vulpinus*) and bigeye thresher shark (*A. superciliosus*), scalloped hammerhead shark (*Sphyrna lewini*), smooth hammerhead shark (*S. zygaena*) and great hammerhead shark (*S. mokarran*) and the pelagic stingray (*Pteroplatytrygon violacea*).

These species have historically sustained the international fin market (Amorim et al. 1998; Hazin et al. 2008; Domingo et al. 2014) and since the 1990s, more precisely after the Brazilian finning ban, became one of the most cheap sea food to be consumed in Brazil. The blue shark, for instance, probably for being the only species that apparently has not yet collapsed, is currently supporting an industry that has grown absurdly in recent decades. Although this species has no capture restrictions in the region, it is currently the main target of longline pelagic fisheries in the South Atlantic, especially in periods where catches of other pelagic fish of high commercial demand (i.e. tunas and billfishes) are lower. Worryingly, both in tuna fishing and swordfish, as in fishing for blue shark, a number of other extremely sensitive sharks, with considerably lower population abundances than the target species are often caught. Finally, Brazil has been recently identified as a global flow channel for shark carcasses (world's largest importer) (Dent and Clarke 2015). Over the last 15 years the country has imported large quantities of shark carcasses from international fleets (>20) which also export fins for international luxury markets channels.

While for oceanic species, despite the limitations, there is some catch information available (mainly due to the industry and their obligations with the government, i.e. fishing statistics, commercial and financial transactions) coastal shark species are far to have their catches known. Large species such as hammerhead sharks, sand-tiger, and also the medium sized daggernose, blacknose and the southern squatinids are at great risk due to coastal fisheries (Barreto et al. 2011; Lessa et al. 2016) and for small and medium-sized species, there is no information about fishing or life history. These species are commonly caught by small-scale fisheries, which in a country with one million registered fishermen in this category, which contributes with 45% of the total fish produced in the country is too big to be ignored (B. Padovani, personal communication). The combined effect of oceanic industrial fishing and the coastal small-scale fisheries can be additive or even synergistic on population depletion due to its spatial complementarity. This relation still needs small-scale fisheries catch data to be assessed.

Rays production in Brazil has always been neglected. In the state of Santa Catarina, main Brazilian elasmobranch producer and one of the last to stop their fishing statistics collection, the ray production increased more than threefold in ten years (from 452,367 kg in 2000 to 1,425,792 kg in 2010). This state reported that more than 2300 tons of elasmobranchs were landed in 2010, whereas over 85% were not identified at species level (UNIVALI/CTTMar 2011). This broad identification levels of landed species makes species-specific regulation very difficult (Bornatowski et al. 2011, 2014), precluding fishing controls. Major concerns however lie in the categorization of ray's species. In Brazilian fisheries some ray species are grouped under a category called "emplastro". Within this label there are at least seven species: bignose banskate (*Sympterygia acuta*, endemic); smallnose fanskate (*S. bonapartii*, endemic), rio skate (*Rioraja agassizi*, endemic); spotback skate (*Atlantoraja castelnaui*, endemic); la plata skate (*A. platana*, endemic); eyespot skate (*A. cyclophora*); and the bluntnose stingray (*Dasyatis say*). Today, the demand of ray meat ("emplastro" group) for the Korean markets is so great that was one of the critical steps to invalidate the Normative Instruction-445 (Brazil 2014), which the main goal was to regulate the use of seafood products.

10.3 The Problem of Mislabeling

Commercial fisheries struggle to apply regulatory and legal mechanisms that depend on reliable species-specific data, and the shark industry faces an even greater obstacle to transparency with sellers changing product names to overcome consumer resistance. Fraudulent representation or mislabeling of fish, including sharks and rays, has been recorded in some countries (e.g. Barbuto et al. 2010; Jacquet and Pauly 2008; Lamendin et al. 2015; Smith and Benson 2001; Wong and Hanner 2008), highlighting the impacts on the economy and ecosystem services (Jacquet and Pauly 2008). The European Union, on the other hand, requires listing the species name on shark products (see Fig. 10.1) to avoid fraud and to aid conservation (Council Regulation No. 104/2000 – December/1999). With these measures rates of seafood fraud appear to have decreased from 2011 to 2015 (Oceana 2016).

While many developing countries depend of fish meat (including shark and ray meat) for their subsistence, many other countries view shark meat as low-quality and therefore commercialize them under generic names to overcome consumer resistance (Vannuccini 1999; Bornatowski et al. 2013, 2015; Dent and Clarke 2015). This measure impedes consumers to link the meat they buy to the animals they know. For instance, the spiny dogfish is sold under names like "saumonette" in France, "schillerlocken" or "seeaal" in Germany, and "spinaroli" in Italy (WildAid 2007). Shortfin mako (*Isurus oxyrinchus*), porbeagle (*Lamna nasus*) and thresher sharks (*Alopias* spp.) are considered to have a highly palatable meat worldwide and sometimes even comparable to swordfish (*Xiphias gladius*) in Europe and USA (WildAid 2007). On the other hand, blue shark (one of the most important shark species in fisheries) is considered one of the less preferred species for human consumption due to its soft and strong flavoured meat.

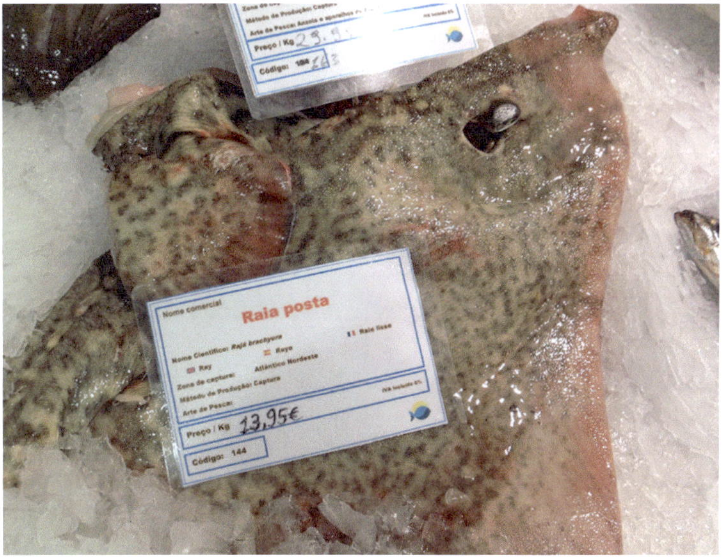

Fig. 10.1 *Raja brachyura* sold in a fish-market in Peniche, Portugal. Note the popular name, scientific name and site of catch (Photo: Hugo Bornatowski)

In Brazil, sharks are sold as "cação" – a popular name attributed for any shark or ray species (Fig. 10.2). People frequently purchase shark meat in pieces as '*fillet'* or '*thermidor*'. The most common dish made with elasmobranch meat is a traditionally dish called "Muqueca". This dish is originally from Espírito Santo, in the Southeast of Brazil, and also from the state of Bahia, in the Northeast.

Curiously, shark meat were not part of the traditional Brazilian diet and is considered to be low-value seafood (values around U$2.00/Kg^{-1}) if comparable to more common fish such as tilapia (*Tilapia* spp. and *Oreochromis niloticus*), flatfish (Paralichthyidae family), salmon (*Salmon salari*), croacker fish (*Cynosion* spp.), snook (*Centropomus* spp.), groupers (Serranidae family) and snappers (Lutjanidae family) and others. However, today the Brazilian market have a high interest in encouraging people to eat "cação" meat, mainly because of their attractive prices (as cited above), unfamiliarity with the product and to avoid meat waste (Ministério da Agricultura, Pecuária e Abastecimento – MAPA). The high interest of Brazil for elasmobranch meat turns it into the perfect "port for meat discard" (finning is prohibited in several regions).

Furthermore, shark meat is traded without labeling and at substantially lower prices under the name of "cação" and not as "shark" or "ray" (Fig. 10.2). This raise a number of questions, mainly in respect to the knowledge of people/consumers, as what are they eating, and why the Brazilian meat market has grown so much in the last years.

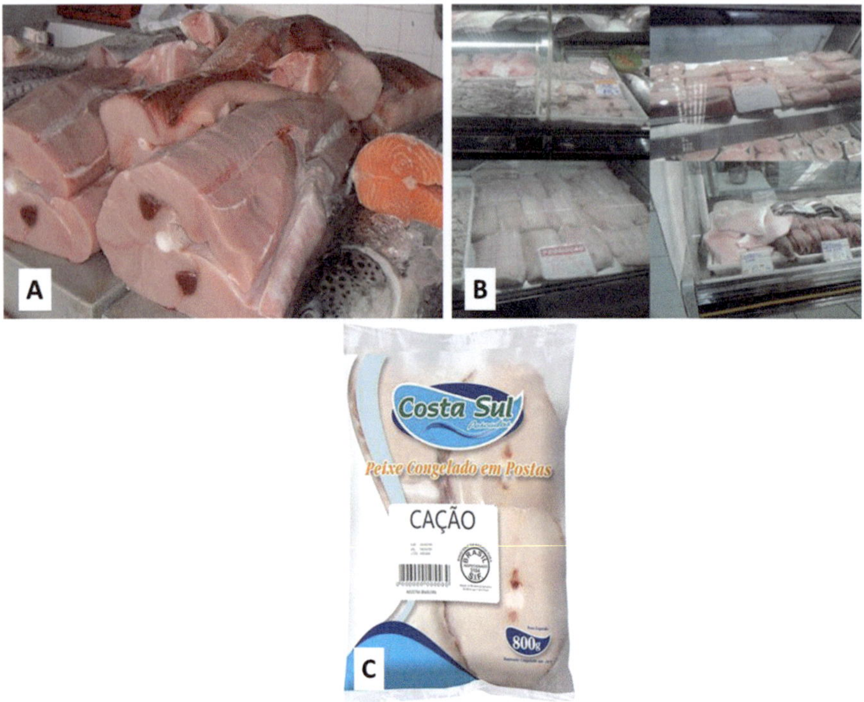

Fig. 10.2 (**a**) Shark meat sold in a market in southeastern Brazil beside of Chilean Salmon (*Salmo salar*) (Credit: Dr. Fernando F. Mendonça.); (**b**) Shark and ray meat sold in a market in southern Brazil (Photo by: Rodrigo Barreto); (**c**) Frozen slices of sharks sold as "cação"

10.4 Decision-Making, Health and Conservation. The Concerns Grow!

According to consumer's knowledge of a large city of southern Brazil, more than 70% of them are often unaware that "cação" refers to sharks, and more than half of respondents claimed to have already eaten "cação" but never eaten sharks or rays (Bornatowski et al. 2015). So, consumers are buying "a pig in a poke". Beside of that, in surveys made in southeastern Brazil (states of São Paulo and Rio de Janeiro), 62% of fish sold as groupers were actually sharks (Estrella et al. 2014). This mislabeling provokes a 25% increase in revenue. Unfortunately, the problem does not stop there and this may be only the tip of the iceberg.

The mislabeling may preclude people from taking health- and/or conservation-related decisions concerning the consumption of elasmobranch meat and thus interfere with efforts to reduce consumption or redirect consumption towards non-threatened species (Barbuto et al. 2010; Bornatowski et al. 2013; Jacquet and Pauly 2008). Furthermore, misleading product descriptions is a crime according to Brazilian Consumer Protection Code (n° 8078/1990) (Brazil 1990). A recent study found that samples labeled under the name "cação" in Brazil were in fact made up of endangered scalloped hammerhead (Carvalho et al. 2015). Another study showed that 55% of "cação" samples were actually large-tooth sawfish, *Pristis perotteti*, a species considered by the IUCN to be critically endangered and for which trade is prohibited in Brazil (Palmeira et al. 2013). The generalization of species in a single name (such as "cação") is therefore puting in risk several threatened species. In 19 October 2016, an online journal published a matter on "Sharks and rays threatened that are irregularly commercialized in Brazil" (http://economia.uol.com.br/noticias/efe/2016/10/19/tubaroes-e-raias-em-extincao-sao-comercializados-irregularmente-no-brasil.htm). More than 16 different species of sharks and rays were sold as "cação", with special concern to the hidden angel shark *Squatina occulta*, and scalloped hammerhead shark *Sphyrna lewini* (23% of samples) classified as critically endangered and vulnerable respectively according IUCN redlist. In southern Brazil, artisanal fisheries capture a lot of sharks, including some threatened species that are also sold as "cação" for consumers in local markets (Figs. 10.3 and 10.4).

This lack of knowledge is also of great concern to human health because shark meat contains high levels of heavy metals (lead and mercury) due to biomagnification (Escobar-Sánchez et al. 2011; Lopez et al. 2013; Pethybridge et al. 2009). A study analyzing specifically blue sharks collected from strategical landing points of central and southern Brazil, found 70% of the samples with mercury levels above recommended (Dias et al. 2008). Brazil has less stringent restrictions for heavy metals in seafood than the European Union, North America, and Asia, allowing products

Fig. 10.3 Scalloped hammerhead sharks (*Sphyrna lewini*) and other *Carcharhinus* spp. captured by artisanal fisheries in Paraná coast. Specimens are without heads and will be cut in slices and sold as "cação" (Photo by Isabella Simões)

Fig. 10.4 Scalloped hammerhead sharks (*Sphyrna lewini*) and other *Carcharhinus* spp. sharks captured by artisanal fisheries in Paraná coast. Specimens landed after evisceration (Photo by Isabella Simões)

Fig. 10.5 Children's tests shark meat to be included in the scholar meals system (Photo by: Ana Chaffin, Macaé-RJ)

with high levels of heavy metals, which would not be marketable elsewhere, to be commercialized legally. The worst is that unlabeled shark meat has been served in public schools from several cities to children from 6 to 17 years old, as subsidized meat by the government (http://macae.rj.gov.br/semed/leitura/noticia/alunos-testam-file-de-cacao-na-merenda-escolar) (Fig. 10.5).

10.5 Actions and Suggestions

We therefore emphasize that a small list of actions need to be taken if developing countries, specially Brazil, to overcome major conservation barriers. They are:

1. Use of scientific names on commercialized products, and use of exclusive popular names in labels. A correct species-specific identification would aid in the solution to the mislabeling of shark meat
2. Supermarkets or any other final seller should be held legally responsible for the identification
3. Development of an identification handbook based on morphological characteristics (on muscle, for instance) to be used by regulatory agencies. This would be a cheap method to prevent fraud in identification. Muscle tissue identification methodology is already applied for teleostean fish (http://www.agricultura.gov.br/arq_editor/manual%20pesca.pdf), and could be used for elasmobranchs also.
4. Use of genetic techniques such as DNA barcoding for species-identification
5. Certification programs such as the Marine Stewardship Council (MSC) and eco-labels, may help consumers to choose more sustainable seafood products (Jacquet and Pauly 2008; Lamendin et al. 2015; Sampson et al. 2015; Von der Heyden et al. 2010). The establishment of species-specific commodity codes can facilitate trade monitoring of protected species, and also inform end consumers. This kind of information should be available to consumers
6. Restructuration of fisheries monitoring programs, and creation of a national program of fishery statistics with wide spatial and temporal coverage, with extensive species-catch monitoring throughout the Brazilian coast
7. Trained personnel to obtain reliable data following the correct identification of elasmobranchs species. Without knowing fishery catches (item 6) and correct species-specific information, any monitoring will fail.

References

Amorim AF, Arfelli CA, Fagundes L (1998) Pelagic elasmobranchs caught by longliners off southern Brazil during 1974–97: an overview. Mar Freshw Res 49:621–632

Barbuto M, Galimberti A, Ferri E, Labra M, Malandra R, Galli P, Casiraghi M (2010) DNA barcoding reveals fraudulent substitutions in shark seafood products: the Italian case of "palombo" (*Mustelus* spp.) Food Res Int 43:376–381

Barreto RP (2015) Historia de vida e vulnerabilidade dos tubarões oceânicos do Atlantico Sul. PhD Thesis, Universidade Federal Rural de Pernambuco, Brazil
Barreto RR, Lessa RP, Hazin FH, Santana FM (2011) Age and growth of the blacknose shark, *Carcharhinus acronotus* (Poey 1860) off the northeastern Brazilian coast. Fish Res 110:170–176
Barreto R, Ferretti F, Mills J, Amorim A, Andrade H, Worm B, Lessa R (2016) Trends in the exploitation of South Atlantic shark populations. Conserv Biol 30:792–804
Baum JK, Myers RA, Kehler DG, Worm B, Harley SJ, Doherty PA (2003) Collapse and conservation of shark populations in the northwest Atlantic. Science 299:389–392
Bonfil R (1994) Overview of world elasmobranch fisheries. Fisheries technical papers. Rome
Bornatowski H, Vitule JRS, Abilhoa V, Corrêa MFM (2011) Unconventional fishing for large sharks in the state of Paraná southern Brazil: a note of concern. J Appl Ichthyol 27:1108–1111
Bornatowski H, Braga RR, Vitule JRS (2013) Shark mislabeling threatens biodiversity. Science 340:923
Bornatowski H, Braga RR, Vitule JRS (2014) Threats to sharks in a developing country: the need for effective and simple conservation measures. Natureza Conserv 12(1):11–18
Bornatowski H, Braga RR, Kalinowski C, Vitule JRS (2015) "Buying a Pig in a Poke": the problem of Elasmobranch meat consumption in Southern Brazil. Ethnobiol Lett 6(1):196–202
Brazil (1990) Lei n. 8.078, de 11 de setembro de 1990. Dispõe Sobre a Proteção do Consumidor e dá Outras Providências. Diário Oficial da República Federativa do Brasil, Brasília, DF, 12 set. 1990. Available at: http://www.planalto.gov.br/ccivil_03/ leis/l8078.htm. Accessed 15 Sep 2016
Brazil (2014) Ministério do Meio Ambiente. Portarias 445, de 17 de Dezembro de 2014, Diário Oficial da União http://pesquisaingovbr/imprensa/jsp/visualiza/indexjsp?data=18/12/2014&jornal=1&pagina=110&totalArquivos=144. Accessed 18 Oct 2016
Camhi M, Fowler S, Musick J, Bräutigam A, Fordham S (1998) Sharks and their relatives, ecology and conservation. Occas. Paper IUCN Spec. Surv. Comm. 39 p
Carvalho DC, Palhares RM, Drummond MG, Frigo TB (2015) DNA barcoding identification of commercialized seafood in south Brazil: a governmental regulatory forensic program. Food Control 50:784–788
Clarke S, Yokawa K, Matsunaga H, Nakano H (2011) Analysis of North Pacific shark data from Japanese commercial longline and research/training vessel records. Western and Central Pacific Fisheries Commission, Pohnpei, Micronesia
Cortés E, Brooks L, Scott G (2002) Stock assessment of large coastal sharks in the U.S. Atlantic and Gulf of Mexico. National Oceanic and Atmospheric Administration, National Marine Fisheries Service Southeast Fisheries Science Center, Sustainable Fisheries Division Contribution SFD-2/03-177
Dent F, Clarke S (2015) State of the global market for shark products. FAO Fisheries and Aquaculture technical paper no. 590. FAO, Rome
Dias ACL, Guimarães JRD, Malm O, Costa PAS (2008) Mercúrio total em músculo de cação *Prionace glauca* (Linnaeus, 1758) e de espadarte *Xiphias gladius* Linnaeus, 1758, na costa sul-sudeste do Brasil e suas implicações para a saúde pública. Cad Saúde Pública 24(9):2063–2070
Domingo A, Forselledo R, Miller P, Jiménez S, Mas F, Pons M (2014) General description of longline fisheries. ICCAT manual, 312. Available from http://wwwiccates/Documents/SCRS/Manual/CH3/CHAP_3_1_2_LL_ENGpdf. Assessed 05 October 2016
Dulvy NK, Harisson LR, Carlson JK et al (2014) Extinction risk and conservation of the world's sharks and rays. elife 3:e00590
Escobar-Sánchez O, Galván-Magaña F, Rosíles-Martínez R (2011) Biomagnification of mercury and selenium in blue shark Prionace glauca from the Pacific Ocean off Mexico. Biol Trace Elem Res 144:550–559
Estrella F, Raposo G, Pascolli J, Gonzalez JG, Motta FS, Moura RL (2014) Comercialização de pescado nas cidades de São Paulo e Rio de Janeiro. https://wwwsosmaorgbr/wp-content/uploads/2014/09/ESTUDO-PESCADO-2014_Relatorio-Finalpdf. Assessed 02 Nov 2016

Ferretti F, Worm B, Britten GL, Heithaus MR, Lotze HK (2010) Patterns and ecosystem consequences of shark declines in the ocean. Ecol Lett 13:1055–1071

Garcia VB, Lucifora LO, Myers RA (2008) The importance of habitat and life history to extinction risk in sharks, skates, rays and chimaeras. Proc R Soc B 275:83–89

Hazin FHV, Broadhurst MK, Amorim AF, Arfelli CA, Domingo A (2008) Catches of pelagic sharks by subsurface longline fisheries in the South Atlantic Ocean during the last century: a review of available data with emphasis on Uruguay and Brazil. In: Camhi MD, Pickitch EA (eds) Sharks of the open ocean: biology, fisheries and conservation. Blackwell Publishing, Oxford, pp 213–227

Holden MJ (1974) Problems in the rational exploitation of elasmobranch populations and some suggested solutions. In: Harden-Jones FR (ed) Sea fisheries research. Elek Science, London, pp 117–138

Hutchings JA (2000) Collapse and recovery of marine fishes. Nature 406:882–885

Instituto Chico Mendes de Conservação da Biodiversidade (ICMBio) (2016) Avaliação do risco de extinção dos elasmobrânquios e quimeras no Brasil: 2010–2012. Available in: http://www.icmbio.gov.br/cepsul/especies-ameacadas.html. Accessed 20 Oct 2016

Jacquet JL, Pauly D (2008) Trade secrets: renaming and mislabeling of seafood. Mar Policy 32:309–318

Lamendin R, Miller K, Ward RD (2015) Labelling accuracy in Tasmanian seafood: an investigation using DNA barcoding. Food Control 47:436–443

Lessa RP, Santana FM, Rincón G, Gadig OBF, El-Deir ACA (1999) Biodiversidade de elasmobrânquios do Brasil. Recife Ministério do Meio Ambiente (MMA), Projeto de Conservação e Utilização Sustentável da Diversidade Biológica Brasileira (PROBIO)

Lessa RP, Batista V, Santana FM (2016) Close to extinction? The collapse of the endemic daggernose shark (*Isogomphodon oxyrhynchus*) off Brazil. Global Ecol Conserv 7:70–81

Lopez SA, Abarca NL, Meléndez R (2013) Heavy metal concentrations of two highly migratory sharks (*Prionace glauca* and *Isurus oxyrinchus*) in the Southeastern Pacific waters: comments on public health and conservation. Trop Conserv Sci 6:126–137

Mace GM, Collar NJ, Gaston KJ, Hilton-Taylor C, Akcakaya HR, Leader-Williams N, Milner-Gulland EJ, Stuart SN (2008) Quantification of extinction risk: IUCN's system for classifying threatened species. Conserv Biol 22:1424–1442

Musick JA (1999) Ecology and conservation of long-lived marine animals. In: Musick JA (ed) Life in the slow lane: ecology and conservation of long-lived marine animals. American Fisheries Society Symposium, Bethesda, pp 1–10

Nelson JS (2006) Fishes of the World, 4th edn. Wiley, New York

Oceana (2016) Seafood frau campaign. http://oceanaorg/our-campaigns/seafood_fraud/campaign. Assessed 2 Oct 2016

Palmeira MCA, Rodrigues-Filho LFS, Sales JBL, Vallinoto M, Schneider H, Sampaio I (2013) Commercialization of a critically endangered species (largetooth sawfish, *Pristis perotteti*) in fish markets of northern Brazil: authenticity by DNA analysis. Food Control 34:249–252

Pethybridge H, Cossa D, Butler CV (2009) Mercury in 16 Demersal sharks from Southeast Australia: biotic and abiotic sources of variation and consumer health implications. Mar Freshw Res 68:18–26

Priede IG, Froese R, Bailey DM, Bergstad OA, Collins MA, Dyb JE, Henriques C, Jones EG, King N (2006) The absence of sharks from abyssal regions of the world's oceans. Proc R Soc B 273:1435–1441

Sampson GS, Sanchirico JN, Roheim CA et al (2015) Sustainability: secure sustainable seafood from developing countries. Science 348(6234):504–506

Smith PJ, Benson PG (2001) Biochemical identification of shark fins and fillets from the coastal fisheries in New Zealand. Fish Bull 99(2):351–355

Smith SE, Au DW, Show C (1998) Intrinsic rebound potentials of 26 species of Pacific sharks. Mar Freshw Res 49:663–678

Snelson FF Jr, Roman BL, Burgess GH (2008) The reproductive biology of pelagic elasmobranchs. In: Camhi MD, Pikitch EK, Babcock EA (eds) Sharks of the open ocean: biology, fisheries and conservation. Blackwell Publishing, Oxford, pp 24–53

Stevens JD, Bonfil R, Dulvy NK, Walker PA (2000) The effects of fishing on sharks, rays, and chimeras (chondrichthyans), and the implications for marine ecosystems. ICES J Mar Sci 57:476–494

Stobutzki IC, Miller MJ, Heales DS, Brewer DT (2002) Sustainability of elasmobranchs caught as bycatch in a tropical prawn (shrimp) trawl fishery. Fish Bull 100:800–821

UNIVALI/CTTMar (2011) Boletim estatístico da pesca industrial de Santa Catarina – Ano 2010. Universidade do Vale do Itajaí, Centro de Ciências Tecnológicas da Terra e do Mar, Itajaí

Vannuccini S (1999) Shark utilization and trade. Food and Agriculture Organization technical paper, n.389, Rome, Italy

Von der Heyden S, Barendse J, Seebregts AJ, Matthee CA (2010) Misleading the masses: detection of mislabelled and substituted frozen fish products in South Africa. ICES J Mar Sci 67:176–185

Vooren CM (1997) Demersal elasmobranchs. In: Seeliger U, Odebrechet C, Castello JP (eds) Subtropical convergence environments. The coast and sea in the southwestern Atlantic. Springer, Heidelberg, pp 141–146

Walker T (1998) Can shark resources be harvested sustainably? A question revisited with a review of shark fisheries. Mar Freshw Res 49:553–572

WildAid (2007) The end of the line? Global threats to sharks. http://www.wildaid.org. Accessed 10 Mar 2015

Wong EHK, Hanner R (2008) DNA barcoding detects market substitution in North American seafood. Food Res Int 41:828–837

Chapter 11
West Indian Manatee *Trichechus manatus* in South America: Distribution, Ecology and Health Assessment

Ana C.O. de Meirelles, Vitor L. Carvalho, and Miriam Marmontel

Abstract The West Indian manatee inhabits exclusively tropical waters of Atlantic Ocean, from Virginia, in United States, to Alagoas, in Northeastern Brazil, in a fragmented way. In South America, low species density and difficulty to detect the animals that inhabit areas with murky waters, associated with a lower financial investment in research in the past, lead to a delay in manatee research. However, in the last decade, an increase in manatee research could be noted in the continent. Nevertheless, a significant part of this information is available only on theses, conference abstracts, reports, etc. Thus, this chapter aims to gather and update important information on the species, mainly on distribution, spatial ecology, feeding habits, genetics and health.

Keywords Sirenia • Abundance • Spatial ecology

11.1 Introduction

The West Indian manatee, *Trichechus manatus*, is one of the four only extant species of the Sirenia order. The IUCN Red list considers the species as Vulnerable, with a decreasing population, due to habitat loss and anthropogenic causes (Deutsch et al. 2008). Many authors and the Society for Marine Mammalogy accept the existence of two subspecies (Committee on Taxonomy 2016): Florida manatee, *T. m. latirostris,* corresponding to animals in the United States.; and Antillean

A.C.O. de Meirelles (✉) • V.L. Carvalho
Associação de Pesquisa e Preservação de Ecossistemas Aquáticos (Aquasis), Av. José Alencar, 150, Sesc IParana, 61627-010 Caucaia, Ceará, Brazil
e-mail: cameirelles@yahoo.com.br

M. Marmontel
Instituto de Desenvolvimento Sustentável Mamirauá, Estrada do Bexiga, 2584, Fonte Boa, 69553-225 Tefé, Amazonas, Brazil

M.R. Rossi-Santos, C.W. Finkl (eds.), *Advances in Marine Vertebrate Research in Latin America*, Coastal Research Library 22,
DOI 10.1007/978-3-319-56985-7_11

manatee, *T. m. manatus,* including manatees that inhabit the tropical and subtropical western Atlantic coasts, from Bahamas to Brazil, including the Caribbean Sea and Gulf of Mexico. However genetic (García-Rodríguez et al. 1998; Vianna et al. 2006; Santos et al. 2016), skull morphology and cytogenetic (Barros 2013; Barros et al. 2017) studies indicate the necessity to revisit this taxonomic classification. Therefore, in this chapter we chose not to use the subspecific classification.

The species inhabits exclusively tropical waters of Atlantic Ocean, from Virginia (37° N, 76° W), in United States, to Alagoas (10° S, 36° W), in Northeastern Brazil, in a fragmented way. In the United States, West Indian manatee is concentrated in Florida, due to warm water temperature. During summer months, some animals disperse to other states and can be found in Georgia, South Carolina, North Carolina, Virginia and rarely in New Jersey. In tropical countries, the species range is not totally understood, however it seems more related to displacements from resting areas to foraging and freshwater sites.

As an exclusively herbivore taxa, manatees occur in rivers, estuaries, lagoons, channels and coastal shallow waters, where animals can feed on a variety of submerged, floating or emergent plants (Marsh et al. 2011). In the United States, during winter, animals concentrate in warm water refuges, natural or artificial, with clear waters. These two factors propitiated early studies with the species there. In other countries, mainly in South America, low species density and difficulty to detect the species, that inhabits areas with murky waters, associated with a lower financial investment in research in the past, lead to a delay in manatee research. In the last decade, an increase in manatee research could be noted in South America. Nevertheless, a significant part of this information is available only in theses, conference abstracts, reports, etc. Thus, in this chapter we present a review on the research conducted with West Indian manatee in South America, discussing its importance to species conservation and management.

11.2 Distribution

In South America, *T. manatus* inhabits rivers, swamps, coastal lagoons, estuaries and coastal areas. Dark waters in these habitats hamper manatee detection, rendering aerial surveys an ineffective method to identify the species distribution. Thus, indirect evidences, such as interviews, feeding traces and presence of feces are the most used approaches to study manatees in the region, associated with direct observations. Aerial surveys were only used with good results in Brazil (Alves et al. 2013, 2016). More recently, the side scan sonar (SSS) emerged as an important detection tool to study manatees (Gonzalez-Socoloske et al. 2009; Gonzalez-Socoloske and Olivera-Gomez 2012). Detection of manatees using sonars initially aimed to avoid collision with boats, the main cause of manatee mortality in U.S.A. (eg, Bowles et al. 2004). However, as demonstrated by Gonzalez-Socoloske et al. (op. cit.) through tests in clear waters of Florida and in the dark waters of a lagoon in Mexico, SSS is a cost-effective option to detect manatees in turbid and dark waters. In South

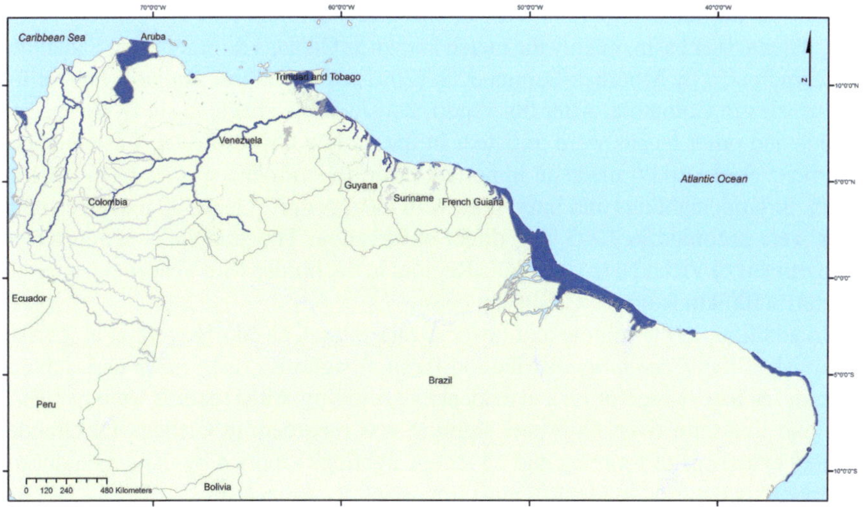

Fig. 11.1 West Indian manatee distribution (in dark *blue*) in South America, including main rivers, bays and estuaries. *Blue dots* are isolated records of the species

America, this tool is already being used in some countries, with encouraging results (eg, Colombia, Arévalo-González et al. 2014; Arévalo-González and Castelblanco-Martínez 2015; Brazil, Choi-Lima et al. 2014; FMA 2015).

Below we review the manatee records in each country of South America (Fig. 11.1). This is not intended to provide an exhaustive review on manatee records in these countries; rather it aims to gather important and update information that not always is available in scientific articles.

11.2.1 Colombia

In Colombia, the West Indian manatee inhabits mainly riverine environments, preferring lower and medium courses of rivers Magdalena, Atrato, Sinú, Cauca, San Jorge, Meta, Orinoco; and marshes, like the Ciénaga (swamps) de Paredes, de Santa Helena and Grande de Santa Marta (Caicedo-Herrera et al. 2004; Montoya-Ospina et al. 2001). Some manatee captures and sightings were recorded in the mouth of Sinú and Magdalena rivers; however, the presence of the species in the Caribbean Sea is not common in the country (Montoya-Ospina and Caicedo-Herrera 1995). In 2004, Caicedo-Herrera et al. compiled important information on the species distribution in the "Programa Nacional de Manejo y Conservación de Manatíes en Colombia" (National Program for manatee conservation and management in Colombia). This is the most complete source of information on *T. manatus* records in the country.

One of the most important sites of manatee occurrence in the country is the Magdalena river basin, mainly the marsh known as Ciénaga de Paredes, in Santander state, indicated by Montoya-Ospina et al. (2001) as the largest available habitat for the species in Colombia. After this report, considerable efforts to study the species in this and other basins were initiated. In the middle Orinoco river, Castelblanco-Martínez et al. (2009) made an important effort to study the species from 2001 to 2005, through sightings and interviews with local people. Over 800 manatee sightings were recorded in 1003 h of direct observation. The study was conducted in Department of Vichada, in Orinoquía Region, in the border with Venezuela, approximately 1100 km from the Caribbean coast.

In addition, in the middle and lower Atrato river, Caicedo-Herrera et al. (2004) established the *T. manatus* distribution through sightings, interviews and indirect records, as feces observation and evidence of foraging. Most records were in *ciénagas* and in Atrato river. Only one sighting was recorded, in Ciénaga La Grande. Ninety evidences of foraging and 17 recent sightings reported by fishermen during interviews were used to identify most important occurrence sites.

11.2.2 Venezuela

In eastern Venezuela, *T. manatus* occurs in the Gulf of Paria, the middle Orinoco river and its tributaries, the Orionoco Delta and the Venezuelan Llanos (Boher and Porras 1991; Lefebvre et al. 2001; Mondolfi and Muller 1979; O'Shea et al. 1988). In the west of the country, the species occurs in the Maracaibo system, comprised of the Gulf of Venezuela, Tablazo Bay and the Maracaibo Strait, which connect Lake Maracaibo in the interior of the Maracaibo Basin, to the Caribbean Sea. Additionaly, there are records of manatees in Limón river and channels south of São Carlos Island, and Ciénaga Juan Manuel de Aguas Claras y Aguas Negras.

According to O'Shea et al. (1988) interviews in Venezuela indicated that a small manatee population exists in Maracaibo Lake and the swamps bordering the lake. Interviews also revealed the presence of manatees in Manatí Lake and rivers Escalate, Catumbo and Bravo. After this research, there was little effort to study manatee in the system. Manzanilla-Fuentes (2007) reported 27 sightings of manatee in Tablazo Bay, including solitary animals and groups of up to four animals. Some stranded and rescued animals were also reported for the area (Boede and Mujica-Jorquera 2016; Ramírez-Carroz and Manzanilla-Fuentes 2008).

O'Shea et al. (1988) indicated that the species did not occur in the Caribbean coast due to inadequate ecological characteristics, representing a major discontinuity in the species distribution (Lefebvre et al. 2001). However, some rare records were reported in the mouth of Neverí river (Ojeda et al. 1993) and in Puerto Cabello (Delgado 1995).

Manatees seem to be more abundant in eastern Venezuela. For this reason, there are more studies published for this region. In the middle Orinoco, in the border with Colombia, the species has been studied for several years by the Colombian

Fundación Omacha (Bermudez-Romero et al. 2004; Castelblanco-Martinez 2004; Castelblanco-Martínez et al. 2001, 2005). In the central lower Orinoco, Rivas-Rodríguez et al. (2012) studied the species in Caicara del Orinoco. In adition, an effort was initiated in 2007 in lower Orinoco, through interviews, sighting records and identification of feeding areas, between Las Bonitas and Mapire, including Caura river (Rivas-Rodriguez et al. 2012). The authors record the presence of manatee in 46 localities, expanding the knowledge of the species occurrence in the Orinoco river originally reported by Mondolfi and Müller (1979) and Correa-Viana et al. (1990). More recently, Rodulfo (2012) studied manatees in Caño Araguao, in the Delta of Orinoco, through interviews, indirect evidences of the species presence (e.g, foraging traces, feces), and sightings. Most records were in distributaries (caños), followed by the river mouth. Foraging areas were recorded in *canõ* Araguao, Río Grande de Araguao, confluence of Rio Grande with *caño* Anacasi, *canõ* Janeida, Canal Sorocobo, Barra de Araguao and Isla Jeocubanoco.

11.2.3 Guyana

Most information published on *T. manatus* in Guyana is from the 1960s and 1970s (eg, Bertram 1963; Bertram and Bertram 1963, 1973). The occurrence of the species on the coast seems to be associated to river mouths. Bertram (1963) suggested that the high water turbidity in the coast probably provided poor feeding grounds to the species. Most records in the country are from rivers, as Courantyne, Nanniekrek, Canje, Berbice, Demerara, Abary, Wuini, Barima, Sebai, Kiatuna, Essequibo, Arapiako, Pomeroon and Akawini. At that time, the species seemed to be more abundant in the Canje and Abary rivers, mainly in upper reaches, on the wet savannahs. In Abary river, manatees were caught by the Department of Drainage and Irrigation and released in channels to eat aquatic plants, as a weed control agent.

This practice started in Guyana after Allsop (1960) observed that the ornamental pools of the Botanic Gardens, in Georgetown, where manatees were kept since the 1880s, were always clear of aquatic plants. In 1916, manatees were used to maintain sugar-estate canals clear of aquatic plants and in 1952 four manatees were introduced in a canal of a water reservoir of the Georgetown Water and Sewage Works with the same purpose and were replaced by others through the years (National Science Research Council 1974). The same procedure was done in other canals from 1959 to 1962 by Drainage and Irrigation Department, when 80 manatees were used to control aquatic vegetation. In 1974, a Workshop to discuss "An International Centre for manatee research" was held in Georgetown. However, the meeting report discusses mainly the importance of using manatees to control "aquatic weeds" in the country, and recommends the species exploration for this purpose (National Science Research Council 1974). In 1991, Haigh published a review on the use of manatees as a means of biological control of aquatic weeds in Guyana. In the article, the author also proposed guidelines on the use of species.

Fig. 11.2 An adult West Indian manatee kept in Georgetown Sewerage and Water Commisioners, in French Guiana, to control aquatic vegetation. This practice started in the 1880s and hundreds of manatees were used in the country for this purpose (Photo: Nataly Castelblanco-Martínez)

There is no recent effort to identify the species distribution in Guyana and almost every report is related to the use of species to control aquatic vegetation. It seems the species is still used in the country to control the aquatic vegetation in water bodies. The Botanical Gardens still maintains manatees in the waterways, and although some articles (eg, Converse et al. 1994) indicated that the population that inhabits the pools is wild, historical information indicates that those animals were introduced in those facilities. Castelblanco-Martínez (2015) reported that in Georgetown at least 30 manatees are inadequate kept in artificial ponds in the Botanical Gardens, the Guyana National Park; and in the Georgetown Sewerage and Water Commissioners, a water treatment plant (Fig. 11.2).

11.2.4 Suriname

There is little information on West Indian manatee in Suriname. Most of published data indicate that the species occurs mainly in rivers and river mouths. Murie (1870) reported the capture of a young male in Maroni (or Marowijne) river, in the border between Suriname and French Guiana. The animal was transported to England, but died from exposure to cold weather. Husson (1978) published a review on the species records in the country indicating the presence of *T. manatus* in Saramacca river. In 1974, Dekker described his trips to Suriname in 1966 and 1971 to capture manatees for a zoo in Amsterdam. According to the author, at that time manatees could be found west of Paramaribo and the Suriname river, in upperwaters of the Commewijne river, and affluents of Suriname river, and in the Nanni swamp, next to the Guyana border. In 1971, he captured 11 manatees at the mouth of large rivers, as Nanni. An interesting fact is that he described that some animals were imprisoned in the drainage canals of plantations, indicating the probable use of the species to control aquatic vegetation. Also in 1978, Duplaix and Reichart compiled information on the species in Suriname in a US Fish and Wildlife Service research report.

A few manatee records in Suriname are reported in articles, book chapters or project reports dedicated to other species, or to a specific area. Mohadin (1993) reported the occurrence of manatees in the Cusewijne river, in the Boven-Cusewijne (or Bove-Coesewijne) Nature Reserve. Duplaix et al. (2001) indicated that T. manatus was hunted to extinction in the area, but still occurred in Nanni Creek and Nanni lake, in the northwestern portion of the country.

In 2012, Pool indicated that in Suriname, four major rivers flow to Atlantic Ocean: Marowijne, Suriname, Coppename and Corantijn, and the estuaries formed by these rivers with contributing rivers are the prime habitat of manatees in the country. As in Guyana, the species was not observed in coastal waters of Suriname. The author indicated the lack of appropriate habitat or insufficient research in the area.

11.2.5 French Guiana

According to Bertram and Bertram (1962, 1963), there was little suitable habitat for manatees farther eastern from Suriname, where "the coastal plain disappears and the mountains come down to the sea". After this, information on manatee in French Guiana came only from Whitehead (1977), who indicated that the species was rarely found to the south of Cayenne. Best and Teixeira (1982) interviewed a manatee hunter in Amapá, a Brazilian state at the border with French Guiana, and reported the presence of the species in rivers Approuague, Mahury, Laughan and Ouanary, as well in some of the smaller rivers in eastern portion of the country.

Only in the twenty-first century, robust studies on manatees started in the country. De Thoisy et al. (2003) conducted interviews and boat surveys in 12 cities and villages along the coast and large river mouths, including fishermen on the Brazilian side of the Oiapoque river, in Amapá state. The results indicated that the species is still present all along French Guiana coast and in most estuaries.

In larger rivers, as Oiapoque, Approuague and Maroni, manatee sightings were reported up to 80 km upriver. Also, the presence of the species was recorded in the rocky coastal areas and around an offshore island 5 km from the continent. The Coswine, a large estuarine mangrove area in the northwestern portion of the country (border with Suriname) was identified as an important habitat for manatees, with the greatest number of sightings. Spiegelberger and Ganslosser (2012) studied the habitat and feeding banks of the species in this estuary, indicating that the area is able to support a large manatee population, with sufficient food supply and free from human disturbance at that time (data collected in 2001).

In 2012, Castelblanco-Martínez visited five regions in French Guiana to recognize the ecological scenario where manatees inhabit: Coswine, Iracoubo, Kourou, Cayenne (Cayenne river and beaches near the city) and Kaw (including part of Oiapoque river). Through shore and boat surveys and surveys using side-scan sonar and hydrophone, the author recorded the presence of the species in Kaw river, Les Vagues beach (near Cayenne city), Counami and Iracoubo region. Interviews with fishermen revealed the presence of manatees in Kourou and Coswine.

11.2.6 Brazil

An expressive amount of information on manatees was published in Brazil in the last decade. Data from significant effort through interviews, strandings and sightings (fixed-based platform, boat and aerial surveys) were recorded in theses, articles, books and conference abstracts. The most recent review on the species biology and conservation in the country became available in a book published by the NGO Aquasis (2016).

There is a significant quantity of historical information on the presence of *T. manatus* in Brazil, since the Pedro Alvares Cabral expedition, in 1500. Most records are from the states of Bahia and Espírito Santo (~19°S), apparently the southernmost limit of the species distribution at that time. According to Whitehead (1977), human predation and possibly habitat destruction exterminated the southerly populations of manatee in Brazil. Domning (1981) reported the species distribution in the northern coast, indicating its occurrence in Amapá and Maranhão coasts, but its extermination from Pará coast and Marajó Island. The author also indicated the Mearim river, in Maranhão, as an important habitat to *T. manatus*, where a large population could be found. In the 1980s, the Brazilian Government coordinated an effort to collect data on manatee distribution in the country through interviews. After this first effort, Albuquerque and Marcovaldi (1982) indicated the former species distribution from Amapá to Espírito Santo. However, it was extirpated from Espírito Santo, Bahia and Sergipe states. Lima et al. (1992) and Luna (2001) complemented the interview efforts and reported a fragmented distribution of the West Indian manatee in Brazil.

Alves et al. (2013) conducted one aerial survey in northeastern Brazilian coast, from Alagoas to Piaui states, covering 4,026 km, recording 36 sightings of manatee (41 individuals). The species was not observed in the Potiguar basin, at the northwestern coast of Rio Grande do Norte state (RN), where the authors described extremely harsh conditions with strong coastal currents associated to intensive exploration and coastal degradation due to oil industry and salt and shrimp farming. However, the presence of the species in this area was recorded through interviews (Choi 2011) and aerial surveys (Petrobras 2014), and this absence reported by Alves et al. (2013) can be related to other factors, as poor visibility associated with low manatee density. In the same aerial survey, the authors reported the absence of manatees in most of Pernambuco coast, and the west coast of Ceará, where Lima (1997) indicated that the species was not present. In Ceará west coast, we believe that the absence of the species is related to the lack of favorable ecological characteristics, which makes it the major discontinuity area in the manatee distribution in Brazil. Lima (op. cit) also indicated that the species was not present in the southern portion of Alagoas state. In 1994, Brazilian Government started a translocation program in Alagoas and Paraíba states, to release rehabilitated calves stranded in the country's coast (Lima 2008; Normande et al. 2015; Normande et al. 2016) to re-colonize areas of historic distribution.

In the states of Ceará and Rio Grande do Norte, interviews (Alves 2007; Aquasis 2006; Choi 2011), aerial (Alves et al. 2013; Aquasis 2006; Costa 2006; Petrobras 2014), boat (Aquasis 2006; Costa 2006) and land-based surveys (Alves 2003, 2007; Aquasis 2006; Moretz-Sohn 2013) and beach monitoring for stranding records (Meirelles 2008; Silva 2010) were performed to identify species distribution and critical areas for conservation. Aerial (24 h effort) and boat surveys (157 h of effort in Ceará coast; Aquasis 2006; Costa 2006), from Fortim to Icapuí, in the eastern portion of the coast, indicated the presence of the species only in Icapuí municipality (Fig. 11.3), mainly in Retiro Grande, Picos, Quitérias and Cajuais Bank. Aerial surveys reported by Petrobras (2014) between Aquiraz, in Ceará, and Touros, in Rio Grande do Norte, in an area of 6,000 km^2, indicated the presence of the species from

Fig. 11.3 An adult manatee sighted during a boat survey performed by Aquasis in Peroba beach, Icapuí municipality, in Ceará, northeastern Brazil. In open coastal areas, manatees lift the entire head out of the water to breathe, avoiding the entrance of water into the nostrils (Photo: Juaci Oliveira/Aquasis' Collection)

Aquiraz to Touros, with a concentration of manatees in the coast of Icapuí municipality, and an absence of records between São Miguel de Touros and Porto do Mangue (RN) and west to Aquiraz.

Choi (2011) through interviews (n = 678) in 77 localities between Beberibe, in east Ceará coast, and Touros, in northwestern Rio Grande do Norte state, indicated the absence of manatees west to Aracati (Beberibe and Fortim) and in Guamaré. The author also reported a high occurrence index in Icapuí and Aracati municipalities. Also, according to interviews, manatees occur only in coastal areas, but used to be observed in the six estuaries of the studied region. In these regions, manatees drink freshwater from the underground aquifers which emerge in the bottom of the sea (Aquasis 2006; Meirelles et al. 2016a).

Interviews were also performed in the border of Ceará and Piauí states, where the species inhabits the estuary of Timonha and Ubatuba rivers and adjacent coastal areas (Aquasis 2008; Choi et al. 2009), a huge estuary with a low human impact (Fig. 11.4). Choi-Lima et al. (2014) also tested the side-scan sonar in this estuary, indicating the viability of the use of this tool to detect manatee in the area (Fig. 11.5). The species is also present in the estuary of Canárias Island, in the border of Piauí and Maranhão states. In Maranhão, Alvite (2008) monitored the presence of manatees in Gato Island, in Tubarão bay, and in Guarapiranga beach, in São José de Ribamar, through fixed-platforms.

Fig. 11.4 Two manatees sighted in the estuary formed by Timonha and Ubatuba rivers, in the border of Ceará and Piauí states, northeastern Brazil. Compared to other estuaries in the region, this one is relatively well preserved. In this area, manatees feed on seagrass, saltmarsh, algae and mangrove
Note that in calm waters, animals slightly bring the snout out of the water to breathe (Photo: Aquasis' Collection)

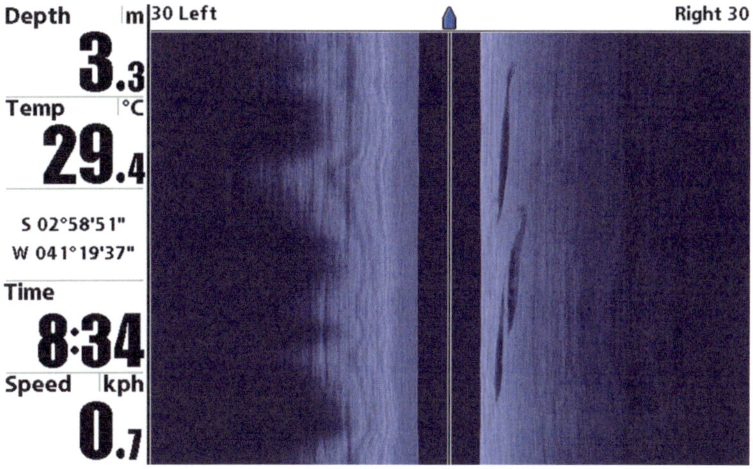

Fig. 11.5 Full-screen capture from Side-scan-sonar Humminbird® 998 cx in Timonha and Ubatuba rivers estuary, during boat surveys performed by Aquasis in 2012 and 2013. Note on the *right side* the presence of three manatees in the bottom of the river

Manatees are also present in the Mamanguape river estuary, in Paraíba state, where the species is abundant (Lima 1997; Silva et al. 1992). Recently, the side-scan sonar was also tested in this estuary as a tool to detect manatees (FMA 2015).

In the northern region, in the state of Amapá, in the border with French Guiana, West Indian manatee is present in the coast and in the estuary. Barbosa (2013) reported the species in coastal protected areas as Estação Ecológica Maracá-Jipioca, in Parque Nacional do Cabo Orange and Reserva Biológica Parazinho. Interviews indicated the species is more commonly observed in Bailique archipelago and Amapá city. Most animals were reported in estuarine systems, followed by river and coastal areas.

The species also occurs in the Oiapoque river where, according to De Thoisy et al. (2003), it can be observed 80 km upriver. Manatees are also present in the river mouth.

In the state of Pará, the species inhabits areas around islands and indentations. There are records in Algodoal and Marajó Islands (Emin-Lima et al. 2010; Sousa et al. 2013), in Salvaterra and Soure municipalities.

11.3 Spatial Ecology

The initiative by few research groups effectively conducting research in South America has produced important biological information on manatees, however, data on movements are scarce. Tagging manatees for telemetry monitoring allows to estimate home range, document movements and seasonal migrations, identify patterns of habitat selection and use, and understand population dynamics aspects, all crucial information to support conservation actions and protected area design. In most cases, logistics, financial and personnel resources impose limitations to the use of sophisticated methodologies, but this has been changing over the last decade, both by the introduction of new technologies and collaborative research initiatives.

The West Indian manatee occurs along almost 15,000 km of coastal area of six countries in South America (Brazil, French Guiana, Guyana, Suriname, Venezuela and Colombia), where no more than 2,000 individuals are believed to exist (Self-Sullivan and Mignucci-Giannoni 2012; Alves et al. 2016). So far, only Colombia and Brazil have implemented captures and/or releases of wild-ranging or captive-raised West Indian manatees, to use telemetry monitoring.

The first release of a captive-raised West Indian manatee in Brazil took place in 1994, and by August 2011 30 animals had been released along the coasts of the Brazilian northeastern states (Normande et al. 2015). Prior to release animals are adapted with a belt around the caudal peduncle, where one or more transmitter types are attached: a VHF-only, a platform-type (PTT) that uses an ARGOS link, and a global positioning system (GPS) that also has an ARGOS link (Normande et al. 2015). All satellite transmitters have built-in VHF transmitters, making it possible to track the target manatee in the field. Satellite data are obtained through the ARGOS service. Twenty-one (eight females and 13 males) were telemetry-monitored between November 2008 and June 2013, with a mean tracking period per animal of 107.48 +− SD 73 days (Normande et al. 2016).

Tracking data of released and reintroduced animals documented a wide manatee distribution in the region, from Rio Grande do Norte to Sergipe states and use coastal corridors for exploratory movements (Lima et al. 2012). Manatees may engage in long movements up to hundreds of kilometers for feeding and reproduction (Lima 2008; Lima et al. 2012; Normande et al. 2016).

Movement pattern was described as dichotomous. Some individuals present sedentary habits, with low rates of daily travel, strong site fidelity and reduced home range size. Wide-ranging manatees have higher daily travel rates and larger home ranges. Home range size (kernel 95%) of 21 (13 males and eight females) manatees released in Alagoas or Paraíba after rehabilitation varied from 4.24 to 30.96 km^2, lower than home ranges reported for other areas (Normande et al. 2016). Home-range size was not significantly influenced by gender, age at release or release site. Centers of activity were strongly associated with freshwater and feeding sites and reefs and estuaries as shelters (Normande et al. 2016).

In Colombia, manatees maintained in semi-captivity in the Sinú river were recaptured and had belt-mounted VHF and floating satellite transmitters adapted to their caudal peduncle. Researchers documented rapid movements towards and between river mouths during the dry season, in search of plants, a behavior similar to wild manatees in the area (Góngora-Correa 2009). In 2009, a female manatee was rescued in the Colombian department of Córdoba, and after a successful rehabilitation process, was released in Ciénaga Grande de Lorica. As of 2015 more than three years passed since her release and she is monitored with VHF technology with the help of the community (Caicedo-Herrera et al. 2015).

Free-ranging manatees were captured for the first time in Brazil, in the state of Ceará, using a small mesh, 200-m long, 7-m-deep seine net launched from a specially designed 10-m long vessel. Six manatees were captured and monitored by satellite telemetry between May 2012 and April 2013 (Petrobras 2014). Home range of four animals were estimated and varied from 41.33 km^2 (female with calf) to 155.56 km^2 (adult male). The studied animals did not use the estuaries of the area. Areas up to 5 m deep were used by the majority of the monitored individuals. Only the adult male used regions with up to 10 m deep. At the time of this writing, four manatees had been captured (two previously captive females with their native calves) in the coast of Alagoas state.

In order to monitor manatees at lower costs, a group of researchers recently developed satellite transmitters with GPS locator in Brazil, that use the Globalstar system and a SPOT Messenger module; a VHF transmitter is also part of the hardware. Tests conducted with captive manatees demonstrated good buoyancy of the housing, transmission mechanism and GPS location, with excellent precision of received coordinates. The use of lower cost materials available in the national market should turn the technology more accessible to users in South America (Borges et al. 2016).

11.4 Abundance

To estimate *T. manatus* abundance in South America is not an easy task. The species occurs mainly in riverine dark waters that, associated with their evasive behavior, makes their detection a challenge.

In Colombia, Arévalo-González et al. (2014) estimated manatee abundance in Ciénagas de Paredes through boat surveys, side-scan sonar surveys and interviews in 2002 and 2010. By boat survey, the maximum number of simultaneous sightings was 17. Using SSS and Distance sampling, a population of 12 manatees was estimated to occur in Los Pozos.

In Suriname, Pool (2012) indicated two manatee abundance estimates for the country: 500–600 (Duplaix and Reichart 1978) and 200–300 with a maximum of 500–600 at best (Wassink 1997). However, no details are presented on these studies' methods.

In Brazil, Lima (1997) and Luna (2001) proposed manatee abundance in northeastern and northern region, respectively, based on interviews. Despite the unusual method, without accuracy, these estimates were the only available in the country for years.

More recently, Alves et al. (2016) conducted aerial surveys from Canárias Island, on the mouth of Parnaíba river (Piauí) to São Francisco river mouth, covering 2,590 km^2, using strip transect method. Through a Bayesian approach, the authors estimated manatee abundance as 1,104 animals (CI 95%: 485–2,221), an inaccurate result that could be improved with additional efforts in a smaller spatial scale.

In the states of Ceará and Rio Grande do Norte, a monitoring project related to federal environmental licensing process conducted by IBAMA estimated manatee abundance through linear transects in aerial surveys from Aquiraz (CE) to Touros (RN) (387 km) (Petrobras 2014). Distance sampling estimated 193 manatees in the area (CI 95%: 93–378).

11.5 Feeding

The manatee is an exclusively herbivorous animal. As the species may occur in a wide variety of habitats, such as marine, freshwater and estuarine environments, the animals feed on a variety of plants, which may be submerged, floating or emergent, and the choice between these types will depend on the availability and nutritional value of each (Marsh et al. 2011).

Trichechus manatus feeding preference is related to species availability, nutritional value, and palatability (Hartman 1979; Best 1981). In South America, the species inhabits a variety of habitats, as rivers, swamps, estuaries and coastal areas. Thus, in each of these habitats animals' diet varies. In Table 11.1, a review on manatee diet in the continent is presented.

An interesting fact when observing the table is the restriction in terms of items quantity in some areas of Brazil, as the states of Ceará and Alagoas, where manatees eat basically *Halodule wrightii* as it has a higher nutritional value than algae (Meirelles et al. 2016a). This fact is alarming, since impact on seagrass meadows can severely affect manatees inhabiting these areas.

11.6 Genetics

Garcia-Rodriguez et al. (1998) compared the mitochondrial DNA control region among eight locations across the western Atlantic Ocean, from Florida to Brazil. Low levels of genetic diversity were observed in Florida and Brazil (at the limits of species' range). Also, the study indicated three mtDNA lineages: Florida and West Indies; from the Gulf of Mexico to the Caribbean rivers of South America; and northeastern Atlantic coast of South America. With a larger sample, Vianna et al. (2006) corroborated these findings. According to Santos et al. (2016), through mtDNA Bayesian phylogeny and chronological calibration with sirenian fossils, separation time between manatee populations from the northern Atlantic region and for those of Caribeean and Gulf of Mexico was estimated in 590,000 years.

Recently, a detailed karyotype analysis revealed differences between chromosomal pair 4 and 10 from *T. manatus* from Brazil and those from Puerto Rico and Caribbean (Barros 2013; Barros et al. 2017). The same authors studied 3D geometric morphometrics of manatee skulls and identified important differences between

Table 11.1 Food items described for West Indian manatee in South America

Country	Region	Food itens	Method	Source
Colombia	Puerto Carreño, Vichada	*Paspalum* sp., *Paspalum fasciculatum*, *Paspalum repens*, *Alchornea castanaeifolia*, *Eichhornia crassipes*, *Ceratopteris thalictroides*, *Mikonia* sp.	Indirect evidence and interviews	Gómez-Camelo et al. (2004)
	Puerto Carreño, Vichada	*Paspalum repens* and *P. fasciculatum*	Interviews, indirect evidence	Castelblanco-Martínez (2004)
	Brazo de Loba and Panceguita, Magdalena ríver	*Eichhornia* sp. and *Paspalum* spp.	Interviews, indirect evidence	Aguilar-Rodríguez et al. (2004)
	Ciénaga de Paredes, caño Peruétano	*Paspalum fasciculatum*, *P. repens* and *Polygonum ferrugimeum*	Interviews, indirect evidence	Castelblanco-Martínez (2004)
Venezuela	Llano Venezoelano	*Eichhornia* sp.		Linares (1998)
	Lower Orinoco	*Paspalum repens* and *Eichhornia crassipes*	Indirect evidence	Rívas-Rodríguez et al. (2012)
	Centroocidental and southwestern Maracaibo lake	*Eichhornia* sp. and *Pistia stratioides*	Interviews	Montiel-Villalobos and Barrios-Garrido (2005)
	Northwestern Maracaibo Lake	*Rhizophora mangle*, *Avicennia germinans*, *Balanus balanoides*	Interviews	Montiel-Villalobos and Barrios-Garrido (2005)
	Orinoco Delta	*Paspalum repens*, *Eichhornia crassipes*, *Eichhornia azurea*, *Pistia stratiotes*, *Hydrocotile umbellata*, *Ludwigia helminthorrhiza*, *Limnobium laevigatum*	Interviews	Rodulfo (2012)

(continued)

Table 11.1 (continued)

Country	Region	Food itens	Method	Source
Guyana	Irrigation canals	*Cabomba aquatica, Nelumbo speciosa, Paspalum repens, Utricularia foliosa, Luziola spruceana, Paspalum vergatum and Hymenachne amplexicaule.*		Haigh (1991)
Suriname		*Montrichardia arborescens, Nymphaea odorata*	Indirect evidence	Lefebvre et al. (2001); Dekker (1974)
French Guiana	Coswine swamps	*Rhizophora* sp., *Rhizophora racemosa, Montrichardia arborescens* and *Scleria pterota*	Indirect evidence	Spiegelberger and Ganslosser (2012)
	Kaw river	*Crinum erubescens, Echinochloa polystachya*	Indirect evidence	Castelblanco-Martínez and De Thoisy (2012)
Brazil	Amapá	*Montrichardia* sp., *Hymenachne* sp., *Avicennia* sp., *Eichhornia crassipes*	Interviews	Barbosa (2013)
	Northern Brazil	*Avicennia nitida, Rhizophora mangle, Laguncularia racemosa, Montrichardia arborescens, Spartina brasiliensis, Eichhornia crassipes, Eleocharis interstincta*	Interviews	Best and Teixeira (1982)
	Marajó Island, Pará	Seagrass, *Spartina brasiliensis*	Interviews	Sousa et al. (2013)
	Gato Island, Maranhão	*Spartina alterniflora*	Observation	Alvite (2008)
	Maranhão	*Spartina* sp., magrove	Stable isotopes (carbon and nitrogen)	Ciotti (2012)
	Timonha and Ubatuba rivers estuary, Piauí	*Spartina* sp., mangrove, seagrass, algae and fish	Interviews	Aquasis (2008)
	Timonha and Ubatuba rivers estuary, Piauí	*Halodule wrightii, S. alterniflora*, mangrove, algae	Stable isotopes (carbon and nitrogen)	Ciotti (2012)

(continued)

Table 11.1 (continued)

Country	Region	Food itens	Method	Source
	East coast of Ceará	Mainly *H. wrightii*	Interviews, direct observation, stomach contents	Aquasis (2006); Vasconcelos (2013)
	East coast of Ceará	*H. wrightii*	Stable isotopes (carbon and nitrogen)	Ciotti (2012)
	Paraíba	Red algae	Stomach contents and feces samples	Borges et al. (2008)
	Paraíba	Mainly algae, but also *H. wrightii*	Stable isotopes (carbon and nitrogen)	Ciotti (2012)
	Alagoas	Mainly *H. wrightii* but also algae	Stable isotopes (carbon and nitrogen)	Ciotti (2012)

skull shape from animals from Brazil and those from Caribbean and Florida, indicating an urgent reassessment of species taxonomy.

In Colombia, Satizábal et al. (2012) tested the Male-biased dispersal (MBD) hypothesis in the country's rivers (Sinú, northern Magdalena, San Jorge, Ciénaga de Paredes, Meta and Orinoco). D-loop analyses indicated high levels of structuring between Sinú, San Jorge and Magdalena rivers, as also between Northern Magdalena and Ciénaga de Paredes. Some evidence of structuring between the Orinoco and Meta rivers was also observed, despite the small number of samples. Microsatellite data suggested no structure. These results indicate male-biased gene flow dispersal in manatees that inhabit Colombia rivers.

In Brazil, Luna et al. (2012) reported three haplotypes (M01, M03 and M04) for the species, with only one (M01) recorded in animals from Ceará to Alagoas. The authors indicated that rehabilitated calves from the states north of Ceará should not be released in current release areas (Paraíba and Alagoas) due to risk of genetic swamping. However, according to Santos et al. (2016), this is not a plausible explanation to restrict release sites in Brazil, since M04 differs from M01 by just one mutation that can happen spontaneously in one generation.

Also in Brazil, in the east coast of Ceará and northwestern Rio Grande do Norte, where a high frequency of newborn live calves strand in the country (Meirelles 2008; Parente et al. 2004; Silva 2010), Silva (2015) studied genetic structuring and kinship between stranded manatees. There was no population structuring between CE and RN. The majority of animals in which it was possible to define kinship had some level of genetic kinship (34.4% of total). This is an indication of a small manatee population in the studied region.

11.7 Health Status

Knowledge on the health of populations of *T. manatus* in South America comes mainly from animals rescued after stranding and/or bycatch, maintained in captivity for rehabilitation and release purposes or for display in zoos and aquariums (Fig. 11.6). Although there are records of maintenance of the species in captivity since the late nineteenth century in Brazil and Guyana, rescue activities were established only in the 1980s in most South American countries (Adimey et al. 2012; Lima et al. 2001). Currently, there are about 11 institutions holding captive manatees in the continent, in Brazil, Guyana, Venezuela and Colombia, serving different purposes (Adimey et al. 2012; Carlos Javier Silva, personal communication; Carvalho and Borges 2016).

In Guyana and Suriname, there are current and historical records of manatees maintained in captivity for the control of aquatic vegetation from irrigation canals, without assessing the risk of this activity for the health of the animals. Dozens of manatees were captured in these countries and sent to aquariums and zoos in Europe and Asia, where they participated in breeding programs (Adimey et al. 2012; Dekker 1974; Haigh 1991). We emphasize the serious concern about these crossbreedings between animals from different populations, even under the control of studbooks, since recent studies indicate that the population from Guyana to Brazil represents a distinct significant evolutionary unit from the Caribbean population and the region of origin of these animals, west of the mouth of the Amazon River, is a hybridization zone (Santos et al. 2016).

Fig. 11.6 A 2-year old manatee calf (named “Tico”) in the management area of a rehabilitation tank of Marine Mammal Rehabilitation Centre/Aquasis, in Ceará, northeastern Brazil. The stranding of newborn calves is a serious problem in the country. Natural and human-related causes are involved in the separation of females and their offspring (Photo: Aquasis’ Collection)

Little information has been published on non-infectious diseases, infectious and parasitic diseases, serological surveys and contaminants that affect the health of native West Indian manatees in South America. In general, publications on *T. manatus* mortality in the continent categorize the origin of these events as anthropogenic (bycatch, hunting, collision with vessels, injury by projectile, solid waste ingestion, etc.), natural (disease, senility, etc.), perinatal and undetermined events, but final pathological diagnoses are unsufficiently detailed. In the Orinoco River, Colombia, Castelblanco et al. (2009) reported that more than 85% of manatee deaths were related to human activities, especially calves and juveniles. In Northeastern Brazil, the perinatal causes are predominant (75–80%), followed by those of anthropogenic origin (Fig. 11.7) (Meirelles 2008; Parente et al. 2004; Silva et al. 2016). Hunting is still a threat on the Northern coast, representing more than 94% mortality (Luna et al. 2008). In Guyana and Suriname, mortality events are poorly documented, but are mainly related to bycatch and hunting, while in Venezuela there are still sporadic records of deaths from boat collisions (Quintana-Rizzo and Reynolds 2010). In French Guyana, at least 10 deaths associated with bycatch and hunting were reported between 2000 and 2002 (De Thoisy et al. 2003).

Medical exams in wild-caught animals are important to determine the influence of environmental change on the health of individuals and populations, hence, they can be used as a tool for conservation (Stamper and Bonde 2012). This activity is still in its infancy in South America, since there are data only on six individuals captured on the coast of Ceará, Northeastern Brazil (Petrobras 2014). The construction of a long-term database is essential for an accurate assessment of the health of populations of manatees in the continent.

Fig. 11.7 Aquasis' rescue team carrying out biometric measures of an adult male manatee accidentally caught in fishing nets in Ponta Grossa beach, Icapuí, northeastern Brazil (Photo: Aquasis' Collection)

Rehabilitation programs are well-established as a strategy for *T. manatus* conservation in Brazil, where it is estimated that more than 100 newborns have been rescued and sent to specialized centers, since the mid-1980s (Carvalho and Borges 2016). Of this total, more than 40 animals were released into the wild, with a success rate exceeding 75% and at least six records of births (Normande et al. 2015, 2016.). In Venezuela, manatee rescues are casual, with most of the animals intended for zoos, with only one record of release. By 2014, the country had seven specimens in captivity, of which three were born in the captive environment (Boede and Mujica-Jorquera 2016). In Colombia, although there is a considerable number of captive animals in artificial lakes, which are kept under inadequate clinical and nutritional support, rescue and rehabilitation programs are not intended for the release of the animals (Caicedo-Herrera et al. 2004).

Several studies have been conducted to determine hematological reference parameters for *T. manatus*, both in captive and free ranging animals. This information is essential for the diagnosis of diseases and to guide the appropriate treatments (Bossart et al. 2001). However, most of these studies were performed on Florida animals, except for the studies of Converse et al. (1994), Silva et al. (2007, 2009) and Boede and Mujica-Jorquera (2016), which showed blood parameters of captive manatees in Guyana, Brazil and Venezuela, respectively. Few native specimens were evaluated in South America, with no further parameters available for these populations. The levels of thyroid hormones were evaluated for captive manatees from Brazil and native animals from Colombia, showing that they are influenced by diet and that despite the low metabolism of the species, the levels found were similar to other marine and terrestrial mammals (Ortiz et al. 2000).

Information on microorganisms that constitute the normal microbiota or act as agents of infectious diseases of manatees has been collected from native and captive animals from Brazil. About 27 species and 15 genera of bacteria and 16 species of yeasts were isolated from the microbiota of natural cavities of these animals, with the detection of strains resistant to multiple antibiotics (beta-lactam, macrolide and aminoglycoside) and antifungals (azole derivatives), respectively (Attademo 2014; Petrobras 2014; Sidrim et al. 2015a; Sidrim et al. 2016; Vergara-Parente et al. 2003a). There were also differences in the composition of colonizing bacteria between captive and native manatees, demonstrating the influence of environmental factors on microbial communities (Petrobras 2014; Vergara-Parente et al. 2003a).

Diagnoses of infectious diseases are uncommon in manatees. Among bacterial diseases, abscesses caused by *Staphylococcus* sp., *Proteus* sp. and *Escherichia coli*; ear infections by *Pseudomonas* sp., *Proteus mirabilis* and *Staphylococcus* sp.; colitis by *Klebsiella pneumoniae* and *Morganella morganii* and systemic infections by *Salmonella panama*, *Pseudomonas* sp. and *Klebsiella* sp. stand out (Attademo 2014; Carvalho and Silva 2016; Lazzarini et al. 2014; Montoya-Ospina et al. 2001; Vergara-Parente et al. 2001, 2003b). Infections by *Flavobacterium meningosepticum* and *Xanthomonas maltophilia* were reported as cause of death of three neonates in Colombia (Montoya-Ospina et al. 2001). Fungal diseases are uncommon, with phaeohyphomycosis reports caused by *Bipolaris hawaiiensis* and outbreak of superficial skin fusariosis in calves kept in captivity (Sidrim et al. 2015b; Carvalho et al. 2015). Regarding viral diseases, Attademo et al. (2013) detected the presence of rotavirus in captive and semicaptive manatee feces without associated clinical signs.

As to parasitic diseases, nasal trematode *Pulmonicola cochleotrema* was the only helminth diagnosed in stranded animals and specimens reintroduced in Northeastern Brazil (Borges et al. 2016; Carvalho et al. 2009). Among the protozoa, cryptosporidiosis is prevalent in both captive and pre-release stage animals in Brazil, with associated intestinal clinical signs (Borges 2007, 2016). A case of myocarditis caused by *Toxoplasma gondii* was detected in a captive manatee that died in Guyana (Dubey et al. 2003). Giardia lamblia infection has been detected in a captive calf (Carvalho and Silva 2016). Garzón (1997) conducted parasitological and microbiological evaluation of *T. manatus* in Colombia.

Non-infectious diseases are poorly described. In the last decade we have seen a higher prevalence of malformations in newborns stranded in Northeastern Brazil, including fatal cases of hydrocephalus, atresia of the right ventricle, and other less severe locomotor disorders, such as spinal fusion in the lumbar region, permanent folding of the caudal fin and scapular-humeral arthropathy. Neoplasias are rare, with a diagnosis of uterine leiomyoma in adult female stranded in Ceará, Brazil (Carvalho and Silva 2016). The diagnosis of disseminated malignant lymphoma was conducted in a specimen exported from Guyana to an aquarium in Denmark, which still had recurrent cutaneous papillomatosis due to immunosuppression (Hammer et al. 2005). Brito (2015) conducted histopathological evaluation of 38 dead manatees in the Northeast coast of Brazil, and diagnosed disorders such as stomatitis, cardiac fibrosis, myocarditis, liver fibrosis, pancreatic atrophy and metritis. Two specimens rescued in the 1990s in Colombia had suspected dwarfism due to a slow development, during more than four years in semicaptivity (Millian-Sánchez et al. 1995).

Serological surveys for the detection of antibodies against *Toxoplasma gondii*, *Leptospira interrogans*, *Brucella abortus* and Morbillivirus have been conducted in captive South American specimens of *T. manatus*. Silva et al. (2001) detected a positive animal for *T. gondii* in Brazil. Later, Attademo (2014) observed prevalence of 10.9% of seropositivity against the same parasite and 9.2% of positivity for anti-*Leptospira* antibodies. Native individuals captured in the country were seronegative for the surveyed agents (Petrobras 2014). Anti-Morbillivirus antibodies were not detected in captive manatees in Guyana (Duignan et al. 1995). Jiménez-Marrero et al. (1998) have measured levels of IgG from 20 semicaptive animals in Colombia, whose parameters can be used in immunological studies and for the diagnosis of immunosuppression.

Anzolin et al. (2012) performed the main toxicological study of manatees in Brazil and found high levels of the metals aluminum, lead, cadmium and tin in specimens maintained in captivity and semicaptivity in Paraíba state. Lailson-Brito et al. (2010) and Cavalheiro et al. (2016) found low concentrations of mercury and organochlorines (DDT, HCH, PCBs and Mirex) in tissues of manatees stranded dead in Ceará and Rio Grande do Norte.

References

Adimey N, Mignucci-Giannoni AA, Da Silva VMF, Auil-Gomez NE, Alvite CMC, Morales-Vela B, Lima RP, Rosas FCW (2012) Manatee rescue, rehabilitation, and release efforts as a tool for species conservation. In: Hines EM, Reynolds J III, Aragones LV, Mignucci-Giannoni AA, Marmontel M (eds) Sirenian conservation: issues and strategies in developing countries. University Press of Florida, Gainesville, pp 204–217

Aguilar-Rodríguez B, Castelblanco-Martinez DN, Trujillo F (2004) Factores de Riesgo y Estado de Conservación del Manatí Antillano Trichehus manatus manatus en el Área de Influencia de Magangue Bolívar (Colombia). In: Abstracts of 11va Reunión de Trabajo de especialistas en Mamíferos Acuáticos de América del Sur y 5to Congreso de la sociedad Latinoamericana de Especialistas en Mamíferos Acuáticos, Quito

Albuquerque C, Marcovaldi GM (1982) Ocorrência e distribuição do Peixe-boi-marinho no Litoral Brasileiro (Sirenia, Trichechidade, *Trichechus manatus*, Linnaeus 1758). In: Abstracts of the Simpósio Internacional sobre a Utilização de Ecossistemas Costeiros: Planejamento, Poluição e Produtividade, Rio Grande

Allsopp WHL (1960) The manatee: ecology and use for weed control. Nature 188(4752):762

Alves MDO (2003) Monitoramento de peixe-boi marinho, *Trichechus manatus manatusLinnaeus*, 1758, no município de Icapuí, litoral leste do Ceará. Undergratuation thesis. Universidade Federal do Ceará.

Alves MDO (2007) Peixe-Boi-Marinho, *Trichechus manatus manatus*: Ecologia e Conhecimento Tradicional no Ceará e Rio Grande do Norte, Brasil. Master's thesis, Universidade Federal de Pernambuco

Alves MDO, Schwamborn R, Borges JCG, Marmontel M, Costa AF, Schettini CAF, Araújo ME (2013) Aerial survey of manatees, dolphins and sea turtles off northeastern Brazil: correlations with coastal features and human activities. Biol Conserv 161:91–100

Alves MDO, Kinas PG, Marmontel M, Borges JCG, Costa AF, Schiel N, Araújo ME (2016) First abundance estimate of the Antillean manatee (*Trichechus manatus manatus*) in Brazil by aerial survey. J Mar Biol Assoc UK 96(04):955–966

Alvite CMC (2008) Indicadores populacionais e ecológicos de peixes-bois-marinhos (*Trichechus manatus manatus*) em duas áreas de manguezais e marismas no Maranhão. Master's thesis, Universidade Federal do Maranhão

Anzolin DG, Sarkis JES, Diaz E, Soares DG, Serrano IL, Borges JCG, Souto AS, Taniguchi S, Montone RC, Bainy ACD, Carvalho PSM (2012) Contaminant concentrations, biochemical and hematological biomarkers in blood of West Indian manatees *Trichechus manatus* from Brazil. Mar Pollut Bull 64:1402–1408

Aquasis (2006) Status de Conservação e Plano de Ação preliminar para o peixe-boi-marinho (*Trichechus manatus manatus*) no litoral leste do Ceará. Relatório Final do Projeto "Status do Peixe-Boi no Litoral Leste do Ceará"- Convênio MMA/FNMA n.059/02.2006, 174 p

Aquasis (2008) Refúgio de vida silvestre peixe-boi-marinho: estudos socioambientais complementares e consolidação da proposta de criação de Unidade de Conservação. Relatório técnico final do projeto "Refúgio de Vida Silvestre Peixe-boi-marinho" – SOS Mata Atlântica/ Fundação Avina/FMA

Aquasis (2016) Peixe-boi-marinho: biologia e conservação no Brasil, 1st edn. Bambu Editora e Artes Gráficas Ltda, São Paulo

Arévalo-González K, Castelblanco-Martínez DN (2015) Effect of the environmental type on manatee detection when using side-scan sonar. In: Abstracts of the 21nd biennial conference on the biology of marine mammals, San Francisco, EUA, 13–18 December 2015

Arévalo-González GK, Castelblanco-Martínez DN, Sanchez-Palomino P, López-Arévalo HF, Marmontel M (2014) Complementary methods to estimate population size of Antillean manatees (Sirenia:Trichechidae) at Ciénagas de Paredes, Santander, Colombia. J Threat Taxa 6(6):1–11

Attademo FLN (2014) Detecção da infecção de rotavírus, coronavírus, enterobactérias, *Leptospira spp.*, *Brucella abortus e Toxoplasma gondii* em peixe-boi marinho (*Trichechus manatus*) em cativeiro no Brasil. PhD thesis, Universidade Federal Rural de Pernambuco

Attademo FLN, Luna FO, Boaviagem AC, Gregori F, Bernardes NTCG, Da Silva JCR (2013) Detecção de rotavirus em peixes-boi marinho (*Trichechus manatus manatus*) mantidos em cativeiro no Brasil. In: Reunião da Sessão WDA América Latina, 1, São Paulo

Barbosa DA (2013) Conhecimento de moradores dos limites e entorno de sete unidades de conservação no estado do Amapá sobre a ocorrência e as possíveis ameaças ao peixe-boi (*Trichechus* sp.). Undergrad thesis, Universidade Federal do Amapá

Barros HMDR (2013) Variação geográfica de Trichechidae (Mammalia: Sirenia): análise morfométrica e citogenética. Ph.D. thesis, Universidade Federal de Pernambuco, 96 p

Barros HMDR, Meirelles ACO, Luna FO, Marmontel M, Cordeiro-Estrela P, Santos N, Astua D (2017) Cranial and chromosomal geographic variation in manatees (Mammalia: Sirenia: Trichechidae) with the description of the Antillean manatee karyotype in Brazil. J Zool Syst Evol Res 55(1):73–87

Bermudez-Romero AL, Castelblanco-Martínez DN, Trujillo-González F (2004) Patrones de presencia y uso del hábitat de *Trichechus manatus manatus* en el rio Orinoco dentro de la zona de influencia de Puerto Carreño, Vichada. In: Diazgranados MC, Trujillo-González F (eds) Estudios de fauna silvestre en ecosistemas acuáticos en la Orinoquia Colombiana. Instituto de Estudios Ambientales para el Desarrollo, Departamento de Ecología y Desarrollo, Pontificia Universidad Javeriana, Bogotá, Colombia, pp 133–158

Bertram C (1963) In search of mermaids: the manatess of Guiana. London, Peter Davies

Bertram GCL, Bertram CKR (1962) Manatees of Guiana. Nature 196(861):329–1331

Bertram GCL, Bertram CKR (1963) The status of manatee in the Guianas. Oryx 7(2–3):90–93

Bertram GCL, Bertram CKR (1973) The modern Sirenia: their distribution and status. Biol J Linn Soc 5:297–338

Best RC (1981) Foods and feeding habits of wild and captive Sirenia. Mammal Rev 11(1):3–29

Best RC, Teixeira DM (1982) Notas sobre a distribuição e "status" aparentes dos peixes-bois (Mammalia: Sirenia) nas costas amapenses brasileiras. Boletim da Fundação Brasileira para Conservação da Natureza 17:41–47

Boede EO, Mujica-Jorquera E (2016) Rescue and handling of Antillean manatees *Trichechus manatus manatus* in Venezuela 1992–2014. Int Zoo Yearbook 50:193–202

Boher S, Porras J (1991) Nuevos registros de manatíes, *Trichechus manatus*, en la costa del Mar Caribe venezolano. Acta Cient Venez 41(1):287

Borges JCG (2007) *Cryptosporidium* spp. (Tyzzer, 1907) em peixes-boi marinhos (*Trichechus manatus*) (Linnaeus, 1758) e peixes-boi amazônicos (*Trichechus inunguis*) (Natterer, 1883) no Brasil. Master's thesis, Universidade Federal Rural de Pernambuco

Borges JCG (2016) *Cryptosporidium* spp. (Tyzzer, 1907), *Giardia* sp. (Leeuwenhoek, 1681) e identificação de helmintofauna em mamíferos aquáticos no Brasil. 2016.105f. PhD thesis, Universidade Federal Rural de Pernambuco

Borges JCG, Araújo PG, Anzolin DG, Miranda GEC (2008) Identificação de ítens alimentares constituintes da dieta dos peixes-bois-marinhos (*Trichechus manatus*) na região nordeste do Brasil. Biotemas 21(2):77–81

Borges JCG, Jung LM, Carvalho VL, Moreira ALO, Attademo FLN, Ramos RAN, Alves LC (2016a) *Pulmonicola cochleotrema* (Digenea: Opisthotrematidae) in Antillean manatees (*Trichechus manatus manatus*) from the north-eastern region of Brazil. J Mar Biol Assoc UK 96:1–4

Borges JCG, Mantovani JE, Dantas RC, Santos S, Marmontel M, Vergara-Parente JE (2016b) Desenvolvimento de nova tecnologia aplicada ao onitoramento satelital de peixes-bois marinhos (*Trichechus manatus manatus*). In: XI Congreso de la Sociedad Latinoamericana de Especialistas en Mamíferos Acuáticos 2016, Valparaíso. Mamíferos aquáticos da América Latina, 2016

Bossart GD, Reidarson TH, Dierauf LA, Duffield DA (2001) Clinical Pathology. In: Dierauf LA, Gulland F (eds) Handbook of marine mammal medicine. CRC Press, Boca Raton, pp 383–436

Bowles AE, Jaffe JS, Yack T, Simonet F (2004) Design for a manatee finder: sonar techniques to prevent manatee-vessel collisions. Hubbs-Seaworld Research Institute for Florida Fish and Wildlife Conservation Commission, Florida Marine Research Institute, Manatee Avoidance Technology Program, p 33

Brito APD (2015) Avaliação histopatológica de mamíferos marinhos encalhados no litoral do nordeste brasileiro. Master's thesis, Universidade Federal Rural do Semi-Árido

Caicedo-Herrera D, Trujillo F, Rodríguez CL, Rivera MA (2004) Programa Nacional para la Conservación y Manejo de los Manatíes (*Trichechus* sp) en Colombia. Fundación Omacha-Ministerio de Ambiente, Vivienda y Desarrollo Territorial, Bogotá, 170 p

Caicedo Herrera D, Mona-Sanabria Y, Espinosa-Forero R (2015) Rescue, rehabilitation and release to the wild of manatees *Trichechus manatus* in the Colombian Caribbean: a conservation program. Sirenews 63:21

Carvalho VL, Borges JCG (2016) Reabilitação. In: Meirelles ACO, Carvalho VL (eds) Peixe-boi-marinho: Biologia e Conservação no Brasil. Bambu, São Paulo, pp 109–130

Carvalho VL, Silva FMO (2016) Sanidade. In: Meirelles ACO, Carvalho VL (eds) Peixe-boi-marinho: Biologia e Conservação no Brasil. Bambu, São Paulo, pp 77–92

Carvalho VL, Meirelles ACO, Motta MRA, Maia DCBSC, Campello MVM, Bevilaqua CML (2009) Occurrence of *Pulmonicola cochleotrema* (syn. Cochleotrema cochleotrema) (Digenea: Opisthotrematidae) in Antillean manatees (*Trichechus manatus manatus*) in Brazil. Lat Am J Aquat Mamm 7:47–52

Carvalho VL, Maia DCBSC, Brilhante RSN, Meirelles ACO, Guedes GMM, Rocha MFG (2015) Surto de fusariomicose em peixes-bois-marinhos (*Trichechus manatus manatus*) cativos. In: Abstract of the 51° Congresso da Sociedade Brasileira de Medicina Tropical, Fortaleza

Castelblanco-Martínez DN (2004) Peixe-boi *Trichechus manatus manatus* na Orinoquia Colombiana: Status de Conservação e uso de hábitat na época seca. Master's thesis, Universidade Federal do Amazonas

Castelblanco-Martínez DN (2012) Relative abundance and conservation status of manatees (*Trichechus manatus manatus*) in French Guyana – preliminary analysis of the study area. Report presented to Park National de La Guadeloupe, France

Castelblanco-Martínez DN (2015) Captive manatees of Guyana – action plan for captures and health assessment. Report presented to Park National de La Guadeloupe, France

Castelblanco-Martínez DN, De Thosy B (2012) Rapid evaluation of methods to monitoring manatees *Trichechus manatus manatus* in French Guiana. In: Abstract of 15ª Reunión de Trabajo de Expertos en Mamíferos Acuáticos de América del Sur (RT 15ª) y 9° Congreso de la Sociedad Latinoamericana de Especialistas en Mamíferos Acuáticos, Puerto Madryn, Argentina

Castelblanco-Martínez D, Kendal S, Fuentes L (2001) Uso de hábitat del manatí y observaciones de manatí (*Trichechus manatus*) en una zona del Orinoco medio, Colombia. In: Abstracts of Congreso Internacional del Manejo de Fauna Silvestre en Amazonía y Latinoamérica. Cartagena de Indias, Colombia, p 80

Castelblanco-Martínez D, Gómez I, Bermúdez A (2005) Ecología y conservación del manatí antillano *Trichechus manatus manatus* en la zona comprendida entre Puerto Carreño, Colombia y Puerto Ayacucho, Venezuela (2004–2005). Informe Técnico, Sinerian International INC, Frederichskburg, USA

Castelblanco-Martínez DN, Bermúdez-Romero AL, Gómez-Camelo IV, Rosas FCW, Trujillo F, Zerda-Ordonez E (2009) Seasonality of habitat use, mortality and reproduction of the vulnerable Antillean manatee *Trichechus manatus manatus* in the Orinoco River, Colombia: implications for conservation. Oryx 43(2):235–242

Cavalheiro KM, Meirelles ACO, Carvalho VL, Fragoso ABL, Silva FJL, Lailson-Brito J, Bisi TL (2016) Mercúrio total em músculo e fígado de peixe-boi-marinho *Trichechus manatus* (Linnaeus, 1758) do nordeste do Brasil. In: Abstract of the XIV Congresso Brasileiro de Ecotoxicologia – ECOTOX, Curitiba

Choi KF (2011) Áreas prioritárias para a conservação do peixe-boi-marinho *Trichechus manatus* no Ceará e no Rio Grande do Norte. 2011. 264 f. Master's thesis, Universidade Federal do Ceará

Choi KF, Campos TM, Meirelles ACO, Campos AA, Fernandes MB (2009) Desenho da área de um refúgio de vida silvestre para a conservação do peixe-boi-marinho. Natureza & Conservação 7(2):82–89

Choi-Lima KF, Umezaki J, Meirelles ACO, Silva CPN, Abessa DMS, Wedekin LL, Sousa-Lima R (2014) Active and passive acoustics for abundance estimation of Antillean manatees, *Trichechus manatus*. In: Abstract of the 16th Reunión de Trabajo de Especialistas en Mamíferos Acuáticos de América del Sur, Cartagena, 01–05

Ciotti LL (2012) Isótopos estáveis de carbono e nitrogênio aplicados ao estudo da ecologia trófica do peixe-boi- marinho (*Trichechus manatus*) no Brasil. Master's thesis, Universidade Federal do Rio Grande

Committee on Taxonomy (2016) List of marine mammal species and subspecies. Soc Mar Mammal. http:// www.marinemammalscience.org. Accessed 1 July 2016

Converse LJ, Fernandes BS, MacWilliams PS, Bossart GD (1994) Hematology, serum chemistry and morphometric reference values for Antillean manatees (*Trichechus manatus manatus*). J Zoo Wildl Med 25:423–431

Correa-Viana M, O'Shea T, Ludlow M, Robinson J (1990) Distribución y abundancia del manatí, *Trichechus manatus* en Venezuela. Biollania 7:101–123

Costa AF (2006) Distribuição espacial e status do peixe- boi-marinho, *Trichechus manatus manatus* (Sirenia: Trichechidae) no litoral leste do Estado do Ceará. 2006. 131f. Master's thesis, Universidade Federal do Ceará

De Thoisy B, Spiegelberger T, Rousseau S, Talvy G, Vogel I, Vié J (2003) Distribution, habitat, and conservation status of West Indian manatee *Trichechus manatus* in French Guiana. Oryx 37(4):431–436

Dekker D (1974) On the natural history of manatees (*Trichechus manatus*) from Suriname for the Amsterdam Zoo. Aquat Mamm 2:1–3

Delgado JR (1995) Estatus del manatí en Venezuela. In: Clemente (ed) Delfines y otros mamíferos acuáticos de Venezuela. Una política para su conservación, Caracas, Venezuela, pp 159–165

Deutsch CJ, Self-Sullivan C, Mignucci-Giannoni A (2008) *Trichechus manatus* The IUCN Red List of Threatened Species 2008: http://dx.doi.org/10.2305/IUCN.UK.2008.RLTS.T22103A9356917.en. Downloaded on 21 Sept 2016

Domning DP (1981) Distribution and status of the *Trichechus* spp. near the mouth of the Amazon River, Brazil. Biol Conserv 19:85–97

Dubey JP, Zarnke R, Thomas NJ, Wong SK, Van Bonn W, Briggs M, Davis JW, Ewing R, Mense M, Kwok OCH, Romand S, Thulliez P (2003) Toxoplasma gondii, Neospora caninum, Sarcocystis neurona, and Sarcocystis canis-like infections in marine mammals. Vet Parasitol 116(4):275–296

Duignan PJ, House C, Walsh MT, Campbell T, Bossart GD, Duffy N, Fernandes PJ, Rima BK, Wright S, Geraci JR (1995) Morbillivirus infection in manatees. Mar Mamm Sci 11(4):441–451

Duplaix N, Reichart HA (1978) History, status and protection of the Caribbean manatee *Trichechus m. manatus* in Suriname. Rare animal relief effort and United States fish and wildlife service, unpublished report, 23 pp + x append

Duplaix N, Lingaard M, Sikimin C (2001) A survey of Kaburi Creek, West Suriname, and its conservation implications. Report to the Oceanic Society, 34 pp. Survey and status report, Suriname

Emin-Lima R, Rodrigues ALF, Sousa MEM, Arcoverde DL, Santos GMA, Martins BML, Silva Júnior JS, Siciliano S (2010) Os mamíferos aquáticos associados aos manguezais da costa norte brasileira. In: Pessôa LM, Tavares WC, Siciliano S (eds) Mamíferos das Restingas e Manguezais do Brasil. Rio de Janeiro, pp 45–58

FMA (Fundação Mamíferos Aquáticos) (2015) Fundação Mamíferos Aquáticos utiliza tecnologia de sonar para pesquisar população do peixe-boi marinho na Paraíba. In: Magalhães K (ed) Revista A Bordo – Projeto Viva o Peixe-boi marinho. Edição 6, Recife-PE, 32 p

Garcia-Rodriguez AI, Bowen BW, Domning D et al (1998) Phylogeography of the West Indian manatee (*Trichechus manatus*): how many populations and how many taxa? Mol Ecol 7:1137–1149

Garzón F (1997) Caracterización microbiológica y parasitológica de las heces fecales del manatí antillano (Trichechus manatus) en determinados sectores de la región Caribe colombiana. Undergrad thesis, Universidad Nacional de Colombia

Gómez-Camelo IV, Zerda Ordóñez E, Castelblanco Martínez DN (2004) Áreas de distribución y alimentación del manatí *Trichechus manatus manatus* en época de aguas altas en la zona de influencia Puerto Carreño, Vichada, Colombia. Acta Biológica Colombiana 9(2):144–115

Góngora-Correa N (2009) Liberación y monitoreo de manatíes (*Trichechus manatus*) a la cuenca baja del río Sinú, Colombia, como una estrategia de conservación. Undergrad thesis, Fundación Universidad Jorge Tadeo Lozano

Gonzalez-Socoloske D, Olivera-Gomez LD (2012) Gentle giants in dark water: using side-scan sonar for manatee research. Open Remote Sens J 5(1):1–14

Gonzalez-Socoloske D, Olivera-Gomez LD, Ford RE (2009) Detection of free-ranging West Indian manatees *Trichechus manatus* using side-scan sonar. Endanger Species Res 8:249–257

Haigh MD (1991) The use of manatees for the control of aquatic weeds in Guyana. Irrig Drain Syst 5:339–349

Hammer AS, Klausen B, Knold S, Dietz HH, Dutoit SJH (2005) Malignant lymphoma in a West Indian manatee (*Trichechus manatus*). J Wildl Dis 41(4):834–838

Hartman DS (1979) Ecology and behavior of the manatee (*Trichechus manatus*) in Florida. Special publication number 5, American Society of Mammalogists, 153 p

Husson AM (1978) The mammals of suriname. EJ Brill, Leiden

Jiménez-Marrero NM, Méndez-Matos I, Montoya-Ospina RA, Williams EH, Brinkley-Williams L, Mignucci-Giannoni A (1998) Valores de Referencia de Inmunoglobulina G en Tres Poblaciones del Manati (*Trichechus manatus*): Puerto Rico, Colombia y Florida. Caribb J Sci 34:313–314

Lailson-Brito J, Meirelles AC, Vidal LG, Ferraz DR, Furtado L, Azevedo A (2010) Mercúrio e organoclorados (DDT, PCB, HCB, HCH e MIREX) em tecidos do peixe-boi-marinho, *Trichechus manatus*, da costa nordeste do Brasil. In: Abstracts of the XIV Reunião de Trabalho de Especialistasem Mamíferos Aquáticos da América do Sul, Florianópolis

Lazzarini SM, Ribeiro DC, Vergara-Parente JE (2014) Sirenia (Peixe-boi-da-Amazônia, peixe-boi-marinho). In: Cubas ZS, Silva JCR, Catão-Dias JL (eds) Tratado de animais selvagens: Medicina veterinária, 2nd edn. São Paulo, Roca, pp 936–972

Lefebvre LW, Marmontel M, Reid JP, Rathbun GB, Domning DP (2001) Status and biogeography of the West Indian manatee. In: Woods CA, Sergile FE (eds) Biogeography of the West Indies: patterns and perspectives, 2nd edn. CRC Press, Boca Raton, pp 425–474

Lima RP (1997) Peixe-boi-marinho (*Trichechus manatus*): Distribuição, status de conservação e aspectos tradicionais ao longo do litoral nordeste do Brasil. Master's thesis, Universidade Federal de Pernambuco

Lima RP (2008) Distribuição espacial e temporal de peixes-bois (*Trichechus manatus*) reintroduzidos no litoral nordestino e avaliação da primeira década (1994/2004) do programa de reintrodução. 2008. 178f. Tese (Doutorado em Oceanografia) – Universidade Federal de Pernambuco, Recife

Lima RP, Paludo D, Silva KG, Soavinski RJ, Oliveira EMA (1992) Distribuição, ocorrência e status de conservação do peixe-boi-marinho (*Trichechus manatus*, Linnaeus, 1758) ao longo do litoral nordeste do Brasil. Coletânia de trabalhos de conservação e pesquisa de sirênios no Brasil 1:47–72

Lima RP, Caldas ST, Candisani L (2001) Peixe-boi: a história da conservação de um mamífero brasileiro. DBA Artes Gráficas

Lima RP, Alvite CMC, Reid JP, Bombassaro Júnior AA (2012) Distribuição espacial e temporal de peixes-bois (*Trichechus manatus*) reintroduzidos no litoral nordeste do Brasil. Nat Res Forum 2(2):63–80

Linares OJ (1998) Mamíferos de Venezuela. AUDUBON, Venezuela, pp 199–201

Luna FO (2001) Distribuição, status de conservação e aspectos tradicionais do peixe-boi-marinho (*Trichechus manatus manatus*) no litoral norte do Brasil. Master's thesis, Universidade Federal de Pernambuco

Luna FO, Lima RP, Araújo JP, Passavante JZO (2008) Status de conservação do peixe-boi marinho (*Trichechus manatus manatus* Linnaeus, 1758) no Brasil. Revista Brasileira de Zoociências 10:145–153

Luna FO, Bonde RK, Attademo FL, Saunders N, Jonathan W, Meigs-Friend G, Passavante JZ, Hunter ME (2012) Phylogeographic implications for release of critically endangered manatee calves rescued in Northeast Brazil. Aquat Conserv Mar Freshw Ecosyst 22:665–672

Manzanilla-Fuentes AG (2007) Estado actual de la población de manatíes (*Trichechus manatus*) en la Bahía del Tablazo, estado Zulia, Venezuela. Master's thesis, Universidad Nacional Experimental de Los Llanos Ezequiel Zamora, Guanare, Venezuela

Marsh, H, O'shea TJ, Reynolds III JE (2011) Ecology and conservation of the Sirenia: Dugongs and Manatees. Conservation biology 18. Cambridge University Press, Nova York

Meirelles ACO (2008) Mortality of the Antillean manatee, *Trichechus manatus manatus*, in Ceará State, northeastern Brazil. J Mar Biol Assoc UK 88(6):1133–1137

Meirelles ACO, Marmontel M, Sousa-Lima RS (2016a) Biologia. In: Meirelles ACO, Carvalho VL (eds) Peixe-boi-marinho: Biologia e Conservação no Brasil. Aquasis, 1st edn. Bambu Editora e Artes Gráficas, São Paulo, pp 29–49

Millián-Sánchez SL, Montoya-Ospina RA, Mignucci-Giannoni AA, Williams EH (1995) Dwarfism in two antillean manatees from Colombia. In: Abstracts of the XI biennial conference on the biology of marine mammals, Orlando

Mohadin K (1993) Conservation of freshwater ecosystems in Suriname. In: Outober PE (ed) The freshwater ecosystems of Suriname. Monographie biologicae, 70. Springer-Science, Dordrecht

Mondolfi E, Müller K (1979) Proyecto de FUDENA para la investigación y conservación del manatí en Venezuela. 2do. Informe sobre los resultados obtenidos en las inspecciones realizadas en los caños del golfo de Paria, en el Delta del Orinoco, el bajo Orinoco y el bajo Apure, Caracas, Venezuela, 76 pp

Montiel-Villalobos M, Barrios-Garrido H (2005) Observaciones sobre la distribución y situación actual del Manatí *Trichechus manatus* (Sirenia: Trichechidae) en el sistema del Lago Maracaibo. Anartia 18:1–12

Montoya-Ospina RA, Caicedo-Herrera D (1995) Status of the Antillean Manatee (*Trichechus manatus manatus*) in Colombia: a preliminary report. In: Abstract of the Eleventh biennial conference on the biology of marine mammal (Society of Marine Mammalogy), Orlando, 14–18 December

Montoya-Ospina RA, Caicedo-Herrera D, Millán-Sánchez SL, Mignucci-Giannoni AA, Lefebvre LW (2001) Status and distribution of the West Indian manatee, *Trichechus manatus manatus*, in Colombia. Biol Conserv 102:117–129

Moretz-Sohn CD (2013) Sazonalidade no uso de habitats pelo peixe-boi marinho (*Trichechus manatus*, Linnaeus, 1758) na praia de Picos, Icapuí-CE. Undergrad thesis, Universidade Federal do Ceará

Murie J (1870) Notice of a memoir on the manatee. Proc Zool Soc London 1870:747–748

National Science Research Council (Guyana) (1974) An international centre for Manatee research. Report of a workshop held 7–13 February 1974, Georgetown, Guyana

Normande IC, Luna FO, Malhado ACM, Borges JCG, Viana Junior PC, Attademo FLN, Ladle RJ (2015) Eighteen years of Antillean manatee *Trichechus manatus manatus* in Brazil: lessons learnt. Oryx 49(2):338–344

Normande IC, Malhado ACM, Reid J, Viana PC, Savaget PVS, Correia RA, Luna FO, Ladle RJ (2016) Post-release monitoring of Antillean manatees: an assessment of the Brazilian rehabilitation and release programme. Anim Conserv 19:235–246

Ojeda M., Porras PJ, Gonzalez H.,OJasti J. (1993). Estado Actual del Manatí Antillano *Trichechus manatus en Venezuela*. Informe MARNR, PROFAUNA, Dirección de Manejo de Fauna Silvestre, Venezuela. 17 pp

O'Shea T, Correa-Viana M, Ludlow M, Robinson J (1988) Distribution, status and traditional significance of the West Indian manatee *Trichechus manatus* in Venezuela. Biol Conserv 46(4):281–301

Ortiz RM, Mackenzie DS, Worthy GA (2000) Thyroid hormone concentrations in captive and free-ranging West Indian manatees (*Trichechus manatus*). J Exp Biol 203(23):3631–3637

Parente CL, Vergara-Parente JE, Lima RP (2004) Strandings of Antillean manatees, *Trichechus manatus manatus*, in northeastern Brazil. Latin Am J Aquat Mamm 3(1):69–75

PETROBRAS (2014) Subprojeto de Monitoramento de Sirênios: Monitoramento Remoto por Telemetria Satelital e Censo Populacional por Meio de Sobrevoo. Relatórios dos Programas e Projetos Ambientais

Pool M (2012) River seismic survey in manatee habitat in Suriname. Sirenews 58:14–17

Quintana-Rizzo E, Reynolds JE (2010) Plan de manejo regional para el manatí de las Indias Occidentales (*Trichechus manatus*). Informe Técnico del Programa Ambiental del Caribe 48. PNUMA, Kingston

Ramírez-Carroz S, Manzanilla-Fuentes AG (2008) Registros de manatí (*Trichechus manatus*) en el Golfo de Venezuela. Revista Mesoamericana de la Conservación 1(1):24–25

Rivas-Rodríguez B, Ferrer-Pérez A, Colonnello G (2012) Distribución, uso de hábitat y status poblacional del manatí *Trichechus manatus* en el tramo central del Bajo Orinoco, Venezuela. Memoria de la Fundación La Salle de Ciencias Naturales 173–174:155–172

Rodulfo SC (2012): Distribución, alimentación y amenazas a las que está siendo sometido el manatí antillano *Trichechus manatus manatus* (Linnaeus 1758) en el Caño Araguao, Reserva de Biosfera Delta del Orinoco. Master's thesis, Universidad de Oriente, Nueva Esparta, Venezuela

Santos FS, Barros HMDR, Schetino MAA, Lima SC (2016) Genética. In: Meirelles ACO, Carvalho VL (ed) Peixe-boi-marinho: biologia e conservação no Brasil. Aquasis, 1st edn. Bambu Editora e Artes Gráficas, São Paulo, pp 63–75

Satizábal P, Mignucci-Giannoni AA, Duchêne S, Caicedo-Herrera D, Perea-Sicchar CM, García-Dávila CR, Trujillo F, Caballero SJ (2012) Phylogeography and sex-biased dispersal across riverine manatee populations (*Trichechus inunguis* and *Trichechus manatus*) in South America. PLoS One 7(2):e52468

Self-Sullivan C, Mignucci-Giannoni AA (2012) West Indian manatees (*Trichechus manatus*) in the Wider Caribbean region. In: Hines EM, Reynolds JR, Aragones LV, Mignucci-Giannoni AA, Marmontel M (eds) Sirenian conservation – issues and strategies in developing countries. University Press of Florida, Gainesville, pp 36–46

Sidrim JJC, Carvalho VL, Castelo-Branco DSCM et al (2015a) Yeast microbiota of natural cavities of manatees (*Trichechus inunguis* and *Trichechus manatus*) in Brazil and its relevance for animal health and management in captivity. Can J Microbiol 61(10):763–769

Sidrim JJC, Carvalho VL, Maia DCBS et al (2015b) *Bipolaris hawaiiensis* as an emerging cause of cutaneous phaeohyphomycosis in an Antillean manatee *Trichechus manatus manatus*. Dis Aquat Org 113:69–73

Sidrim JJC, Carvalho VL, Maia DCBS et al (2016) Antifungal resistance and virulence among Candida spp. from captive Amazonian manatees and West Indian manatees: potential impacts on animal and environmental health. Eco Health 13:328

Silva RB (2010) Aspectos biológicos, físico-ambientais e antrópicos de encalhes de peixes-bois-marinhos, *Trichechus manatus manatus*, no Ceará. 2010. 101f. Undergrad thesis, Universidade Federal do Ceará

Silva SM (2015) Caracterização populacional e estimativa de parentesco entre peixes-bois-marinhos *Trichechus manatus* (LINNAEUS, 1758) usando marcadores microssatélites. Master's thesis, Universidade Federal do Pará

Silva KG, Paludo D, Oliveira EMA, Lima RP, Soavinski RJ (1992) Distribuição e ocorrência do peixe-boi-marinho (*Trichechus manatus*) no estuário do rio Mamanguape, Paraíba-Brasil. Periódico Peixe-boi/IBAMA- FMM 1(1):6–18

Silva JCR, Marvulo MFV, Picanço MC et al. (2001) Pesquisa de anticorpos anti-*Toxoplasma gondii, Leptospira interrogans* e *Brucella abortus* em peixes-bois-marinhos (*Trichechus manatus manatus*) mantidos em cativeiro. In: Abstracts of V Congresso e X Encontro da Associação Brasileira de Veterinários de Animais Selvagens, São Paulo

Silva FMO, Vergara-Parente JE, Gomes JKN et al (2007) Contribution for the definition of serum chemistry values in captive adults Antillean Manatees (*Trichechus manatus manatus* Linnaeus, 1758). J Vet Med Ser A 54:119–122
Silva FMO, Vergara-Parente JE, Gomes JKN et al (2009) Blood chemistry of Antillean Manatees (*Trichechus manatus manatus*): age variations. Aquat Mamm 35:253–258
Silva CPN, Meirelles ACO, Umezaki J (2016) Ameaças. In: Meirelles ACO, Carvalho VL (eds) Peixe-boi-marinho: Biologia e Conservação no Brasil. São Paulo, Bambu, pp 93–108
Sousa MEM, Martins BML, Fernandes MEB (2013) Meeting the giants: the need for local ecological knowledge (LEK) as a tool for the participative management of manatees on Marajó Island, Brazilian Amazonian coast. Ocean Coast Manag 86:53–60
Spiegelberger T, Ganslosser U (2012) Habitat analysis and exclusive bank feeding of the Antillean manatee (*Trichechus manatus manatus* L. 1758) in the Coswine Swamps of French Guiana. South Am Trop Zool 18(1):1–12
Stamper MA, Bonde RK (2012) Health assessment of captive and wild-caught West Indian manatees (*Trichechus manatus*). In: Hines EM, Reynolds III J, Aragones LV, Mignucci-Giannoni AA, Marmontel M (eds) Sirenian conservation: issues and strategies in developing countries. University Press of Florida, Gainesville, pp 139–147
Vasconcelos AMO (2013) Dieta de *Trichechus manatus* (Linnaeus, 1758), no litoral leste do Ceará, Brasil. Undergrad thesis, Universidade Federal do Ceará. Fortaleza
Vergara-Parente JE, Marcondes MCC et al. (2001) Identificação de *Pseudomonas* sp. e *Proteus mirabillis* como agentes etiológicos de otite externa em filhotes de peixes-bois marinhos em cativeiro. In: Abstracts of the VII Encontro de Pesquisadores da UECE, Fortaleza
Vergara-Parente JE, Sidrim JJC et al. (2003a) Bacterial flora of upper respiratory tract of captive Antillean manatees. Aquat Mamm 29:124–130
Vergara-Parente JE, Sidrim JJC et al (2003b) Salmonellosis in an Antillean manatee (*Trichechus manatus manatus*) calf: a fatal case. Aquat Mamm 29:131–136
Vianna JA, Bonde RK, Caballero S, Giraldo JP, Lima RP, Clark A, Marmontel M, Morales-Vela B, De Souza MJ, Parr L, Rodríguez-Lopez MA, Mignucci-Giannoni AA, Powell JA, Santos FR (2006) Phylogeography, phylogeny and hybridization in trichechid sirenians: implications for manatee conservation. Mol Ecol 15(2):433–447
Wassink W (1997) Zeekoeien van Suriname. Afdeling Natuurbeheer, Dienst's Lands Bosbeheer. Ministerie van Natuurlijke Hulpbronnen
Whitehead PJP (1977) The former southern distribution of new world manatees (*Trichechus* spp.). Biol J Linn Soc 9: 165–189

Chapter 12
Ecology and Conservation of Cetaceans of Costa Rica and Panama

Laura J. May-Collado, Marisol Amador-Caballero, José Julio Casas, Mónica P. Gamboa-Poveda, Frank Garita-Alpízar, Tim Gerrodette, Rocío González-Barrientos, Gabriela Hernández-Mora, Daniel M. Palacios, José D. Palacios-Alfaro, Betzi Pérez, Kristin Rasmussen, Lissette Trejos-Lasso, and Javier Rodríguez-Fonseca

Abstract The cetacean species richness of Costa Rica and Panama represents about 36% of the species described worldwide. Within the economical coastal waters of these countries bottlenose dolphins, the pantropical spotted dolphins, the Guiana dolphin, and humpback whales are the most commonly observed. Their high sight fidelity and predictability has resulted in a rapid increase on the whale-watching industry, perhaps second to incidental mortality, when unregulated whale-watching can negatively affect coastal cetacean populations. In this Chapter, we provide a summary of

L.J. May-Collado (✉)
Department of Biology, University of Vermont, Burlington, VT, USA
e-mail: lmaycoll@uvm.edu

M. Amador-Caballero • J. Rodríguez-Fonseca
PROMAR, San José, Costa Rica
e-mail: marisol.acaballero@gmail.com; javierf2006@yahoo.com

J.J. Casas
Universidad Marítima Internacional de Panamá, Panama City, Panama

PANACETACEA, Saint Paul, MN, USA
e-mail: josejuliocasas@yahoo.es

M.P. Gamboa-Poveda • B. Pérez • K. Rasmussen • L. Trejos-Lasso
PANACETACEA, Saint Paul, MN, USA
e-mail: monicagamboapo@gmail.com; betziperez@yahoo.com; Krill@aol.com; sselita18@gmail.com

F. Garita-Alpízar
Fundacion VIDA, Santa Ana, Costa Rica

PANACETACEA, Saint Paul, MN, USA
e-mail: frankgarita@gmail.com

T. Gerrodette
National Marine Fisheries Service, La Jolla, CA, USA
e-mail: Tim.Gerrodette@noaa.gov

M.R. Rossi-Santos, C.W. Finkl (eds.), *Advances in Marine Vertebrate Research in Latin America*, Coastal Research Library 22,
DOI 10.1007/978-3-319-56985-7_12

the state of knowledge of the ecology and conservation status of the cetaceans of Costa Rica and Panama. Along this line, we describe the consensus of our small scientific community on what relates to future scientific research and conservation.

The cetacean species richness of Costa Rica and Panama represents about 36% of the species described worldwide. The dolphin family, Delphinidae is the most diverse group of cetaceans in these countries, with some species having year-round resident populations in coastal habitats like the bottlenose dolphins (*Tursiops truncatus*), the pantropical spotted dolphins (*Stenella attenuata),* and the Guiana dolphin (*Sotalia guianensis*). Other dolphin species are found primarily in oceanic and neritic waters like the rough-toothed dolphins (*Steno bredanensis*), striped dolphins (*Stenella coeruleoalba*), spinner dolphins (*S. longirostris*), and common dolphins (*Delphinus delphis*) (Rodriguez-Fonseca 2001; May-Collado et al. 2005). Among baleen whales, the humpback whale (*Megaptera novaeangliae*) is by far the most common species, and both Costa Rica and Panama possess important habitats used by these whales during their breeding season (Steiger et al. 1991; Calambokidis et al. 2000; Rasmussen et al. 2007, 2012).

The economic value of cetaceans in Costa Rica and Panama has increased in the past 20 years particularly due to whale watching activities, an activity that has become highly profitable across Latin America (Hoyt and Iñiguez 2008). Profitable whale watching can mutually benefit humans and wild animal populations, resulting in successful conservation. However, high observation intensities may nevertheless negatively affect animals. In Costa Rica and Panama whale watching is boat-based, and cetaceans are exposed daily to various levels of engine noise, which has been shown to elicit avoidance behaviors and stress, and mask important acoustic signals. For example bottlenose dolphins of Bocas del Toro, Panama, are exposed daily to high levels of underwater noise associated with dolphin-watching boats. In response, these dolphins modify their communicative signal frequency and duration presumably to decrease signal masking (e.g., May-Collado and Wartzok 2008, 2015; May-Collado and Quinones-Lebron 2014). Although cetaceans have evolved ways to compensate for masking by natural ambient noise, noise from human activities represent a new challenge that these animals may not be able to circumvent. The only option may be to avoid contact with these noises in which case noise acts

R. González-Barrientos • G. Hernández-Mora
Ministerio de Agricultura y Ganadería, Servicio Nacional de Salud Animal (SENASA), Heredia, Costa Rica
e-mail: cristyrocio@gmail.com; gabbytica@gmail.com

D.M. Palacios
Marine Mammal Institute and Department of Fisheries and Wildlife, Oregon State University, Corvallis, OR, USA
e-mail: daniel.palacios@oregonstate.edu

J.D. Palacios-Alfaro
Fundacion Keto, San Jose, Costa Rica
e-mail: pala1611@gmail.com

much like other pollutants in rendering habitats unsuitable. Because of the potentially harmful effects of noise, the International Whaling Commission (1995) and International Fund for Animal Welfare (1996) have recognized boat-based whale watching as potentially detrimental to cetaceans and their environment and have developed guidelines in an attempt to reduce the impact of the industry.

In Costa Rica, several coastal communities have become economically dependent on whale-watching activities. For instance the communities of Sierpe, Uvita, and Drake located in the South Pacific coast of Costa Rica specialize in taking tourists to observe humpback whales that arrived from the southern hemisphere from August to November. In Panama, whale watching activities occur near Isla Contadora, Archipelago de Las Perlas, and Isla Iguana to name a few areas off the Pacific coast, but the most active whale watching activities in Panama occurs in the Archipelago of Bocas del Toro, located in north Caribbean coast of Panama, where resident bottlenose dolphins are the main target (May-Collado et al. 2012, 2014, 2015). While both countries have developed regulations for whale watching operators to reduce the impact of the activity, these guidelines emphasize the regulation of distance at which boats can approach the animals. However there is growing evidence that engine and propeller noise are the main causes of disturbance (e.g., Morisaka et al. 2005; Nowacek et al. 2001; May-Collado and Wartzok 2008), highlighting the importance of designing experiments that allow to measure noise levels and incorporate this information into local regulations (IWC 1995; IFAW 1996).

Despite their economic value, some cetaceans remain affected by incidental mortality in artisanal fisheries and purse-seine tuna fishery in Costa Rican waters. In Costa Rica dolphin estimates of annual mortality rates due to artisanal fisheries is about 9.5% of the population size for all small cetaceans combined (Palacios and Gerrodette 1996). Anthropogenic mortality rates should not exceed 1–2% of population size, based on the low reproductive rate of cetaceans (Perrin et al. 1994; IWC 1996). In addition, anecdotal information indicates the direct killing of small dolphins for shark bait in both the Pacific and Caribbean coast of Costa Rica, and it has been documented in Guatemala as well by Quintana-Rizzo (2014). The frequency of these incidents and the number of animals being killed is unknown at this moment.

Historically, whaling in Panamanian waters occurred between 1883 and 1886. The *Gay Head*, a commercial whaling vessel from New Bedford, visited the Gulf of Panama three times, capturing and processing a total of 50 humpback whales (Best 2008). This is the first reference of the presence of commercial whaling in the coast of Panama. In addition, the *Olympic Challenger*, a Norwegian factory vessel registered in Panama, was proved to be involve in "pirate whaling" (Birnie 1985) in Panamanian waters.

The impact of both historical and current killing of cetaceans is unknown largely because it occurs far from the coast and under great secrecy. However, this is a source of mortality that needs to be addressed and taken into account in future models along with incidental mortality to better assess the impact of these activities in the local cetacean populations.

In this Chapter we will provide a summary of the state of knowledge of the ecology and conservation status of the cetaceans of Costa Rica and Panama. Along this line, we will identify what our small scientific community considers are the most important threats to these animals and their habitat, highlighting recommendations for their conservation.

12.1 Pacific Coast

The list of cetacean species for the Pacific coast of Costa Rica includes about 28 species. However, information about the ecology and conservation status for most of these species remains unknown. In the past 20 years research has focused primarily on three coastal species: bottlenose dolphins, spotted dolphins, and humpback whales (e.g., Acevedo-Gutierrez and Burkhart 1998; Cubero-Pardo 1998; Calambokidis et al. 2000; Garcia and Dawson 2003; May-Collado and Morales-Ramirez 2005; May-Collado and Forcada 2012; Herrera-Miranda et al. 2016). We will start providing a general background on what we know so far about these three species.

The coastal subspecies of the spotted dolphin (*Stenella attenuate graffmani*) is one of the 16 dolphin species inhabiting Costa Rican and Panamanian waters (May-Collado et al. 2005; May-Collado 2009) and possibly the most common dolphin off the Pacific coast of both Costa Rica and Panama. In Costa Rica, coastal spotted dolphin are commonly observed throughout the Pacific coast, however most of what we know about these dolphins comes from a handful of locations: Cuajiniquil and Gulf of Papagayo in the north Pacific and from Drake Bay, Osa Peninsula, and Gulf Dulce and Isla del Caño in the south Pacific (e.g., Cubero 1998; Oviedo 2007; Oviedo et al. 2014; May-Collado et al. 2005; Martinez-Fernandes et al. 2011, 2014).

In the north Pacific, coastal spotted dolphins appear to be resident year round (Martinez-Fernandez et al. 2011; May-Collado et al. 2005; May-Collado and Forcada 2012). The gulf is characterized by dry and rainy seasons. During the dry season strong northeast trade winds generate the richest coastal upwelling of the country, and spotted dolphin's abundance seem to be associated to it. May-Collado and Forcada (2012) found that the abundance of coastal spotted dolphins increased linearly with water depth and transparency, and non-linearly with the dissolved oxygen concentration. Also, they found that high variability in the relative dolphin abundance occurred during the dry season, a time where a higher number of spotted dolphin groups were observed primarily in foraging activities (May-Collado and Morales-Ramirez 2005). Therefore seasonal changes in relative abundance for these dolphins probably are associated with food availability. Similar patterns were described for the spotted dolphins of the Gulf Dulce by Cubero-Pardo (1998), but no estimates of relative abundance were made, hindering further comparisons.

There is some evidence that at least some coastal spotted dolphins show high site fidelity (May-Collado 2001). Thus, there is concern about the impact coastal human activities may have on these animals. For instance, Montero-Cordero y Lobo (2010) studied the impact of dolphin watching activities on coastal spotted dolphins that inhabit the waters between Drake Bay and Caño Island. They found that the highest density of dolphin groups accompanied by tour boats was near Caño Island. They also found that when operators did not comply with conduct guidelines, the spotted dolphins responded negative towards the boat, by swimming away. What this study suggest is that even dolphin species that live in large group sizes are affected by intense dolphin watching activities and that negative effects can be minimized if operators comply with regulations.

While coastal spotted dolphins are found along the Pacific coast, bottlenose dolphins do not showed a continuous in their distribution. For instance, bottlenose dolphins are rarely observed in the Gulf of Papagayo and surrounding areas, whereas in the gulfs of Nicoya and Dulce they appear to be more common. In Gulf Dulce spotted and bottlenose partition the gulf by differences in habitat use and physical characteristics of the gulf such as bathymetry and topographic characteristics. For instance, bottlenose dolphins are primarily associated with all major river drainages at the inner basin and sill area of the gulf, likely representing a key critical habitat where they main prey is located (Acevedo and Buckhart 1998; Herrera-Miranda et al. 2016). Dietary information is not available yet, but observations on their feeding behavior suggest bottlenose dolphins feed primarily on demersal fish while coastal spotted dolphins seem to feed mainly on pelagic fish (Oviedo 2007).

May-Collado et al. (2005) distribution maps indicate that the Gulf of Nicoya is potentially an important habitat for spotted and bottlenose dolphins. The gulf and surrounding areas are considered the most important fishing ground in Costa Rica (Vargas 1995). To put this in perspective, between the two main fishing ports (located in Puntarenas and Quepos), just in 2014 they accounted for 7511 tons of fish (including sharks). This is about 62% of total catches in the country for the same period (INCOPESCA 2016). This high productivity likely explains the regular presence of spotted and bottlenose dolphins. In addition, other 13 species of cetaceans, including beaked whales, humpback whales, pilot whales, false killer whales, among others have also been observed in this area (May-Collado et al. 2005; Martinez-Fernandez et al. 2014). While high species richness may be associated to food, Palacios-Alfaro pers. comm. (2016) also suggest that the short distance between coast and continental shelf may also increase the probability of observing deep water species.

Although the Gulf of Nicoya and surrounding areas represent important habitats for cetaceans we still know very little about their population dynamics and interactions with human activities in the area. The Central Pacific coast of Costa Rica has the highest fishing effort in the country, mainly due to artisanal fishing activities that use of bottom lines and gillnets, purse seine, trawling, and long lines at small and medium scales. Serious concerns about the potential impact of the fishery industry in the local cetaceans stems from previous work by Palacios and Gerrodette (1996) who estimated that the Costa Rican cetacean bycatch in gillnets was considerably high in comparison with other Central American countries, such as Panama, and by anecdotal accounts of fisherman witnessing bottlenose dolphin calves being trapped in sardine nets (Jose D. Palacios Alfaro pers. comm. 2016).

An indirect impact of fisheries is creating dependence for resources. For example, Palacios-Alfaro (2007a, b) found that dolphins can potentially change their foraging patterns in areas where shrimp fishing occurrence is high. He observed pantropical spotted dolphins and bottlenose dolphins regularly following trawlers to feed on the discarded demersal and pelagic fish. Some dolphin groups consistently approached trawlers and followed them for up to 11 h, suggesting that at least some dolphins might have develop a dependency on this resource.

The third most common cetacean in the Pacific coast of Costa Rica is the humpback whale. Humpback whales make seasonal migrations from high-latitude feeding grounds to low latitude breeding areas where whales give birth and mate. Costa Rica receives humpback whales from two main feeding grounds: off California-Oregon-Washington in the Eastern North Pacific (northern whales) between December and April (Steiger et al. 1991; Acevedo and Smultea 1995; Calambokidis et al. 2000; Rasmussen et al. 2012) and off Chile and Antarctica between July to November (Acevedo et al. 2007; Rasmussen et al. 2007).

Southern humpback whales have been previously documented off the Pacific coast of Central America during the austral wintering season (Townsend 1935; Acevedo and Smultea 1995; Florez-González et al. 1998; Rasmussen et al. 2007; Best 2008). Central America is unique because it harbors the northernmost breeding area of any southern hemisphere humpback whale population, with whales migrating approximately 8300 km from the feeding areas (Acevedo et al. 2007; Rasmussen et al. 2007). As mentioned before these whales have been linked to feeding areas off Chile and Antarctica (Acevedo et al. 2007; Rasmussen et al. 2007) as well as to adjacent breeding areas off Colombia (Florez-González et al. 1998), and are part of the IWC-designated "Breeding Stock G" (BSG). In Panama, one of the most studied breeding grounds for southern humpback whales is the Gulf of Chiriquí. This gulf lies in the western part of Panama, and is bordered by the Azuero Peninsula to the east, and Punta Burica to the west. Here, humpback whales are frequently sighted, predominantly in shallow waters near island groups and close to shore. Approximately half of the sightings contained calves, many of which were small in size, suggesting the females are calving in these waters or in waters very close by. Similar patterns are found in Costa Rica, where 64% of the sightings between Caño Island and Drake Bay are mothers with calves (Palacios-Alfaro et al. 2014) (Figs. 12.1, 12.2, 12.3, 12.4, 12.5, 12.6, 12.7, 12.8, 12.9, 12.10, 12.11, 12.12, 12.13, 12.14).

Although, research on humpbacks whales in Costa Rica was initially focused on northern whales, over the last years a perceived decrease in the number of these whales is raising concerns among local scientists. For this reason whale-watching activities are now only targeted towards southern whales. According to the National Oceanographic Atmospheric Administration (NOAA) while most humpback whale populations are recovering, the Central American humpback whales remain endangered. Therefore, future research should focus on determining if the number of northern whales breeding in Costa Rica is indeed decreasing, and if so what are the factors responsible for this decrease.

The Gulf of Chiriqui is also an important habitat for six other species of cetacean. The most common dolphin is the spotted dolphin both in number of sightings and abundance, followed by bottlenose dolphins, which are sighted both in estuarine waters and offshore waters near islands. Bryde's whales (*Balaenoptera edeni*) are seen occasionally near the Islas Contreras and the Islas Secas. Less common are false killer whales (*Pseudorca crassidens*), sighted twice near the Islas Secas; one

A. Boat passing by masking elements of singing humpback whale

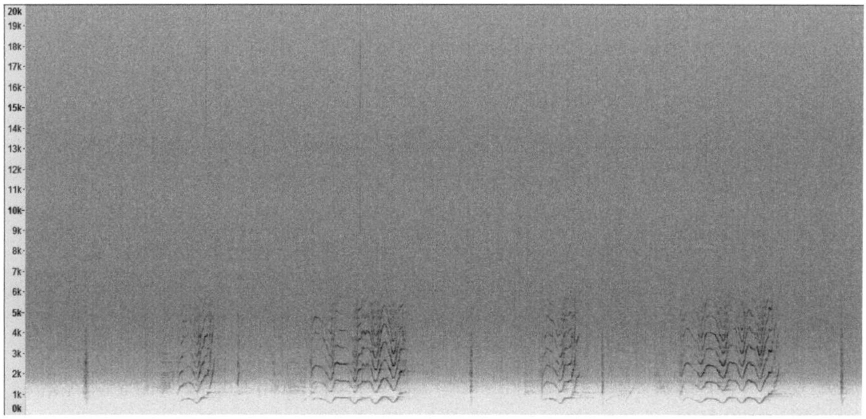

B. Humpback whale signing in the absence of engine noise

Fig. 12.1 Southern hemisphere humpback whale male singing near Isla del Caño, Costa Rica in 2016. The island, reefs, and humpback whales attract hundreds of tourists to the island each year. Tour operators in Uvita, Drake bay and Sierpe (and likely other locations along the coast) offer tours to the island on daily basis. In this figure we show an example of the noise emitted by a passing boat and its overlap in frequency with some elements of the whale song. These recordings were made using underwater recorders (uRUDAR-mk2 from Cetacean Research Technology) during a pilot project funded by Conservation International Costa Rica (**a**) Boat passing by masking elements of singing humpback whale. (**b**) Humpback whale signing in the absence of engine noise (Credits to Laura J May-Collado, Ana Guzman, and Marco Quesada Alpizar)

minke whale (*Balaenoptera acutorostrata*) sighted near Isla Coiba, and one group of rough toothed dolphins (*Steno bredanensis*) sighted near the Islas Paridas. Other areas were cetacean research is starting to generate information about their occurrence is Isla Iguana, Gulf Montijo, and Islas Perlas.

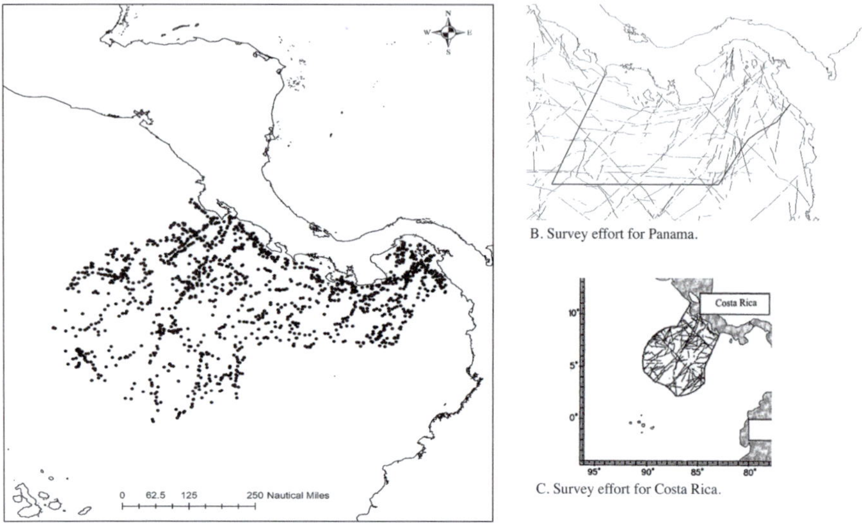

B. Survey effort for Panama.

C. Survey effort for Costa Rica.

A. Distribution of cetacean sightings in the Pacific EEZ of Costa Rica and Panama.

Fig. 12.2 Overall distribution of cetacean sightings during SWFSC-NOAA surveys between 1979 and 2016 in Costa Rica and Panama. (**a**) Distribution of cetacean sightings in the Pacific EEZ of Coasta Rica and Panama. (**b**) Survey effort for Panama. (**c**) Survey effort for Costa Rica (Data provided by Dr. Tim Gerrodette)

Fig. 12.3 Spotted and bottlenose dolphins feeding on fish discharged by shrimp fishing boats in in Central Pacific waters of Costa Rica (Photos by Jose D. Palacios)

12.2 Caribbean Coast

In the Caribbean only nine of the 29 expected species for both Costa Rica and Panama have been confirmed (May-Collado 2009). These species are the pilot whale (*Globicephala macrorhynchus*), killer whales (*Orcinus orca*), false killer whales (*Pseudorca crassidens*), Guiana dolphins, spotted dolphins (*Stenella frontalis*), rough-toothed dolphins, bottlenose dolphins, sperm whales (*Physeter macrocephalus*), and the dwarf sperm whale (*Kogia sima*). This low number of confirmed

Fig. 12.4 Southern humpback whales migrate during the austral wintering season to Panama and Costa Rica and other parts of Central America to breed and give birth (Photo by Kristin Rasmussen)

species is likely the product of the lack of consistent and standard survey efforts in these countries for the Caribbean. The diversity is likely higher as suggested stranding data, where an unidentified beaked whale of the genus *Mesoplodon* and a baleen whale of the genus *Balaenoptera* where documented (see the stranding section).

The most commonly observed cetacean in the Caribbean coast of both Costa Rica and Panama is the bottlenose dolphin. In Costa Rica, bottlenose dolphins have been documented along the entire Caribbean coast, in the north, sightings of bottlenose dolphins have been made in the mouth of the Colorado River, Samay Lagoon, and the Parismina River, in the central Caribbean from the Matina River to the Cieneguita River, and in the south in Gandoca-Manzanillo Wildlife Refuge. In Panama, it is presumed that the only resident population of bottlenose dolphins is located in the Archipelago of Bocas del Toro. However, just like in Costa Rica the Caribbean coast of Panama remains vastly unexplored. Most of what is known about Caribbean bottlenose dolphins comes from a handful of long-term studies in Gandoca-Manzanillo, Costa Rica (Gamboa Poveda and May-Collado 2006; Gamboa-Poveda 2009) and Bocas del Toro, Panama (May-Collado et al. 2012–2015). However, recent efforts in the central Caribbean coast of Costa Rica are changing our perception of bottlenose habitat use and residency. Amador-Caballero, Garita-Alpízar, and Rodríguez-Fonseca surveyed the Central Caribbean between 2015 and 2016. They found that bottlenose dolphins was most commonly foraging in mouth of the river Matina. The coloration of these dolphins suggests that they might transient oceanic animals. Of a total of 74 dolphins photographed between only 7.35% of the animals showed regular use of the area. In this area the second most sighted species was the Atlantic spotted dolphins, which was observed primarily

Fig. 12.5 Cetacean species observed within coastal waters of the Pacific coast of Costa Rica and Panama from top to bottom spinner dolphins (Photo by Monica Gamboa), spotted dolphins (Photo by Laura May-Collado), false killer whales (Photo by Kristin Rasmussen), common dolphins (Photo by Jose David Palacios), Bryde's whale (Photo by Monica Gamboa), killer whale, and rough-toothed dolphins (Photos by Jose David Palacios).

Fig. 12.6 Cetaceans observed in the north Caribbean of Costa Rica and Panama from top to bottom: false killer whales, bottlenose dolphins, and Atlantic spotted dolphins

Fig. 12.7 Guiana dolphin's home range involved the nearshore waters of Gandoca-Manzanillo (Costa Rica) and Changuinola (Panama) where they are regularly expose to illegal nets aimed to trap sea turtles (Photo by Laura J May-Collado and Monica Gamboa-Poveda)

Fig. 12.8 Mixed species interactions of Guiana dolphins and bottlenose dolphins occur daily in Gandoca-Manzanillo. Interactions are particularly intense during social activities where bottlenose dolphins tend to separate and harass one or two Guiana dolphin. The *first photograph* shows a bottlenose dolphin (with the tall and falcate dorsal fin) following three Guiana dolphins (short and triangular dorsal fin) (Photo by Monica Gamboa-Poveda) and the *second photograph* two bottlenose dolphins have already separated a Guiana dolphin from the rest of the group (Photo by Laura J May-Collado)

Fig. 12.9 Bottlenose dolphin of Bocas del Toro belong to the coastal ecotype and are highly isolated from other Caribbean populations including the neighboring dolphins of Gandoca-Manzanillo, Costa Rica located only 35 km north (Photo by Betzi Perez)

in traveling activities and interacting with passing boats. They also observed in two occasion rough-tooted dolphins and false-killer whales highlighting the need for long-term monitoring in the area.

In Gandoca-Manzanillo Wildlife Refuge dolphin sightings are common within and nearby the refuge. Here the majority of the sightings are Guiana dolphins followed by bottlenose dolphins (Gamboa-Poveda 2009). Other species such as the dwarf sperm whale, Atlantic spotted dolphins, and rough-tooted dolphins have been observed occasionally within the refuge (Palacios-Alfaro 2007b). While the bottlenose dolphin is observed throughout the Caribbean coast, the Guiana dolphin

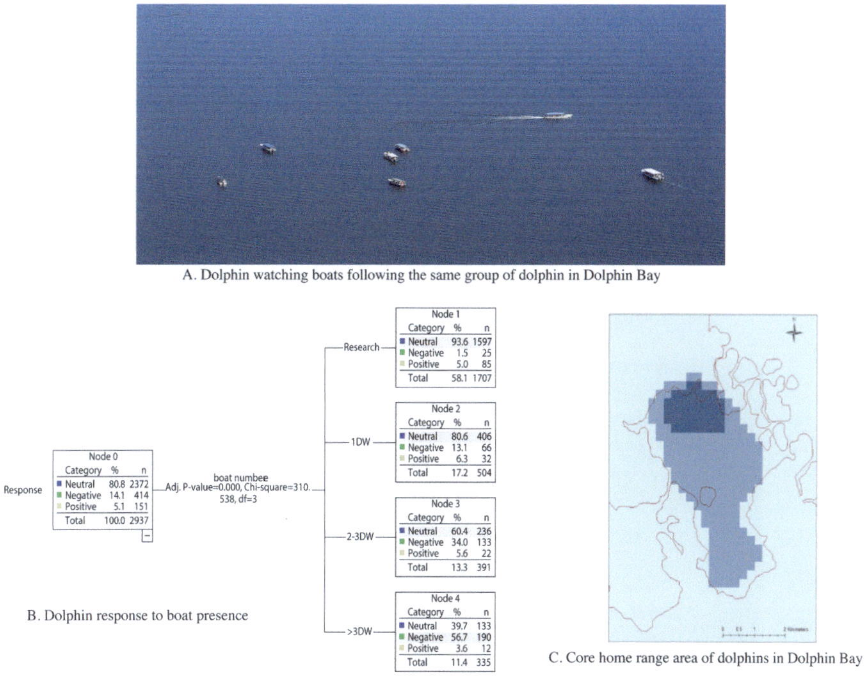

Fig. 12.10 The impact of boat-based dolphin watching in the resident population of bottlenose dolphins in Bocas del Toro. (**a**) highlights the high congregation of boats following the same group of dolphins, (**b**) is A decision tree analysis evaluate of dolphin response (negative, positive, neutral) to number of dolphin watching boats present simultaneously during an encounter (cross validation estimate = 0.311, std. error = 0.013, X2 = 310.5 df = 3, $p < 0.0001$), (**c**) shows the core home range area of dolphins within Dolphin Bay were the majority of encounters with dolphin watching boats occur

appears to be only found in the nearshore waters of Gandoca-Manzanillo and the Changuinola area in the Panamanian side. In both sites, biological productivity is highly influenced by large rivers, in the Costa Rican side by the Sixaola River and in the Panamanian side by the Changuinala River. However, its distribution may had been more ample in the past. For instance, the first documentation of Guiana dolphins for Panama was in Colon, near the Caribbean entrance to the Panama Canal, (Bossenecker 1978) but since this report there has been not sightings of these dolphins in Colon. This highlights the needs for standardized and long-term surveys in the region that can help us understand not only the diversity of cetaceans in it but also the threats to which they may be exposed to.

A special aspect of the dolphins inhabiting Gandoca-Manzanillo is the daily formation of interspecific groups. Gamboa-Poveda (2009) found that the majority of the dolphin groups observed within the refuge belong to the Guiana dolphins (50%), followed by mixed species groups (27.2%) and bottlenose dolphins (22.8%). The mixed groups of Guiana and bottlenose dolphins are observed on daily basis

Fig. 12.11 Adult male striped dolphins with (**a**) Facial scar completely healed, stranded at Camaronal Beach in 2010. (**b**) Ulcerative lesion with active borders, stranded at Tambor Beach, 2011

Fig. 12.12 Adult female dwarf sperm whale with bad body condition at the Pacific exit to the Panama Canal in 2010 (Photos by Lissette Trejos)

Fig. 12.13 Marine mammal specialist and veterinarian Lissette Trejos-Lasso working on a recent stranding of 60 rough-toothed dolphins in the Pacific coast of Panama. Using limited resources Lissette and her support team spent hours rescuing the healthy looking animals and necropsying the deceased

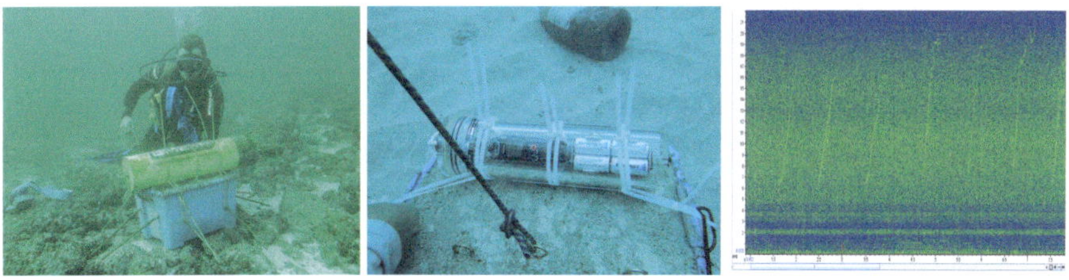

Fig. 12.14 Underwater acoustic recorders used to monitor the acoustic habitat of Cano and Murcielago Island in Costa Rica, both locations are protected areas used by humpback whales and dolphins. The last photograph shows spotted dolphin whistles recorder at 2 a.m. The project is funded and coordinated by Conservation International Costa Rica, and lead by Dr. May-Collado

Table 12.1 List of cetacean species confirmed for both Costa Rica and Panama

	Pacific Coast		Caribbean Coast	
Species	Costa Rica	Panama	Costa Rica	Panama
Balaenoptera edeni	x	x		
Balaenoptera musculus	x			
Balaenoptera physalus	x			
Balaenoptera borealis	x			
Delphinus delphis	x	x		
Feresa attenuata	*x*	x	x	
Globicephala macrorhynchus	x	x		
Grampus griseus	x	x		
Kogia sima	x	x	x	
Lagenodelphis hosei	x			
Megaptera novaeangliae	x	x		
Mesoplodon densirostris	x			
Mesoplodon peruvianus				
Orcinus orca	x	x		
Peponocephala electra	x	x		
Physeter macrocephalus	x	x	x	
Pseudorca crassidens	x	x	x	
Sotalia guianensis			x	x
Stenella attenuata	x	x		
Stenella frontalis			x	x
Stenella coeruleoalba	x	x		
Stenella longirostris	x	x		
Steno bredanensis	x	x	x	x
Tursiops truncatus	x	x	x	x
Ziphius cavirostris	x	x		

primarily during socializing activities (Forestell et al. 1999; Acevedo-Gutierrez et al. 2005; Gamboa-Poveda 2009; May-Collado 2010). Because of the commonality of these social associations and the intermedia morphological characteristics of the dorsal fin of some individuals, Forestell et al. (1999) hypothesized that hybridization between the species might be occurring. However, to this point not genetic study has taken place to determine if such hybrids exist. Another interesting aspect of these mixed species groups, is that sometimes, social interactions between species turned into aggressive interactions. This happens when bottlenose dolphins target one or two Guiana dolphins by separating them from the rest of the group and by chasing them and forcing them into intercourse activities. It is during these antagonistic events that 'hybrid' whistles are produced presumably by the Guiana dolphin as an attempt to match the whistles of its attackers and maybe reduce the intensity of the attack (May-Collado 2010).

Photo-identification work by Gamboa-Poveda (2009) suggested that overall residency rate is low for both dolphin species. Only 14% of Guiana dolphins and 12% of bottlenose dolphins seem to regularly use the refuge. This result is likely biased due to survey effort been limited to the refuge, as shown by recent boat surveys extending from Gandoca to Changuinola. During this recent surveys, Guiana dolphin were observed south the Changuinola River, suggesting that their distribution is larger than previously thought. Interestingly this does not seem to be the case for the bottlenose dolphins from Gandoca-Manzanillo, whose seem to include offshore waters.

Bottlenose dolphins have been documented in several parts of the Caribbean coast of Panama including the Laguna of Chiriquí, but only those from the Archipelago of Bocas del Toro are considered residents. In Bocas del Toro, the population of bottlenose dolphins is small, highly predictable, and it appears to be genetically isolated. Barragan-Barrera et al. (2015) described a unique mtDNA haplotype for this population, not found anywhere else in the Caribbean suggesting some extent of genetic isolation. Microsatellite data also support the isolation of this population even from neighboring Costa Rican bottlenose dolphins. Based on preliminary estimates, the population is considered to be between 80–100 animals, and it seems to be organized into two communities one that has low residency rate (<38.5%) and wider distribution, and a smaller communities that has more restricted movement patterns and higher residency rate (38.5–100%) (May-Collado et al. 2015a).

This small community seems to be restricted in great part to a small semi-enclosed bay called Dolphin Bay, where due to its high predictability it is in constant interactions with dolphin-watching boats. May-Collado et al. (2012) found that the number of boats interacting with these animals has increased exponentially in the Bay; up to 40 boats have been counted to interact with the same group of dolphins within a period of an hour (May-Collado et al. 2015). The increase in the number of boats is correlated with an increase in underwater engine noise and dolphins seem to respond to this noise by modifying their whistles presumably to reduce masking and by switching to avoidance behaviors (Kassamali-Fox et al. 2015; May-Collado and Wartzok 2008 2015, Sitar et al. 2016). Data shows that these dolphins do not respond acoustically in the same way to other watercraft, for example transport and other passing boats that are not directly targeting the animals, rarely elicit changes in their acoustic behavior (May-Collado 2015). Interestingly, interactions with dolphin watching boats vary depending on the animal behavior at the moment of the interaction. For example, foraging dolphins interrupted by dolphin watching boats lower the frequency and increase the duration of their communicative signals but when the animals were socializing not significant changes in the acoustic structure of their signals were noted.

12.3 Cetacean Strandings

In Costa Rica cetacean stranding information is available since 1996. Between 1966 and 1999 a total of 35 stranding events (244 individuals) were recorded. These strandings included predominantly single animals from 13 species of the families Delphinidae, Kogiidae, Physeteridae and Balaenopteridae (Table 12.1) (Rodríguez-Fonseca and Cubero-Pardo 2001). In addition, in 1976 a stranding of 200 individual melon headed whales (*Peponocephala electra*) (~80 died) was reported at Tambor Beach at Cobano, Peninsula of Nicoya in Puntarenas province. During these 33 years, the causes of death in the stranded animals were not determined (Janzen and Wilson 1983). Between 2000 and 2016, 96 stranding events involving 98 individuals were described largely by trained scientists. Most of these strandings (97.9%) consisted of single animals belonging to 16 species within the families Delphinidae, Kogiidae, Physeteridae, Ziphiidae and Balaenopteridae (see Table 12.1). In 2002, another mass stranding was reported at Cedros Beach, 24 kilometers apart from the site of the earlier mass stranding at Tambor Beach in 1976. This time 37 individuals (4 died) of rough-toothed dolphins stranded (Loaiza 2002).

In Costa Rica systematic postmortem examinations started to be performed in 2001, first by the Pathology Department of the School of Veterinary Medicine of the National University of Costa Rica; and later on (after 2009) by the National Veterinary Service of the Ministry of Livestock and Agriculture of the country. Currently, histopathological analyses are routinely performed for each stranding including tissue sections from all organs using standard procedures as described by Kiernan (2003). Since 2001 a total of 70 animal necropsies have been performed in 11 species of odontocetes including, striped dolphins, pantropical spotted dolphins, spinner dolphins, common dolphins, rough-toothed dolphins, bottlenose dolphins, Risso's dolphins, dwarf sperm whale and Cuvier's beaked whale (*Ziphius cavirostris).* All these strandings occurred in the Pacific coasta, to this day only two necropsies have been performed in animals stranded in the Caribbean coast, including one specimen of beaked whale belonging to the genus *Mesoplodon* and one sperm whale. By far the most common stranded species was the striped dolphin that represented 72% of all the specimens analyzed during this period of time. The main cause of death in this species was neurobrucellosis associated with infection caused by *Brucella ceti* (Table 12.2).

Causes of death varied among stranded animals and among species. There is variation in the presence of gastrointestinal parasites, parasitic pneumonias, and skin lesions. For instance, about 32% of stranded striped dolphins presented a localized unilateral or bilateral ulcerative lesion at the level of the mandibular condyle, directly below the eye or slightly frontal or caudal, irregular to oval in shape and with variable levels of chronicity. Some animals presented deep ulcerative lesion with active borders and secondary infection with different species of *Vibrio,* in the other hand other animals presented with completely healed wounds with prominent scar tissue (see figure). Histopathology revealed moderate to severe chronic ulcerative granulomatous dermatitis, negative for special stains intended to demonstrate

Table 12.2 List of cetacean species stranded in Costa Rica and Panama

Common name	Species	Costa Rica	Panama
Orca	*Orcinus orca*	Yes	–
Short finned pilot whale	*Globicephala macrorhynchus*	Yes	–
Melon headed whale	*Peponocephala electra*	Yes	–
False killer whale	*Pseudorca crassidens*	Yes	–
Guiana dolphin	*Sotalia guianensis*	Yes	–
Common dolphin	*Delphinus delphis*	Yes	–
Bottlenose dolphin	*Tursiops truncatus*	Yes	Yes
Spotted dolphin	*Stenella attenuata*	Yes	Yes
Spinner dolphin	*Stenella longirostris*	Yes	–
Striped dolphin	*Stenella coeruleoalba*	Yes	Yes
Rough toothed dolphin	*Steno bredanensis*	Yes	Yes
Risso's dolphin	*Grampus griseus*	Yes	–
Beaked whale	*Mesoplodon spp.*	Yes	–
Cuvier's beaked whale	*Ziphius cavirostris*	Yes	–
Sperm whale	*Physeter macrocephalus*	Yes	Yes
Dwarf sperm whale	*Kogia sima*	Yes	Yes
Fin whale	*Balaenoptera physalus*	Yes	–
Sei whale	*Balaenoptera borealis*	Yes	–
Bryde's whale	*Balaenoptera edeni*	Yes	Yes
Humpback whale	*Megaptera novaeangliae*	Yes	Yes

etiological agents like fungi, protozoans and bacteria. More exhaustive studies are needed to rule out viral agents like herpes and pox virus. These lesions appear to be traumatic in origin but the exact etiology has not been confirmed. There is no tendency for the presentation of these scars and active lesions between gender, age or time of the year of the stranding.

In Costa Rica the incidence of *Brucella ceti* in stranded cetaceans, particularly striped dolphins is considered high. Brucellosis is a zoonotic disease caused by bacteria of the *Brucella* genera. Traditionally, this bacterial disease has been associated in terrestrial animals such as ruminants, with reproductive problems such as abortions, testicular inflammation, lower milk productivity, and partial or complete infertility. Marine brucellosis was first described affecting stranded cetaceans in 1996 in the Scottish coast in United Kingdom (Foster et al. 1996). First, this bacterium was named as *Brucella maris*, then *B. cetaceae* and by 2007, *B. ceti* was recognized as a novel species of the genera affecting cetaceans. In 2004, this bacterial disease was identified as an important causing agent of meningoencephalomyelitis in striped dolphins stranded along the Pacific coast of Costa Rica. Isolation of *B. ceti* was first obtained in the country from a stranded striped dolphin in 2006 from cerebrospinal fluid (CSF), but the presence of this bacterium since then, has been also demonstrated from placenta, umbilical cord, amniotic and allantoic fluids, multiple fetal organs, milk, cardiac valve, atlanto occipital joint fluid, lung and lung nematodes (*Halocercus spp.*) (Hernández-Mora et al. 2008; González-Barrientos et al. 2010; Oliveira et al. 2011; Guzmán- Verri et al. 2012; Hernández-Mora et al. 2013). Serological diagnoses was first described under field and laboratory conditions, using Rose Bengal test as screening assay, (Rapid slide agglutination test, ID VET -France) and confirmed using an indirect ELISA designed for diagnosis of brucellosis in toothed-whales (Hernández-Mora et al. 2009). From 2004 to 2016, six stranded cetacean species in Costa Rica have been described as seropositive for this bacterial disease including striped and spotted dolphins, common dolphins, rough-tooted dolphins, and Cuvier's beaked whale. However, striped dolphins are the only species for which neurobrucellosis was diagnosed as the related cause of death and for which *B. ceti* was isolated (Hernández-Mora et al. 2013).

The most relevant macroscopic findings observed in striped dolphins infected with *B. ceti* include meningeal hyperemia and elevated levels of cerebrospinal fluid, resulting in secondary hydrocephalus in some cases, generalized hyperplastic lymphadenopathy and less frequently valvular endocarditis, placentitis and atlanto-occipital and scapulo-humeral osteoarthritis (González-Barrientos et al. 2010). Central nervous system histopathological lesions observed in striped dolphins with neurobrucellosis consist in a moderate to severe diffuse non-suppurative meningo-encephalomyelitis. These lesions are especially severe on basal structures of the brain like spinal cord, medulla oblongata and pons and less severe in upper regions like the thalamus and cerebral cortex.

Parasites in cetaceans are relatively common and some have been considered a cause of cetacean strandings (Gibson et al. 1998). In 2007, *Taxoplasmosis gondii* was isolated from a stranded striped dolphin found in the Pacific Coast of Costa Rica. This discovery suggest that cetaceans can become infected when this parasite is washed from land to the sea (Dubey and Beattie 1988). In a systematic survey of the parasitic fauna of stranded cetaceans of Costa Rica by Oliveira et al. (2011) it was found that parasites was present primarily in striped dolphins with a prevalence of parasites of 89.5%. Other cetaceans such as bottlenose dolphins, spinner dolphins, pantropical spotted dolphins and Cuvier's beaked whale also presented parasitosis. Fourteen helminth taxa were found and morphologically identified, including six species of cestodes, four nematodes and four digeneans. The nematodes and cestodes were the most prevalent parasite groups (90.9%), followed by the digeneans (22.7%). The most prevalent species were *Anisakis spp.* (90.9%), followed by *Tetrabothrius forsteri* (63.6%), *Halocercus lagenorhynchi* (54.5%) and *Tetraphyllidean plerocercoids* (40.9%). Additionally, *Xenobalanus globicipitis* a commensal crustacean species was also identified (Oliveira et al. 2011).

The pathologies related to the different parasitic infection within these animals include four infected cetaceans with *Anisakis* spp. presented mild gastritis. Moderate to severe granulomatous pneumonia was described in a striped dolphin infected with *Halocercus lagenorhynchi* and spotted and spinner dolphins were infected with *Halocercus* sp. Necrosis, fibrosis and partial destruction of the kidneys was found of Cuvier's beaked whale with *Crassicauda anthonyi,* similar to reported in Australia and Puerto Rico (Robson 1984; Mignucci-Giannoni et al. 1998). Because the related tissue damage demonstrated in crassicaudosis, this is a parasitic disease threatening the health and population recovery of cetaceans (Lambertsen 1992).

In addition to brucellosis and parasites, viruses can be an important cause of cetacean strandings. Among the most important viruses is the cetacean morbillivirus (CeMV) considered as the most pathogenic virus in these animals (Bellière et al. 2011).The reason for this is that CeMV may trigger high mortality events of cetaceans characterized by pneumonia, nonsuppurative meningo-encephalitis and prominent lymphoid cell depletion (Domingo et al. 1992). Strandings caused by CeMV were particularly dramatic between 1990 and 1992 in the Western Mediterranean cetacean species (mainly striped dolphins) and in the east coast of United States from 2010 to 2016 (Colegrove et al. 2016). In Costa Rica CeMV has not been detected yet.

In Panama, regular documentation of cetacean stranding initiated in 2009. In the Panamanian Pacific coast the most common species known to strand are the humpback whale, bottlenose dolphin, pantropical spotted dolphin, and sperm whale. Other species that have stranded include the dwarf sperm whale, false-killer whale, Bryde's whale, striped dolphins and rough-tooted dolphins. Strandings of dwarf sperm whales were associated to parasite infections in kidneys by *Crassicauda anthonyi* and stomach parasites by *Anisakis*, and in the muscle by *Diphilobrotium*. In 2016, a mass stranding of rough-tooted dolphins of 60 animals occurred in the Península of Azuero. The cause is suspected to be zoonotic diseases but not lab analysis were made to confirm it, however, parasite of the genus *Anisakis* where found in their stomach. In the Caribbean coast of Panama, the most common stranded species is the bottlenose dolphin, due primarily to collisions with dolphin-watching boats (Trejos-Lasso and May-Collado 2015). A sperm whale stranded on the Caribbean entrance to the Panama Canal also died due to boat collision (Trejos-Lasso unpublished data).

In recent years, Panamanian scientists have received training courses including one organized by the International Whaling Commission. About 40 people has been trained to assist in cetacean stranding and rescue events. Gear and tools have been donated to attend also whales trapped in fishing nets with the support of the National Coast Guard of Panama, and a protocols are being develop to create a national network that can respond to individual and mass strandings.

12.4 Conservation Threats and Concerns

Conservation efforts to protect cetaceans in Costa Rica and Panama have grown in the past 10 years largely due to increasing economic value of these animals in touristic activities, which consequently prompted the creation of guidelines to protect them. In addition, a number of non-profit organizations have developed workshops and training opportunities for tour operators and coastal communities, encouraging compliance to conduct guidelines. However, *in situ* regulation remains scarce resulting in the 'overuse' of some populations as is the case of the bottlenose dolphins of Bocas del Toro.

In both countries a number of marine protected areas have been created and some expanded. Many of these areas cover a relatively small area given the ranges of most local cetaceans and there is little to non-vigilance to enforce their protection. This is an important issue that needs to be addressed through education of neighboring communities and by funneling more resources for vigilance. In this respect, Panama faces many challenges when it comes to whale watching activities. While lack of compliance in some areas is due to operators limited opportunities for training and education (Sitar et al. 2016) in others operators purposefully ignored regulations, like those that sell tours to swim with humpback whales and dolphins, an activity prohibited in the country.

Although, scientific studies have increased in the past 20 years they remained bias to a few species and limited in space and time. Long-term data such as that of the Gulf of Chiriquí, Bocas del Toro, and the Osa Peninsula have been key to help us asses the demographic trends of these populations as well as any migratory shifts that may occur. Both of these aspects could be affected by climate change, or anthropogenic pressures. For example the Gulf of Chiriquí is an important nursery area for humpback whales with a high proportion of calves seen every year, ensuring that the proper conservation measures are taken will be crucial to the continued recovery of this humpback whale population, as well as the other species seen in these waters.

In Panama, southern humpback whales that travel by the Canal of Panama may be vulnerable to ship collisions. Guzman et al. (2013) using satellite data from adults and calves found that their movement coincides with major commercial maritime routes. In contrasts, humpback whales that use the Gulf of Chiriqui face other more important threats such as entanglement of fishing gear, vessel traffic, pollution and other anthropogenic pressures. Unlike the Gulf of Panama, this area is not used by large vessels transiting through the Panama Canal, so large ship strikes are not considered a major issue (Guzman et al. 2013). However, commercial fishing vessels are found in these waters as well as sport fishing and other pleasure craft, including a small but growing whale-watching industry. In the Osa Peninsula, Costa Rica the non-profit organization Keto has been monitoring touristic activities in the area for over a decade. Their research finds that since diving and whale watching activities were incorporated to local tour offers, the revenue of these communities went up 500%. This rapid economic growth inevitable translates into a larger tourism boat fleet that without control can significantly change the acoustic space of cetaceans.

For this reason the incorporation of acoustic monitoring to current cetacean survey protocols in Costa Rica and Panama can significantly improve population assessments and evaluate the impact of boat traffic. In an ongoing study established by Conservation International Costa Rica, acoustic monitoring is providing novel information about cetacean presence, habitat use, and daily activity, in relation to boat traffic. This type of assessments are key particularly in countries like Costa Rica and Panama where resources are limited. Acoustic monitoring is proven cost efficient methods that ensures large amounts of data of high quality.

In addition to long-term acoustic monitoring, incorporating molecular data to monitoring protocols is urgently needed. The population genetic studies by Barragan-Barrera et al. (2015) in Bocas del Toro, highlights the importance of such information. Her research indicates that this small bottlenose dolphin population is highly isolated form others in the Caribbean, and thus should be treated it as a separate conservation unit, and likely considered threatened in Panama. Another example is the case of coastal spotted dolphins. Escorza-Trevino et al. (2002) genetic work showed significant differences between Costa Rican spotted dolphins and those from the rest of Central America. Molecular data is relevant for conservation and management strategies, especially for cetaceans, which habitat overlaps with important fishing areas and tourism. Finally, we need infrastructure and capacity building for our stranding networks. As described in the stranding section, most strandings are associated with pathogens highlighting the importance of trained scientists, access to laboratories, and novel methods that can provide a better understanding of the causes of death in both countries. Helping to produce conservation policies that reflect cetacean population's health status.

While Costa Rica and Panama are leading the way in marine mammal research and conservation efforts in Central America, the lack of resources and funding can quickly hamper these efforts. Our small scientific community worries about the impact of rapidly growing boat-based whale watching industry and associated noise along our coasts, and how increasing noise levels can affect the soundscape of important habitats such as breeding grounds for migratory whales. We also worry about the lack of estimations of local cetacean mortality in fishing nets and direct killing (for shark bait) and, the effect that global warming will have in tropical habitats. Finally, we worry about the limited sources for funding whether governmental or private that allow us to continue filling up the gaps on the basic ecology of these animals, this information is the basis to make the right decisions on management and policy making.

References

Acevedo A, Burkhart S (1998) Seasonal distribution of bottlenose (*Tursiops truncatus*) and pantropical spotted (*Stenella attenuata*) dolphins (Cetacea: Delphinidae) in Golfo Dulce, Costa Rica. Rev Biol Trop 46: 91–101

Acevedo A, Smultea ME (1995) First records of humpback whales including calves at Golfo Dulce and Isla del Coco, Costa Rica, suggesting geographical overlap of northern and southern hemisphere populations. Mar Mamm Sci 11:554–560

Acevedo J, Rasmussen K, Félix F, Castro C, Llano M, Secchi E, Saborío MT, Aguayo-Lobo A, Haase B, Scheidat M (2007) Migratory destinations of humpback whales from the Magellan Strait feeding ground, Southeast Pacific. Mar Mamm Sci 23:453–463

Acevedo-Gutierrez A, DiBerardinis A, Larkin S, Larkin K, Forestell P (2005) Social interactions between Tucuxis and bottlenose dolphins in GandocaManzanillo, Costa Rica. Lat Am J Aquat Mamm 4:49–54

Barragán-Barrera DC, May-Collado LJ, Islas-Villanueva V, Caballero S (2015) Isolated in the Caribbean: low genetic diversity of a bottlenose dolphin in Bocas del Toro, Panama. International Whaling Commission. SC/66a/SM/13

Bellière EN, Esperón F, Fernández A, Arbelo M, Muñoz MJ, Sánchez-Vizcaíno JM (2011) Phylogenetic analysis of a new cetacean morbillivirus from a short-finned pilot whale stranded in the Canary Islands. Res Vet Sci 90:324–328

Best PB (2008) Nineteenth-century evidence for the Golfo de Panama as a migratory destination for southern humpback whales, including the first mention of singing. Mar Mamm Sci 24:737–742

Birnie P (1985) The role of developing countries in nudging the international whaling commission from regulating whaling to encouraging non-consumptive uses of whales. Ecology LQ 12:937–975

Bossenecker PJ (1978) The capture and care of *Sotalia guianensis*. Aquat Mamm 6:13–17

Calambokidis J, Steiger GH, Rasmussen K, Urban RJ, Balcomb KC, Guevara P, Salinas ZM, Jacobsen JK, Baker CS, Herman LM, Cerchio S, Darling JD (2000) Migratory destinations of humpback whales that feed off California, Oregon and Washington. Mar Ecol Prog Ser 192:295–304

Colegrove KM, Venn-Watson S, Litz J, Kinsel MJ, Karen AT, Fougeres E et al (2016) Fetal distress and in utero pneumonia in perinatal dolphins during the Northern Gulf of Mexico unusual mortality event. Dis Aquat Org 119:1–16

Cubero-Pardo P (1998) Patrones de comportamiento diurnos y estacionales de *Tursiops truncatus* y *Stenella attenuata* (Mammalia: Delphinidae) en el Golfo Dulce, Costa Rica. Rev Biol Trop 46: 103–110

Domingo M, Visa J, Pumarola M, Marco AJ, Ferrer L, Rabanal R, Kennedy S (1992) Pathologic and immunohistochemical studies of morbillivirus infection in striped dolphins (*Stenella coeruleoalba*). Vet Pathol 29:1–10

Dubey JP, Beattie CP (1988) Toxoplasmosis of animals and man. CRC Press, Boca Raton, 220 p

Escorza-Trevino S, Archer FI, Rosales M, Lang A, Dizon AE (2005) Genetic differentiation and intraspecific structure of eastern tropical Pacific spotted dolphins, Stenella attenuata, revealed by DNA analyses. Conserv Genet 6:587–600

Florez-González L, Capella J, Haase B, Bravo GA, Félix F, Gerrodette T (1998) Changes in winter destinations and the northernmost record of Southeastern Pacific humpback whales. Mar Mamm Sci 14:189–196

Forestell PH, Wright AM, DiBerardinis A, Larkin S, Schot V (1999) Sex and the single Tucuxi: observations of interspecies mating between free-ranging bottlenose (Tursiops truncatus) and tucuxi (Sotalia fluviatilis) dolphins in Costa Rica. In: 13th biennial conference on the biology of marine mammals, November 28–December 3, 1999

Foster G, Jahans KL, Reid RJ, Ross HM (1996) Isolation of *Brucella* species from cetaceans, seals and an otter. Vet Rec 138:583–586

Gamboa-Poveda MP (2009) Tamaño poblacional, distribución y uso de hábitat de dos especies simpátricas de delfines en el refugio nacional de vida silvestre Gandoca-Manzanillo, Costa Rica. Universidad Nacional de Costa Rica, Tesis de Maestria, 78pp

Gamboa-Poveda MP, May-Collado LJ (2006) Insights on the occurrence, residency, and behavior of two coastal dolphins from Gandoca-Manzanillo, Costa Rica: Sotalia guianensis and Tursiops truncatus (Family Delphinidae). International Whaling Commission. SC/58/SM4

Garcia C, Dawson SM (2003) Distribution of Pantropical spotted dolphins in Pacific coastal waters of Panama. Lat Am J Aquat Mamm 2:29–38

Gibson DI, Harris EA, Bray RA, Jepson PD, Kuiken T, Baker JR, Simpson VR (1998) A survey of the helminth parasites of cetaceans stranded on the coast of England and Wales during the period 1990–1994. J Zool 244:563–574

González-Barrientos R, Morales JA, Hernández-Mora G, Guzmán Verri C, Chaves Olarte E, Moreno E (2010) Pathology of striped dolphins (*Stenella coeruleoalba*) infected with Brucella ceti. J Comp Pathol 142:347–352

Guzmán-Verri C, González-Barrientos R, Hernández-Mora G, Morales JA, Baquero-Calvo E, Chaves-Olarte E, Moreno E (2012) *Brucella ceti* and brucellosis in cetaceans. Front Cell Infect Microbiol 2:3

Guzman HM, Gomez CG, Guevara CA, Kleivane L (2013) Potential vessel collisions with Southern Hemisphere humpback whales wintering off Pacific Panama. Mar Mam Sci 29:629–642
Hernández-Mora G, González-Barrientos R, Morales JA, Chaves-Olarte E, Guzmán-Verri C, Baquero-Calvo E, De-Miguel MJ, Marín CM, Blasco JM, Moreno E (2008) Neurobrucellosis in stranded dolphins, Costa Rica. Emerg Infect Dis 14:1429–1433
Hernández-Mora G, Manire CA, González-Barrientos R, Baquero-Calvo E, Guzmán-Verri C, Chaves-Olarte E, Moreno E (2009) Serological diagnosis of *Brucella* infections in Odontocetes. Clin Vaccine Immunol 16:906–915
Hernández-Mora G, Palacios-Alfaro JD, González-Barrientos R (2013) Other wildlife reservoirs: brucella in aquatic environments. Rev Sci Tech Off Int Epiz (OIE) 32:89–103
Herra-Miranda D, Pacheco-Polanco JD, Oviedo L, Iniguez M (2016) Spatial analysis of the critical habitats of bottlenose dolphins (*Tursiops truncatus*) and humpback whales (*Megaptera novaeangliae*) in Golfo Dulce, Costa Rica: considerations for a marina construction project. Rev Mar Cost 8:9–27
Hoyt E, Iñíguez M (2008) The State of Whale Watching in Latin America. WDCS, Chippenham, UK; IFAW, Yarmouth Port, USA and Global Ocean, London, 60pp
INCOPESCA (2016) Estadísticas de desembarcos periodo 2014. Disponible en Setiembre 2016
Janzen DH, Wilson DE (1983) Mammals. In: Janzen DH (ed) Costa Rican natural history. Chicago University, Chicago, pp 439–456
Kassamali-Fox A, Christiansen F, Quinones-Lebron S, Rusk A, May-Collado LJ, Kaplin B (2015) Using Markov chains to model the impacts of the dolphin watching industry on the dolphin community of Dolphin Bay, Bocas del Toro, Panama. International Whaling Commission. SC/66a/WW11
Kiernan JA (2003) Histological and histochemical methods: theory and practice. Oxford University Press, New York, p 1e477
Lambertsen RH (1992) La crassicaudose: une maladie parasitaire menacant la sante et la reconstitutution despopulation de gros mysticetes. Rev Sc et techn 11:1131–1141
Loaiza VN (2002) Buscan causa de delfines varados. La Nación, San José Costa Rica 2 April
Martinez-Fernandez D, Montero-Cordero A, May-Collado LJ (2011) Cetaceos de las aguas costeras del Pacifico norte y sur de Costa Rica. Rev Biol Trop 59:283–290
Martinez-Fernandez D, Montero-Cordero A, Palacios-Alfaro D (2014) Areas de congregación de cetáceos en el Pacifico norte de Costa Rica: recomendaciones para la gestión del recurso. Rev Biol Trop 62:99–108
May-Collado LJ (2001) Ecología y Comportamiento del delfín manchado costero (*Stenella attenuata graffmani*) del Pacífico Norte de Costa Rica. Universidad de Costa Rica, Tesis de Maestria, 78 p
May-Collado LJ (2009) Chapter 42: marine mammals. In: Wehrmann I, Cortes J (eds) Marine biodiversity of Costa Rica, Central America. Springer Academic Press, NewYork
May-Collado LJ (2010) Changes in whistle structure of two dolphin species during interspecific associations. Ethology 116:1065–1074
May-Collado LJ (2015) The acoustic behavior of Bocas del Toro dolphins varies with watercraft activity International Whaling Commission. SC/66a/WW6
May-Collado LJ, Forcada J (2012) Small-scale estimation of relative abundance for the coastal spotted dolphins (*Stenella attenuata*) in Costa Rica: the effect of habitat and seasonality. Rev Biol Trop 60:133–142
May-Collado LJ, Morales Ramírez A (2005) Presencia y Patrones de comportamiento del delfín manchado costero, *Stenella attenuata graffmani* (Cetacea: Delphinidae) en el Golfo de Papagayo, Costa Rica. Rev Biol Trop.l. 53:265–276
May-Collado LJ, Quinones-Lebron S (2014) Dolphin changes in whistle structure with watercraft activity depends on their behavioral state. J Acoust Soc Am EL 135:193–198
May-Collado LJ, Wartzok D (2008) A comparison of bottlenose dolphin whistle in the Western Atlantic Ocean: insights on factors promoting whistle variation. J Mammal 89:205–216

May-Collado LJ, Wartzok D (2015) The effect of dolphin watching boat noise levels on the whistle acoustic structure of dolphins in Bocas del Toro, Panama. International Whaling Commission 2015. SC/66a/WW5

May-Collado LJ, Gerrodete T, Calambokidis J, Rassumen K, Sereg I (2005) Distribution of Cetaceans sightings in the EEZ of Costa Rica. Rev Biol Trop 53:249–263

May-Collado LJ, Barragán-Barrera DC, Quiñones-Lebrón SG, Aquino-Reynoso W (2012) Dolphin watching boats impact on habitat use and communication of bottlenose dolphins in Bocas delToro, Panama during 2004, 2006–2010. International Whaling Commission. SC/64/WW2

May-Collado LJ, Quiñones-Lebrón SG, Barragán-Barrera DC, Palacios JD, Gamboa-Poveda MP (2014) The dolphin watching industry of Bocas del Toro continues impacting the resident bottlenose dolphin population. International Whaling Commission, Slovenia. SC/65b/WW06

May-Collado LJ, Quiñones-Lebrón SG, Barragán-Barrera DC, Palacios JD, Gamboa-Poveda MP, Kassamali-Fox A (2015) The Bocas del Toro dolphin watching industry relies on a small community of bottlenose dolphins: implications for management. International Whaling Commission. SC/66a/WW10

Montero-Cordero A, Lobo J (2010) Effect of tourist vessels on the behaviour of the pantropical spotted dolphin, *Stenella attenuata*, in Drake Bay and Cano Island, Costa Rica. J Cetacean Res Manag 11:285–291

Morisaka T, Shinohara M, Nakahara F, Akamatsu T (2005) Effects of ambient noise on the whistles of Indo-Pacific bottlenose dolphin populations. J Mammal 86:541–546

Nowacek SM, Wells RS, Solow AR (2001) Short-term effects of boat traffic on bottlenose dolphins, *Tursiops truncatus*, in Sarasota Bay Florida. Mar Mamm Sci 17:673–688

Oliveira JB, González-Barrientos RC, Hernández-Gamboa J, Hernández-Mora G (2011) Parasites of cetaceans stranded on the Pacific coast of Costa Rica. Vet Parasitol 182:319–328

Oviedo L (2007) Dolphin sympatric ecology in a tropical fjord: habitat partitioning by bathymetry and topography as a strategy to coexist. J Mar Biol Assoc UK 87:1327–1335

Oviedo L, Herra-Miranda D, Pacheco-Polanco JD, Figgener C, Marquez-Artavia A, Iniguez M (2014) Diversidad de cetáceos en el paisaje marino costeros de Golfo Dulce, Peninsula de Osa, Costa Rica. Rev Biol Trop 63:395–406

Palacios DM, Gerrodette T (1996) Potential impact of artisanal gillnet fisheries on small cetacean populations in the Eastern Tropical Pacific. National Marine Fisheries Service Southwest Fisheries Science Center, Administrative Report LJ-96–11.16 p

Palacios-Alfaro JD (2007a) Presencia y comportamiento de dos especies de delfines en el Pacifico. Central de Costa Rica. Tesis de Bachillerato. Universidad Nacional de Costa Rica, 70pp

Palacios-Alfaro JD (2007b) Fish selection of Pantropical spotted dolphins (*Stenella attenuata*) and Bottlenose dolphin (*Tursiops truncatus*) when interacting with deep shrimp trawlers in the central Pacific of Costa Rica. In: 17th biennial conference on the biology of marine mammals. Cape Town, 29 November–3 December 2007

Palacios-Alfaro JD, Martinez Fernandez D, Sanchez Godinez C, Venegas R. (2014) Distribution and behavior of humpback whales (*Megaptera novaeangliae* Borowski, 1781) (Breeding stock G) in southern Pacific of Costa Rica. International Whaling Commission. SC/64/SH16

Perrin WF, Caldwell DK, Caldwell MC (1994) Atlantic spotted dolphin – Stenella frontalis (Gray, 1828). In: Ridgway SH, Harrison SR (eds) Handbook of marine mammals, The first book of dolphins, vol 5. Academic Press, London, pp 173–190

Quintana-Rizzo E (2014) Harpooning and entanglement of wild dolphins in the Pacific coast of Guatemala. Lat Am J Aquat Mamm 9:179–182

Rasmussen K, Palacios DM, Calambokidis J, Saborío MT, Dalla Rosa L, Secchi ER, Steiger GH, Allen JM, Stone GS (2007) Southern hemisphere humpback whales wintering off Central America: insights from water temperature into the longest mammalian migration. Biol Lett 3:302

Rasmussen K, Calambokidis J, Steiger GH (2012) Distribution and migratory destinations of humpback whales off the Pacific coast of Central America during the boreal winters of 1996–2003. Mar Mamm Sci 28:E267–E279

Rodríguez-Fonseca J (2001) Diversidad y distribución de los Cetáceos de Costa Rica (Cetacea:Delphinidae, Physeteridae, Ziphiidae y Balaenopteridae). Rev Biol Trop 49(Supl. 2):135–143
Rodríguez-Fonseca J, Cubero-Pardo P (2001) Cetacean strandings in Costa Rica (1966–1999). Rev Biol Trop 49:667–672
Sitar A, May-Collado LJ, Wright AJ, Peters-Burton E, Rockwood L, ECM P (2016) Boat operators in Bocas del Toro, Panama display low levels of compliance with national whale-watching regulations. Mar Policy 68:221–228
Steiger GH, Calambokidis J, Sears R, Balcomb KC, Cubbage JC (1991) Movement of humpback whales between California and Costa Rica. Mar Mamm Sci 7:306–310
Townsend CH (1935) The distribution of certain whales as shown by logbook records of American whaleships. Zoologica 19:1–50
Trejos-Lasso L, May-Collado LJ (2015) Bottlenose dolphins tursiops truncatus strandings in Bocas del Toro caused by boat strikes and fishing entanglement. International Whaling Commission. SC/66a/WW7
Vargas JA (1995) The Gulf of Nicoya Estuary, Costa Rica: past, present, and future cooperative research. Helgoländer Meeresun 49:821–828

Chapter 13
Anthropogenic Noise and Guiana Dolphins (*Sotalia guianensis*) in Brazil: Ecological and Conservation Concerns

Fábio de Sousa Pais, Rafaela Prado Cardoso, Marcos R. Rossi-Santos, Leonardo L. Wedekin, Flávio J.L. Silva, Emygdio L.A. Monteiro-Filho, and Dalila Teles Martins Leão

Abstract The *Sotalia guianensis* dolphin occurs in the western Atlantic Ocean, from the south of Brazil to Central America. It is a coastal species, much susceptible to human impacts, as anthropogenic noises, which in general causes potential impacts in

F. de Sousa Pais (✉)
Instituto de Biologia, Universidade Federal da Bahia,
Rua Barão de Geremoabo, 147, Campus de Ondina, 40170-290 Salvador, Bahia, Brazil
e-mail: paisfs@gmail.com

R.P. Cardoso
Instituto Oceanográfico, Universidade de São Paulo, Praça do Oceanográfico,
191, Cidade Universitária, 05508-120, São Paulo, São Paulo, Brazil

M.R. Rossi-Santos
Instituto de Biologia, Universidade Federal da Bahia,
Rua Barão de Geremoabo, 147, Campus de Ondina, 40170-290 Salvador, Bahia, Brazil

Centro de Ciências Agrárias, Ambientais e Biológicas, Universidade Federal Recôncavo Baiano,
Rua Rui Barbosa, 710, Campus Universitário, 44380-000 Cruz das Almas, Bahia, Brazil

L.L. Wedekin
Departamento de Ecologia, Universidade de São Paulo,
Rua do Matão, 321, Cidade Universitária, 05508-090, São Paulo, São Paulo, Brazil

F.J.L. Silva
Departamento de Turismo, Universidade Federal do Rio Grande do Norte,
Avenida Airton Senna, 4241, Neópolis, Natal, Rio Grande do Norte, Brazil

E.L.A. Monteiro-Filho
Instituto de Pesquisas Cananéia, IPEC, Rua Tristão Lobo, 199, Centro, 11990-000
Cananéia, São Paulo, Brazil

Departamento de Zoologia, Universidade Federal do Paraná,
Centro Politécnico, Jardim das Américas, 8153-1970 Curitiba, Paraná, Brazil

D.T.M. Leão
Departamento de Fisiologia, Universidade Federal do Rio Grande do Norte,
Avenida Senador Salgado Filho, sn, Campus Universitário Lagoa Nova, 59078-970 Natal,
Rio Grande do Norte, Brazil

M.R. Rossi-Santos, C.W. Finkl (eds.), *Advances in Marine Vertebrate Research in Latin America*, Coastal Research Library 22,
DOI 10.1007/978-3-319-56985-7_13

marine mammals' populations, because the emission and reception of sound signals plays a complex and vital role in this populations' ecology through many different ways. This study reviews the acoustic repertoire and potential anthropic impacts on populations of Guiana dolphin (*S. guianensis*) through its occurrence, and described and compared the anthropogenic noises with the acoustic repertoire of three Brazilian populations: Cananéia estuary – São Paulo State, Caravelas estuary – Bahia State and Pipa beach – Rio Grande do Norte State. Data was collected similarly in these three sites during boats surveys using hydrophones and audio recorders. Results showed overlapping of the *S. guianensis* repertoire and noises produced from boats, dredge activities and wood-barges, leading to harm on individuals and populations. This review has gathered important information supporting marine activity management, and can be used to evaluate acoustic impacts over cetaceans population.

Keywords *Sotalia guianensis* • Guiana dolphin • Soundscape • Acoustic • Anthropogenic noise • Cetaceans

13.1 Introduction

Aquatic ecosystems are very sensitive environments, with diversity and abundance of interdependent organisms. Oceanic, coastal and estuarine areas suffer intense pressure from anthropogenic activities, therefore the importance of better understanding the impacts in these ecosystems, with the purpose of assisting in the management guidelines and conservation issues, so that these environments are able to maintain their natural ecological balance.

Acoustic ecology studies the relationship of sounds between organisms and individuals, and between organisms and the environment (Erbe et al. 2016). The term "noise" refers to diffuse sounds that don't provide significant biological information. Sounds emitted by aquatic mammals are called "acoustic signals", as they contain specific information. The raising number of sound sources in marine waters, with wide range of frequencies and intense energy levels, lead to ecosystems with greater ambient noise.

The variability of sound sources results in acoustic environments with singular characteristics in different space-time scales that are denominated "soundscapes" (Pijanowski et al. 2011). They can also be interpreted as the acoustic environments of aquatic ecosystems, composed by sounds of natural (biotic and abiotic) or anthropic sources. Examples of natural sound sources are: earthquakes, wind, rain, waves, fishes, shrimps, etc. Anthropic sound sources occur as result of several human activities, such as: shipping, boat traffic, dredging, wind turbines, sonars, oil and gas exploitation, port activities and seismic activities (Richardson et al. 1995; Popper and Hawkins 2016).

Such anthropogenic noise can potentially cause negative impacts on aquatic mammal populations, since the emission and detection of sound signals play a complex and vital role in their ecology through various aspects, such as group cohesion, individual recognition, escape from predators, prey detection and reproductive behavior (Stewart 2008; Tyack 2000a, b). Thus, underwater noises are able to prevail over the sounds emitted by aquatic mammals, becoming a threat for individuals and population.

Cetaceans are aquatic mammals with complex social structures that have large cognitive abilities. Exploitation of marine resources by anthropogenic activities have the potential to affect aquatic organisms in many ways, interfering with the maintenance of social relations and the balance of ecosystems, leading to population declines and, in extreme cases, extinction of species (Whitehead et al. 2000).

A methodology that has been growing in a scientific outlook is the study of sounds in aquatic environments, bringing new ways to research about animals' life cycle, social behavior and interactions with the whole ecosystem.

The growing interest and funding of research about the sound in the marine environment are of great importance for achieving sustainable management of ocean especially in coastal regions (Erbe et al. 2016). Bioacoustics is a field of study dedicated to understand the mechanisms of sound production and perception of animals and how these animals communicate in their acoustic habitats.

The application of bioacoustics techniques in behavioral ecology and Cetacean conservation studies are recent in Brazil, having started in the 1990's, but only after the year 2000 studies began to be published (*eg* Monteiro-Filho and Monteiro 2001). Most studies in Brazil have focused on the Guiana dolphin (*Sotalia guianensis*), the principal species aimed in this study.

13.1.1 Guiana Dolphin

The dolphin *Sotalia guianensis* (Van Bénedén 1864) is a small Cetacean, from the Order: Cetartiodactyla (Suborder: Odontoceti, Family: Delphinidae), popularly known in the Brazilian coast as *'boto'*, '*golfinho'*, '*boto-comum'*, '*golfinho-cinza'* or '*boto-cinza'* (Flores et al. 2010) (Fig. 13.1).

Fig. 13.1 *Sotalia guianensis*, often observed in pairs in the residence area of Pipa beach, Rio Grande do Norte State, Brazil. Like other cetaceans, they have a highly developed sense of hearing and depend greatly on it for feeding, to communicate and to locate obstacles and are therefore particularly sensitive to disturbing noises (Photo: Marcos Rossi-Santos)

Fig. 13.2 On photo-identification studies, photos capturing details like format, scars and size of the triangular dorsal fin of *Sotalia guianensis* can distinguish each one of the individuals of a population (Photo: Marcos Rossi-Santos)

Previously, the genus *Sotalia* was monospecific, *Sotalia fluviatilis*, represented by animals that inhabit fluvial areas, in the Amazon region, and those that occur along the coast of South and Central America. The taxonomy of the species was revised and through tridimensional morphometric analysis (Monteiro-Filho et al. 2002) and genetic analysis from mitochondrial DNA (Cunha et al. 2005), recently there was split into two species. Currently the fluvial dolphin is categorized as *Sotalia fluviatilis* and the marine dolphin as *Sotalia guianensis*.

Guiana dolphins present longevity about 30 years (Rosas et al. 2003), are shorter than 220 cm (Flores and da Silva 2009) and show no apparent sexual dimorphism (Flores 2002). Its dorsal portion is described with grayish and the ventral can vary between pink and white (da Silva and Best 1996), and its dorsal fin has a triangular shape and small size, typical of the species (da Silva and Best 1996) (Fig. 13.2).

Its distribution ranges from the State of Santa Catarina (27° 35′ S), Brazil (Simões-Lopes 1988) to Nicaragua (14° 35′ N) (Carr and Bonde 2000) and possibly Honduras (da Silva and Best 1996). In the Brazilian coast, the occurrence of Guiana dolphin has been recorded since the 1980s and 1990s, through stranding (*eg* Geise and Borobia 1987, 1988; Simões-Lopes 1987) and visual observations (*eg* Bittencourt 1984; Andrade et al. 1987; Simões-Lopes 1988; Simões-Lopes and Ximenez 1990).

This species inhabits preferably coastal and protected areas such as estuaries and bays, usually shallow regions with presence of mangroves, these considered to be high primary production environments (Wedekin et al. 2004; Monteiro-Filho et al. 2008; Cremer et al. 2012). The diet of *S. guianensis* includes fish, squid and shrimp, although with variations along the Brazilian coast according to the environment inhabited by the animal (Daura-Jorge et al. 2011). The most common described prey species are schools of fish like *Sardinella brasiliensis, Trichiurus lepturus* and *Micropogonias furnieri* (Rosas et al. 2010).

The gestation time of the Guiana dolphin is about 11–12 months for only one calf. The calves demand long parental care, about 5–6 years, when they begin to

Fig. 13.3 Mother and calf of *Sotalia guianensis* interacting in protected waters of the Caravelas estuary, Bahia State, Brazil. Parental care lasts usually over 5 years (Photo: Marcos Rossi-Santos)

Fig. 13.4 Group of *Sotalia guianensis* surfing a small wave in Pipa beach, Rio Grande do Norte State, Brazil (Photo: Marcos Rossi-Santos)

reach sexual maturity (about 6–7 years old) (Perrin and Reilly 1984; Ramos 1997; Ramos et al. 2000; Rosas 2000) (Fig. 13.3).

They are animals that usually form groups with an average size of 1–30 individuals (Daura-Jorge et al. 2005), although groups of over 200 individuals (Simão et al. 2000.; Flach et al. 2008) and by 400 individuals (Lodi and Hetzel 1998) have already been reported on the Ilha Grande region, Rio de Janeiro State. Large aggregations of this species can be characterized by cooperation for food (Flores and Silva 2009) (Fig. 13.4).

Several studies indicate that *S. guianensis* have a high degree of fidelity to the areas used by them, therefore areas that animals use several times or stay for a long time are usually characterized as breeding grounds and feeding areas (*eg* Simão et al. 2000; Gamboa-Poveda and May-Collado 2006; Araújo et al. 2007; Rossi-Santos et al. 2007; Wedekin 2007; Hardt et al. 2010; Oshima et al. 2010; Rossi-Santos et al. 2010; Batista et al. 2014).

Habitats of Guiana dolphins are close to the coast and their fidelity to these areas make this species vulnerable to anthropogenic impacts, such as: whale watching tours, boat traffic, buildings and coastal works, water pollution, driftnet fishing nets and bycatch (*eg* Simões-Lopes and Ximenez 1990; Wedekin et al. 2002; Valle and Melo 2006; Oliveira 2007; Di Benedito and Rosas 2008; Zappes et al. 2009; Meirelles et al. 2010).

In this context, *S. guianensis* is currently one of the most studied cetacean species in Brazil and since 2014 is classified as "vulnerable" by Instituto Chico Mendes de Conservação da Biodiversidade (ICMBio 2016). Despite this increase in research efforts on this species, it is still classified as '*insufficiently known*' species (*deficient data*) by the International Union for Conservation of Nature and Natural Resources (IUCN 2016).

Many issues still need to be elucidated in the relation of noises from anthropogenic sources and its implications to the life cycle of various aquatic animals. To investigate this, studies are needed on the auditory sensitivity and acoustic repertoire of species, the use of sounds in social, behavioral changes associated with noise, temporary and permanent shifts in hearing thresholds and masking, when the anthropogenic noise, especially at high intensities occur simultaneously in the same frequency bands and time that the acoustic signals of marine animals (Clark et al. 2009).

Thus, this chapter presents a review of the *Sotalia guianensis* and describes the repertoire of populations of this species recorded in three locations along the Brazilian coast (Cananéia estuary – São Paulo State, Caravelas estuary – Bahia State and Pipa beach – Rio Grande do Norte State) and compares them to anthropogenic noises recorded in these areas, bringing discussions on the acoustic ecology of the Guiana dolphin and considerations on conservation.

13.1.2 Marine Acoustic Environment and Sound Propagation

The marine acoustic environment is composed by sounds that are in propagation in an area and may undergo different processes (attenuation and degradation) according to the sound transmission related to oceanographic conditions, hydro and geoacoustics (Erbe et al. 2016). Substrate type, small changes in density gradients mainly due to different temperatures, salinity and pressure determine the stratification and the sound distribution, which can become trapped in isobaric layers and change their characteristics (*eg* Maglio et al. 2015).

Attenuation is a process in which the acoustic signal declines in intensity due primarily to the spherical scattering effect, by the loss of energy during its transmission. Degradation may occur through reverberation, amplitude fluctuations and different attenuations at different frequencies (Rabin et al. 2003).

Anthropogenic alterations in the natural environments such as topography, granulometry, salinity, temperature, and other factors, have the potential to change the propagation characteristics of the acoustic signals by increasing the signals' excess of attenuation or degradation. Therefore, it's extremely important to understand the sound propagation in the specific study site before discussing the spectrum of the soundscape (Erbe et al. 2016).

Experiments on mathematical models simulating the propagation of sound in the ocean at different depths and bottom properties indicated that sound in shallow waters (400 m) generally has intensity levels around 10–12 dB higher than in deep waters (3000 m) and also has lower acoustic signal attenuation rate (less effect of the spherical spreading) (Breitzke and Bohlen 2010).

In more consolidated substrates, *i.e.* superior hardness, the potential noise impact is greater, as in hard bottoms multiple reflections at the water-substrate interface overlap the refractions promoting sound reverberation. Therefore, an intense acoustic signal is potentially harmful especially in cases where the substrate is harder and shallower, because the sound will remain with the same intensity (Breitzke and Bohlen 2010).

A study of Ramos et al. (2012) on the spreading of tonal sounds reproduced in the estuarine environment showed that low frequency signals (60–240 Hz) can be significantly attenuated (reaching up to one fifth of the intensity of the original signal) in small depths sites (~ 5 m). Due to Lloyde mirrors' effect (Urick 1983), based on the reflection of the incident signals at the water-air interface, low frequency sounds (typically generated by ships and large ships) can become very difficult to be heard, hence increasing the risk of aquatic mammals collisions with ships (Gerstein 2002).

13.1.3 Anthropogenic Noise

The steady increase of human activities in oceanic and coastal areas have contributed to the increase of anthropogenic noise in the marine environment, making the acoustic environments gradually more polluted (Andrew et al. 2002; Hildebrand 2009). Continental shelves are regions highly impacted by anthropogenic noise, offshore mainly due to increased seismic and oil activity, and in the coastal area, including bays and estuaries, which are impacted by noises from boats, worksites, industrial and recreational activities (Rossi-Santos and Oliveira 2016).

13.1.3.1 Vessel Traffic Commercial

The traffic of large vessels (Fig. 13.5) every day across the oceans in frequent routes, mainly in the Northern Hemisphere, creates very noisy acoustic environments, making it quite difficult for cetaceans' communication to be effective especially for long

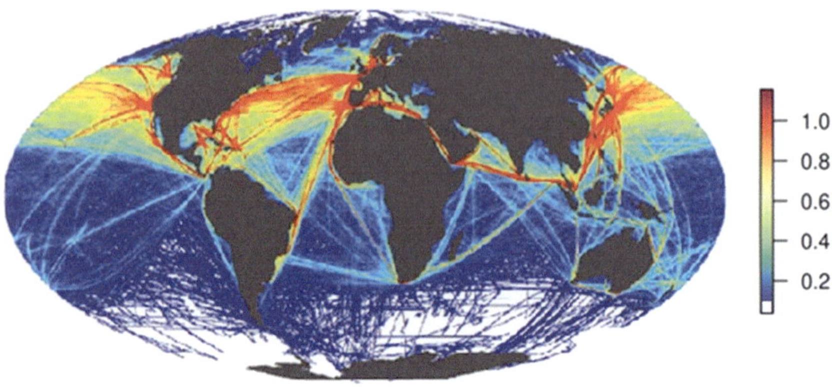

Fig. 13.5 Routes with heaviest traffic, highlighted in *red* in the image are considered to be areas with higher ambient noise due to shipping activity. Scale (0 a 1) reffers to the intensity of the traffic (From: National Center for Ecological Analysis and Synthesis 2016)

distances, as these ships primarily emit low frequencies sounds with high intensities (Southall 2004; Hildebrand 2009; Erbe et al. 2016.)

According to a study of Hester et al. (2008) the ocean noise will intensify as the increasing acidification of the oceans (due to various factors, including the increase of CO_2), causing a reduction on the absorption of low and medium frequency sounds (≤1 kHz to ~10 kHz), allowing it to travel 10–15% further.

The shipping noise is influenced by a combination of vessel size, type of engine, quantity and operation speed (Holt 2008; Houghton et al. 2015.). Reine et al. (2014) measured high levels of intensity produced by ships at the port of New York, the highest value obtained was 188.9 dB re 1 μPa @ 1 m.

The low frequency sounds (below 200 Hz) are generated by the propulsion system (vibration of engine and hull) (Hildebrand 2009), while the majority of medium and high frequencies (above about 2 kHz) are from cavitation, a phenomenon that consists of the formation and explosion of air bubbles due to the accelerated movement of the engine propeller blades (Evans et al. 1992; Leaper et al. 2014).

13.1.3.2 Coastal Boat Traffic

While shipping traffic is the largest global source of noise into the oceans, at a regional scale the fishing, recreational and whale watching boats are often the main sources of impacts on coastal populations of aquatic mammals (The Acoustic Ecology Institute 2008) (Fig. 13.6).

Noise generated by large vessels, composed mainly of lower frequencies, is extensively studied for their influence on the communication of baleen whales (*eg* Rossi-Santos 2015). Smaller boats usually have higher potential to stop feeding

Fig. 13.6 Different types of boats are commonly seen in Pipa beach, as artisanal fishing boats, recreational boats and whale-watching (schooners and fiber boats) creating a noisy environment with a high range of frequencies and intensities that have the potential to impact marine mammals (Photo: Rafaela Cardoso)

activities, communication masking of coastal species and usually cause behavioral changes in these animals (The Acoustic Ecology Institute 2008).

The noise of cavitation, a phenomenon that occurs in the engine propellers, is the main source of medium and high frequencies emitted by vessels and it's increased according to the engine speed (Arveson and Vendittis 2000).

This noise ranges from medium to high frequencies, so has the potential to overlap the sound signals of various species of toothed whales (Bain and Dahlheim 1994) and is the major factor contributing to noise levels in coastal areas. A recent study for a population of killer whale (*Orcinus orca*) pointed the boat's navigation speed as the most important cause in the production of noise (Houghton et al. 2015).

Other studies have shown that in the coastal acoustic environment, noise baseline levels have increased 10–27 dB in intensity in the presence of smaller boats (Erbe 2002). Some species have to send their signals with intensities 20–40 dB above than the environmental noise to have a successful communication (Holt 2008).

13.1.3.3 Whale Watching Activity

Whale and dolphin watching tourism, or only whale watching, represents a specific boat traffic and demands special attention, because in addition to the potential impact of the boat itself, there is also the risk related to the tourism activity, since

Fig. 13.7 Tourism Schooner looking for dolphins in the concentration area very close to the shore of Pipa beach, Rio Grande do Norte State, Brazil (Photo: Marcos Rossi-Santos)

these boats are approaching and remaining in areas inhabited by cetacean populations (Fig. 13.7).

The whale watching is a rising activity around the world, present in over 500 locations, many of them in Brazil (Hoyt 2001). It is currently an important source of income for coastal communities (Garrod and Wilson 2002) also presenting value in other aspects that are not only economic, but social and environmental, such as environmental education, aid to research and coastal management (Lück 2015). Thus, it can be considered as an alternative to non-lethal exploitation and economical solution for the conservation of cetaceans (Hoyt 2001).

Nevertheless, there are several studies warning about the impacts and negative effects of tourism on the animals, considering potential effects in short and long-term for the survival of the species and then the effective sustainability of this activity (*eg* Ritter 2003; Stockin et al. 2008; Tosi and Ferreira 2009; Parsons and Scarpaci 2016). Within the acoustics, the increasing number of vessels targeting this activity can lead to an increase in noise produced and therefore, a noisier acoustic habitat (Popper and Hawkins 2016).

Furthermore, in order to be successful, these activities in most cases occur in areas of residence or concentration of these animals, that are often used for foraging activities (Hastie et al. 2004), for example in Pipa beach (Rio Grande do Norte State).

These conditions of tourism activity generate a greater potential of impact over coastal species with a high degree of site fidelity, such as the Guiana dolphin, possibly causing behavioral changes that affect their distribution and survival (Scheidat et al. 2004; Carrera et al. 2008; Popper and Hawkins 2016).

Therefore, the whale watching activity can be indicated as a threat to marine animals (Coscarella et al. 2003; Popper and Hawkins 2016), requiring increased attention for the subject, control of the activity and management in areas in which it is held, since positive results have been shown in some locations (*eg* Tosi and Ferreira 2009).

13.1.3.4 Dredging and Port Activities

Dredging procedures are usually performed adjacent to the coast, where the depth needs to be increased (as the entrance of the Caravelas estuary, Bahia State, to ensure efficiency of ports operations). Many negative environmental impacts are related to this activity, such as increased turbidity, resuspension of pollutants, habitat destruction, alteration of hydrodynamics, changes on sound propagation and intense sound noise during dredging activities (Torres-Guijarro et al. 2013).

A dredger used on sandy bottoms is the Trailing Suction Hopper Dredger. This type of dredger generates noises from different sources, as illustrated on the picture below (Fig. 13.8).

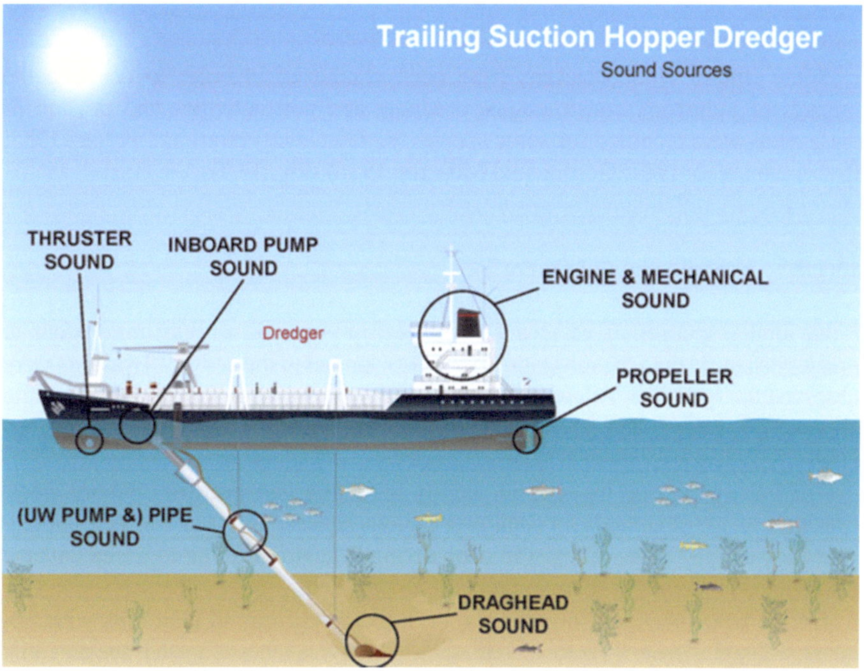

Fig. 13.8 Dredging activity is very common and has many different sources of sound, may causing acoustic impact where it occurs (From: Reine et al. 2014)

Fig. 13.9 Construction of a port using pile driving in the *S. guianensis* occurrence area of Caravelas estuary, Bahia State, Brazil. This activity produces an intense noise during its operation (Photo: Marcos Rossi-Santos)

Some studies with cetaceans have been conducted in order to describe the impacts of industrial, constructions, dredging and port activities and evasion of these areas were recorded for some species (*eg* Glockner-Ferrari and Ferrari 1985; Richardson et al. 1987; Gordon and Moscrop 1996) (Fig. 13.9).

13.1.3.5 Seismic and Oil Operations

Behavioral responses of cetaceans to noise from seismic activity may occur at greater distances than the mitigation measures applied to the activity. Evasions were observed for species of baleen whales and toothed whales, including animals using areas of great depth and also pelagic and coastal animals (Castellote and Llorens 2016).

The noise produced by seismic activity is usually generated in offshore areas, causing leakage of animals for shallower regions consequently leading to stranding. The animals may show changes in deep diving behavior, leading to tissue damage also related to stranding (Fernández et al. 2004).

On the Brazilian coast, noises have been recorded from oil platforms with frequencies ranging from 5 Hz to 48 kHz, such a wide frequency range has the potential of masking acoustic signals of numerous species (Rossi-Santos 2015).(Fig. 13.10).

Fig. 13.10 Oil and gas exploitation in the breeding ground of the humpback whale, *Megaptera novaeangliae*, in the Brazilian continental shelf (From: Rossi-Santos and Oliveira 2016)

13.1.4 Anthropogenic Noise Effects

In the last decade there has been an increase in studies on anthropogenic noise in the oceans and their impact on cetacean ecology in their natural habitats. The issue has been widely discussed in recent years suggesting that usually cause changes in their communication skills (Branstetter and Finneran 2008). This change in the communication capacity can generate, therefore, potential impacts of different magnitudes on the animals, which may have basically behavioral character, acoustic or physiological. Effects can also be divided into short-term, long-term and in extreme cases, fatalities (Nowacek et al. 2007).

In a study on cetacean reactions to anthropogenic noise, Nowacek et al. (2007) characterized the effects for each character mentioned above. The character behavioral effects include changes in patterns of breathing, dip and rise to the surface and foraging (Weilgart 2007). Acoustic character responses include changes in the modulation frequency, maximum frequency of the signal, duration and repetition of vocalizations according to the source of the noise (May-Collado and Wartzok 2008). The physiological effects are the Permanent Thresholds Shifts (PTS) or Temporary Thresholds Shifts (TTS) of hearing by the auditory system (Ketten 2004), changes in heartbeat rate (Lyamin et al. 2015) and stress (Wright et al. 2007).

Changes in behavioral patterns of cetaceans have been widely documented especially as a result of vessel traffic. The main effects on dolphin species are: avoidance of the area when in presence of boats (Watkins 1986; Janik and Thompson 1996;

Nowacek et al. 2007; Filla and Monteiro-Filho 2009), Increase in swimming speed (Nowacek et al. 2001; Erbe 2002; Jahoda et al. 2003), changes in swimming direction (Nowacek et al. 2001; Lemon et al. 2006; Filla and Monteiro-Filho 2009), changes of behavior (Miller et al. 2008; Stockin et al. 2008; Pirotta et al. 2015), changes in group composition (Bejder et al. 1999; Filla and Monteiro-Filho 2009), decrease in the distances between individuals (Nowacek et al. 2001; Valle and Melo 2006; Tosi and Ferreira 2009), changes in respiratory rate and breath timing (Janik and Thompson 1996; Jahoda et al. 2003; Filla and Monteiro-Filho 2009; Hastie et al. 2003; Tosi and Ferreira 2009).

Acoustic changes are intrinsically associated with the masking phenomenon, when the anthropogenic noise overlaps the signals of animals (in the same frequency and greater intensity). According to Bain and Dahlheim (1994), noise generated by human activities can mask sound signals emitted by animals, forcing them to spend more energy to communicate. Thus, according to review by Tyack and Janik (2013), for greater efficiency to overcome ambient noise, the cetaceans can change the intensity of the emitted signals, frequency, duration and repetition of their signals.

Some authors claim that there is occurrence of habituation and/or acclimatization, referring to the body's ability to change the perception of danger through experience, then decreased in behavioral response over time (Weilgart 2007; Wright and Kuczaj 2007). But other authors believe it may be a mistake to say that there is acclimatization only by finding the reduction of behavioral responses (Wright et al. 2007).

Anyway, behavioral change was observed in *S. guianensis* by Pereira et al. (2007) in Baía Norte, Santa Catarina State, where during a period of 10 years in contact with tour boats, reactions shifted from negative to neutral. It was also observed decrease in behavioral responses by Izidoro and Le Pendu (2012) in southern Bahia State (Port of Ilhéus), which was described impartiality of Guiana dolphin front of the approaches of the boats.

Behavioral responses observed in these cases potentially expose individuals to physical injuries such as cuts, hearing damage and accidents with fishing nets (Southall 2004; Zappes et al. 2010).

It is quite likely that aquatic animals are suffering high levels of stress in the presence of noises even when showing minimal behavioral change (Wright and Highfill 2007), as well as the apparent tolerance of noise does not imply that it is not causing long-term impacts in animals (Weilgart 2007).

Clark et al. (2009) pointed out that the animals are exposed to sound sources less dynamic, wider range of frequencies and constantly growing base level of noise in the panorama of modern marine environment. In animal perspective, they are considerably different acoustic habitats, where it is more difficult to communicate when compared to the environment in which they evolved.

13.1.5 S. guianensis *Acoustic Repertoire*

Odontocetes developed a sophisticated biosonar system that provides accurate information even in dark environments, in which case it would not be possible for a normal vision system, because their emitted signals penetrate and pass through objects causing reflections, which are received by them again and so the information is processed and interpreted by the central nervous system (Ary et al. 2016).

As other species of dolphins, *S. guianensis* presents an acoustic repertoire, usually divided into three categories (Richardson et al. 1995): whistles and two types of pulsed sounds (a variety of calls which may contain gargle) and the clicks for echolocation (Norris et al. 1994; Monteiro-Filho and Monteiro 2001; Rossi-Santos et al. 2008).

The whistles have been reported in several studies around the Brazilian coast, being basically pure tones and can have some harmonics and frequency values ranging from 1.05 kHz (Deconto and Monteiro-Filho 2016) with records up to 66.7 kHz (Andrade et al. 2015) (Fig. 13.11a, b) has already been registered harmonics with frequencies up to 136 kHz (May-Collado and Wartzok 2009).

Some studies indicate that the behavior of animals and social group size can interfere in the parameters of whistles (Erber and Simão 2004) and are often associated with social behavior, to maintain vocal or physical contact between the dolphins (Tyack 2008; Andrade et al. 2015), allowing group cohesion and individual recognition (Monteiro-Filho and Monteiro 2001; Figueiredo and Simon 2009) and is therefore considered a social vocalization (Caldwell and Caldwell 1968; Sayigh et al. 1990; Smolker et al. 1993).

Pulsed calls, presents a wide variation in the frequency parameters, duration and intensity, always occurring with harmonics (Fig. 13.11c, d). The frequency parameters of these sounds have been described between 0.7 kHz and 5 kHz, with duration varying between 0.02 and 0.9 s. As the whistles, pulsed calls are associated with social communication, but usually in groups' vocalizations (Monteiro-Filho and Monteiro 2001).

Another sound already registered, 'gargles' are similar to pulsed calls and therefore classified by some authors in the same category (Fig. 13.11e) (Monteiro-Filho et al. 2008; Garcia 2009). Gargles are characterized by a long, low frequency notes, about 0.3 kHz, containing 2 harmonics and average duration of 3 s, reaching 20 s (Monteiro-Filho and Monteiro 2001). Gargles were described only in Cananéia estuary (Monteiro-Filho and Monteiro 2001) and in Caravelas estuary (Garcia 2009). These sounds were associated with groups containing calves and interacting in shallower waters as estuarine areas.

Clicks are basically related with navigation and prey detection, vocalization and echolocation (*eg* Kellogg et al. 1953; Kellogg 1958; Pilleri 1990). They are characterized by a sequence of pulsed sounds with registered frequency range ranging from 0.1 kHz to 43.62 kHz (Deconto and Monteiro-Filho 2016) (Fig. 13.11f) having been described with the highest occurrence in foraging activity (Monteiro-Filho and Monteiro 2001).

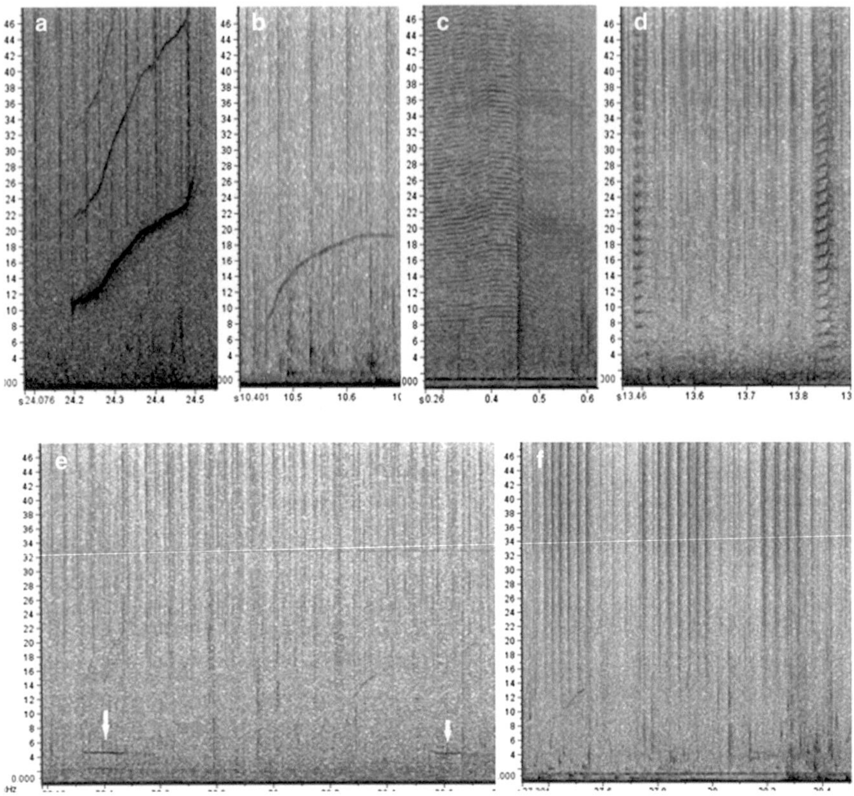

Fig. 13.11 Examples of whistles (**a**, **b**), pulsed calls (**c**, **d**), gargling (**e**) and clicks (**f**) from the *Sotalia guianensis* repertoire. Axis y: frequency range (kHz), axis X: duration (s) (From: Deconto and Monteiro-Filho 2016)

The latest data on frequency range of whistles of *S. guianensis* throughout its distribution in the South Atlantic Ocean (in various locations along the Brazilian coast, Costa Rica and Venezuela) is summarized in the table below (Table 13.1).

The low values at maximum frequency in Caravelas estuary (Bahia State) and Venezuela are explained by the authors possibly due to the equipment used for recording having frequency response up to 24 kHz. The best hearing sensitivity range of this species is described between 64 and 105 kHz (Sauerland and Dehnhardt 1998).

During fishing activities there are evidences on emission of complex compositions and intercalating types of sounds described above resulting in possible phrases (Fig. 13.12), as occurrence was observed in the presence of lone dolphin (Monteiro-Filho and Monteiro 2001).

There is evidence of individual whistles of Guiana dolphin, equivalent to an "acoustic signature", since specific types of whistles would be produced by each animal and can be used for individual identification (Figueiredo and Simão 2009; Lima and Le Pendu 2014).

Table 13.1 Recent researches describing the frequency range of **whistles** by the Guiana dolphin (*Sotalia guianensis*) on its occurrence in Brazilian coast, in Costa Rica and Venezuela

Whistles (*S.guianensis*)	Cananéia estuary (SP)	Guanabara Bay (RJ)	Sepetiba Bay (RJ)	Paraty Bay (RJ)	Caravelas estuary (BA)	Pipa beach (RN)	Costa Rica	Venezuela
Max. Freq. (kHz)	46.8	66.7	44.9	42.84	24	47.5	48.4	2.88
Min. Freq. (kHz)	1.05	1.7	1.87	1.03	1.39	0.42	1.4	5.31
References	Deconto and Monteiro-Filho (2013, 2016)	Andrade et al. (2015)	Andrade et al. (2014)	Andrade et al. (2014)	Garcia (2009)	Martins et al. (2016), Albuquerque and Souto (2013)	May-Collado (2013)	Barrios-Garrido et al. (2016)

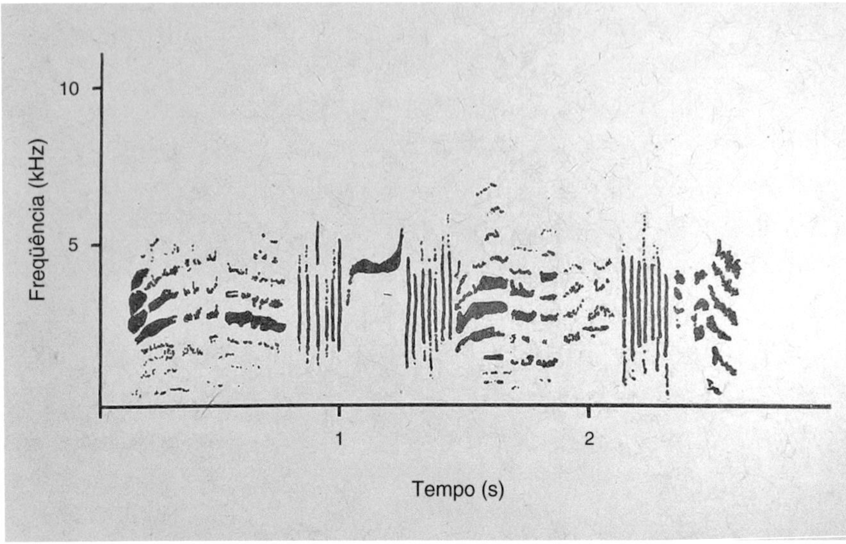

Fig. 13.12 Considering the short intervals between the different sounds and their lack of superposition, it is possible that they were emitted by the same animal and therefore represent a phrase emitted by an individual of *Sotalia guianensis* (From: Monteiro-Filho et al. 2008)

13.2 Material and Methods

13.2.1 Study Areas

Three regions were studied on the Brazilian coast from south to north: Cananéia estuary (25° 04′S, 47° 56′W) in São Paulo State; Caravelas estuary (17° 30′S, 39° 30′W) in Bahia State and Pipa beach (06° 10′S, 35° 05′W) in Rio Grande do Norte State (Fig. 13.13). The sampled areas were chosen because they represent typical examples of how the coastal and estuarine regions have been shared between aquatic mammals and increasing human activities.

13.2.1.1 Cananéia Estuary (CAN)

The Cananéia estuary (CAN) is located on the southern coast of São Paulo State being part of a lagoon-estuarine system of about 180 km. It is an area of intense fishing activity, characterized by mangroves, beaches and predominantly muddy bottom (Monteiro-Filho and Monteiro 2001). The hydrodynamics of this estuary is influenced mainly by the action of tidal waves, with contribution of fluvial waters from rivers, shallow depths (up to 23 m), causing water with high turbidity throughout the year (Schaeffer-Novelli et al. 1990).

Fig. 13.13 Occurrence areas of *Sotalia guianensis* (highlighted area) and study sites on the Brazilian coast: (PIP): Pipa beach – Rio Grande do Norte State; (CVL): Caravelas estuary – Bahia State and (CAN): Cananéia estuary – São Paulo State (Adapted from: Marchetto 2010)

The region of Cananéia has a high density and intense use by *Sotalia guianensis* (Bisi 2001). The region has a total abundance of 195 animals and most of the Guiana dolphin groups occur with two individuals (Havukainen et al. 2011).

The recordings used in this study were obtained between the years 2005 and 2008, and can be found in Monteiro-Filho and Monteiro (2001); (Monteiro-Filho et al. 2008) and Rezende (2008). Data were recorded in the waters of the Trapandé Bay, south opening of Cananéia estuary and region adjacent to the Cananéia Island, Comprida Island and Cardoso Island. This area is the region of the estuary with highest occurrence of *S. guianensis* (Oliveira 2007) (Fig. 13.14).

13.2.1.2 Caravelas Estuary (CVL)

The Caravelas estuary (CVL) is located in the southern portion of Bahia State and is the second largest mangrove forest in northeastern Brazil, comprising an area of approximately 66 km^2 (Herz 1991).

Caravelas is the main city of the National Marine Park Abrolhos Bank. The region is characterized by a pronounced increase in the extension of the continental shelf, protected by banks of coral reefs, with a predominance of sandy beaches and

Fig. 13.14 Cananéia estuary is characterized by shallow waters and a muddy *bottom*, what makes the sound even more important for the population of Guiana dolphin that inhabit there, as the water is turbid almost over the all year (Photo: Caio Louzada, IPEC)

mangrove areas. The bottom is heterogeneous, composed of sandy areas, mud and calcareous algae (Leão 1994).

Since 2002, when the port of Caravelas was built, an increase of human activities could be observed in the region. The main activity is the wood-barges traffic, which to cross some areas the responsible company needs to dredge sand from the bottom of the estuary regularly (Rossi-Santos et al. 2003).

According to Cantor et al. (2012), the population of *Sotalia. guianensis* is relatively small, ranging from 57–124 animals with constant abundance over the years. For the species, fidelity and residence to the study area have also been described (Rossi-Santos et al. 2007) (Fig. 13.15).

Dolphins sightings are frequent off the mouth of the Caravelas River area considered extremely important to the species (Rossi-Santos et al. 2010), where these animals mainly present foraging behavior (Daura-Jorge et al. 2007). The bottom of this region is predominantly muddy, there is evidence of contact of the animals with the floor, probably a foraging strategy (Rossi-Santos and Wedekin 2006), being observed preference of animals to areas with flat bottoms (Rossi-Santos et al. 2010).

The recordings analyzed in this study were collected between the years 2005 and 2008 at the mouth of the Caravelas estuary and can be found in Garcia (2009).

Fig. 13.15 *Sotalia guianensis* in Caravelas estuary, an area of intense human activities and dolphins sightings (Photo: Marcos Rossi-Santos)

13.2.1.3 Pipa Beach (PIP)

Pipa beach (PIP) is located in the municipality of Tibau do Sul, a Sustainable Use Reserve, in Rio Grande do Norte State. The region called Baía dos Golfinhos is about 0.4 km^2 area and is bounded by vertical cliffs that reach heights of 40 m, in some areas are covered by sand dunes, increasing the overall height to 70 m high (Carrera et al. 2008) (Fig. 13.16).

The southern coast of Baía dos Golfinhos is well protected from strong southeast winds and currents (Beaufort scale <1.0), while the northern coast is more susceptible to these factors. Its maximum depth reaches 7 m (Garri 2006), with temperatures ranging from 19 to 28 °C and salinity between 36 and 37‰ (Souto et al. 2006) and slightly turbidity in the water column. The bottom is composed mainly of sand except the edges, composed of sedimentary rocks. The region is also influenced by Guaraíra estuary, located about 5 km northwards.

Pipa beach is characterized as a growing tourist region, with particular attention to the increase in whale watching activities aimed at the *S. guianensis* in Baía dos Golfinhos. Due to its unique features, the region is largely inhabited by Guiana dolphin that primarily use for foraging activity (Araújo et al. 2003). According to Carrera et al. (2008), the average number of animals using this region is about 2–4 individuals, varying according to the presence or absence of vessels.

The acoustic data used in this study was collected at Baía dos Golfinhos during five campaigns in April 2009 and can be found in Martins (2010).

Fig. 13.16 Vertical cliffs, some with dunes on the top, *S. guianensis* and tourism boats in Baía dos Golfinhos in Pipa beach, Rio Grande do Norte State, Brazil (Photo: Rafaela Cardoso)

13.2.2 Data Collection

The bioacoustic data used in this study was collected between the years 2005 and 2009 by different researchers, grouped in this chapter for a better evaluation of acoustic conditions that the Guiana dolphin are exposed to in three different areas along the Brazilian coast.

The data was collected in a similar way in the three areas sampled through field activities, using different equipment that gradually improved the ability to capture the sounds in the aquatic environment. Initially tape deck (usually Sony TCD – 5 M), digital recorder Sony DAT D8, both recorders with frequency responses of about 24 kHz, replaced by a Panasonic PV-GS 120 digital camera (frequency response of 48 kHz), and as from 2008, a M-Audio Microtrack II Professional recorder, with the capacity to record higher frequencies (frequency response up to 96 kHz). Hydrophones used were basically two HTI SSQ 94 and HTI 90 min (Fig. 13.17).

At laboratory, the sounds collected in field trips were transferred to the computer and processed through the RAVEN 1.3 Pro software (Cornell University). The following parameters were obtained: minimum and maximum frequency, initial and final frequency, delta frequencies and duration of the signal.

Currently, knowledge of the acoustic repertoire of aquatic mammals evolved significantly due to the use of equipment with higher frequency response and sampling rate.

Fig. 13.17 Data collection in Pipa beach, Rio Grande do Norte State, Brazil. Standard methods are essential to be applied in data collection and analysis in order to future comparisons (Photo: personal archive of Marcos Rossi-Santos)

It is important that the recording system is adequate for the target species, in order to be able to sample the entire acoustic repertoire. (Oswald et al. 2004; Simão et al. 2007).

13.3 Results

The data recorded describes the acoustic repertoire of the Guiana dolphin, especially whistles (Table 13.2) and pulsed calls (Table 13.3), also fishing boats, tourism boats, wood-barges, a water pump and dredging activity (Table 13.4).

Regarding *S. guianensis* whistles, values of three populations were similar in general, but with some small differences, for example, the amplitude of minimum frequencies in Caravelas (2.5 kHz) is higher than in Cananéia (4.6 kHz) and Pipa (5.6 kHz) and the higher frequency have also been registered in this location (24 kHz).

The acoustic signals related to pulsed sounds showed wider variations among the three populations. The highest amplitudes of minimum and maximum frequencies, as well as whistles, were recorded in Caravelas (0.18–24 kHz), and this site also had the highest number of harmonics (38 harmonics). The greater value of minimum

Table 13.2 Values of minimum, maximum, initial, final and delta frequencies and duration of the **whistles** from *Sotalia guianensis* populations from Cananéia estuary (CAN), Caravelas estuary (CVL) and Pipa beach (PIP)

Whistles		Min. Freq. (kHz)	Max. Freq. (kHz)	Initial Freq. (kHz)	Final Freq. (kHz)	Delta Freq.	Duration (s)
CAN	Min.-Max.	4.6–12.4	8.5–19.9	4.7–12.6	8.5–19.9	Ø – 15.2	0.038–0.476
	Mean	8.2	16.4	8.5	16.3	–	0.231
CVL	Min.-Max.	2.5–14.3	10.4–24.0	3.7–23.3	2.5–24.0	0.6–18.4	0.032–0.564
	Mean	10.1	17.2	10.1	16.8	7.1	0.265
PIP	Min.-Max.	5.6–11.9	10.5–18.4	5.6–15.6	10.5–18.4	Ø – 12.8	0.056–0.680
	Mean	8.6	15.7	9.7	16.4	–	0.315

Table 13.3 Values of minimum, maximum and mean frequencies, mean duration and minimum and maximum number of harmonics of the **pulsed calls** from *Sotalia guianensis* populations from Cananéia estuary (CAN), Caravelas estuary (CVL) and Pipa beach (PIP)

Pulsed Calls	Min. Freq. (kHz)	Max. Freq. (kHz)	Mean (kHz)	Mean duration (s)	Min.- Max. Harmonics
CAN	0.42	5.00	3.00	0.1–2.1	1–15
CVL	0.18	24.00	6.65	0.12	1–38
PIP	0.78	17.20	0.98	0.20	3–24

Table 13.4 Values of minimum and maximum frequencies of the anthropogenic noises in Cananéia estuary (CAN), Caravelas estuary (CVL) and Pipa beach (PIP)

Source	Min. Freq. (kHz)	Max. Freq. (kHz)
Diesel engine boats (CAN; CVL; PIP)	0.38	35.71
Petrol engine boats (CAN; CVL; PIP)	0.84	46.80
Wood-barges (CVL)	0.02	20.00
Dredging activity (CVL)	0.15	17.30
Water Pump (PIP)	1.84	8.14

frequency occurred in Pipa (0.78 kHz), with a minimum number of harmonics higher than other regions (3 harmonics).

The sounds produced by boats (diesel and petrol) in the three regions were synthesized in order to obtain a range of values (frequencies) for comparison of these impacts. The minimum and maximum values of frequencies (kHz) of all recorded anthropic activities are presented in the Table 13.4.

A water pump in Pipa beach used to be turned on every day in the late afternoon to fill up a swimming pool of a resort, creating noise from 1.84 to 8.14 kHz.

Regarding to the frequency range, small petrol engine boats (Fig. 13.18) emitted the highest frequencies, reaching 46.8 kHz, followed by diesel engine boats (35.71 kHz) (Fig. 13.19) while other noise sources had significantly lower maximum frequencies 20 kHz (wood-barges) (Fig. 13.20), 17.30 kHz (dredging) (Fig. 13.21) and 8.14 kHz (water pump).

Fig. 13.18 Petrol engine boats frequently used for whale watching tourism in Pipa beach, Rio Grande do Norte State, Brazil. Conservation of animals and the environment where they live while keeping economical and educational benefits of this industry, is one of the greatest challenges of this activity (Photo: Fábio Pais)

Fig. 13.19 Fishing boats (diesel engine) traditionally used in the Cananéia estuary, São Paulo State, Brazil. Fishers' local knowledge is an additional tool to obtain information about cetaceans, regarding their local particularities, fishing interactions, and behavior (Photo: Caio Louzada, IPEC)

Fig. 13.20 Wood-barge in presence of a group of *Sotalia guianensis* with a calf in Caravelas estuary. These protected environments are used for this species to ensure the survival of their calves (Photo: Marcos Rossi-Santos)

Fig. 13.21 Dredging activity in Caravelas estuary happens once a year in order to make the estuary's mouth deeper for wood-barges crossings (Photo: Marcos Rossi-Santos)

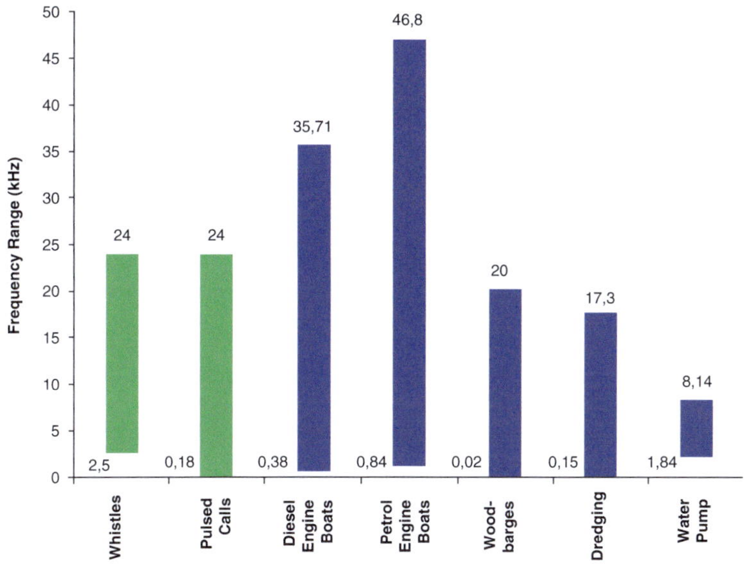

Fig. 13.22 Minimum and maximum frequencies values of ***Sotalia guianensis*** **repertoire** and the **anthropogenic noise** from the three study sites: Cananéia estuary, Caravelas estuary and Pipa beach

Based on the data presented above, in order to better visualize and compare the frequency ranges emitted by human activities and the repertoire of Guiana dolphin, a graphic representation was generated (Fig. 13.22), considering the lowest and highest values of whistles and pulsed calls between the three studied areas.

The image above shows the overlapping range of frequencies emitted by anthropogenic activities and the Guiana dolphin. Biological sounds (green bars) are fully in the spectrum of anthropogenic noise frequencies (blue bars).

13.4 Discussion

This study aimed in presenting a literature review on the acoustic ecology of the Guiana dolphin (*Sotalia guianensis*) and acoustic impacts to which they are subject, bringing a description of the acoustic repertoire of three populations of the species along the Brazilian coast and the noise produced by human activities in these areas, contributing to the conservation of the species in Brazil.

Noises generated by anthropogenic activities are associated with several impacts on cetacean populations around the world. The coastal ecosystems already pose many threats, for example, water pollution, overexploitation of natural resources, increase in traffic of commercial and recreational boats and climate change (De Fontaubert et al. 1996).

Fig. 13.23 Examples of petrol engine boat (*left*) and a diesel engine boat (*right*) in Pipa beach, Rio Grande do Norte State, Brazil. Both boats are in the same site for dolphin watching, looking for the Guiana dolphin (Photo: Marcos Rossi-Santos)

The interaction of these threats with continuous exposure to anthropogenic noise, presents a major risk to cetacean species that inhabit and use these coastal areas, particularly even more alarming for animal populations which already presents a classification status as vulnerable or endangered (Perry 1998), as is the case of *Sotalia guianensis* in Brazil.

By comparing the sounds emitted by the Guiana dolphin in the three areas with anthropogenic noise (Fig. 13.22), there is an overlapping of frequency ranges for all populations, indicating a possible masking on the acoustic signals of this species by anthropogenic noise.

In all study regions sounds of small boat traffic were recorded, with the highest frequency (46.80 Hz) generated by vessels with outboard petrol engine, while lower frequencies were recorded for internal diesel engine boats (35.71 Hz) (Fig. 13.23).

One explanation for this difference between the engines is that the diesel engine generates less direct vibration in water and minor amount of bubbles by the propellers' blades, causing noises of lower intensities and at lower frequencies (Seppänen and Nieminen 2004). Another reason is that the number of revolutions per minute in internal diesel engines is less than of the outboard petrol engines, generating less noise and possibly less impact on aquatic life (Martins et al. 2016).

Rezende (2008) observed different reactions of the Guiana dolphin to the boats in the Cananéia estuary region, for wooden boats equipped with diesel engine approaches, the animals showed acoustic avoidance behavior (decreased signaling

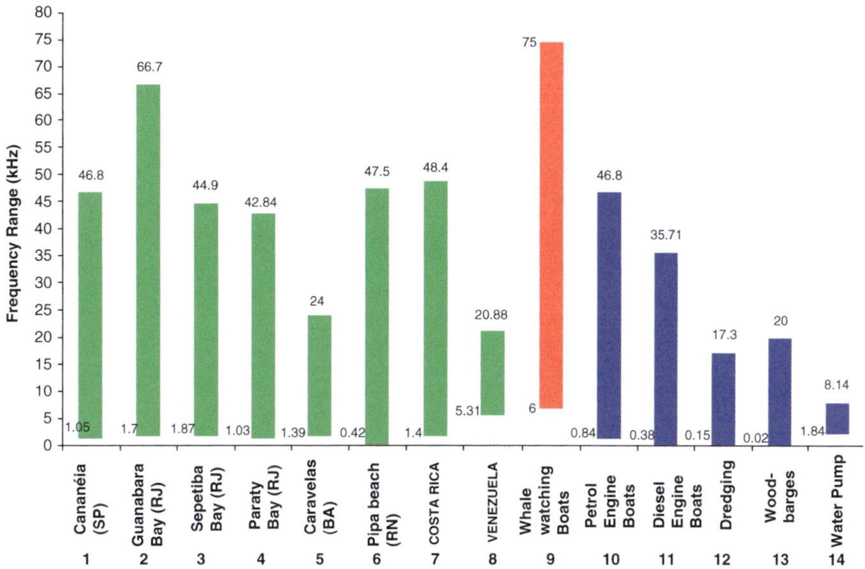

Fig. 13.24 Frequency range of whistles from eight *Sotalia guianensis* populations (*green bars*), anthropogenic noise evaluated on this study (*blue bars*) and whale watching activity (*red bar*). *1*. Deconto and Monteiro-Filho 2013 (max. Freq.), 2016 (min. Freq.). *2*. Andrade et al. 2015. *3*. Andrade et al. 2014. *4*. Andrade et al. 2014. *5*. Garcia 2009. *6*. Martins et al. 2016 (max. Freq.); Albuquerque and Souto 2013 (min. Freq.). *7*. May-Collado 2013. *8*. Barrios-Garrido et al. 2016. *9*. Hildebrand et al. 2006. *10*; *11*; *12*, *13* e *14*. Present study

rate as a possible way to avoid masking) in only 1.4% of the time, while in 100% of the encounters of animals with aluminum boats equipped with petrol engines this avoidance behavior was recorded.

Therefore, it can be suggested that internal diesel engines are relatively less harmful to the Guiana dolphin, mainly because these animals can shift their signals to higher frequencies and also for the reason that they have fewer hearing sensitivity at low frequencies (Martins et al. 2016).

The graphic below (Fig. 13.24) compares whistle data presented on Table 13.1, referring to the range of frequencies of eight *S. guianensis* populations and anthropogenic noises of this study, together with results obtained for whale watching activity from a study made by Hildebrand et al. (2006).

Recent studies, with results pictured above, were carried out using modern equipment able to record broadband sounds, both the acoustic repertoire of the Guiana dolphin (66.7 kHz, Andrade et al. 2015) and anthropogenic noise (75 kHz, Hildebrand et al. 2006). Thus it is clear that dolphins produce sounds with higher frequencies than most impactful human activities described, but are still potentially impacted by noise generated by whale watching boats, mainly petrol powered.

Ocean noise stimulated several studies evaluating behavioral changes and the acoustic repertoire of several cetacean species around the world. When in the pres-

Fig. 13.25 Jet-skis crossing fastly the Cananéia estuary in presence of *Sotalia guianensis*, São Paulo State, Brazil. These type of boat reach high speeds and represent a great concern for the species that inhabit the areas, especially because its sound is not much propagated under the water, so the animals can't hear the fast boat coming in their direction, consequently can't escape from possible accidents (Photo: Caio Louzada, IPEC)

ence of tourism boats, dolphins reduced time spent in foraging activities from 10 min to 6 min and when stopped, animals took 54% more time than normal to return to the activity (Stockin et al. 2008).

The presence of a boat reduced by 49% the foraging activity on bottlenose dolphins (*Tursiops truncatus*), and the greater the amount of vessels in the region the greater the impact on feeding, also considering that the physical presence of vessels, and not only the noise generated by them, showed an important role in these disturbances (Pirotta et al. 2015). In Mississipi, bottlenose dolphins changed their behavior in less than 1 min in the presence of jet-skis, thus Miller et al. (2008) concluded that the increase of this activity can cause long-term impacts on dolphins' populations (Fig. 13.25).

Martins et al. (2016) in a study of Guiana dolphin in Pipa beach, in the presence of boats they found changes in all acoustic parameters of the signals. The main changes in whistles and pulsed calls were characterized by the increasing of frequencies and the amount of signals emitted. On the clicks, the main change was an increase in the duration of the signals.

In another study on the responses of dolphins to noise generated by boats, more than half of the reactions considered extreme (as animals clearly avoiding the area, prolonged separation of females and calves and prolonged disruption of mating behavior) occurred at intensity levels of only 100–130 dB re 1 μPa @ 1 m (Southall et al. 2007).

Studies with bottlenose dolphins (*Tursiops truncatus*) described changes in whistles of individuals according to the noise in the environment, producing signals of higher frequencies in noisy environments (Heiler et al. 2016) and may have wide variation on its tones and longer duration when compared to the environment with only one boat (May-Collado and Wartzok 2008), confirming the results obtained by Martins et al. (2016) to the Guiana dolphin in Pipa beach.

Studies with killer whales (*Orcinus orca*) have also showed changes in vocalizations, increasing the intensity of signals with increasing noise caused by boats, which can have as a consequence a greater energy expenditure, higher stress level and also an inefficient communication between individuals, may affect then other activities (for example, reproduction and socialization) (Holt et al. 2009).

May-Collado and Quiñones-Lebrón (2014) in a study comparing changes in the acoustic repertoire of bottlenose dolphins with the presence of small boats concluded that such changes vary according to the behavior of animals. It was described that when the dolphins were in social interactions or swimming, in the presence of boats, the acoustic parameters showed no significant change, while when the activity was foraging, showed a reduction on the frequencies and the signals duration has increased.

Whereas the main activity described for *S. guianensis* in the three areas of study described in this paper (Cananéia estuary, Caravelas estuary and Pipa beach) is the foraging activity, we can suggest a high sensitivity of the animals to the presence of the vessel generated noise, especially petrol engines ones, indicating a high potential of impact over the three populations studied.

A study about the economic value of the Guiana dolphin observation tours in Cananéia estuary came to the conclusion that this is an activity of great importance in the income of the community (Filla et al. 2012), similarly as in Pipa beach, a region that presents a fast growth of this kind of tourism, also reflecting the local economy.

The Guiana dolphin watching tourism is a rising activity both in the estuary of Cananéia and in Pipa beach. For the activity to prosper it must be carried out in a sustainable manner. The sustainability depends basically on management of visitors (respecting the carrying capacity of visitors in each region), fluctuations in the local and regional economy, and especially the maintenance of the resource or species, which is the target of the tourism (Sousa-Lima and Clark 2008).

Filla and Monteiro-Filho (2009) showed that the Guiana dolphins of Cananéia estuary have different responses according to the approaching of tourist boats. When boats proceeded properly, behavioral responses were positive (the animals approached the boat, swam around, followed the boat or foraged, for example) or neutral (continued their activity) (Fig. 13.26). In 100% of approaches considered inadequate, the results were negative reactions from animals, such as abandonment of activity, group separation, diving in different directions and evacuation of the area (Filla and Monteiro-Filho 2009).

Albuquerque and Souto (2013) showed that in 54% of the time that the Guiana dolphins were using the area of the Baía dos Golfinhos (Pipa beach, Rio Grande do Norte state) boats were present, and in 95% of cases the vessels approached in a distance less than 50 m away from the animals (Fig. 13.27).

Fig. 13.26 Correct approach by a whale watching boat in Cananéia estuary, São Paulo State, Brazil. An approach was considered as being correct when, on seeing the animals at 500 m distant, the skipper immediately reduced speed and approached slowly and laterally, not interfering with their activities. On reaching about 100 m, the skipper put the engine into neutral and drifted freely (Photo: Caio Louzada)

Fig. 13.27 Whale watching boat proceeding an *direct approach with intention of stopping* in front of mother and calf of *S. guianensis* in Pipa beach, Rio Grande do Norte State, Brazil. The skipper deliberately approached to within less than 50 m from the animals, on some occasions getting to within less than 3 m (Photo: Fábio Pais)

Studies reported that tourism boats at a speed of 5 knots in shallow waters have the potential to reduce the communication of dolphins in 26% at a distance of up to 50 m (Jensen et al. 2009). By increasing the speed to 10 knots, cavitation noise may rise by up to 10 dB re 1 μPa @ 1 m (Gervaise et al. 2012), also increasing considerably the frequency and the distance at which noise is propagated (Jensen et al. 2009). For comparison, speeds of 2.5 knots have been described as suitable for areas of high concentration of cetaceans, as it showed no masking factor in odontocetes species analyzed (Jensen et al. 2008).

Erbe (2002) estimated the intensity of noises emitted by whale watching tourism boats at 145–169 dB re 1 μPa @ 1 m and Hildebrand et al. (2006) described noises intensities from 110–145 dB re 1 μPa @ 1 m, with frequencies from 6 to 75 kHz. Values that are probably enough to mask the signals emitted by the *S. guianensis.*

Carrera et al. (2008) reported that in the presence of tourism boats in Pipa beach, the Guiana dolphin reduced foraging activity, with avoidance of the area also reported, especially for younger animals, considered by some authors as protective behaviors (*eg* Santos-Jr et al. 2006).

Previous studies in the same location also showed behavioral changes of Guiana dolphins in the region due to the presence of vessels, causing the decrease in distance between individuals, which the authors suggested to be an attempt to facilitate communication between them (Valle and Melo 2006) (Fig. 13.28).

Fig. 13.28 *Sotalia guianensis* very close from each other in Baía dos Golfinhos, Pipa beach, Rio Grande do Norte State, Brazil. This behavior can be described for some researches as a probable technique to improve their communication in a noisy place (Photo: Fábio Pais)

For the whale watching activity to become less harmful to the exploited species, some measures should be implemented.

A concept to be reinforced in the whale watching tourism industry is that the main motivation of tourists to opt for this activity is not to get as close as possible to the animal, but to get a better knowledge, observe them in their natural environment without disturbing them, and receive information on cetaceans and their habitat (Orams 2000; Andersen and Miller 2006; Oliveira 2007; Filla et al. 2008, 2012; Lück 2015).

Studies have been conducted in the context of evaluating the satisfaction of the tourists about the number of boats in the area carrying out the activity and found that more than two boats approaching the animals have negative influences to the activity (Hoyt 2001; Avila-Foucat et al. 2013). It was observed that with an increase in the number of boats, there is also an increase in the proportion of dolphins with their activities interrupted (Stockin et al. 2008), making clear the need to control the amount of vessels on this activity.

A management plan should include training of operators and travel agents, so they can know the potential impacts of the activity on dolphin populations and know how an appropriate procedure can minimize or even avoid impacts over them.

It is recommended the spreading of information on the noise generated by different boats and type of engines, also correlating to its speed, and how each situation can affect the acoustic ecology of animals, so conductors promotes reduction of noise aiming at a lower acoustic impact on dolphin populations.

Tosi and Ferreira (2009), evaluated the behavior of Guiana dolphin on the presence of tourism boats in Pipa beach after the creation of a Municipal Coastal Reserve, a management action proposed by Ferreira et al. (2005), which led to control in the number of tourism boats using the region. The study concluded that a simple conservation measure such as controlling vessel traffic results in a decrease of negative behavioral responses, demonstrating the effectiveness of regulations and supervision in the dolphins' ecology in areas which whale watching activity is in development (Tosi and Ferreira 2009) (Fig. 13.29).

In 2008, a proposal to implement a zone with regulation for human activities in Cananéia estuary was created aiming to the protection of the *S. guianensis* (Filla et al. 2008). Initiatives like this are extremely important as an aid in the management of coastal areas.

Thus shows the importance of proper management of the whale watching activity both in the Cananéia estuary and in Pipa beach, so that the conservation of Guiana dolphin allow that tourism remains an important part of the local economy.

Recordings of dredging activities and wood-barges, both recorded in the Caravelas estuary, when analyzed showed that the minimum frequency of the Guiana dolphin's repertoire (0.18 kHz) is lower than of these anthropic activities (0.02 kHz) while maximum frequency of the dolphins (24 kHz) is higher than of the anthropogenic noise (20 kHz).

Martins et al. (2016) in Pipa beach found frequency values of whistles reaching 47.5 kHz, above the limit presented in this study, which shows the possibility of Guiana dolphin using signals with higher frequencies for communication if they

Fig. 13.29 Non-controlled number of boats used for whale watching in Pipa beach, Rio Grande do Norte State, Brazil. An important management for whale watching tourism activity is the maximum number of boats appropriate for each place. In some cases the exceeded number of boats interrupting dolphins' activities can lead to ecological impacts to the population (Photo: Fábio Pais)

need. Therefore, it is concluded that the ecology of Guiana dolphins are affected by wood-barges and dredging activity on the use of signals, but especially in pulsed calls, for the reason of its lower frequencies indicating a potential to masking.

Wood-barges activity are very often in Caravelas estuary, exposing the population of Guiana dolphins to high levels of noise intensity very often, which may lead to impacts on the hearing thresholds (TTS or even PTS) depending on the acoustic effect suffered by the noise, which can suffer, for example, reverberation due to local morphology (stratification of the water column, bottom hardness and low depths) (Breitzke and Bohlen 2010) (Fig. 13.30).

A study of the noise of dredging activities showed intensity levels over 150 dB re 1 μPa @ 1 m (Reine et al. 2014), and this noise was perceptible about 7 km distance from the source (depending on the dredger type and environment conditions) (Clarke et al. 2002).

Considering the Caravelas estuary setting, with shallow and low hydrodynamics areas, it is likely that the average noise and high frequencies (> 1 kHz) of dredging activity is perceived by the animals that inhabit the region and is potentially impactful. It is noteworthy that this activity is held annually in the mouth of the Caravelas estuary generating an intense acoustic impact, which can cause long-term effects.

The water pump in Pipa beach has the potential to impact the pulsed calls and whistles, but only at night (when it was turned on) and especially when the dolphins

Fig. 13.30 Since 2002, Caravelas estuary has been under significant increase of anthropogenic impacts. Daily, trucks transporting eucalypt trunks to the wood-barges and port activities are contributing to a noisier soundscape, may affecting the Guiana dolphins' ecology in Caravelas estuary, Bahia State, Bazil (Photo: Marcos Rossi-Santos)

used signals with lower frequencies (1.84–8.14 kHz). Due to the prevail of smooth surface of the sea, typical of Baía dos Golfinhos, and mainly at night, when the winds are generally weaker, the acoustic impact of the water pump is probably high due to low sound attenuation by the Lloyde's mirror effect.

Coastal environments have generally low depths and are environments with major trends to noise pollution of medium and high frequencies, because they suffer lower loss of the acoustic signal due to spherical spreading (Jensen et al. 2009; Rabin et al. 2003).

The three regions analyzed in this study have some characteristics in common: vital use for Guiana dolphin as priority areas for foraging, shallow depths, near the coast and intensive use by human activities. Cananéia and Caravelas estuaries have more turbid waters, factor that increases the importance of using acoustic signals by animals, since the water transparency affects visual interactions (Van Parijs and Corkeron 2001).

The use of areas for feeding and the site fidelity presented by *Sotalia guianensis*, in the study areas reinforces the vulnerability of this species due to possible changes in their habitats (Batista et al. 2014). It is likely that short and long-term impacts on breeding and foraging activities will have serious consequences for these populations (Perry 1998).

This study could point the evidence of potential anthropogenic impacts on three *Sotalia guianensis* populations along the Brazilian coast from the bioacoustics approach. This methodology is an important tool for coastal management and understanding of Guiana dolphin's ecology, than can be used for impact assessment, distribution and population data for aquatic species.

Nowadays, with the decrease on equipment costs, establishment of partnerships between researchers around the world, software tools available as free share-ware, the potential for studies of acoustic ecology is growing. However, the quality of the results still need to improve with the establishment of standard protocols for data collection, analysis and description of acoustic results (Erbe et al. 2016).

13.5 Conclusion

Dolphins are highly socialized animals, using sounds for communication and foraging, as well as for many other aspects in their life, directly being influenced by the acoustic habitat. We can conclude that the anthropogenic noises described here have the potential to affect the Guiana dolphin's ecology on the studied areas.

Sotalia guianensis is a marine ecosystem sentinel, species that reflect the quality of the marine coast habitat that it live. The discussions in this chapter contribute to the species knowledge and the influence caused by anthropogenic noise on these animals, allowing for an evaluation of potential acoustic impacts of human activities on marine ecosystems.

Nowadays, anthropogenic noise represents one of the biggest problems facing oceans conservation. Bioacoustics studies are extremely important to support management actions, offering data about acoustic habitats and ecology of aquatic species, helping to stablish preservation measurements, restrictions and limits of acoustic parameters, for the management of use and exploitation of aquatic ambient by humans.

References

Andersen MS, Miller ML (2006) Onboard marine environmental education: Whale watching in the San Juan Islands, Washington. Tour Mar Environ 2(2):111–118. doi:10.3727/154427306779436327

Andrade L, Siciliano S, Capistrano L (1987) Movimentos e atividades do boto *Sotalia guianensis* (Cetacea, Delphinidae) na baía de Guanabara - Rio de Janeiro. In: Reunião de trabalho de especialistas em mamíferos aquáticos da América do Sul, 2. Fundação Brasileira para a Conservação da Natureza, Rio de Janeiro, pp 49–55

Andrade LG, Lima IMS, Macedo HS, Carvalho RR, Lailson-Brito J, Flach L, Azevedo AF (2014) Variation in Guiana dolphin (*Sotalia guianensis*) whistles: using a broadband recording system to analyze acoustic parameters in three areas of southeastern Brazil. Acta Ethologica 18(1):47–57

Andrade LG, Lima IMS, Bittencourt L, Bisi TL, Lailson-Brito J Jr, Azevedo AF (2015) High-frequency whistles of Guiana dolphins (*Sotalia guianensis*) in Guanabara Bay, southeastern Brazil. J Acoust Soc Am 137(1):15–19

Andrew RK, Howe BM, Mercer JA, Dzieciuch MA (2002) Ocean ambient sound: Comparing the 1960's with the 1990's for a receiver off the California coast. Acoust Res Lett 3(2):65–70

Araújo JP, Passavante JZO, Souto AS (2003) Behavior of the estuarine dolphin, *Sotalia guianensis*, at Dolphin Bay - Pipa - Rio Grande do Norte - Brazil. Trop Oceanogr 31(2):101–112

Araújo JP, Araújo ME, Souto A, Parente CL, Geise L (2007) The influence of seasonality, tide and time of activities on the behavior of *Sotalia guianensis* (Van Bénéden) (Cetacea, Delphinidae) in Pernambuco, Brazil. Revista Brasileira de Zoologia 24(4):1122–1130. doi:10.1590/S0101-81752007000400032

Arveson PT, Vendittis DJ (2000) Radiated noise characteristics of a modern cargo ship. J Acoust Soc Am 107(1):118–129

Ary W, Cranford TW, Berta A, Krysl P (2016) Functional morphology and symmetry in the odontocete ear complex. In: Popper AN, Hawkins A (eds) The effects of noise on aquatic life II, Advances in Experimental Medicine and Biology. Springer, New York. doi:10.1007/978-1-4939-2981-8_7

Avila-Foucat VS, Vargas AS, Jordan AF, Flores OMR (2013) The impact of vessel crowding on the probability of tourists returning to whale watching in Banderas Bay, Mexico. Ocean Coast Manag 78:12–17. doi:10.1016/j.ocecoaman.2013.03.002

Bain D, Dahlheim M (1994) Effects of masking noise on detection thresholds of killer whales. In: Loughlin T (ed) Marine mammals and the Exxon Valdez. Academic Press, San Diego, pp 243–256

Barrios-Garrido H, De Turris-Morales K, Nash CM, Delgado-Ortega G, Nínive ER (2016) Acoustic parameters of Guiana Dolphin (*Sotalia guianensis*) Whistles in the Southern Gulf of Venezuela. Aquat Mamm 42(2):127–136. doi:10.1578/AM.42.2.2016.127

Batista RLG, Alvarez MR, dos do MSS R, Cremer MJ, Schiavetti A (2014) Site fidelity and habitat use of the Guiana dolphin, *Sotalia guianensis* (Cetacea: Delphinidae), in the estuary of the Paraguaçú River, northeastern Brazil. North-Western J Zool 10(1):93–100

Bejder L, Dawson SM, Harraway JA (1999) Responses by Hector's dolphins to boats and swimmers in Porpoise Bay, New Zealand. Mar Mamm Sci 15(3):738–750. doi:10.1111/j.1748-7692.1999.tb00840.x

Bisi TL (2001) Estimativa da densidade populacional do Boto-cinza (Cetacea, Delphinidae) na região estuarina lagunar de Cananéia, SP. Dissertation, Universidade Estadual Paulista "Júlio de Mesquita Filho"

Bittencourt ML (1984) Contribuições para a identificação de cetáceos na costa e águas interiores do Brasil. Arquivo de Tecnologia do Paraná 27:529–547

Branstetter BK, Finneran JJ (2008) Comodulation masking release in bottlenose dolphins (*Tursiops truncatus*). J Acoust Soc Am 124:625–633

Breitzke M, Bohlen T (2010) Modelling sound propagation in the Southern Ocean to estimate the acoustic impact of seismic research surveys on marine mammals. Geophys J Int 181(2):818–846. doi:10.1111/j.1365-246X.2010.04541.x

Caldwell MC, Caldwell DK (1968) Vocalization of naive captive dolphins in small groups. Science 159:1121–1123

Cantor M, Wedekin L L, Daura-Jorge F G, Rossi-Santos M R, Simões-Lopes P C (2012) Assessing population parameters and trends of Guiana dolphins (*Sotalia guianensis*): An eight-year mark-recapture study. Mar Mamm Sci 28(1)63–83. doi.org:10.1111/j.1748-7692.2010.00456.x

Carr T, Bonde RK (2000) Tucuxi (*Sotalia fluviatilis*) occurs in Nicarágua, 800 km of its previously known range. Mar Mamm Sci 16:447–452

Carrera ML, Favaro EGP, Souto A (2008) The response of marine tucuxis (*Sotalia fluviatilis*) towards tourist boats involves avoidance behavior and a reduction in foraging. Anim Welf 17:117–123

Castellote M, Llorens C (2016) Review of the effects of offshore seismic surveys in cetaceans: are mass strandings a possibility. In: Popper AN, Hawkins A (eds) The effects of noise on aquatic life II, Advances in Experimental Medicine and Biology, p 875. doi:10.1007/978-1-4939-2981-8_16

Clark CW, Ellison WT, Southall BL, Hatch L, Van Parijs SM, Frankel A, Ponirakis D (2009) Acoustic masking in marine ecosystems: intuitions, analysis, and implication. Mar Ecol Prog Ser 395:201–222. doi:10.3354/meps08402

Clarke D, Dickerson C, Reine K (2002) Characterization of underwater sounds produced by dredges. Dredging 119:64. doi:10.1061/40680(2003)64

Coscarella JM, Dans SL, Crespo EA, Pedraza SN (2003) Potential impact of unregulated dolphin-watching activities in Patagonia. J Cetacean Res Manag 5(1):77–84

Cremer MR, Pinheiro PS, Simões-Lopes PC (2012) Prey consumed by Guiana dolphin *Sotalia guianensis* (Cetacea, Delphinidae) and franciscana dolphin *Pontoporia blainvillei* (Cetacea, Pontoporiidae) in an estuarine environment in southern Brazil. Iheringia Serie Zoologia 102(2):131–137

Cunha HA, Silva VMF, Lailson-Brito J Jr, Santos MCO, Flores PAC, Martin AR, Azevedo AF, Fragoso ABL, Zanelatto RC, Solé-Cava AM (2005) Riverine and marine ecotypes of *Sotalia* dolphins are different species. Mar Biol 148:449–457

da Silva VMF, Best RC (1996) *Sotalia fluviatilis*. Mamm Species 527:1–17

Daura-Jorge FG, Wedekin LL, Piacentini VDQ, Simões-Lopes PC (2005) Seasonal and daily patterns of group size, cohesion and activity of the estuarine dolphin, *Sotalia guianensis* (P.J. van Bénéden) (Cetacea, Delphinidae), in southern Brazil. Revista Brasileira de Zoologia 22(4):1014–1021. doi:10.1590/S0101-81752005000400029

Daura-Jorge FG, Rossi-Santos MR, Wedekin LL, Simões-Lopes PC (2007) Behavioral patterns and movement intensity of *Sotalia guianensis* (P. J. van Bénéden) (Cetacea, Delphinidae) in two different areas on the Brazilian coast. Revista Brasileira de Zoologia 24(2):265–270

Daura-Jorge FG, Wedekin LL, Simões-Lopes PC (2011) Feeding habits of the Guiana dolphin, *Sotalia guianensis* (Cetacea: Delphinidae), in Norte Bay, southern Brazil. Sci Mar 75(1):163–169. doi:10.3989/scimar.2011.75n1163

de Albuquerque NS, da Souto AS (2013) Motorboat noise can potentially mask the whistle sound of estuarine dolphins (*Sotalia guianensis*). Ethnobiol Conservation 2(5):1–15

De Andrade LG, Lima IMS, Bittencourt L, Bisi TL, Lailson-Brito J Jr, de Azevedo AF (2015) High-frequency whistles of Guiana dolphins (*Sotalia guianensis*) in Guanabara Bay, southeastern Brazil. J Acoust Soc Am 137(1):15–19. doi:10.1121/1.4902428

de Filla GF, de Monteiro-Filho ELA (2009) Monitoring tourism schooners observing estuarine dolphins (*Sotalia guianensis*) in the Estuarine Complex of Cananéia, south-east Brazil. Aquat Conserv Mar Freshwat Ecosyst 19:772–778. doi:10.1002/aqc

de Filla GF, Atem ACG, Bisi TL, de Oliveira LV, Domit C, Gonçalves M, de Monteiro-Filho ELA et al (2008) Proposal for creation of a "zoning with regulation of use in the Cananéia estuarine-lagoon complex" aiming the conservation of the estuarine dolphin, *Sotalia guianensis* (van Bénéden) (Cetacea, Delphinidae). Pan-Am J Aquat Sci 3(1):75–83

de Filla GF, de Oliveira CIB, Gonçalves JM, de Monteiro-Filho ELA (2012) The economic evaluation of estuarine dolphin (*Sotalia guianensis*) watching tourism in the Cananéia region, south-eastern Brazil. Int J Green Econ 6(1):95–116. doi:10.1504/IJGE.2012.048427

de Flores PAC, Bazzalo M, Caballero S, de Santos MCO, Rossi-Santos MR, Trujillo F, Bolaños-Jimenez J, Cremer MJ, May-Collado LJ, de Silva FJL, Montiel-Villalobos MG, de Azevedo AF, Meirelles ACO, Flach L, Barrios-Garrido H, Simões-Lopes PC, Cunha HA, Van Waerebeek K (2010) Proposed english common name for the neotropical delphinid *Sotalia guianensis* (P.J. Van Bénedén, 1864). Lat Am J Aquat Mamm 8(1–2):179–181

De Fontaubert AC, Downes DR, Agardy TS (1996) Biodiversity in the seas: implementing the convention on biological diversity in marine and coastal habitats. IUCN, Gland and Cambridge, p 82

de Meirelles ACO, Ribeiro ÁC, Silva CPN, Soares Filho AA (2010) Records of Guiana dolphin, *Sotalia guianensis*, in the State of Ceará, Northeastern Brazil. Lat Am J Aquat Mamm 8(1–2):97–102. doi:10.5597/lajam00157

de Oshima JEF, de Santos MCO, Bazzalo M, de Flores PAC, do Pupim FN (2010) Home ranges of Guiana dolphins (*Sotalia guianensis*) (Cetacea: Delphinidae) in the Cananéia estuary, Brazil. J Mar Biol Assoc UK 90(8):1641–1647. doi.org/10.1017/S0025315410001311

Deconto LS, Monteiro-Filho ELA (2013) High initial and minimum frequencies of whistles in the southeast and south of Brazil. J Acoust Soc Am 134(5):3899–3904

Deconto LS, de Monteiro-Filho ELA (2016) Day and night sounds of the Guiana dolphin, *Sotalia guianensis* (Cetacea: Delphinidae) in southeastern Brazil. Acta Ethol 19:61–68. doi:10.1007/s10211-015-0223-y

Di Benedito APM, Rosas FCW (2008) Mortalidade. In: Monteiro-Filho ELA, Monteiro KDKA (eds) Biologia, ecologia e conservação do Boto-Cinza. Páginas & Letras Editora, São Paulo, pp 211–222

do Valle AL, FCC M (2006) Alterações comportamentais do golfinho *Sotalia guianensis* (Gervais, 1953) provocadas por embarcações. Biotemas 19(1):75–80

Erbe C (2002) Underwater noise of whale-watching boats and potential effects on Killer whales (*Orcinus orca*), based on an acoustic impact model. Mar Mamm Sci 18:394–418. doi:10.1111/j.1748-7692.2002.tb01045.x

Erbe C, McCauley R, Gavrilov A (2016) Characterizing marine soundscapes. In: Popper A, Hawkins A The effects of noise on aquatic life II, cap 31, pp 265–271

Erber C, Simão SM (2004) Analysis of whistles produced by the Tucuxi dolphin *Sotalia fluviatilis* from Sepetiba Bay, Brazil. An Acad Bras Cienc 76(2):381–385

Evans P G H, Cadwell PJ, Lewis E (1992) An experimental study of the effects of pleasure craft noise upon bottlenose dolphins in Cardigan Bay, West Wales. In: Evans PGH European research on cetaceans, European Cetacean Society, San Remo, pp 43–46

Fernández A, Arbelo M, Deaville R, Patterson IAP, Castro P, Baker JR, Degollada E, Ross HM, Herráez P, Pocknell AM, Rodríguez E, Howie FE, Espinosa A, Reid RJ, Jaber JR, Martin V, Cunningham AA, Jepson PD (2004) Pathology: whales, sonar and decompression sickness (reply). Nature 428:U1–U2

Ferreira RG, Santos E Jr, Pansard K, Ananias S (2005) Reserva Faunística Costeira de Tibau do Sul em defesa de um turismo sustentável no município. Report presented to IBAMA, IDEMA and GRPU for delimitation of a Faunal Reserve

Figueiredo LD, Simão SM (2009) Possible occurrence of signature whistles in a population of *Sotalia guianensis* (Cetacea, Delphinidae) living in Sepetiba Bay, Brazil. J Acoust Soc Am 126(3):1563–1569. doi:10.1121/1.3158822

Flach L, Flach PA, Chiarello AG (2008) Aspects of behavioral ecology of *Sotalia guianensis* in Sepetiba Bay, southeast Brazil. Mar Mamm Sci 24(3):503–515. doi:10.1111/j.1748-7692.2008.00198.x

Flores PAC (2002) Tucuxi *Sotalia fluviatilis*. In: Perrin WF, Würsig B, Thewissen GM (eds) Encyclopedia of marine mammals. Academic Press, San Diego, pp 1267–1269

Flores PAC, da Silva VMF (2009) Tucuxi and Guiana dolphin. In: Perrin WF, Wursig B, Thewissen JGM (eds) Encyclopedia of marine mammals. II. Academic Press, Burlington, pp 1188–1192

Gamboa-Poveda M, May-Collado LJ (2006) Insights on the occurrence, residency, and behavior of two coastal dolphins from Gandoca-Manzanillo, Costa Rica: *Sotalia guianensis* and *Tursiops truncatus* (Family Delphinidae). Science 58(4)

Garcia FCP (2009) Caracterização do repertório sonoro da população do boto-cinza (*Sotalia guianensis*) da região do estuário do rio Caravelas, Brasil. Dissertation, Universidade Federal do Rio Grande do Norte

Garri R (2006) Comportamento de mergulho do boto-cinza, *Sotalia guianensis*, na Enseada do Curral, Praia da Pipa - RN, Brasil. Possíveis adaptações cardíacas ao mergulho. Dissertation, Universidade Federal do Rio Grande do Norte

Garrod B, Wilson J (2002) Nature on the edge: marine ecotourism in coastal peripheral communities. Visions of Sustainability, ATLAS Annual Conference, Estoril, Portugal. doi:10.1016/j.annals. 2003.12.003

Geise L, Borobia M (1987) New Brazilian records for *Kogia*, *Pontoporia*, *Grampus*, and *Sotalia* (Cetacea, Physeteridae, Platanistidae, and Delphinidae). J Mammal 68:873–875. doi:10.2307/1381571

Geise L, Borobia M (1988) Sobre a ocorrência de cetáceos no litoral do Rio de Janeiro, entre 1968 e 1984. Revista Brasileira de Zoologia 4:341–346

Gerstein ER (2002) Peixes-boi, bioacústica e barcos. Am Sci 90:154–165

Gervaise C, Simard Y, Roy N, Kinda B, Ménard N (2012) Shipping noise in whale habitat: characteristics, sources, budget, and impact on belugas in Saguenay–St. Lawrence Marine Park hub. J Acoust Soc Am 132(1):76. doi:10.1121/1.4728190

Glockner-Ferrari DA, Ferrari MJ (1985) Individual identification, behavior, reproduction and distribution of humpback whales, *Megaptera novaeangliae*, in Hawaii, Report. Marine Mammal Commission, Washington, DC, p 36

Gordon J, Moscrop A (1996) Underwater noise pollution and its significance for whales and dolphins. In: Simmonds M, Hutchinson JD (eds) The conservation of Whales and Dolphins. Wiley, New York, pp 381–320

Hardt FAS, Cremer MJ, Tonello AJ Jr, Simões-Lopes PCA (2010) Residence patterns of the guiana dolphin *Sotalia guianensis* in Babitonga Bay, south coast of Brazil. Lat Am J Aquat Mamm 8(1–2):117–121. doi:10.5597/lajam00160

Hastie GD, Wilson B, Tuff LH (2003) Bottlenose dolphins increase breathing synchrony in response to boat traffic. Mar Mamm Sci 19(1):74–84. doi:10.1111/j.1748-7692.2003.tb01093.x

Hastie GD, Wilson B, Wilson LJ, Parsons KM, Thompson PM (2004) Functional mechanisms underlying cetacean distribution patterns: Hotspots for bottlenose dolphins are linked to foraging. Mar Biol 144:397–403

Havukainen L, de Monteiro-Filho ELA, de Filla GF (2011) Population density of *Sotalia guianensis* (Cetacea: Delphinidae) in the Cananéia region, Southeastern Brazil. Int J Trop Biol 59(3):1275–1284. doi:10.15517/rbt.v0i0.3398

Heiler J, Elwen SH, Kriesell HJ, Gridley T (2016) Changes in bottlenose dolphin whistle parameters related to vessel interaction, surface behavior and group composition. Anim Behav 117:167–177. doi:10.1007/s13398-014-0173-7.2

Herz R (1991) Manguezais do Brasil. Editora da Universidade de São Paulo, São Paulo

Hester KC, Peltzer ET, Kirkwood WJ, Brewer PG (2008) Unanticipated consequences of ocean acidification: a noisier ocean at lower pH. Geophys Res Lett 35:L19601

Hildebrand JA (2009) Anthropogenic and natural sources of ambient noise in the ocean. Mar Ecol Prog Ser 395:5–20

Hildebrand J, McDonald M, Calambokidis J, Balcomb K (2006) Whale watch vessel ambient noise in the Haro Strait. (Available from Univ. California San Diego, Scripps Institution of Oceanography, Joint Institute for Marine Observations, 9500 Gilman Dr., La Jolla, CA 92093)

Holt M (2008) Sound exposure and southern resident killer Whales (*Orcinus orca*): a review of current knowledge and data gaps. NOAA Technical Memorandum NMFS-NWFSC-89, pp 77

Holt M, Noren DP, Veirs V, Emmons CK, Veirs S (2009) Speaking up: killer whale (*Orcinus orca*) increase their call amplitude in response to vessel noise. J Acoust Soc Am 125(1):27–32

Houghton J, Holt MM, Giles DA, Hanson MB, Emmons CK, Hogan JT, VanBlaricom GR (2015) The relationship between vessel traffic and noise levels received by killer whales (*Orcinus orca*). PLoS One 10(12):1–20. doi:10.1371/journal.pone.0140119

Hoyt E (2001) Whale-watching 2001: woldwide tourism numbers, expenditures, and expanding socioeconomic benefits. International Fund for Animal Welfare, Yarmouth Ports. doi.org/10.2307/4444572

ICMBio Instituto Chico Mendes de Conservação da Biodiversidade. Portaria (2016) Mamíferos Aquáticos do Brasil: Plano de Ação, versão II, 2ª edn. Instituto Brasileiro do Meio Ambiente e dos Recursos Naturais Renováveis – Brasília, Brasil

IUCN (2016) Secchi E 2012 *Sotalia guianensis*. The IUCN Red List of Threatened Species 2012: e.T181359A17583662. http://dx.doi.org/10.2305/IUCN.UK.2012.RLTS.T181359A17583662.en. Accessed 25 May 2016

Izidoro FB, Le Pendu Y (2012) Estuarine dolphins (*Sotalia guianensis*) (Van Bénéden, 1864) (Cetacea: Delphinidae) in Porto de Ilhéus, Brazil: group characterization and response to ships. J Zool 8(2):10

Jahoda M, Lafortuna CL, Biassoni N et al (2003) Mediterranean fin whale's (*Balaenoptera physalus)* response to small vessels and biopsy sampling assessed through passive tracking and timing of respiration. Mar Mamm Sci 19(1):96–110. doi:10.1111/j.1748-7692.2003.tb01095.x

Janik VM, Thompson PM (1996) Changes in surfacing patterns of bottlenose dolphins in response to boat traffic. Mar Mamm Sci 12:597–602

Jensen FH, Wahlberg M, Bejder L, Madsen PT (2008) Noise levels and masking potential of small whale-watching and research vessels around two delphinid species. Int J Anim Sound Rec 17:166–168
Jensen FH, Bejder L, Wahlberg M, Soto NA, Johnson M, Madsen PT (2009) Vessel noise effects on delphinid communication. Mar Ecol Prog Ser 395:161–175. doi.org/10.3354/meps08204
Kellogg WN (1958) Echo ranging in the porpoise. Science 128:982–988
Kellogg WN, Kohler R, Morris HN (1953) Porpoise sounds as sonar signals. Science 117:239–243
Ketten DR (2004) Marine mammal auditory systems: a summary of audiometric and anatomical data and its implications for underwater acoustic impacts. In: Fütterer DK (ed) Polarforschung, vol 2/3, pp 79–92
Leão ZMAN (1994) Os recifes de coral do sul da Bahia. In: Hetzel B, Castro CB (eds) Corais do Sul da Bahia. Editora Nova Fronteira, Rio de Janeiro, pp 151–159
Leaper R, Renilson M, Ryan C (2014) Reducing underwater noise from large commercial ships: current status and future directions. J Ocean Technol 9:50–69
Lemon M, Lynch TP, Cato DH et al (2006) Response of travelling bottlenose dolphins (*Tursiops aduncus*) to experimental approaches by a powerboat in Jervis Bay, New South Wales, Australia. Biol Conserv 27(3):63–372. doi:10.1016/j.biocon.2005.08.016
Lima A, Le Pendu Y (2014) Evidence for signature whistles in Guiana dolphins (*Sotalia guianensis*) in Ilhéus, northeastern Brazil. J Acoust Soc Am 136(6):3178. doi:10.1121/1.4900829
Lodi L, Hetzel B (1998) Grandes agregações do boto-cinza (*Sotalia fluviatilis*) na Baía da Ilha Grande, Rio de Janeiro. Bioikos 12(2):26–30
Lück M (2015) Education on marine mammal tours – But what do tourists want to learn? Ocean Coast Manag 103:25–33. doi:10.1016/j.ocecoaman.2014.11.002
Lyamin OI, Korneva SM, Rozhnov VV, Mukhametov LM (2015) Cardiorespiratory responses to acoustic noise in belugas. Adv Exp Med Biol 875:665–672. doi:10.1007/978-1-4939-2981-8
Maglio A, Soares C, Bouzidi M, Zabel F, Souami Y, Pavan G (2015) Mapping shipping noise in the Pelagos Sanctuary (French part) through acoustic modelling to assess potential impacts on marine mammals. Sci Rep Port-Cros Natl Park 29:167–185
Marchetto C (2010) Estimativas de abundância de boto-cinza (*Sotalia guianensis*) (Cetacea, Delphinidae) no complexo estuarino de Paranaguá utilizando modelos de captura-recaptura. Dissertation, Universidade do Vale do Rio dos Sinos
Martins DTL (2010) Caracterização do repertório acústico do boto-cinza, *Sotalia guianensis*, e impacto de embarcações no Nordeste do Brasil. Dissertation, Universidade Federal do Rio Grande do Norte
Martins DT, Rossi-Santos MR, de Silva FJL (2016) Effects of anthropogenic noise on the acoustic behaviour of *Sotalia guianensis* (Van Bénéden, 1864) in Pipa, North-eastern Brazil. J Mar Biol Assoc UK 1–9
May-Collado LJ, Quiñones-Lebrón SG (2014) Dolphin changes in whistle structure with watercraft activity depends on their behavioral state. J Acoust Soc Am 135(4):193–198. doi:10.1121/1.4869255
May-Collado LJ, Wartzok D (2008) A comparison of bottlenose dolphin whistles in the Atlantic Ocean: factors promoting whistle variation. J Mammal 89(5):1229–1240
May-Collado LJ, Wartzok D (2009) A characterization of Guyana dolphin (*Sotalia guianensis*) whistles from Costa Rica: the importance of broadband recording systems. Journal of the Acoustic Society of America 125(2):1202–1213. doi:10.1121/1.3058631
May-Collado LJ (2013) Guyana dolphins (*Sotalia guianensis*) from Costa Rica emit whistles that vary with surface behaviors. J Acoust Soc Am 134(4):359–365
Miller LJ, Solangi M, Kuczaj SJ (2008) Immediate response of Atlantic bottlenose dolphins to high-speed personal watercraft in the Mississippi Sound. J Mar Biol Assoc U K 6:1139–1143
Monteiro-Filho ELA, Monteiro KDKA (2001) Low-frequency sounds emitted by *Sotalia fluviatilis guianensis* (Cetácea: Delfhinidae) in an estuarine region in southeastern Brazil. Can J Zool 79:59–66
Monteiro-Filho ELA, Monteiro LR, Reis SF (2002) Skull shape and size divergences in dolphins of the genus *Sotalia*: A tridimensional morphometric analysis. J Mammal 83(1):125–134

Monteiro-Filho ELA, Monteiro LR, Reis SF (2008) Sistemática e distribuição de *Sotalia*. In: Monteiro-Filho ELA, Monteiro KDKA (eds) Biologia, ecologia e conservação do Boto-Cinza. Páginas & Letras Editora, São Paulo

National Center for Ecological Analysis and Synthesis, NCEAS (2016). http://www.nceas.ucsb.edu/globalmarine/impactbyactivity. Accessed 13 Oct 2016

Norris KS, Würsig B, Wells RS (1994) Aerial behavior. In: Norris KS et al (eds) The Hawaiian spinner dolphin. University of California Press, Berkeley, pp 103–120

Nowacek SM, Wells RS, Solow AR (2001) Short-term effects of boat traffic on bottlenose dolphins, *Tursiops truncatus*, in Sarasota Bay, Florida. Mar Mamm Sci 17(4):673–688

Nowacek DP, Thorne LH, Johnston DW, Tyack PL (2007) Responses of cetaceans to anthropogenic noise. Mammal Rev 37(2):81–115

Oliveira F (2007) Conhecimento tradicional e etnoconservação de cetáceos em comunidades caiçaras no município de Cananéia, litoral sul de São Paulo. Dissertation, Universidade de São Paulo

Orams MB (2000) Tourists getting close to whales, is it what whale-watching is all about? Tour Manag 21:561–569

Oswald JN, Rankin S, Barlow J (2004) The effect of recording and analysis bandwidth on acoustic identification of delphinids species. J Acoust Soc Am 116:3178–3185

Parsons ECM, Scarpaci C (2016) Recent advances in Whale-watching. Tour Mar Environ 11(4):251–262. doi.org/10.3727/154427316X14580612748687

Pereira MG, Bazzalo M, Flores PAC (2007) Reações comportamentais na superfície de *Sotalia guianensis* (Cetacea, Delphinidae) durante encontros com embarcações na Baía Norte de Santa Catarina. Rev Bras Zoociências 9(2):123–135

Perrin WF, Reilly SB (1984) Reproductive parameters of dolphins and small whales of the family Delphinidae. Reports of the International Whaling Commission, Special Issue 6, pp 97–133

Perry C (1998) A review of the impact of anthropogenic noise on cetaceans. Report for the Environmental Investigation Agency, London, pp 1–27

Pijanowski BC, Villanueva-Rivera LJ, Dumyahn SL, Farina A, Krause BL, Napoletano BM, Gage SH, Pieretti N (2011) Soundscape ecology: the science of sound in the landscape. Bioscience 61(3):203–216

Pilleri G (1990) Adaptation to water and the evolution of echolocation in the Cetacea. Ethol Ecol Evol 2:135–163

Pirotta E, Merchant ND, Thompson PM, Barton TR, David Lusseau D (2015) Quantifying the effect of boat disturbance on bottlenose dolphin foraging activity. Biol Conserv 181:82–89

Popper AN, Hawkins A (2016) The effects of noise on aquatic life II, Advances in Experimental Medicine and Biology 875. Springer, New York, p 1292

Rabin LA, McCowan B, Hooper SL, Owings DH (2003) Anthropogenic noise and its effect on animal communication: an interface between comparative psychology and conservation biology. Int J Comp Pshychology 16:172–192

Ramos RMA (1997) Determinação de idade e biologia reprodutiva de *Pontoporia blainvillei* e da forma marinha de *Sotalia fluviatilis* (Cetacea: Pontoporiidae e Delphinidae) no norte do Rio de Janeiro. Dissertation, Universidade Estadual do Norte Fluminense

Ramos RMA, Di Benedito APM, Lima NRW (2000) Growth parameters of *Pontoporia blainvillei* and *Sotalia fluviatilis* (Cetacea) in northern Rio de Janeiro, Brazil. Aquat Mamm 26:65–75

Ramos A, Amorim MCP, Fonseca PJ (2012) Propagation of Lusitanian toad fish sounds in estuary shallow Waters. Adv Exp Med Biol 730:173. doi:10.1007/978-1-4419-7311-5_39

Reine KJ, Clarke D, Dickerson C (2014) Characterization of underwater sounds produced by hydraulic and mechanical dredging operations. J Acoust Soc Am 135(6):3280–3294. doi.org/10.1121/1.4875712

Rezende F (2008) Alterações acústico comportamentais. Cap. 04. In: Monteiro-Filho ELA, Monteiro KDKA (eds) Biologia, ecologia e conservação do boto-cinza. Paginas & Letras Editora e Gráfica LTDA, São Paulo

Richardson WJ, Davis RA, Evans CR, Ljungblad DK, Norton P (1987) Summer distribution of Bowhead Whales, *Balaena mysticetus*, Relative to Industry Activities in the Canadian Beaufort Sea, 1980-84. Arctic 40(2):93–104

Richardson WJ, Greene JR, Malme CI, Thomson DH (1995) Marine mammals and noise. Academic Press, San Diego and London. p 576

Ritter F (2003) Boat-related behaviours of cetaceans as a tool for the development of species-specific whale watching guidelines. Eur Cetacean Soc Conf Guide and Abstr 17(1):135

Rosas FCW (2000) Interações com a pesca, mortalidade, idade, reprodução e crescimento de *Sotalia guianensis* e *Pontoporia blainvillei* (Cetacea, Delphinidae e Pontoporiidae) no litoral sul do Estado de São Paulo e litoral do Estado do Paraná, Brasil. Dissertation, Universidade Federal do Paraná

Rosas FCW, Barreto AS, Monteiro-Filho ELA (2003) Age and growth of the estuarine dolphin (*Sotalia guianensis*) (Cetacea, Delphinidae) on the Paraná coast, southern Brazil. Fish Bull 101:377–383

Rosas FCW, Marigo J, Laeta M, Rossi-Santos MR (2010) Natural history of dolphins of the genus *Sotalia*. LAJAM 8(1–2):57–68

Rossi-Santos MR (2015) Oil Industry and noise pollution in the Humpback Whale (*Megaptera novaeangliae*) soundscape ecology of the Southwestern Atlantic Breeding Ground. J Coast Res 31(1):184–195. doi:10.2112/JCOASTRES-D-13-00195.1

Rossi-Santos MR, Oliveira G (2016) Combining cetacean soundscape ecology and Niche modeling to contribute in the mapping of the Brazilian continental shelf. In: Finkl CW, Makowski C (eds) Seafloor Mapping along Continental Shelves, Coastal Research Library 13. Elsevier, Amsterdam. doi:10.1007/978-3-319-25121-9_6

Rossi-Santos MR, Wedekin LL (2006) Evidence of bottom contact behavior by Estuarine Dolphins (*Sotalia guianensis*) on the Eastern Coast of Brazil. Aquat Mamm 32(2):140–144. doi.org/10.1578/AM.32.2.2006.140

Rossi-Santos M, Wedekin LL, Engel MH (2003) Behavioral ecology of the *Sotalia guianensis* dolphins in the extreme southern Bahia region. Revista de Etologia 5:200–201

Rossi-Santos MR, Wedekin LL, Monteiro-Filho ELA (2007) Residence and site fidelity of *Sotalia guianensis* in the Caravelas River Estuary, eastern Brazil. J Mar Biol Assoc UK 87(01):207. doi.org/10.1017/S0025315407055683

Rossi-Santos MR, da Silva JM, Silva FL, Monteiro-Filho ELA (2008) Descriptive parameters of pulsed calls for the spinner dolphin, *Stenella longirostris*, in the Fernando de Noronha. J Mar Biol Assoc U K 88(6):1093–1097

Rossi-Santos MR, Wedekin LL, de Monteiro-Filho ELA (2010) Habitat use of the Guiana Dolphin, *Sotalia guianensis* (Cetacea, Delphinidae), in the Caravelas River Estuary Eastern Brazil. Lat Am J Aquat Mamm 8(1–2):111–116. doi.org/10.5597/lajam00159

Santos-Jr E, Pansard KC, Yamamoto ME, Chellappa S (2006) Comportamento do boto-cinza, *Sotalia guianensis* (Van Benédén) (Cetacea, Delphinidae) na presença de barcos de turismo na Praia de Pipa, Rio Grande do Norte, Brasil. Rev Bras Zool 23(3):661–666

Sauerland M, Dehnhardt G (1998) Underwater audiogram of a tucuxi (*Sotalia fluviatilis guianensis*). J Acoust Soc Am 103:1199–1204

Sayigh LS, Tyack PL, Wells RS, Scott MD (1990) Signature whistles of free-ranging bottlenose dolphins *Tursiops truncatus*: stability and mother–offspring comparisons. Behav Ecol Sociobiol 26:247–260

Schaeffer-Novelli Y, Mesquita HSL, Cintrón-Molero G (1990) The Cananéia lagoon estuarine system, São Paulo, Brazil. Estuaries 13:193–203

Scheidat M, Castro C, Janira G, Williams R (2004) Behavioural responses of humpback whales (*Megaptera novaeangliae*) to whalewatching boats near Isla de la Plata, Machalilla National Park, Ecuador. J Cetacean Res Manag 6(1):1–6

Seppänen J, Nieminen M (2004) Measurements and descriptions of underwater noise in Finland. Geophysica 40(1–2):23–38

Simão SM, Pizzorno JLA, Perry VN, Siciliano S (2000) Aplicação da técnica de fotoidentificação do boto-cinza, *Sotalia fluviatilis* (Cetacea, Delphinidae), da Baía de Sepetiba. Floresta e Ambiente 7(1):31–39

Simão SM, Moreira SC, Barbosa MMC (2007) Ultrasound vocalizations of *Sotalia guianensis* population of Sepetiba Bay, RJ, Brazil. VII Encontro de Tecnologia Em Acústica Submarina, Rio de Janeiro, pp 332–336. doi:10.13140/2.1.3531.3444

Simões-Lopes PC (1987) Sobre a ampliação da distribuição do gênero *Sotalia* Gray, 1886 (Cetacea, Delphinidae) para as águas do Estado de Santa Catarina - Brasil. In: Reunião de trabalho de especialistas em mamíferos aquáticos da América do Sul, 2. Fundação Brasileira para a Conservação da Natureza, Rio de Janeiro, pp 87–88

Simões-Lopes PC (1988) Ocorrência de uma população de *Sotalia fluviatilis* Gervais, 1853, (Cetacea, Delphinidae) no limite sul de sua distribuição, Santa Catarina, Brasil. Biotemas 1(1):57–62

Simões-Lopes PC, Ximenez A (1990) O impacto da pesca artesanal em área de nascimento do boto cinza, *Sotalia fluviatilis*, (Cetacea, Delphinidae), Santa Catarina, Brasil. Biotemas 3(1):67–72

Smolker RA, Mann J, Smuts BB (1993) Use of signature whistles during separations and reunions by wild bottlenose dolphin mothers and infants. Behav Ecol Sociobiol 33:393–402

Sousa-Lima RS, Clark CW (2008) Modeling the effect of boat traffic on the fluctuation of Humpback Whale singing activity in the Abrolhos National Marine Park, Brazil. Can Acoust 36(1):174–181. doi:10.1227/01.NEU.0000255476.06247.F1

Southall BL (2004) Shipping noise and marine mammals: a forum for science, management, and technology. Final Report of the National Oceanic and Atmospheric Administration (NOAA) International Symposium, pp 1–40

Southall BI, Bowles AE, Ellison WT, Finneran JJ, Gentry RL, Greene CRJ, Kastak D, Ketten DR, Miller JH, Nachtigall PE, Richardson WJ, Thomas JA, Tyack P (2007) Marine mammal noise exposure criteria: initial scientific recommendations. Aquat Mamm 33(4):411–521

Souto A, Araújo JP, Geise L, Araújo ME (2006) The surface behavior of the estuarine dolphin in Baía dos Golfinhos, RN, Brazil: a field and comparative study. Zoociências 8:185–193

Stewart RH (2008) Introduction to physical oceanography. Department of Oceanography, Texas A & M University, p 345

Stockin K, Lusseau D, Binedell V, Wiseman N, Orams MB (2008) Tourism affects the behavioural budget of the common dolphin *Delphinus sp*. in the Hauraki Gulf, New Zealand. Mar Ecol Prog Ser 355:287–295

The Acoustic Ecology Institute (2008) Special Report: Ocean Noise 1–29

Torres-Guijarro S, Sobreira-Seoane M, Santos-Domínguez D, Pena A (2013) Evaluation of underwater dredging noise. Inter-Noise: Noise control for quality of life, Innsbuck, Austria, p 7

Tosi CH, Ferreira RG (2009) Behavior of estuarine dolphin *Sotalia guianensis* (Cetácea, Delphinidae), in controlled boat traffic situation at southern coast of Rio Grande do Norte, Brazil. Biodivers Conserv 18:67–78. doi:10.1007/s10531-008-9435-z

Tyack PL (2000a) Function aspects of cetacean communication. In: Mann J, Connor RC, Tyack PL, Whitehead H (eds) Cetacean societies: fields studies of dolphins and whales. The University of Chicago Press, Chicago, pp 270–307

Tyack PL (2000b) Dolphins whistle a signature tune. Science 289:1310–1311

Tyack PL (2008) Implications for marine mammals of large-scale changes in the marine acoustic environment. J Mammal 89(3):549–558

Tyack PL, Janik VM (2013) Effects of noise on acoustic signal production in marine mammals. In: Brumm H (ed) Animal communication and noise, Animal Signals and Communication, vol 2. Springer, New York, pp 251–271

Urick RJ (1983) Principles of underwater sound, 3rd edn. McGraw-Hill, New York, p 423

Van Parijs SM, Corkeron PJ (2001) Boat traffic affects the acoustic behavior of Pacific humpback dolphins, *Sousa chinensis*. J Mar Biol Assoc U K 81:533–538

Watkins WA (1986) Whale reactions to human activities in Cape Cod waters. Mar Mamm Sci 2(4):251–262

Wedekin L L (2007) Preferência de hábitat pelo boto-cinza, *Sotalia guianensis* (Cetacea, Delphinidae) em diferentes escalas espaciais na costa sul do Brasil. Dissertation, Universidade Federal do Paraná

Wedekin LL, Daura-Jorge FG, Simões-Lopes PC (2002) Desenhos de unidades de conservação marinhas com cetáceos: estudo do caso do boto-cinza *Sotalia guianensis*, na Baía Norte de Santa Catarina, sul do Brasil. In: Anais Do III Congresso Brasileiro de Unidades de Conservação, pp 56–62

Wedekin LL, Daura-Jorge FG, Simoes-Lopes PC (2004) An aggressive interaction between bottlenose dolphins (*Tursiops truncatus*) and estuarine dolphins (*Sotalia guianensis*) in southern Brazil. Aquat Mamm 30:391–397

Weilgart LS (2007) A brief review of known effects of noise on marine mammals. Int J Comp Psychol 20:159–168

Whitehead H, Reeves RR, Tyack PL (2000) Science and the conservation, protection, and management of wild cetaceans. In: Mann J, Connor RC, Tyack P, Whitehead H (eds) Cetacean Societies: Field studies of dolphins and whales. The University of Chicago Press, Chicago and London, pp 308–332

Wright AJ, Highfill L (2007) Considerations of the effects of noise in marine mammals and other animals. Int J Comp Psychol 20:89–316

Wright AJ, Kuczaj S (2007) Noise-related stress and marine mammals: An introduction. Int J Comp Psychol 20:3–8

Wright AJ, Soto NA, Baldwin AL, Bateson M, Beale CM, Clark C, Deak T, Edwards EF, Fernández A, Godinho A, Hatch LT, Kakuschke A, Lusseau D, Martineau D, Romero LD, Weilgart LS, Wintle BA, Notarbartolo-di-sciara G, Martin V (2007) Do marine mammals experience stress related to anthropogenic noise? Int J Comp Psychol 20:274–316

Zappes CA, Andriolo A, Silva FO, de Monteiro-Filho ELA (2009) Potencial conflicts between fishermen and *Sotalia guianensis* (Van Bénéden, 1864) (Cetacea, Delphinidae) in Brazil. Sitientibus Serie Ciências Biológicas 9(4):208–214

Zappes CA, Nery MF, Andriolo A, Simão SM (2010) Ethnobiology and photo-identification: identifying anthropic impacts on boto-cinza dolphin *Sotalia guianensis* in Sepetiba Bay, Brazil. Rev Bras Biociências 8(2):221–224

Chapter 14
The Atlantic Goliath Grouper: Conservation Strategies for a Critically Endangered Species in Brazil

Maurício Hostim-Silva, Áthila A. Bertoncini, Maíra Borgonha, Jonas Rodrigues Leite, Matheus O. Freitas, Felippe Alexandre Daros, Leonardo S. Bueno, Ana Paula C. Farro, and Christopher C. Koenig

Abstract Commonly known as the Atlantic Goliath Grouper, *Epinephelus itajara* (Lichtenstein 1822) is a marine fish in the family Epinephelidae. Because of slow growth, late maturity, large adult size (up to 2 m long and weighing 400 kg), and shallow coastal habitats, the species is highly vulnerable to anthropogenic activities, such as fishing and activities that cause habitat loss and reduce water quality. This endangered species (IUCN) is at great risk of being lost from Brazilian waters.

M. Hostim-Silva (✉) • A.P.C. Farro
Universidade Federal do Espírito Santo – UFES,
Rodovia BR 101 Norte, Km. 60, Bairro Litorâneo, S/N São Mateus, ES 29932-900, Brazil
e-mail: mhostim@gmail.com

Á.A. Bertoncini
Universidade Federal do Estado do Rio de Janeiro – UNIRIO,
296 Pasteur Av., Urca, RJ 22290-240, Brazil

M. Borgonha
Universidade Federal Fluminense – UFF, Outeiro São João Batista Centro,
Niterói, RJ 24020-141, Brazil

J.R. Leite • M.O. Freitas
Rede Abrolhos, Instituto de Pesquisas Jardim Botânico do Rio de Janeiro, Pacheco Leão St, 920, Rio de Janeiro, RJ 22460-000, Brazil

Instituto Meros do Brasil, 67 Benjamin Constant St, Curitiba, PR 80060-020, Brazil

F.A. Daros
Instituto Meros do Brasil, 67 Benjamin Constant St, Curitiba, PR 80060-020, Brazil

Pós-graduação em Ciências e Tecnologia Ambiental (PGCTA) - Universidade do Vale do Itajaí (UNIVALI) , Rua Uruguai, 458 – Centro, Itajaí, SC 88302-901, Brazil

M.R. Rossi-Santos, C.W. Finkl (eds.), *Advances in Marine Vertebrate Research in Latin America*, Coastal Research Library 22,
DOI 10.1007/978-3-319-56985-7_14

14.1 Presentation

Commonly known as the Atlantic Goliath Grouper, *Epinephelus itajara* (Lichtenstein 1822) is a marine fish in the family Epinephelidae. Because of slow growth, late maturity, large adult size (up to 2 m long and weighing 400 kg), and shallow coastal habitats, the species is highly vulnerable to anthropogenic activities, such as fishing and activities the cause habitat loss and reduced water quality. This endangered species (IUCN) is at risk of being lost from Brazilian waters.

Atlantic Goliath Grouper, *Epinephelus itajara*, occur in tropical and subtropical waters of the Atlantic Ocean. They live in estuarine and coastal areas, including mangroves, coral reefs, rocky shores, submerged rocks, and artificial structures, such as shipwrecks and piers. *Epinephelus itajara* feed on crustaceans and fish, such as stingrays and catfish, and sometimes even young turtles. Besides their large size, they are considered gentle, allowing close contact with divers.

Epinephelus itajara can live up to 40 years and they become reproductively mature at seven years old, and just over one meter in total length. They form spawning aggregations in specific times and places. However, researchers have shown that spawning aggregations have disappeared from many traditional sites, but the scale of illegal fishing activities on Goliath Grouper populations in Brazilian waters is still unknown.

Concern about declining Goliath Grouper populations throughout the Atlantic and Caribbean has induced many researchers to focus on the conservation of this species which is classified as critically endangered by the International Union for Conservation of Nature (Craig 2011). The species has been protected since 1990 in the southeastern United States. In Brazil the capture of Goliath Grouper has been prohibited since 2002 by IBAMA Ordinance No. 121/2002. Through this instrument, Goliath Grouper has become the first marine fish to have a specific law establishing the prohibition of its capture, transportation, and commerce for a period of 5 years. In 2007, Ordinance No. 42/2007 from IBAMA was prolonged for 5 more years.

In 2012, combined efforts from *Projeto Meros do Brasil* ("Brazilian Atlantic Goliath Grouper Project") research and the Ministry of Environment (through the ICMBio) generated subsidies for a new renovation for 3 more years, prioritizing the development of more comprehensive studies, and to give time for the population

L.S. Bueno
Instituto Meros do Brasil, 67 Benjamin Constant St, Curitiba, PR 80060-020, Brazil

Instituto COMAR – Conservação Marinha do Brasil,
104 Helena Degelmann St, Joinville, SC 89218-580, Brazil

C.C. Koenig
The Florida State University Coastal and Marine Laboratory,
3618 Coastal Highway, St Teresa, FL 32358, USA

recover in Brazil, as determined by the Interministerial Normative Instruction (INI) No. 13/2012. In 2016, a new Normative Instruction was launched, protecting Goliath Grouper from capture, transport, and commerce until 2023.

The following presents the research topics that *Projeto Meros do Brasil* has been dedicated to in the past 10 years. The aim is to produce and disseminate modern scientific knowledge about the populations of Brazilian Goliath Grouper, establishing important national and international partnerships to access new technologies, especially the non-lethal ones, generating data that can be used for the conservation of the species in Brazil.

14.2 The *Projeto Meros do Brasil* ("Brazilian Atlantic Goliath Grouper Project")

In 2002, observing the declining of Goliath Grouper populations over the Brazilian coast, a group of researchers from the South of the country founded the *Projeto Meros do Brasil* ("Brazilian Atlantic Goliath Grouper Project"), in order to raise funds and facilitate researches and conservation efforts of Goliath Grouper.

In 2007 and in 2012, the *Projeto Meros do Brasil* was funded by the state-owned oil company *Petrobras*, through its program *Petrobras Ambiental*, by public notice process. *Petrobras* sponsorship encouraged and is still supporting the development of new research in Brazil, not only for the preservation of the Atlantic Goliath Grouper, but also for coastal environments associated with the species, such as mangroves, coral reefs, and rocky shores. In 2013, the *Instituto Meros do Brasil* was founded, with the objective of continuing, improving and implementing management and governance measures of the Project.

Nowadays, the *Projeto Meros do Brasil* works in research and conservation through studies of fisheries biology, genetics, marine fish aquaculture, environmental education, participatory research, scientific diving, and knowledge dissemination among seven coastal Brazilian states.

The *Instituto Meros do Brasil* has five institutions as their main partners, working together in different focal points: *Instituto de Pesca* ("Fisheries Institute", located in São Paulo); Rio de Janeiro State Federal University (UNIRIO); Espírito Santo Federal University (UFES/CEUNES); Alagoas Federal University (UFAL); Instituto Recifes Costeiros (IRCOS; "Coastal Reefs Institute", located in Pernambuco); Pará Federal University (UFPA); Instituto Conservação Marinha do Brasil (COMAR) located in Santa Catarina state and Museu de História Natural do Capão da Imbúia (Natural History Museum Of capão da Imbúia) in Paraná.

The Project works in autonomous partnership under technical and scientific cooperation with more than 50 governmental and non-governmental institutions, as well as organized civil society groups in Brazil and abroad, assisting the development of actions in its many Focal Points. Therefore, the Project proposes to unify the research and conservation challenges of a species that is widely distributed through the Brazilian coast.

This chapter present methods and techniques of six lines of research currently in progress, which can be used for other threatened species in Latin America: the knowledge from traditional populations, the use of submarine photography and video for photo-identification, reproductive biology, captive breeding, genetic, and the study of stable isotopes.

The last part of this chapter is a brief of the exchange experience with the Florida State University, a partner of the *Projeto Meros do Brasil*, bringing to Brazil, well-developed techniques already used for Goliath Grouper population's studies in the USA. Such techniques are already being used along the Brazilian coast.

14.3 Methods & Techniques to Study the *E. itajara* in Latin America and in Brazil

14.3.1 The Study of Traditional Knowledge as Subsidy to Conservation of the E. itajara *in Brazil*

In the past three decades, the study of indigenous people knowledge systems has been gaining increasing recognition. This tendency can be observed in the fields of knowledge of ichthyology, marine conservation, and fisheries management (Johannes 2002; Begossi 2010; Leite and Gasalla 2013).

Many of the studies that deal with the nature of the relationships between human beings and ecosystems, including their elements, have their origin in the Ethnoecology. Defined as the field of knowledge that develops the scientific study behind the traditional ecological knowledge (Marques 2001), the investigation in Ethnoecology has become an important instrument in marine conservation. When working with different knowledge systems, the researcher aims to understand the structure or the components of nature and their classifications; patterns, processes, and relationships among the elements, and the natural events and the use of objects and resources by indigenous people (Toledo and Barrera-Bassols 2009).

Part of Ethnoecology is Ethnoichthyology, aiming to investigate specifically the traditional knowledge that indigenous people have regarding fish (Marques 1991; Begossi 1993; Diegues 2000). Generally, studies in Ethnoichthyology try to record the knowledge that traditional and/or indigenous people have relative to morphology, habitat, spatio-temporal distributions, behavior, trophic interactions, feeding, reproduction/spawning, and taxonomy of fish species (Marques 2001; Pinheiro 2004; Mourão and Nordi 2006; Begossi 2010).

The first studies in Ethnoichthyology that provided the knowledge of fishermen regarding the Goliath Grouper, *Epinephelus itajara*, in Brazil started early in the 1990s. As a result of these surveys, clear evidences of the decrease in abundance and size of the species in the Northeastern Brazilian coast was observed (Ferreira and Maida 1995).

Considering the observed decline of populations of the Goliath Grouper in several other areas along the Brazilian coast, in 2002, in a pioneering attitude, representatives from research institutions, the government, and members of the fishing communities met to promote, discuss, and propose important protective measures for the species. As a result, a moratorium was established, prohibiting for 5 years (2002–2007) the capture, transport, and the commerce of Goliath Grouper in Brazil. Surprisingly, the Goliath Grouper became the first marine fish species to have a moratorium in Brazil (Hostim-Silva et al. 2005). Then after, from the starting point of the prohibition, research efforts started addressing the real conservation status of the species, and the proposal of adequate protective measures throughout its occurrence in Brazil.

The first studies regarding Goliath Grouper by the *Projeto Meros do Brasil* were performed in south Brazil between 2002 and 2004 with the support from local fishermen (Gerhardinger et al. 2007a, b), where known and studied spawning aggregations of Goliath Grouper occurred (Bueno et al. 2016). Later, in 2005 and 2007, ethnoichthyological studies were spread to the states of Bahia, Pernambuco, and São Paulo (Gerhardinger et al. 2007a, b). The identification and monitoring of spawning aggregation sites was possible only due the acquisition of bioecological and behavioral information of the Goliath Grouper from the local ecological knowledge, especially the information coming from specialized fishermen, also known as elders (Gerhardinger et al. 2006a, b). The results of these studies with local fishermen showed the existence of old aggregations that were extinct, evidence that revealed a decline of the population of Goliath Grouper, and also prominent threats to the species (Gerhardinger et al. 2006a, b). Such evidence was fundamental to arguments to maintain the moratorium for 5 more years (from 2007 to 2012).

Later, new studies considering the reproductive aspects and threats, based on the knowledge of fishermen using similar methods, became widespread throughout the Brazilian coast (Bertoncini et al. 2012; Ferreira et al. 2014; Jesus et al. 2016).

In regard to the methods and data collecting techniques used in the research with the Goliath Grouper, the information related to the ecological knowledge was obtained preferably from the identified native intellectuals, considered expert-informers (Fig. 14.1), documented by semi-structured interviews, surveys, and mental maps (Hostim-Silva et al. 2005; Gerhardinger et al. 2006a, b, 2007a, b; Bertoncini et al. 2012). The mapping of experts was supported by the use of well-known methods and techniques in Ethnoecological research, such as snowball sampling (Babbie 2012), which is based on the identification and recognition of the main experts in certain topics, who will point out other members of the group they belong to.

Nowadays, works based on traditional/local knowledge about Goliath Grouper by the *Projeto Meros do Brasil* focus on the identification of new spawning aggregation areas. As an example, one of the efforts in which the *Projeto Meros do Brasil* was involved (2009–2011) was the *Projeto Pró-Arribada: Agregações Reprodutivas de Peixes Recifais no Brasil: Subsídios para o Licenciamento Ambiental de Atividades de Exploração e Produção* ("Reproductive Aggregation of Reef Fish in

Fig. 14.1 Use of survey by the researcher of *Projeto Meros do Brasil* in Tamandaré, Pernambuco (Photo: Áthila Bertoncini)

Brazil: Subsidies for Environmental Licensing of Oil and Gas Exploration and Production").

This initiative, resulted from a Federal environmental damage compensation fee, and combined many institutions and researchers through the country, aiming to map spawning aggregation areas of reef fishes along the Brazilian coast.

The Goliath Grouper was one of the main targeted species in this study. At first, the existing data was compiled from fisheries landings monitoring programs, and at the same time, specialized fishermen were interviewed, to investigate the existence of areas and periods when spawning aggregations occurred in order to complement the data and to define the reproductive season of the species.

In this context, the researchers of the *Projeto Meros do Brasil* were assigned to investigate of two areas: one in the South of Brazil (the coast of the Brazilian State of Santa Catarina) (Bertoncini et al. 2012), and another in the East-Northeast (Northern shore of the Brazilian State of Espírito Santo to the Southern shore of the State of Bahia). The area located between Bahia and Espírito Santo, included one of the main reef areas of the Brazilian coast, characterized by the elevated biological importance, considered as a priority for conservation (Mazzei et al. 2016). This coastal zone is known as an area of artisanal fishermen, who perform subsistence fishing or small-scale fishing.

However, the location, as well as the characterization of the Goliath Grouper spawning aggregations is not a simple task, and requires a large sample effort through different seasons and areas yet to be explored. As a result from this study, the information gathered subsidized diving expeditions for 40 different areas, where samples of Goliath Grouper and other Epinephelidae species were found, demonstrating that these areas are in fact potential aggregations sites. The indications from the combination of traditional knowledge and the diving expeditions provided enough support for new field efforts to be continued.

It is important to highlight that a great part of the present interest from government and researchers in spawning aggregations are due the deleterious effects fishing has brought to the main marine fish resources. Additionally, the effects in the exploration and production of oil and gas industry has forced the responsible Environmental Agencies to better understand the dynamics of these spawning aggregations events, and to propose more adequate management plans, in order to guarantee the sustainability of both resources and traditional populations.

14.3.2 Photo-Identification

Photo-identification (photo-iD) is a widely utilized approach on a variety of marine species including marine mammals, sea turtles and cartilaginous fish, providing a single non-invasive mark-recapture technique, which is critical for threatened species such as the Goliath Grouper *Epinephelus itajara*.

Being the largest grouper in the Atlantic Ocean (>2 m total length - TL, > 400 Kg) (Bullock et al. 1992), Goliath Grouper have a series of characteristics, which turn them a potential species to work with photo-iD. They are sedentary, have a high degree of site fidelity – performing yearly spawning migrations and aggregations (Eklund and Schull 2001; Pina-Amargós and Gonzáles-Sansón 2009; Leite et al. 2012) – and have spots and blotches that are unique for each individual. Adults inhabit both natural and artificial reefs, generally in depths less than 50 m (Bullock et al. 1992), within the recreational dive limits (i.e., 40 m), and juveniles are often observed in mangrove areas, considered their nursery grounds.

Investigations on habitat use and abundance of Goliath Grouper in Brazil from a participative survey – which engaged volunteer divers in data-collection (see Giglio et al. 2014b) – provided the first clues to focus efforts on photo-iD research.

Our first investigations (Giglio et al. 2014a) started with a small but intriguing set of images, where scars were used to photo-identify three Goliath Grouper, photographed and re-sighted several times at their home sites along the Brazilian coast, including the longest report for site fidelity, in more than 4 years.

Since 17% of the sightings in the participatory survey (Giglio et al. 2014b), came from Fernando de Noronha Archipelago, a National Marine Park, our efforts were directed to this offshore archipelago.

Fernando de Noronha is located 186 nm off northeastern Brazil (03°50′ S, 32°25′ W), and is under the influence of the northern branch of the South Equatorial Current. The leeward seascape is mainly composed of descending slopes with large scattered boulders and patch reefs, whereas the windward side of the archipelago is characterized by extensive algal-vermetid ridges along rocky shorelines (Maida and Ferreira 1997).

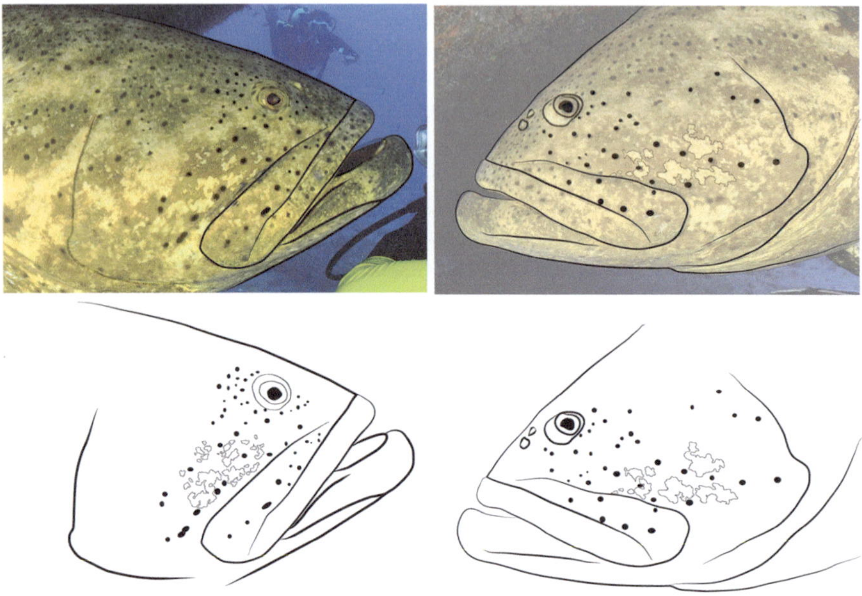

Fig. 14.2 Sketch of the *black dots* that cover the head of Goliath Grouper (*Epinephelus itajara*). Each image was mapped and analyzed in order to identify the individuals at the study site (Photo: Guilherme Cavalcante Vasconcelos)

Our databank of Photographic records came from the Participative Survey Program of *Projeto Meros do Brasil* and from photo/video companies. Recognition of individual animals used natural marks, mainly spots over the head of the Goliath Grouper (Fig. 14.2). The free computer-aided photo-identification I3S (available from http://www.reijns.com/i3s/) was used, although manual inspection of the photo database was performed, considering the many different angles available.

Between April 2004 and May 2013, a total of six different Goliath Grouper (Fig. 14.3) were registered in 84 different days. From April 2004 to November 2006, only four different Goliath Grouper were in the area, disappearing in the following years. The fifth specimen was observed only from December 2007 on, lasting in the records until January 2012, being then replaced by the sixth specimen in October 2012, registered for the last time in May 2013.

Fernando de Noronha National Marine Park provided so far, the relevant data that allows long time analyses. Such factor that can be related to its strong touristic potential (specially for SCUBA diving), allied to consolidated companies that provide photographic services to tourists, where Goliath Grouper sightings figure as a major attraction (see Shideler and Pierce 2016).

Results showed that the six Goliath Grouper populated the archipelago at 11 different dive sites (Fig. 14.4a). *Ilha do Meio* and *Caverna da Sapata* sheltered most of the sightings, accounting to frequencies of occurrence of 46% and 40%, followed by *Pontal Norte* (4%) and *Ilhéu do Frade* (2%), while the remaining sites had smaller contributions (Fig. 14.4b).

Fig. 14.3 The six different Goliath Grouper observed in Fernando de Noronha National Marine Park, from April 2004 to May 2013 (Photos by: Rafael Antunes, Zaira Matheus and Andreza dos Santos)

The first Goliath Grouper (pink circle) showed up in April 2004, and was last observed in April 2006, performing 25 months, being registered other three times between dates. The second (light green circle) was first observed in May 2004, and never seen again after August 2005, 16 months, also with three registers between dates. The third Goliath Grouper (blue circle) showed up in July 2005, leaving in September of the same year, 2 months. The fourth (orange circle) appeared a single time, on November 2006.

On the other hand, the fifth Goliath Grouper (purple circle), showed up on December 2007, and was registered on 63 different dates, up to the last observation on January 2012, being then *replaced* in our registers by the sixth specimen (dark green) in October 2012, registered for the last time in May 2013, 8 months.

Caverna da Sapata and *Ilha do Meio* were the most visited places by the fifth specimen along 50 months in Fernando de Noronha, performing small migrations detected by the indexing system of *Photoshop Adobe Lightroom*®, used by collaborator photographers.

Regarding migrations, the fifth Goliath Grouper migrated distances of ~14 km in 2008, from *Caverna da Sapata* (MAY, 8th) – 56 days – to *Pontal do Norte* (JUL, 3rd) and again – 36 days – to *Caverna da Sapata* (AUG, 8th). Then after in 2011, ~11 km within at least 6 days, between *Ilha do Meio* (DEC 5th), to *Caverna da Sapata* (DEC, 11th), and again at *Ilha do Meio* (DEC 17th), in 2011. The shortest

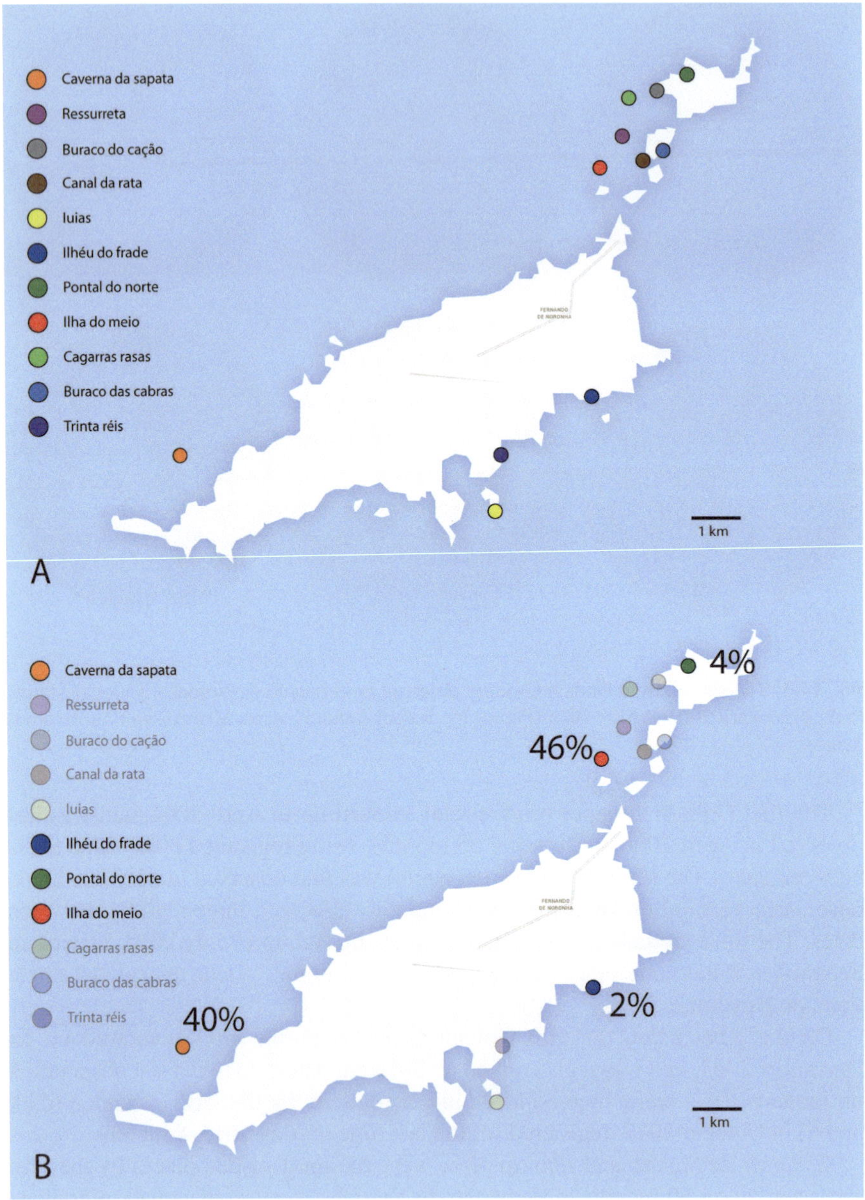

Fig. 14.4 Fernando de Noronha Archipelago, and the 11 sites where Goliath Grouper (*Epinephelus itajara*) were spotted (**a**); percentage contribution of most relevant sites (**b**)

movement time was detected in 2012, from *Ilha do Meio* (DEC, 20th) to *Caverna da Sapata* (DEC 24th), comprising no more than 4 days to perform the ~11 km that separate both areas.

Our future efforts focus on gathering as much as possible of photographic material from diving operators, enthusiast underwater photographers and photo companies, as well as, pre-digital pictures from underwater photographers in order to better refine data on Goliath Grouper movements at this offshore Brazilian Archipelago. Besides that, efforts have shown promising results for tagging smaller Goliath Grouper in estuaries. Since some conventional marks are lost between recaptures, the photo-Id provides a trustful tool in order to prevent the lost of data.

14.3.3 Telemetry and Acoustics

Telemetry is an established method used to track the movements, habitat use, and survival of animals using active tracking or stationary receiver in aquatically environments such lakes, rivers, estuaries, and marine waters (Voegeli et al. 2001).

Telemetry studies had been applied to monitor migration dynamics, survival, habitat use and ecology of tagged animals (Goulette et al. 2014). Since the 1990s, tracking of mobile fauna was restricted to manual monitoring via vessel-based hydrophone surveys. Tagged animals could be monitored for some period of time after implantation, and potentially located and followed again (Welch et al. 2003). With the development of the technology, it is available to researchers to use autonomous hydrophones allowing continuous *in situ* monitoring, which permits the study of migration routes, key habitats using presence, and absence of detections during logger periods of time (Welch et al. 2003). In Brazil, the acoustic telemetry in marine fish started in latter 1990s–2000s, primary being applies to sharks and rays. Nowadays, the use of this technology is growing, and being used in more groups of fish. This technology can be very useful to generate data of endangered species, bring a large possibility to collect many types of important data in a long-term monitoring.

Pioneer studies involving telemetry of the Goliath Grouper took place in Florida, with most of the work performed by the Coleman & Koenig Lab at the Florida State University (Ellis et al. 2013), which brought new findings about the behavior of the Goliath Grouper. The referred study demonstrates the importance of the collaborative partnership network, such as the one used to collect information, with the Florida Atlantic Coast Telemetry (FACT). This initiative is a collaborative partnership of 24 different organizations using passive acoustic telemetry to document site fidelity, habitat preferences, seasonal migration patterns, and reproductive strategies of valuable sportfish, sharks, and marine turtles over all East cost of Florida, South Georgia, and the Bahamas. Ellis et al. (2013) bring up results about the distance travelled by the Goliath Grouper to get to spawning aggregations, reconsidering the idea that the Goliath Grouper was a very slow moving and sedentary fish.

In Brazil, the first efforts to study Goliath Grouper behavior using acoustic telemetry were done in 2013, by the Projeto Meros do Brazil (Project Atlantic

Goliath Grouper of Brazil), in the southern coast of the country. Goliath Grouper telemetry in Brazil starts with adults, aiming to monitor newly confirmed spawning aggregations (Bueno et al. 2016). There are many challenges to work with a such large and critically endangered species. The main one was to catch and handle these animals. An important step was established with the cooperation partnership between the Project and the Coleman & Koenig Lab at the Florida State University, when researchers from Brazil got the opportunity to experience a full –scale methodology. Later, they brought the information regarding the methods to be adapted and implemented under the Brazilian environment. In this research, the acoustic telemetry system by Vemco was used with internal tags (Fig. 14.5), surgically implanted into the abdominal cavity. Besides the acoustic tag, the fish would also get two (pit and ear tag) or three additional tags (dart tag) (Fig. 14.6). The receivers were installed on 4 artificial reefs attached to the reef structure (Fig. 14.7).

The acoustic telemetry data of the Goliath Grouper in South Brazil has shown good results, demonstrating their behavior at a spawning aggregations site. There were 42% of tagged Goliath Grouper showing they lived on the spawning sites after the summer, and returning during the spring, spending 8 to 10 months in unknown sites. This pattern confirms the visual census data presented by Bueno et al. (2016).

To improve the telemetry studies of the Goliath Grouper in Brazil, it is necessary to expand the covered area, numbers of tagged fish, and to develop new studies bringing different questions, such as 3D–positioning monitoring. The implementation of a cooperative system such as FACT will be the best scenario in order to have a large-scale monitoring system, but few studies with telemetry on marine fish are in progress in Brazil at this moment. Without large-scale coverage, only limited spatial and temporal details in the movement of the Goliath Grouper can be observed.

Fig. 14.5 Receiver model VR2W attached on the artificial reef in Parana state, South Brazil (Photo: Áthila Bertoncini)

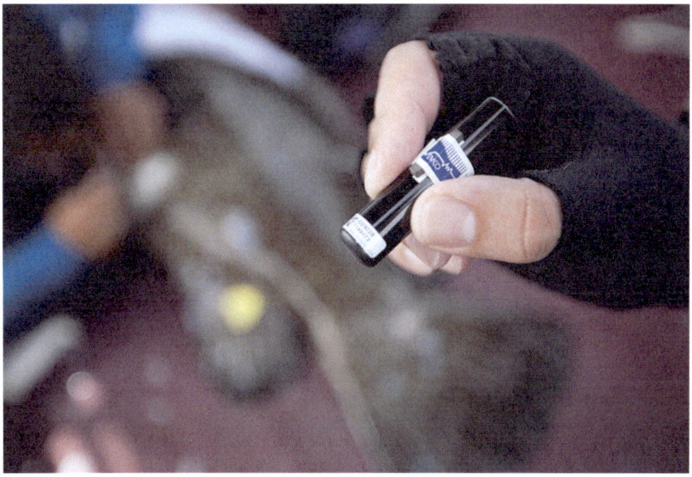

Fig. 14.6 Acoustic tag used on the Goliath Grouper in South Brazil, model V16 by Vemco (Photo: Áthila Bertoncini)

Fig. 14.7 Goliath Grouper handling on board. 1- surgical suture to acoustic tag; 2- ear tag; 3- pit tag; 4- dart tag, implanted under water by a diver before the capture (Photo: Áthila Bertoncini)

14.3.4 Reproduction

The reproductive biology and processes are one of the most well-known topics in fish studies. Fishes comprise the majority of vertebrates, representing all reproductive patterns and adaptations known for vertebrates. In regard to the determination and differentiation of gender among the teleosts, a variety of patterns have been described including hermaphroditism, unisexuality (parthenogenesis), bisexuality (gonochorism), or a combination of these. The evolutionary success of fishes is due to their complex and diversified reproductive patterns. Spawning patterns can be exemplified as: Semelparous (spawn once and die), and Iteroparous, with its many variations such as Isocronal, or Total Spawn (generate a batch of eggs per year), Heterochronal or Serial Spawn (spawn some batches during the reproductive season), and Indeterminate (multiple spawning in one season). Fertilization of eggs can be internal or external (external is most common). In oviparous external fertilization, the type of embryonic nutrition is lecithotropic, while those species with internal fertilization exhibit ovoviviparity which may include lecithotrophy and/or matrotrophy (common to Poeciliids). The most observed process among oviparous species, is the lack of parental care post-fertilization, with liberation of pelagic eggs (e.g., *Centregraulis*, *Anguilla*, Lutjanidae, Epinephalidae), pelagic eggs in gelatinous masses (*Scorpaena*), individual demersal release (*Gadus*) or in masses (*Onchorhynchus*). For species with present parental care, eggs release can be as demersal masses and guard of nests (*Micropterus* spp. and Pomacentrids), the release of larvae (Hemiramphids), or release of juveniles (some Poeciliids such as *Poecilia vivipara*).

Among Hermaphrodites, four general patterns are known: *Simultaneous*, *Protogynous*, *Protandrous*, and *bidirectional sex changer* (Pandian 2011). Here we will focus on the *Protogynous* pattern because it is the most common in the family *Epinephelidae*. Due to a great plasticity and flexibility in such systems, the identification of sexuality among hermaphrodites is a difficult task (Pandian 2011), with the histological analysis of the gonads being the most powerful tool to gender determination (Sadovy and Shapiro 1987; Sadovy de Mitcheson and Liu 2008). This analysis often involves sacrificing the fish.

Many species in the family Epinephelidae are considered threatened or endangered by the IUCN (Sadovy de Mitcheson et al. 2013) necessitating the use of non-lethal techniques to identify gender and gonadal maturation stages. Among these methods, it is important to mention: the use of differentiated color pattern for males and females in reproductive aggregations (Sadovy de Mitcheson and Colin 2012), the use of a borescope (Kynard and Kieffer 2002), sexing based on plasma sex hormone and vitellogenin levels (Craig et al. 2009a, b; Norris and Lopez 2011), examination of the external morphology of urogenital ducts (Vecsei et al. 2003; Koenig et al. 2016) (Fig. 14.8), use of ultrasound imaging (Colombo et al. 2004), gonad biopsy through removed tissue with a hand-operated vacuum pump (Koenig et al. 2016) (Fig. 14.9), ELISA for a model protein – vitellogenin (Specker and Anderson 1994), and microsurgery (gonadectomy) (McLean 1994).

The reproductive biology of Goliath Grouper in southern Atlantic Ocean studies were first performed in the state of Santa Catarina, Southern Brazil, using local ecological knowledge of fishermen (Gerhardinger et al. 2009), and some gonad sample of individuals seized during official investigations of illegal fishing. Freitas et al. (2015), through a partnership between Conservation International Brazil, *Rede Abrolhos* and *Meros do Brasil Project*, determined the size of sexual maturation, sex ratio, and reproductive patterns of the species under the largest and most diverse reef complex of southern Atlantic Ocean, the Abrolhos Bank. Samples were acquired

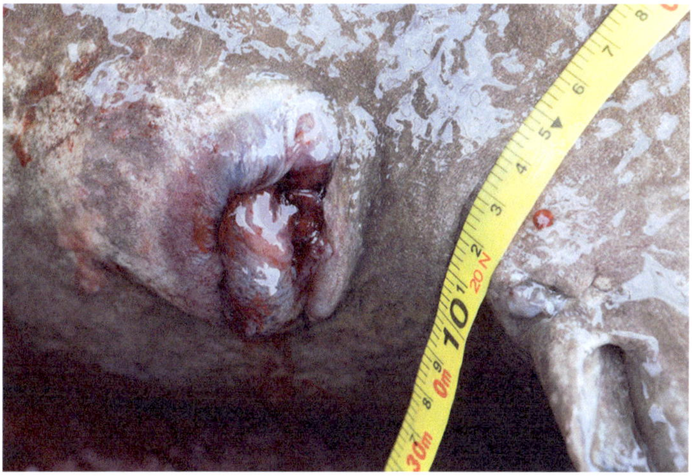

Fig. 14.8 Macroscopic sex determination and oviduct features of the Goliath Grouper females captured in the state of Paraná, southern Brazil (Photo: Áthila Bertoncini)

Fig. 14.9 Gonad biopsies to determine sex and reproductive condition of captured females. A polyethylene catheter is inserted through the oviduct into the lumen of the ovary to remove ovarian tissue with a hand-operated vacuum pump (Photo: Áthila Bertoncini)

from donations of fishermen and official seizures. Suggesting that the species reproduces during the southern Summer (from January to March).

Through the partnership established between the *Projeto Meros do Brasil/Instituto Meros do Brasil* and the Florida State University (Coastal and Marine Laboratory), was implemented in Brazil the protocol using non-lethal sampling (Koenig et al. 2016), generating important understanding of the reproductive biology and the reproductive aggregations of Goliath Grouper in the southern Brazil (Bueno et al. 2016).

The future perspective for Goliath Grouper in Brazil is to amplify these methods to other areas along the coast, empowering the researchers to use hormones and blood plasmatic steroids to determine the gender and the stage of sexual maturity.

14.3.5 Genetics and Molecular Advances for Atlantic Goliath Grouper's Conservation

Atlantic Goliath Grouper has been overfished throughout its range from southern Brazil to the southeastern United States, including the Caribbean and the Gulf of Mexico. Although the species was considered endangered by the International Union for Conservation of Nature (IUCN) since 2007, genetic information about its stocks are still incipient. Some efforts to increase the knowledge about population and evolutionary aspects were applied using different molecular strategies for the *E. itajara* and related species. We present here an overview on these researches and their main results, as well as some progress in forensic genetics and the perspectives to genomic studies of the Atlantic Goliath Grouper.

Considering that some basic information about the phylogeny, diversity and connectivity of the populations are essential for species conservation programs, studies focusing on these aspects can be a good starting point for obtaining knowledge about the real conservation status of fish stocks.

Currently, several molecular tools are available for this purpose. We can highlight mitochondrial and nuclear markers that have been widely applied in conservation genetic studies, and that with the advent of the genomic and new technologies started to be called "traditional markers".

The first molecular study involving the genus *Epinephelus* had a phylogenetic focus. Craig et al. (2001) provided evidence for a monophyletic *Epinephelinae* and a paraphyletic *Cephalopholis* and *Epinephelus,* and suggested that the evaluated grouper lineages were not limited to biogeographical provinces. This study used mitochondrial DNA sequences of the16S ribosomal DNA region, but based on a limited number of taxa. Aiming to expand it, a new research using more markers and species was performed (Craig and Hastings 2007). In this one, two nuclear and two mitochondrial genes from 155 species of serranids and other acanthomorphs were evaluated. The authors showed *Serranidae* as polyphyletic and demonstrated the paraphyly of the genera *Cephalopholis, Epinephelus,* and *Mycteroperca.*

Until 2008, data on genetic diversity and population structure of *Epinephelus* were available only for: hawaiian grouper, *E. quernus* (Rivera et al. 2004); red grouper, *E. morio* (Zatcoff et al. 2004); *E. marginatus* (Maggio et al. 2006); and *E. labriformis* (Craig and Hastings 2007).

Silva-Oliveira et al. (2008) presented the first data about diversity and population connectivity for Atlantic Goliath Grouper. In this study, 116 individuals from the Northern Brazilian coast were evaluated, using 499 bp of the mtDNA control region (D-loop) from five localities (Bragança and Ajuruteua, in the state of Pará; Parnaiba, in the state of Piauí; Fortaleza, in the state of Ceará; and Natal, in the state of Rio

Grande do Norte). The analyses showed that the genetic diversity of *E. itajara* is lower if compared with other already studied endangered species of this genus. There was no structure among the sampling units, but a significant association between genetic variability and geographic distribution for three of these populations (Ajuruteua, Parnaíba and Natal). Genetic diversity was higher in populations from the Amazon region, what might be related to the vast preserved mangrove habitats in this area.

This study focused on a small part of the species distribution in South Atlantic Ocean. On the following year, information about Pacific Ocean and other spot in Atlantic Ocean were published. Craig et al. (2009a, b) tested the hypothesis that Pacific and West Atlantic populations of Atlantic Goliath Grouper constitute a single species by analyzing nuclear (S7 intron) and mitochondrial DNA sequence data (cytochrome b and 16S rRNA gene). They found genetic differences for all markers in the sampled populations: Panama, Belize, Florida and Brazil (Caravelas, in the state of Bahia). These data indicated that Atlantic Goliath Grouper in the West Atlantic is subdivided into discrete populations, and in the Pacific and Western Atlantic it is represented by two or more distinct species. They resurrected the species *Epinephelus quinquefasciatus* (Bocourt 1868) for Pacific Goliath Grouper and recommended that these distinct populations/species required separate management and conservation strategies.

Despite the extensive distribution of Atlantic Goliath Grouper in the Brazilian coast, genetic studies were still scarce and more effort was necessary to improve genetic data from Southwestern stocks.

Benevides et al. (2014) evaluated the genetic diversity of the species and the genetic/evolutionary relationships of its populations along the southwestern Atlantic coast. Samples from 95 specimens of *E. itajara* were collected at ten sites in South America, including French Guiana and Brazil. Eight ISSR primers were used and a total of 92 polymorphic loci were obtained. The results indicated relatively reduced genetic variation, re-emphasizing the low adaptive potential of the species. Populations from Babitonga Bay (in the state of Santa Catarina) presented relatively high degrees of genetic diversity and were evolutionary isolated from all other populations (north, northeast, and southeast Brazil, and French Guiana). The evidences indicated the existence of two Evolutionarily Significant Units comprising *E. itajara* in the Southeast Atlantic Ocean and the authors suggested that conservation prospects for the species took this evidence into account.

Damasceno et al. (2015), using 218 individuals from six locations in Brazilian waters (Bragança, in the state of Pará; Tamandaré, in Pernambuco; Vaza Barris, in Sergipe; Caravelas, in Bahia; Conceição da Barra, in Espírito Santo; and Babitonga, in Santa Catarina) found haplotype diversity ranging from 0.40 to 0.86; and nucleotide diversity from 0.1 to 0.5% for mtDNA control region, which is moderate to high genetic diversity for the species along the coast. Besides evidences of an incipient population structure in Brazilian coast (three to five populations) were found, they recommended these findings to be used for the identification of priority areas for conservation. Interesting to note that this study included unit areas surveyed by previous studies, such as Silva-Oliveira et al. (2008), Craig et al. (2009a, b) and Benevides et al. (2014), but evaluating a different marker.

In addition to the studies involving phylogeny and population genetics aspects, some efforts to increase the available number of specific molecular markers for epinephelids were conducted in the Americas.

Ramírez et al. (2006) characterized nine microsatellite loci and identified 60 genomic regions containing microsatellites in the red hind grouper, *Epinephelus guttatus*. Nine loci were highly polymorphic and were tested for 16 other groupers in the genera *Epinephelus* and *Mycteroperca*.

Silva-Oliveira et al. (2013) performed new nuclear primers for molecular studies of Epinephelidae fishes. EPIC strategy was used and twelve primers were designed for six genomic regions. The results of the study provide a number of appropriate new nuclear DNA markers for future genetic studies in several endangered/overexploited grouper species.

Seyoum et al. (2013) isolated and characterized 29 microsatellites (nuclear markers) for the endangered Atlantic Goliath Grouper (*Epinephelus itajara*) and the Pacific Goliath Grouper (*E. quinquefasciatus*).

Concerning the Forensic Genetics applied to the species conservation, some advance was obtained in Brazil. Since public policies led to a moratorium of Atlantic Goliath Grouper in 2002, capture of the species is controlled. However, only in 2014 it was included as critically endangered (CR) in the list of threatened species (MMA 2014). In 2015, Brazilian government approved a National Conservation Plan for the endangered and socioeconomic important species of Mangrove Ecosystems (PAN Manguezal; MMA 2015), which included the Atlantic Goliath Grouper.

Even with the creation of the PAN, the main difficulty found by the environmental agencies is the supervision of illegal fishing activities. Brazilian coast is considerable length and the environmental organizations lack human and financial resources for law enforcement.

Molecular tools have been used to contribute in this hard task. DNA barcoding, proposed by Hebert et al. (2003), has proved able to help agencies deal with global putative illegal activities and for emitting seafood products authentications (Cawthorn et al. 2012, Di Pinto et al. 2013, Palmeira et al. 2013, Ferrito et al. 2016). The method is based on the sequencing of a unique mtDNA region to identify the correct species.

Examples that can be quoted for *Epinephelus itajara:* Torres et al. (2013) used DNA barcoding for the Atlantic Goliath Grouper and other exploited epinephelids in order to provide procedures for DNA authentication and support illegal fishing activities surveillance.

Damasceno et al. (2016) developed specific primers from COI gene, to be used in multiplex PCR for identification of nine species of Epinephelidae: *Epinephelus itajara, E. quinquefasciatus, E. morio, Hyporthodus flavolimbatus, H. niveatus, Mycteroperca acutirostris, M. bonaci, M. marginata,* and *M. microlepis*. The multiplex PCR technique can be a low cost, rapid, safe, and a practical method to help the regulation and management of endangered marine fish species.

After DNA barcoding, with the advances of Genomics approaches, in the last decade, many important issues in conservation have been best explored for different organisms. Within Genomics, Next-generation Sequencing (NGS) opened up the possibility of working with thousands of molecular markers, even for little-known species. The low cost of NGS, when compared with the amount of data generated by traditional sequencing, has made this technology feasible to widely access the species genome.

Genomic studies focusing on *Epinephelus itajara* were not developed yet. However some researches can be mentioned for the genus. Complete mitochondrial genomes of other *Epinephelus* species were published (Qu et al. 2012; Wang et al. 2016; Wu et al. 2015; Yang et al. 2016).

Zhuang et al. (2013) characterized and compared genomes among 22 groupers species of the family Epinephelidae. Three distinct genome organizations were identified, two of them never previously reported in vertebrates. With this study the authors provide further support for the elevation of the subfamily Epinephelinae to family Epinephelidae; the resurrection of the genus *Hyporthodus*; and the combination of the monotypic genera *Anyperodon* and *Cromileptes* to genus *Epinephelus*, and *Aethaloperca* to genus *Cephalopholis.*

Yu et al. (2016) identified 264,072 raw SNPs (single-nucleotide polymorphisms) by double digest restriction site associated DNA sequencing (ddRADseq), and utilized 3029 of these SNPs to construct a genetic linkage map in orange-spotted grouper (*Epinephelus coioides*). They determined a map containing 24 linkage groups (LGs) with a total genetic distance of 1231.98 cM. Twenty-seven significant growth-related QTLs were identified and 17 genes, including three that have been considered to be involved in fish growth. The authors consider that these QTLs will be useful in marker-assisted selection (MAS) efforts to improve growth-related traits in this economically important fish.

NGS has been considered a very promising advance not only for genetic breeding, but also to answer relevant biological issues in Conservation Genetics.

One good example of discussion about the NGS applications for marine systems comes from Willete et al. (2014), that organized questions mainly into two general categories: (1) population genomics, and (2) local adaptation and demographic history of marine organisms. Some of those questions we can point out as really important for future genetic studies of *Epinephelus itajara*: (a) Is it possible to assess the population structure and genetic variability at a level that can guide a restoration of local stocks?, (b) How might we estimate population size of marine organisms?, (c) How will marine organisms respond to environmental stress, particularly those related to climate change?, and (d) How to infer the spatial connectivity of important commercially populations or threatened species, and how can this support positioning of Fisheries Management Zones (FMZ) or Marine Protected Areas (MPAs)?

Concluding, despite of advances in *Epinephelus itajara* genetics there is still much to do. Data including individuals from different localities and evaluation of more traditional molecular markers associated to next generation sequencing are necessary to figure out more conclusive structure and phylogeography, as well as provide better access to its genome and understanding about the adaptive processes in which the species is involved.

14.3.6 Sclerochronology

Sclerochronology is the study of the physical and chemical patterns in temporal growth rings (daily, sub-daily, or annual) registered in calcified structures of seawater organisms, with the objective to rebuild the life history and its relationship with environmental processes (Pasten 2016). In fish, the most used calcified structures are scales, spines, the rays of the fins, and especially, the otoliths.

Besides being frequently used to estimate age and growth of bony fish (i.e. Francis and Campana 2004; Mendoza 2006; Correia et al. 2009), the otoliths are also used as natural markers in migration studies (Albuquerque et al. 2012; Daros et al. 2016), differentiation of fish stocks (i.e. Correia et al. 2011; Geffen et al. 2011; Silva et al. 2011), determination of spawning site (i.e. Di Franco et al. 2011; Hamer et al. 2011), and in the study of connectivity between spawning site and coastal recruiting zones (i.e. Vasconcelos et al. 2009; Reis-Santos et al. 2012). This is possible because, besides $CaCO_3$ and otoline, other elements are deposited in small concentrations, and these are not chemically transformed or modified after deposition (Campana and Neilson 1985); they can be used as environmental markers (Campana et al. 2000).

The chemical composition of the otoliths proved to be a powerful tool to resolve the population structure of fish characterized by an elevated gene flow, in particular from migratory species, with a long pelagic larvae stage, especially when there is environmental heterogeneity reflected in the chemical composition of the otoliths (Bradbury et al. 2008; Selkoe et al. 2008; Smith and Campana 2010). The knowledge regarding the ecology of fish species is fundamental to identify the population units or stocks, the identification of migratory routes, and the study of connectivity between the spawning zones or nurseries, and the respective recruiting and growth coastal zones. This information is fundamental to a more integrated and rational management of fish resources, as well as for the protection of threatened species due to overfishing (Stephenson 1999), and the creation of protected seawater network (Palumbi et al. 2013; Cook 2014).

However, the extraction of otoliths requires the kill of fish, and when the study is focused on protected or threatened species, the use of non-lethal sampling is preferred (Willmes et al. 2016). Other calcified tissues (i.e. scale, rays and spines, and dorsal scute), other than otoliths are mostly composed by aragonite. These tissues are composed of biological apatite (Willmes et al. 2016), where the incorporation of ions occurs via the blood (Veinott and Evans 1999), which is different from otoliths, where ions are incorporated directly from the endolymph (Campana and Neilson 1995). These structures also provide valuable information in studies, as for example, the chemical signature of the areas of spawning and nursing, age and growth (Fig. 14.10) and ontogenetic migration (Clarke et al. 2007; Brushei and Schull 2009; Smith and Whitledge 2010; Tzadik et al. 2014; Altenritter and Kinnison 2015; Linley et al. 2016), once these tissues present an internal structure in layers, which correspond to periods in time, similar to the otoliths (Erickson 1983). Already tested by other authors, the rays from the dorsal fin of the Goliath Grouper are useful to

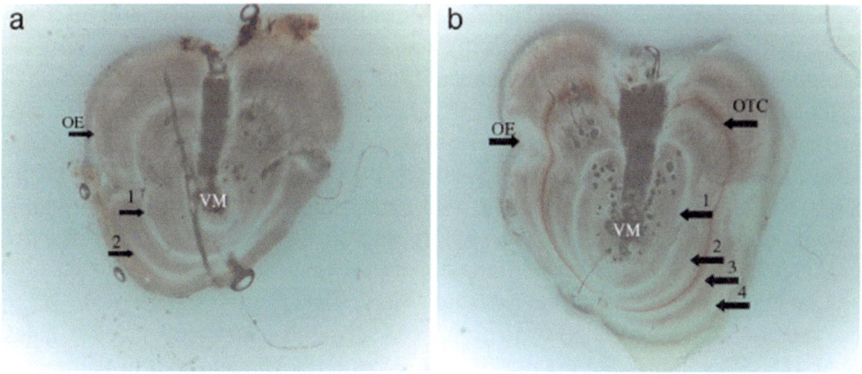

Fig. 14.10 *Epinephelus itajara.* A juvenile Goliath Grouper captured 4 times over 2 years: photographs of dorsal spine cross sections taken from (**a**) date of original capture, October 1999, 365 mm total length (TL); and (**b**) from a subsequent capture event in December 2000, 487 mm TL. This fish was injected with oxytetracycline (OTC) at time of capture in October 1999; note OTC stain visible between 2nd and 3rd annuli in (**b**). *Numbers above arrows* refer to annuli. *VM* vascularized matrix, *OE* outer edge (Brusher and Schull 2009)

determine the age (Brusher and Schull 2009; Artero et al. 2015) and for studies of ontogenetic changes in habitat and diet, using $\delta^{15}N$ (Tzadik et al. 2014). In the future, these structures will be able to demonstrate the life history and connectivity between the environments the Goliath Grouper visits, through the analysis of elementary chemical structures.

14.3.7 Non-lethal Method to Analyze Stable Isotopes

In ecology studies of animal species, especially the ones considered migratory, it is very important to know which geographical regions are being used by certain individuals and their population variations during reproduction and feeding. Normally, fish displacement is observed through external markers, in which the success of markers recovery and the number of retrieved individuals are too low to have a good analysis.

Based on these limitations, in the past decades new techniques emerged to assist researchers to understand a little more about the ecology of animals in nature. These techniques included the use of fatty acid profiles, analysis of molecular genetics, and determination of natural occurrence of stable isotopes in different animal tissues.

There are many researches with stable isotopes in different animal groups. Here, we will focus on the use of stable isotopes in fish, emphasizing non-lethal techniques, and finally, what has been done in this sense for the study of *Goliath Grouper* in the United States of America. In most fish SIA applications, samples of white muscle tissue, liver, gonads, bone, otoliths, brain or the whole fish are used for isotopic determinations (DeNiro and Epstein 1977; Pinnegar and Polunin 1999).

Stable Isotope Analysis (SIA) has had a great impact over many areas of research, since it offers significant advantages in comparison to traditional methods, such as, avoiding sacrifice of extinction endangered species, allowing the comprehension of complex diet networks over large space and time scale, it is cheaper than traditional methods, it enables open sea and deep-water species studies, beyond diet reconstruction of extinct species, what was simply impossible by stomach content analysis. Furthermore, contrary to the traditional analysis of stomach content, stable isotope analysis is not challenged by empty stomachs, and permits a more effective use of collected individuals for different analysis (Vander Zanden et al. 1998).

Stable isotopes use in marine fish studies has a long history. Harrod et al. (2005) examined stable isotope analysis to evaluate the ecological plasticity of eels. Cunjak et al. (2005) demonstrated the use of isotopes as a tracking element of fish movement, combined with use the telemetry and mark-recapture data.

Schlacher et al. (2005) analyzed the relationships between waste water pollution and isotopes in fish, in this study, he observed that δ15N signatures obtained from estuarine fish muscles were effective indicators of nitrogen derived from sewage pollution, since all analyzed fish showed high values of δ15N.

Through the pioneering researches of Smith and Epstein (1971), Minson et al. (1975), and Haines (1976), it was possible to verify that the isotopic signatures (expressed as δ) of the consumers were similar to the isotope signatures of associated plants. Therefore, the analysis of stable isotopes proved to be a useful tool to rebuild diets, characterize trophic relationships, clarify patterns of resource allocation, and construct food webs. As a consequence, the number of studies using this tool in trophic ecology has increased exponentially in the past years.

Based on trophic niche width, isotope variation indicates higher amplitude in the diet and its consistency, while low variation can be attributed to a large population and little consistency, or high amplitude combined with a low population (Vander Zanden et al. 2000). A higher isotopic variability is directly related to a greater diet range, while a low variability can reflect a low diet variety used by a population (Vander Zanden et al. 2000).

Although few studies have used stable isotope analysis to assess diet variability within and between populations, some researchers have suggested that the change in isotopic variability offers a promising tool for studying dietary variation at a population level (Paterson et al. 2006; Syvaranta and Jones 2008).

For a long time muscles (usually white muscles) were used in stable isotope studies, but it was confirmed that the use of tissues that do not require fish sacrifice, such as tissues from scales and fins, can be substituded for muscular tissue. Thus a new moment in the study of stable isotopes arises, using now fish tissues obtained through non-lethal procedures. These sampling methods can be used to investigate rare and endangered species, as well as it allows the analysis of fish scales from scientific collections (Kelly et al. 2006).

Scales can better represent the carbon sources of a fish diet since they are not influenced by lipid dynamics. However, scales seem to be less sensible to short-term changes in diet when compared to muscles, therefore, they should only be used to infer seasonal diet patterns, or to certain phases of the life cycle stages, where the process is slower (Sinnatamby et al. 2007).

Sanderson (2009) observed that $\delta^{15}N$ and $\delta^{13}C$ signatures found in caudal fins were strongly correlated to the signatures found in muscular tissue in space and time, representing a viable alternative of non-lethal analysis. Therefore, it is not necessary to sacrifice a fish subgroup population; instead of that few individuals will need to be sacrificed to calibrate the conversion factor (Kelly et al. 2006).

Thus, the use of fish fins is having a more prominent place as a non-lethal alternative for stable isotope analysis, replacing traditional methods of white muscular tissue sample, undoubtedly reducing the impact of sampling on tropical aquatic systems. By minimizing lethal techniques, researchers may avoid undesirable effects on the system, particularly in those cases where the same sites are being sampled repeatedly, and the possible population-level consequences may result in the removal of individuals for research purposes (Jardine et al. 2011). In addition, this technique allows sampling the same individual over time, which can be very useful in tracking patterns of individual migration or ontogenetic changes in food ecology. More importantly, this powerful tool allows researchers to study fish that are threatened or endangered without scarifying every individual sampled (Graham et al. 2013).

A pioneer study using the determination of $\delta^{15}N$ isotopic values in Goliath Grouper fin rays from two different locations of similar latitudes (East and West coast of Florida, USA), demonstrated that this non-lethal technique is quite promising for the study of near-extinction species (Tzadik et al. 2015).

In Brazil, as in other countries, the Atlantic Goliath Grouper, *Epinephelus itajara* (Lichtenstein 1822) are under threat, with decreasing populations as a result of fishing pressure and habitat degradation. However, life history data for *E. itajara* in Brazil are scarce and, when available, this information is fragmented and apparently insufficient to satisfactorily assess the population status of this species along the Brazilian coast. Due to its endangered status, non-lethal methods are very important for biological sampling in order to cover such knowledge gaps for *E. itajara* and other endangered species in Brazil. To understand the characteristics of a 1-year life history, researchers analyzed a conserved organic matrix that holds the isotopic ratio values over the lifetime of the individual.

Recently, studies conducted at the Florida State University have shown that fishes dorsal fin rays can record the chronology of isotopes through non-lethal sampling. This methodology represents a great advance to study the life history characteristics of endangered species like the Atlantic Goliath Grouper and may be replicated to other grouper species (that are of interest for management).

One of the researches conducted at the Forida State University compared ray fin samples of *E. itajara* collected from three mangrove systems in the north of Espírito Santo state in Brazil (Itaúnas – 18°25′16″S; 39°42′33″W; Conceição da Barra - 18°35′33″S; 39°44′05″W; and Barra Nova – 18°56′58″S; 39°44′41″W) to samples collected in Ten Thousand Islands (USA). In this study, the rays number 5–7 of each

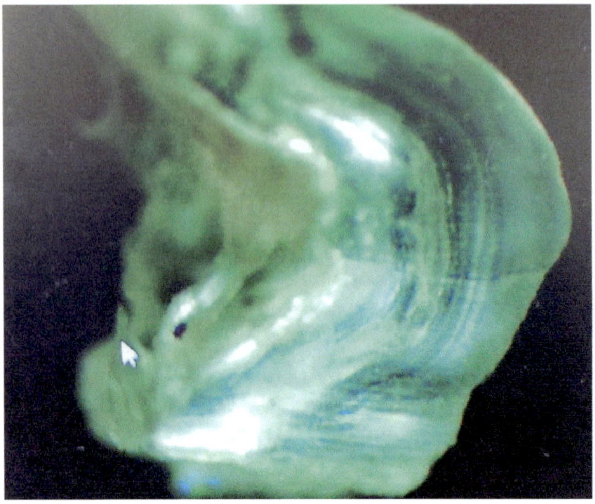

Fig. 14.11 Annuli from a cross section of a dorsal fin ray of Atlantic Goliath Grouper, *Epinephelus itajara*

specimen dorsal fin were extracted to cover its entire structure (including distal pterygiophores) or as close as possible to the base (Fig. 14.11). A set of two sections pad (1.5 mm thick) were cut from the fin rays using a Buehler Isomet© (low-speed saw), as in Tzadik et al. (2015). The aim of these cross sections was to isolate individual rings for stable isotopes analysis. The rings were analyzed for molar mass concentrations of Carbon and Nitrogen (C, N, C:N), and the stable isotopes rates ($\delta^{13}C$ e $\delta^{15}N$). The stable isotopes compositions were expressed in per thousand (‰) using the δ (delta) notation.

All statistical analyzes were performed using MATLAB version R2012b. Age and length distributions among sampling sites were analyzed using Kolmogorov-Smirnov Test. In order to investigate if isotopic ratios were preserved over time, a test was performed using the ^{15}N values of the rings corresponding to a determined age for two age groups within each study site (Fig. 14.12). Age data from Florida samples were used once such information was not available in Brazil yet. The values of $\delta^{13}C$ and $\delta^{15}N$ were plotted against total lengths and age at the moment of capture for all specimens, and the linear least square regressions were calculated for both comparisons.

A total of 53 samples of 5 estuarine habitats were analyzed: 6 from Itaúna, 6 from Conceição da Barra, 22 from Barreiras, 2 from Barra Nova and 17 from Ten Thousand Islands - USA. As a result, the average number of growth annulli from the dorsal fin rays corroborated that individuals captured in Brazil were juveniles, with maximum age of 4 years old.

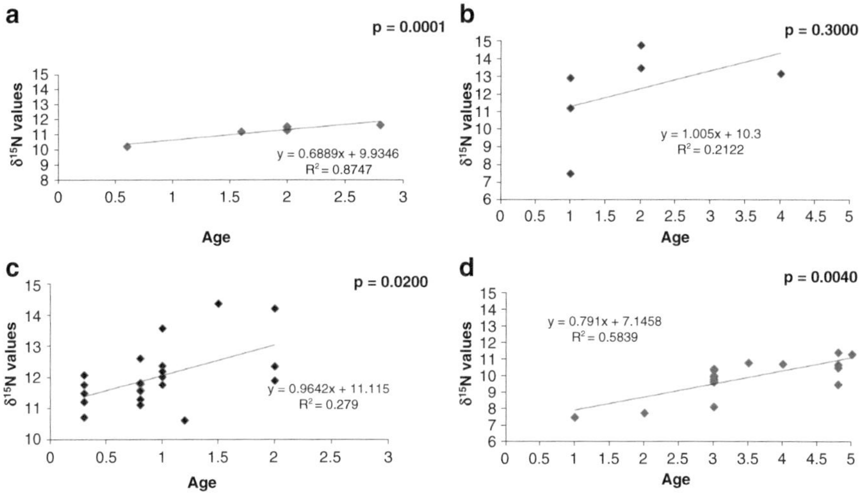

Fig. 14.12 Relationship between Age and $\delta^{15}N$ values for 4 estuarine habitats in Brazil: Itaúnas (**a**), Conceição da Barra (**b**) and Barreiras (**c**); in the United State of America: Ten Thousand Islands (**d**)

14.3.8 Atlantic Goliath Grouper in Aquaculture and in Public Aquariums

In the past years, aquaculture studies with *Epinephelus itajara* are taking place in Brazil through a partnership between the *Instituto de Pesca de São Pau*lo (São Paulo Fishing Institute) and the *Projeto Meros do Brasil.* This research has two main goals related to the species conservation: first, providing captive breeding fingerlings to studies of physiology, development, settlement preferences, among others; second, developing a protocol for Goliath Grouper aquaculture, helping the development of Brazilian mariculture, reducing the fishing impact on wild populations, besides creating minimum criteria to maintain Goliath Grouper in public aquariums. The results found so far, are the first steps to the success of the reproduction of the Goliath Grouper in captivity.

World fish production (fishing + aquaculture) was 167 million tons in 2014, generating US$ 600 billion, seven times higher than cattle industry business (MPA 2015). Aquaculture was responsible for 73 million tons of consumed fish in the world in the year of 2014 (FAO 2016), thus becoming the most growing livestock activity in Brazil and in the world (FAO 2013; SEBRAE 2015). With a growth rate of 5% per year, it is now considered one of the main sources of protein capable to provide the crescent demand for animal protein, because it is more effective, producing up to 250 tons per hectare, and because it generates less environmental impact than present production of cattle, pigs, and poultry (FAO 2013, 2014; MPA 2015). Sustainable fish farming is considered an important strategy to economical development of artisanal fishermen under favorable environmental conditions

(Syukur and Mahrus 2016). Today, mariculture produces 35%, of the whole fish production in the world, with pisciculture producing 23% of it (FAO 2016).

Brazil is considered a future marine fish aquaculture world power due to its long coast line, the favorable weather, high biodiversity, and self-sufficient in grain production (Sanches et al. 2013; Bueno et al. 2015; FAO 2016), but is still placed in the 14th position of world fish producers, with 562 thousand tons produced in 2014 (FAO 2016). Continental fish farming has increasing significantly, but the marine fish aquaculture has not progressed commercially. The difficulty of acquiring fingerlings, and the slow process of licensing seawater use areas has hampered the activity, despite these facts, important basic steps have occurred in the past years.

Some of the characteristics of the Brazilian coast, as quantity of estuaries and bays, together with the lack of expertise in handling offshore environment indicate that the real marine fish aquaculture development in Brazil should start in estuary and salty waters at or near the coast. Therefore euryhaline species or ones with high tolerance to variation of salinity would be ideal for culture. Besides this requirement, adding growth rate, rusticity, quality of the meat, and market price, an ideal target species would be the Atlantic Goliath Grouper *Epinephelus itajara*.

Besides issues related to commercial aquaculture, the development of seawater fish farming of endangered species can contribute as an strategic tool for conservation, helping to reduce captures in nature (providing aquacultured fish), generating agribusiness opportunities, assisting production expansion, creating jobs and income for coastal communities (Sanches 2007), besides enabling the possibility of re-population, and the implementation of semen banks contributing to maintain the genetic diversity of populations (Cloud et al. 1990; Goswami et al. 2016).

Goliath Grouper aquaculture research in Brazil started in 2012, by the Instituto de Pesca de São Paulo, in partnership with the Projeto Meros do Brasil, chaired by Eduardo Gomes Sanches (PhD), in Ubatuba, São Paulo, Brazil. Due to the lack of information regarding the biology of the species in captivity, the knowledge of another Epinephelidae, the dusky grouper *Epinephelus marginatus* (Lowe 1834) was used as the start point, once it had already been well studied by Dr. Sanches. The initial target of the researches was the reproduction of the Goliath Grouper, and setup protocols to maintain the species (nutrition, density, environmental enrichment) in captivity. The focus was set in the sexual inversion and maturation of small-size males (<50 cm), cryopreservation of semen, maturation of small- and medium-size females, hatchery, and nursing.

The Goliath Grouper are protogynous hermaphrodites, presenting a late sexual maturation, with 6–8 years old, more than 110 cm long (Bullock et al. 1992; Sadovy and Eklund 1999), and breed in reproductive aggregations. The use of large breeders is a challenge for the management of the species in captivity, due to the difficulties in: acquiring, capturing, and transporting the individuals, besides the complexity of proceedings and risks to the operators. Due to these factors, the study was concentrated in acquiring sexually viable males. In order to proceed, young individuals (from 30 to 50 cm long) captured by artisanal fishermen in the Vaza-barris estuary, in the Brazilian state of Sergipe, were used.

Fish were fed daily with fresh sardines, with half of the individuals treated with the androgynous hormone 17- alpha-methyltestosterone (in capsules) with 1 mg per kg of each individual. The other half received sardines with placebo, only to evaluate the effectiveness of hormonal treatment. The methodology adopted was successfully used in prior investigations of sexual inversion of the dusky grouper (Sanches et al. 2009).

After 6 months applying the androgynous hormone, in an unprecedented way for the species, it was confirmed the sexual inversion of the treated individuals, with production of semen. Fish who were receiving placebo pills kept the ovotestis as immature female, which was observed after the collection of oocytes, demonstrating their functional condition. Collecting semen from small-size individuals (one of the fish had only 2.0 kg) was a great achievement – in Asia the males of other aquacultured Epinephelidae weight around 40 kg.

Through a partnership with researchers from Embrapa Tabuleiros Costeiros (EMBRAPA, Brazilian Agricultural Research Corporation), the extracted semen was characterized in its volume, sperm density, viability, motility rate, and duration of motility using the Sperm Class Analyzer (SCA). This was the first time this system was used to analyze the semen of marine fish in Brazil, and the first time for Goliath Grouper semen.

After the success in controlled sexual inversion with the acquisition of viable sperm, experiments were performed to find a proper cryopreservation procedure. The freezing preservation of semen brings countless advantages to the development of a cultivation protocol. It allows the optimization for transportation, as well as semen storage in gene banks. These banks present great benefits, such as: maintenance of semen stock of a large number of individuals in cryogenic tanks, which guarantees the genetic viability and the exchange among many research institutes, a continuous supply of semen throughout the reproductive season, and the reduction of costs in maintenance of broodstocks.

The freezing protocol was a successful methodology for the *E. marginatus*. A diluent, with pH 8.2 and 172 mOsm was used, together with the following chemical compound (Diluent (g/L): NaCl, 7.89; KCl, 1.19; $CaCl_2$, 0.22; MgC_{l2}, 0.72531; NaH_2PO_4, 0.0805; $NaHCO_3$, 0.84; pH 8.2; 172 mOsm). The ratio semen:diluent was 1:3 (v/v), with the cryoprotector (dimethyl sulfoxide) at 10%, equilibrium time of 1 min, and freezing speed of −60 °C/min. The freezing equipment used was used a cryogenic tank with nitrogen vapor (CP 100 Taylor-Wharton – Harsco Corp., Theodore, AL, USA), at −196 °C. The cryopreserved semen was stored in a Cryometal tank of 34 l.

The medium-size broodstocks (between 50 and 100 cm) used in experiments of female gonadal maturation were collected in supply channels of two shrimp farms, one in the Brazilian state of Pernambuco, and the other from Bahia. Fish were kept in the laboratory in Ubatuba in recirculating aquaculture systems (RAS) of 3000 l, with skimmers and water sterilization (using ultra-violet filters), with filtering rate of 200% of the total volume per hour. The quality of the water, together with the sterilization system permitted elevated survival rates and no report of diseases.

Individuals were monitored weekly, after eugenol analgesia, performing ovarian biopsy, with part of the oocytes collected and analyzed in stereo-microscope. During 8 months, females presented oocytes smaller than 420 μm. In previous studies with the dusky grouper, the oocytes need to reach a larger size to fit the procedures of hormone induction. Until the present moment, the fertilization of oocytes of Goliath Grouper with cryopreserved semen was no successful.

The successfully semen cryopreservation and the creation of the first sperm bank opens new doors to the protection of the Goliath Grouper, besides consolidating a tool to preserve their gene heritage. The final stages of the research to consolidate a reproduction protocol for the species in captivity is in progress, and the presented results were achieved in short time indicated that soon this tool for conservation of the Atlantic Goliath Grouper will be fully available.

Besides the specific steps of the aquaculture protocol, the development and improvement of methodologies and practices for ideal management of the species were designed and are still being updated. The pioneer work with Goliath Grouper in captivity, from the transportation of individuals from Northeast Brazil to the laboratories in Ubatuba (over 2000 km distance, and over 48 h of travel), the maintenance of fish in marine RAS, the determination of sanitary protocols, and the identification of parasites and pathologies, associated to reports of unknown biological events, such as skeletal deformations (Silva et al. 2014, Sanches et al. 2015), were successfully overcome challenges. The obtained results are part of the data for a complete manual of Goliath Grouper management in aquaculture and public aquariums.

In the last decade, many public aquariums opened in Brazil, and most of them have large groupers in their tanks. Unfortunately, due to the lack of proper supervision and regulation, higher portion of these fish are not kept in adequate environments, shortening their life span under captivity conditions. Efforts using the knowledge generated in this research are being used jointed with the practices demanded by the World Association of Zoos and Aquariums to determine minimum criteria to be adopted on Goliath Grouper and other large-scale bony fish management in captivity in Brazil, detailing specific animal well-being practices, to reduce the mortality, improving the life quality and supporting future related policies.

The definition of techniques for successfully captive bred species is a long-term procedure that involves many researches and experiments. Atlantic Goliath Grouper aquaculture still involves many challenges and only through the interaction of research institutions can be overcome, contributing to the larger objective, the species conservation. In this sense, efforts in captive bred are a priority, the availability of fingerlings and juveniles will facilitate and accelerate the elaboration of aquaculture and captive ideal Goliath Grouper protocols.

14.4 Goliath Grouper's Research and Conservation in USA

14.4.1 Catching, Sampling and Tagging

Sampling the protected Atlantic Goliath Grouper requires non-lethal methods. We have developed these methods in our research on the ecology and recovery of this species in Florida since it was fully protected in 1990. To catch adults we use 20–0 circle hooks attached to 900 lb. test monofilament leader about 6–7 m long. Monofilament leader is superior to stainless steel leader because monofilament will abrade and break if the fish becomes entangled in the reef structure—cable will not break causing extreme stress to the fish and potentially death. We use heavy swiveled longline clips at the upper end of the leader and connect the leader to a loop in the lower end of a 30 m long hand line made of 1.0 cm diam braided handline. Depending on the current velocity, we usually attach about 2.0 kg of lead weight to the juncture between the handline and the monofilament leader. Large pieces of dead bait or live bait are suitable.

Under some circumstances such as fishing on wrecks, we attach a polyball float (#50) to the hand line to suspend the bait above the bottom. The float provides a buffer to the impact of the initial strike exerted by a hooked fish and can keep the fish out of the structure of the reef/wreck. To avoid extreme stress to the fish from barotrauma we limit our depth of fishing to about 30 m.

After bringing the fish to the surface, we pull it on deck, then onto a 2.0 m long stretcher, with a short swivel-handle gaff. We hook the gaff behind the lower jaw through the thin membrane and pull the fish into the boat. The fish is then tied down to the stretcher to keep it from thrashing about on deck—this eliminates the possibility of injury to both the fish and the crew. A running seawater hose is inserted into the fish's mouth to irrigate the gills and the exposed eye is covered with a wet towel to protect it from direct sunlight. We then vent the fish using a stainless steel trocar (9.5 mm diam) & cannula which is designed for venting cows with bloat, but it works well for venting Goliath Grouper. We puncture the fish with the trocar (with the cannula slid over it) at a point just behind the tip of the pectoral fin (while the fin is laid flat against the body) and just ventral to the midline—removal of a few scales is necessary before penetrating through the skin. After puncturing the swimbladder, the trocar is removed leaving the cannula inserted for about 30 s or until all the gas is expelled.

While the vented fish is immobilized on the stretcher with seawater flushing the gills, we sample and tag the fish. For aging the fish we excise fin rays 6 and 7 at their base (or remove them at their articulation with their respective pterygiophores) from the second dorsal fin. Fin rays are kept in labeled bags and put on ice until they can be stored frozen for later processing. Stomach contents are collected by hand—we insert a 15 cm diam, 30 cm long PVC tube into the mouth, then extend our arm through the tube into the fish's stomach to remove contents. We use several types of tags: livestock ear tags, PIT tags (passive integrated transponders), and acoustic tags (Vemco V16 s, 10-year battery). Ear tags are clipped into the base of the posterior part of the dorsal or anal fin; PIT tags are injected into the dorsal musculature just below the juncture of the spinous and soft dorsal fins, and acoustic tags are inserted into the body cavity through a small incision just anterior and lateral to the vent region. We use surgical staples or monofilament sutures to close the wound. The distorted fin ray pattern of regrown amputated fin rays provide a non-specific external tag to alert for the presence of a PIT tag or acoustic tag (if present).

14.4.2 Size, Age, Reproduction and Sound Production

Size structure—Total length (TL, cm) of individual Goliath Grouper is measured directly on captured fish using a tape measure (straight-line distance from tip of upper jaw to end of caudal fin) or from digital videos taken underwater using a video camera equipped with paired green lasers. The two lasers are arranged so that the beams are parallel and 20 cm apart. During scuba dives, lasers and the attached video camera are aimed at the fish when it is perpendicular to the beams. The lasers appear as two green dots on the side of a fish separated by 20 cm. We use dividers on the video/computer screen in the lab to estimate the total length of each laser-measured fish. The method is useful for evaluating whether or not there is size or age selection in the catch. That is, the determination of the age structure of the stock depends on the assumption that there is no bias in sampling.

Age structure—Murie et al. (2009) provides details of aging Goliath Grouper using fin rays. Processing involves: cleaning the soft tissue off the cartilaginous rays, drying the rays, embedding the rays in epoxy, and then sectioning the rays at widths of about 1.0–1.4 mm. After several sections are mounted to a slide, they are read under a compound microscope using a green filter which enhances contrast between translucent and opaque zones of the annuli.

Gonad biopsy and histology—The Goliath Grouper oviduct is very large during the spawning season, so it is easy to access ovarian tissue using a rather large catheter. We biopsy ovarian tissue with a polyethylene catheter (6.3 mm OD, 4 mm ID), an in-line plastic collection cup, and a hand-operated vacuum pump (designed for testing automotive vacuum hoses). The procedure involves inserting a polyethylene catheter into the oviduct (for males we use a 2 mm OD catheter or small forceps) while increasing the vacuum with the pump. The tube is drawn back and forth in the lumen of the ovary so that it continuously removes tissue from the inner wall. The tube is withdrawn while still under vacuum and the ovarian tissue is sucked into the collection cup. We preserved gonad tissue immediately in 10% formalin. After several days of fixing, the tissues are processed using standard histological procedures, then examined for reproductive condition.

Sound production—We use acoustic recordings as additional verification of Goliath Grouper spawning activity at selected sites. Night-time sounds are produced in a lunar pattern only on spawning sites and only during the spawning season (Mann et al. 2009, Koenig et al. 2016). We record Goliath Grouper sounds using a DSG-Ocean (http://loggerheadinstruments.com/) or a H2n recorder and an H2a

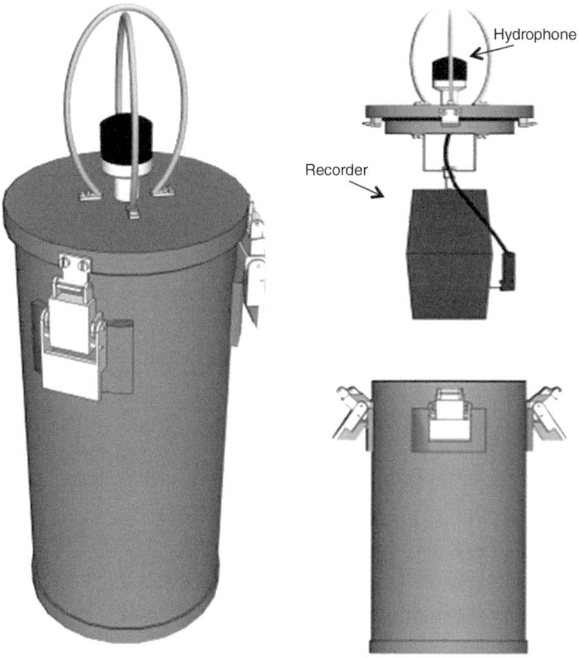

Fig. 14.13 Hydrophone and H2n Zoom recorder in PVC housing, built for recording continuously for 20 h

hydrophone contained in a PVC housing (Fig. 14.13). The DSG-Ocean is a low-power underwater acoustic recorder that records to SD memory cards using a FAT32 file system which makes high-quality acoustic recordings over long time periods. The DSG-Ocean can sample continuously at rates up to 80 kHz or intermittently, for example, 10 s per 10 min. The H2n recorder can only record continuously for up to 20 h on two AA batteries—it cannot be programed to record intermittently. It is therefore useful for overnight recordings on specific sites, but not for long-term recordings (Koenig and Coleman 2016).

14.5 Final Considerations

In the last decade, new technologies have been used to study endangered fish species. The use of non-lethal methods to collect data has allowed researchers to produce important knowledge regarding their habits, as well as following up the stages of these fish life cycle in nature.

The participation of indigenous people, specially fishermen, in producing and validating data acquired in nature is considered a yardstick in studies performed by the Projeto Meros do Brasil (Brazilian Atlantic Goliath Grouper Project, in

Portuguese) since the beginning of its activities. Through partnerships, the researchers get not only scientific data, but also exchange experiences among various social actors while sharing their views regarding the environment in which both human beings and studied species belong together.

Because the research is based on a threatened species, and despite the fact the investigators are covered by legal instruments to collect the fish, non-lethal methods are growing in order to gather information in Brazil. Ethnobiology/ethnoecology, under-water photography/video recording, and telemetry have been used as important tools to identify individuals, defining moving patterns, besides understanding inter- and intra-relationships specifically observed in the species in a natural environment. Recently, other non-lethal technologies were added to the studies, especially the molecular genetics, stable isotopes, and microchemistry, the later used in individuals accidently captured by fishermen.

In Latin America, non-lethal methods normally involve large use of expensive and inaccessible technologies for most countries, taking into consideration the economical instability faced by the domestic economies of the continent nowadays. In order to overcome financial limitations, the use of international partnerships, consolidating a research network, has allowed to maximize financial resources and to expand areas of operation of investigators. The partnerships not only promote an exchange between institutions, but also enable that graduate researchers perform their academic studies in different areas around the world.

In this sense, the Projeto Meros do Brasil has built and strengthened a vast network of collaborators, with projects related to the conservation of marine species in Brazil and together with other countries in Latin America, such as Argentina, Chile, Colombia and Mexico, and also with the United States of America, which has a vast experience in the field of marine conservation.

Together, new technologies and the collaboration of researchers in a network filled many gaps in the comprehension and in the struggle to preserve the Goliath Grouper in the past 10 years in Brazil and in Latin America. The publication of more than 12 scientific articles with a large number of partners is the main product that shows that the Projeto Meros do Brasil is being successful.

In the end, the good news is that the development achieved and the good experiences in studying the species, considered internationally as an emblematic case in marine conservation, has been applied to other species with similar ecological interest, however unknown by the general public. Based on so many innovations and new technologies, the old saying can still be used here: "it is important to know in order to protect".

References

Albuquerque CQ, Miekeley N, Muelbert JH (2012) Whitemouth croaker, *Micropogonias furnieri*, trapped in a freshwater coastal lagoon: a natural comparison of freshwater and marine influences on otoliths chemistry. Neotrop Ichthyol 8(2):311–320. doi: http://dx.doi.org/10.1590/S1679-62252010000200009

Altenritter ME, Kinnison MT (2015) Assessing dorsal scute microchemistry for reconstruction of shortnose sturgeon life histories. Environ Biol Fish 98:2321–2335

Artero C, Murie DJ, Koenig CC et al (2015) Age, growth, and mortality of the Atlantic Goliath Grouper *Epinephelus itajara* in French Guiana. Endanger Species Res 28:275–287

Babbie E (2012) The practice of social research, 13th edn. Wadsworth, Belmont

Begossi A (1993) Ecologia humana: um enfoque das relações homem-ambiente. Interciencia 18(3):121–132

Begossi A (2010) Small-scale fisheries in Latin America: management models and challenges. MAST 9:7–31

Benevides EA, Vallinoto MNS, Fetter Filho AFH et al (2014) When physical oceanography meets population genetics: the case study of the genetic/evolutionary discontinuity in the endangered Atlantic Goliath Grouper (*Epinephelus itajara*; Perciformes: Epinephelidae) with comments on the conservation of the species. Biochem Syst Ecol 56:255–266

Bertoncini AA, Borgonha M, Bueno L et al (2012) Reef fish aggregations in Southern Brazil: Pró - Arribada and Meros do Brasil Initiatives. In: Proceedings of the 65th Gulf and Caribbean Fisheries Institute. GCFI, Santa Marta, pp 287–292

Bradbury IR, Campana SE, Bentzen P (2008) Estimating contemporary early life history dispersal in an estuarine fish: integrating molecular and otolith elemental approaches. Mol Ecol 17:1438–1450

Brusher JH, Schull J (2009) Non-lethal age determination for juvenile Goliath Grouper *Epinephelus itajara* from southwest Florida. Endanger Species Res 7:205–212

Bueno GW, Ostrensky A, Canzi C, De Matos F et al (2015) Implementation of aquaculture parks in Federal Government waters in Brazil. Rev Aquac 7:1–12

Bueno LS, Bertoncini AA, Koenig CC et al (2016) Evidence for spawning aggregations of the endangered Atlantic Goliath Grouper *Epinephelus itajara* in southern. J Fish Biol 89:876–889. doi: 10.1111/jfb.13028

Bullock LH, Murphy MD, Godcharles MF, Mitchell ME (1992) Age, growth, and reproduction of jewfish *Epinephelus itajara* in the eastern Gulf of Mexico. Fish Bull 90:243–249

Campana SE, Chouinard GA, Hanson JM, Fréchet A et al (2000) Otolith elemental fingerprints as biological tracers of the fish stocks. Fish Res 46:343–357

Cawthorn DM, Steinman HA, Witthuhn RC (2012) DNA barcoding reveals a high incidence of fish species misrepresentation and substitution on the South African Market. Food Res Int 46:30–40

Clarke AD, Telmer KH, Shrimpton JM (2007) Elemental analysis of otoliths, fin rays and scales: a comparison of bony structures to provide population and life-history information for the Arctic grayling (*Thymallus arcticus*). Ecol Freshw Fish 16:354–361

Cloud JG, Miller WH, Levanduski MJ (1990) Cryopreservation of sperm as a means to store salmonid germ plan and to transfer genes from wild fish to hatchery populations. Prog Fish Cult 52(1):51–53

Colombo RE, Wills PS, Garvey JE (2004) Use of ultrasound imaging to determine sex of shovelnose sturgeon. N Am J Fish Manag 24(1):322–326

Cook GS (2014) Population connectivity shifts at high frequency within an open-coast marine protected area network. PLoS One 9(7):e103654

Correia AT, Manso S, Coimbra J (2009) Age, growth and reproductive biology of the European conger eel (*Conger conger* Linnaeus, 1758) from the Atlantic Iberian waters. Fish Res 99:196–202

Correia AT, Barros F, Sial AN (2011) Stock discrimination of European conger eel (*Conger conger* L.) using otolith stable isotope ratios. Fish Res 108:88–94

Craig MT (2011) *Epinephelus itajara*. The IUCN Red List of Threatened Species 2011: e.T195409A8961414. Available via IUCN Red list. http://dx.doi.org/10.2305/IUCN.UK.2011-2.RLTS.T195409A8961414.en. Accessed 15 Oct 2016

Craig MT, Hastings PA (2007) Molecular phylogeny of the groupers of the subfamily Epinephelinae (Serranidae) with a revised classification of the Epinephelini. Ichthyol Res 54:1–17

Craig MT, Pondella DJ II, Hafner JC et al (2001) On the status of the serranid fish genus *Epinephelus*: evidence for paraphyly based on 16S rDNA sequences. Mol Phylogenet Evol 19:121–130

Craig MT, Graham RT, Torres RA et al (2009a) How many species of Atlantic Goliath Grouper are there? Cryptic genetic divergence in a threatened marine fish and the resurrection of a geopolitical species. Endanger Species Res 7:167–174

Craig JM, Papoulias DM, Thomas MV et al (2009b) Sex assignment of lake sturgeon (Acipenser fluvescens) based on plasma sex hormone and vitellogenin levels. J Appl Ichthyol 25(Sup 2):60–67

Cunjak RA, Roussel JM, al GMA e (2005) Using stable isotope analysis with telemetry or mark-recapture data to identify fish movement and foraging. Oecologia 144:636–646

Damasceno JS, Siccha-Ramirez R, Morales MJA et al (2015) Mitochondrial DNA evidences reflect an incipient population structure in Atlantic Goliath Grouper (*Epinephelus itajara*, Epinephelidae) in Brazil. Sci Mar 79(4):419–429

Damasceno JS, Siccha-Ramirez R, Oliveira C et al (2016) Molecular identification of Atlantic Goliath Grouper *Epinephelus itajara* (Lichtenstein, 1822) (Perciformes: Epinephelidae) and related commercial species applying multiplex PCR. Neotrop Ichthyol 14(3):e150128. doi: 10.1590/1982-0224-20150128

Daros FA, Spach HL, Correia AT (2016) Habitat residency and movement patterns of *Centropomus parallelus* juveniles in a subtropical estuarine complex. J Fish Biol 88:1796–1810. doi:10.1111/jfb.12944

Di Franco A, De Benedetto G, De Rinaldis G et al (2011) Large scale-variability in otolith microstructure and microchemistry: the case study of *Diplodus sargus sargus* (Pisces: Sparidae) in the Mediterranean Sea. Ital J Zool 78(2):182–192

Di Pinto A, Di Pinto P, Terio V et al (2013) DNA barcoding for detecting market substitution in salted cod fillets and battered cod chunks. Food Chem 141:1757–1762

Diegues AC (2000) Etnoconservação da natureza: Enfoques alternativos. In: Diegues AC (ed) Etnoconservação: Novos rumos para a proteção da natureza nos trópicos. 2a. Edição. NUPAUB-USP: HUCITEC, São Paulo, pp 01–46

Eklund AM, Schull J (2001) A stepwise approach to investigate the movement patterns and habitat utilization of Goliath Grouper, *Epinephelus itajara*, using conventional tagging, acoustic telemetry and satellite tracking. In: Sibert JR, Nielsen JL (eds) Electronic tagging and tracking in marine fisheries. Springer, Dordrecht/London, p 456

Erickson CM (1983) Age determination of Manitoban Walleyes using otoliths, dorsal spines, and scales. N Am J Fish Manag 3(2):176–181

Ferreira BP, Maida M (1995) Projeto Mero: apresentação de resultados preliminares. Bol Téc Cient Cepene 3(1):210–210

Ferreira HM, Reuss-Strenzel GM, Alves J et al (2014) Local ecological knowledge of the artisanal fishers on *Epinephelus itajara* (Lichtenstein, 1822) (Teleostei: Epinephelidae) on Ilhéus coast – Bahia State, Brazil. J Ethnobiol Ethnomed 10(1). doi:10.1186/1746-4269-10-51

Ferrito V, Bertolino V, Pappalardo AM (2016) White fish authentication by COIBar – RFLP: toward a common strategy for the rapid identification of species in convenience seafood. Food Control 70:130–137

Food and Agriculture Organization of the United Nations (2013) A global assessment of offshore mariculture potential from a spatial perspective. Fao Fisheries and Aquaculture Technical Paper 549, Rome, p 202

Food and Agriculture Organization of the United Nations (2014) The state of world fisheries and aquaculture. Opportunities and challenges, Rome, p 243

Food and Agriculture Organization of the United Nations (2016) The state of world fisheries and aquaculture. Contributing to food security and nutrition for all. Rome, p 204
Francis RICC, Campana SE (2004) Inferring age from otolith measurements: a review and a new approach. Can J Fish Aquat Sci 61:1269–1284
Freitas MO, Abilhoa V, Giglio VJ et al (2015) Diet and reproduction of the Goliath Grouper, *Epinephelus itajara* (Actinopterygii: Perciformes: Serranidae), in eastern Brazil. Acta Ichthyol Piscat 45:1–11
Geffen AJ, Nash RDM, Dickey-Collas M (2011) Characterization of herring populations West of the British Isles: an investigation of mixing based on otolith microchemistry. ICES J Mar Sci 68(7):1447–1458
Gerhardinger LC, Bertoncini AA, Hostim-Silva M (2006a) Local ecological knowledge and Goliath Grouper spawning aggregations in the South Atlantic Ocean: Goliath Grouper spawning aggregations in Brazil. SPC Tradit Mar Resour Manag Knowl Inform Bull 20:33–34
Gerhardinger LC, Marenzi RC, Bertoncini AA et al (2006b) Local ecological knowledge on the Goliath Grouper *Epinephelus itajara* (teleostei: serranidae) in southern Brazil. Neotrop Ichthyol 4(4):441–450. doi: https://dx.doi.org/10.1590/S1679-62252006000400008
Gerhardinger LC, Freitas MO, Bertoncini AA et al (2007a) Collaborative approach in the research of Dusky grouper reproductive biology (*Epinephelus marginatus*, LOWE, 1834) (Perciformes: Serranidae, Epinephelinae). Acta Sci 28:219–226
Gerhardinger LC, Medeiros RP, Marenzi RC et al (2007b) Local ecological knowledge on the planning and management of marine protected areas and conservation of fish spawning aggregations: the experience of Meros do Brasil Project. In: Prates AP, Blanc D (eds) Áreas Aquáticas Protegidas como Instrumento de Gestão Pesqueira. Áreas Protegidas do Brasil vol 4. Ministério do Meio Ambiente, Brasília, pp 106–129
Gerhardinger LC, Hostim-Silva M, Medeiros RP et al (2009) Fishers' resource mapping and Goliath Grouper *Epinephelus itajara* (Serranidae) conservation in Brazil. Neotrop Ichthyol 7:93–102
Giglio VJ, Alves JA, Bertoncini AA (2014a) Using scars to photo-identify the Goliath Grouper, *Epinephelus itajara*. Mar Biodivers Rec 7:1–4
Giglio VJ, Alves JA, Gerhardinger LC et al (2014b) Habitat use and abundance of Goliath Grouper *Epinephelus itajara* in Brazil: a participative survey. Nat Conserv 12:118–123
Goswami M, Mishra A, Ninawe NS et al (2016) Bio-banking: an emerging approach for conservation of fish germplasm. Poult Fish Wildlife Sci 4:143. doi:10.4172/2375-446X.1000143
Goulette GS, Hawkes JP, Kocik JF (2014) Opportunistic acoustic telemetry platforms: benefits of collaboration in the Gulf of Maine. Fisheries 39:441–450
Graham CT, Harrison SS, Harrod C (2013) Development of non-lethal sampling of carbon and nitrogen stable isotope ratios in salmonids: effects of lipid and inorganic components of fins. Isot Environ Health Stud 49:555–566
Haines EB (1976) Relation between the stable carbon isotope composition of fiddler crabs, plants, and soils in a salt marsh. Limnol Oceanogr 21:880–883
Hamer PA, Acevedo S, Jenkins GP et al (2011) Connectivity of a large embayment and coastal fishery: spawning aggregations in one bay source local and broad-scale fishery replenishment. J Fish Biol 78:1090–1109
Harrod C, Grey J, McCarthy TK, Morrissey M (2005) Stable isotope analyses provide new insights into ecological plasticity in a mixohaline population of European eel. Oecologia 144:673–683
Hebert PDN, Cywinska A, Ball SL et al (2003) Biological identifications through DNA barcodes. Proc Biol Sci 270:313–321
Hostim-Silva M, Bertoncini AA, Gerhardinger LC et al (2005) The Lord of the Rocks conservation program in Brazil: the need for a new perception of marine fishes. Coral Reefs 24:74. doi:10.1007/s00338-004-0437-3
Jardine TD, Hunt RJ, Pusey BJ, Bunn SE (2011) A non-lethal sampling method for stable carbon and nitrogen isotope studies of tropical fishes. Mar Freshw Res 62:83–90

Jesus L, Pereira G, Cristina S, Fernandes P et al (2016) Conhecimento ecológico local sobre o Mero *Epinephelus itajara* (Lichtenstein, 1822) no Nordeste Paraense Amazônico. Biota Amazônia 6(2):110–119

Johannes RE (2002) The renaissance of community-based marine resource management in Oceania. Annu Rev Ecol Evol Syst 33:317–340

Kelly MH, Hagar WG, Jardine TD, Cunjak RA (2006) Nonlethal sampling of sunfish and slimy sculpin for stable isotope analysis: how scale and fin tissue compare with muscle tissue. N Am J Fish Manag 26:921–925

Koenig CC, Coleman FC (2016) Regional age structure, reproductive biology and trophic patterns of adult Goliath Grouper in Florida. Final Report. NOAA/NMFS Project Project NA11NMF4330123, 162pp

Koenig CC, Bueno LS, Coleman FC et al (2016) Diel, lunar, and seasonal spawning patterns of the Atlantic Goliath Grouper, Epinephelus itajara, off Florida, United States. Bull Mar Sci 92. doi: http://dx.doi.org/10.5343/bms.2016.1013

Kynard B, Kieffer (2002) Use of a borescope to determine the sex and egg maturity stage of sturgeons and the effect of borescope use on reproductive structures. J Appl Ichthyol 18:505–508

Leite MCF, Gasalla MA (2013) A method for assessing fishers' ecological knowledge as a practical tool for ecosystem-based fisheries management: seeking consensus in Southeastern Brazil. Fish Res 145:43–53. doi:http://dx.doi.org/10.1016/j.fishres.2013.02.013

Leite JR, Bueno LS, Alves JA et al (2012) Description of different color displays of Goliath Grouper on aggregations in Santa Catarina, Brazil. In: 12th international coral reef symposium, Cairns, Australia. Book of abstracts 12th ICRS, 2012

Linley T, Krogstad EJ, Nims NK et al (2016) Geochemical signatures in fin rays provide a nonlethal method to distinguish the natal rearing streams of endangered juvenile Chinook Salmon *Oncorhynchus tshawytscha* in the Wenatchee River, Washington. Fish Res 181:234–246

Maggio T, Andaloro F, Arculeo M (2006) Genetic population structure of *Epinephelus marginatus* (Pisces, Serranidae) revealed by two molecular markers. Ital J Zool 73:275–283

Maida M, Ferreira BP (1997) Coral reefs of Brazil: an overview. In: Abstracts Proceedings of the 8th international coral reef symposium 1997 1:263–274

Mann DA, Locascio JV, Coleman FC, Koenig CC (2009) Goliath Grouper *Epinephelus itajara* sound production and movement patterns on aggregation sites. Endanger Species Res 7:229–236

Marques JGW (1991) Aspectos Ecológicos na Etnoictiologia dos Pescadores do Complexo Estuarino-Lagunar Mundaú-Manguaba. Universidade Estadual de Campinas, Alagoas

Marques JG (2001) Pescando Pescadores: Ciência e etnociência em uma perspectiva ecológica. 2a. NUPAUB/USP, São Paulo

Mazzei EF, Bertoncini AA, Pinheiro HT et al (2016) Newly discovered reefs in the southern Abrolhos Bank, Brazil: anthropogenic impacts and urgent conservation needs. Mar Pollut Bull. doi:10.1016/j.marpolbul.2016.08.059

McLean E (1994) Microsurgery: gonadectomy. In: Hochachka PW, Mommsen TP (eds) Biochemistry and molecular biology of fishes, vol 3. Elsevier Science, Amsterdam, pp 63–75

Mendoza RPR (2006) Otoliths and their applications in fishery science. Ribarstvo 64:89–102

Ministério do Meio Ambiente (2014) Lista oficial das espécies da fauna brasileira ameaçadas de extinção. MMA, Brasília, Portaria n 445

Ministério do Meio Ambiente (2015) Plano de Ação Nacional para a Conservação das Espécies Ameaçadas e de Importância Socioeconômica do Ecossistema Manguezal (PAN Manguezal). Brasília Portaria n 9

Mourão JS, Nordi N (2006) Pescadores, peixes, espaço e tempo: uma abordagem etnoecológica. Interciencia 31:358–363

Murie DJ, Parkyn DC, Koenig CC et al (2009) Evaluation of finrays as a non-lethal ageing method for protected Goliath Grouper *Epinephelus itajara*. Endanger Species Res 7:213–220

Norris DO, Lopez KH (2011) The endocrinology of the mammalian ovary. In: Norris DO, Lopez KH (eds) Hormones and reproduction of vertebrates. Elsevier, Amsterdam, pp 59–72

Palmeira CAM, Rodrigues-Filho LFS, Sales JBL et al (2013) Commercialization of a critically endangered species (largetooth sawfish, *Pristis perotteti*) in fish markets of northern Brazil: authenticity by DNA analysis. Food Control 34:249–252
Palumbi SR, Gaines SD, Leslie H et al (2013) New wave: high-tech tools to help marine reserve research. Front Ecol Environ 1(2):73–79
Pandian TJ (2011) Sex determination in fish. CRC Press, Boca Raton, p 276
Pasten GP (2016) Qué es la esclerocronología? http://kronolab.cl/esclerocronologia.html. Accessed in 20 Aug 2016
Pina-Amargós F, González-Sansón G (2009) Movement of Goliath Grouper (*Epinephelus itajara*) in southeast Cuba: implications for its conservation. Endanger Species Res 7:243–247
Pinheiro L (2004) Da ictiologia ao etnoconhecimento: saberes populares, percepção ambiental e senso de conservação em comunidade ribeirinha do rio Piraí, Joinville, Estado de Santa Catarina. Acta Sci 26(3):325–334. doi:10.4025/actascibiolsci.v26i3.1594
Qu M, Zhang X, Ding S (2012) Complete mitochondrial genome of yellow grouper *Epinephelus awoara* (Perciformes, Epinephelidae). Mitochondrial DNA 23(6):432–434
Ramírez MA, Patricia-Acevedo J, Planas S et al (2006) New microsatellite resources for groupers (Serranidae). Mol Ecol Resour 6(3):813–817
Reis-Santos P, Gillanders BM, Tanner SE et al (2012) Temporal variability in estuarine fish otolith elemental fingerprints: implications for connectivity assessments. Estuar Coast Shelf Sci 112:216–224
Rivera MAJ, Kelley CD, Roderick GK (2004) Subtle population genetic structure in the Hawaiian Grouper, *Epinephelus quernus* (Serranidae) as revealed by mitochondrial DNA analyses. Biol J Linn Soc 81:449–468
Sadovy Y, Eklund AM (1999) Synopsis of biological data on the Nassau grouper, *Epinephelus striatus* (Bloch, 1792), and the jewfish, *E. itajara* (Lichtenstein, 1822). NOAA Technical Report *NMFS* 146.
Sadovy Y, Shapiro DY (1987) Criteria for the diagnosis of hermaphroditism in fishes. Copeia 1:136–156. doi:10.2307/1446046
Sadovy de Mitcheson Y, Colin PL (eds) (2012) Reef fish spawning aggregations: biology, research and management, vol 35: Fish and fisheries. Springer, New York
Sadovy de Mitcheson Y, Liu M (2008) Functional hermaphroditism in teleosts. Fish Fish 9:1–43
Sadovy de Mitcheson Y, Craig MT, Bertoncini AA et al (2013) Fishing groupers towards extinction: a global assessment of threats and extinction risks in a billion dollar fishery. Fish Fish 14:119–136. doi:10.1111/j.1467-2979.2011.00455.x
Sanches EG (2007) Piscicultura marinha no Brasil: uma alternativa de produção e conservação. Aqüicultura e Pesca nov/dez:16–22
Sanches EG, Oliveira IR, Serralheiro PC (2009) Inversão sexual da garoupa-verdadeira *Epinephelus marginatus*. Rev Bras Saúde Prod Anim 10:198–209
Sanches EG, Tosta GAM, Souza-Filho JJ (2013) Viabilidade econômica da produção de formas jovens de bijupirá (*Rachycentron canadum*). Bol Inst Pesca 39(1):15–26
Sanches EG, Silva FC, Herrera LA (2015) Anormalidades esqueléticas em meros. Bol Inst Pesca 42:191–198
Selkoe KA, Henzler CM, Gaines SD (2008) Seascape genetics and the spatial ecology of marine populations. Fish Fish 9:363–377
Serviço Brasileiro de Apoio às Micro e Pequenas Empresas (2015) Aquicultura no Brasil. Série estudos mercadológicos, p 76
Seyoum S, Tringali MD, Barthel BL et al (2013) Isolation and characterization of 29 polymorphic microsatellite markers for the endangered Atlantic Goliath Grouper (*Epinephelus itajara*), and the Pacific Goliath Grouper (*E. quinquefasciatus*). Conserv Genet Resour 5:729–732
Shideler GS, Pierce B (2016) Recreational diver willingness to pay for Goliath Grouper encounters during the months of their spawning aggregation off eastern Florida, USA. Ocean Coast Manag 129:36–43. http://dx.doi.org/10.1016/j.ocecoaman.2016.05.002

Silva DM, Santos P, Correia AT (2011) Discrimination of *Trisopterus luscus* stocks in the northern of Portugal using otolith elemental fingerprints. Aquat Living Resour 24:85–91

Silva FC, Leite JR, Hostim-Silva M, Valença AR, Sanches EG (2014) First record of *Neobenedenia melleni* like species (Monogenae: Capsalidae) in Goliath Grouper (*Epinephelus itajara*) in Brazil. Rev Bras Parasitol Ve 23(1):248–250. http://dx.doi.org/10.1590/S1984-29612014025

Silva-Oliveira GC, Rêgo PS, Schneider H et al (2008) Genetic characterization of populations of the critically endangered Atlantic Goliath Grouper (*Epinephelus itajara*, Serranidae) from the Northern Brazilian coast through analyses of mtDNA. Genet Mol Biol 31(4):988–994

Silva-Oliveira GC, Silva ABC, Oliveira Y et al (2013) New nuclear primers for molecular studies of Epinephelidae fishes. Conserv Genet Resour 5:165–168

Sinnatamby RN, Bowman JE, Dempson JB, Power M (2007) An assessment of de-calcification procedures for d13C and d15N analysis of yellow perch, walleye and Atlantic salmon scales. J Fish Biol 70:1630–1635

Smith SJ, Campana SE (2010) Integrated stock mixture analysis for continuous and categorical data, with application to genetic–otolith combinations. Can J Fish Aquat Sci 67:1533–1567

Smith BN, Epstein S (1971) Two categories of 13C/12C ratios for higher plants. Plant Physiol 47:380–384

Smith KT, Whitledge GW (2010) Fin ray chemistry as a potential natural tag for Smallmouth Bass in Northern Illinois Rivers. J Freshw Ecol 25(4):627–635

Specker JL, Anderson TR (1994) Developing an ELISA for a model protein – vitellogenin. In: Hochachka PW, Mommsen TP (eds) Biochemistry and molecular biology of fishes, vol 3. Elsevier Science, Amsterdam, p 718

Stephenson RL (1999) Stock complexity in fisheries management: a perspective of emerging issues related to population sub-units. Fish Res 43:247–249

Syukur A, Mahrus SAR (2016) The potential assessment environment friendly aquaculture of small-scale fishermen as a conservation strategy seagrass beds in coastal areas of Tanjung Luar East Lombok Indonesia. Int J Fish Aquat Stud 4(2):22–27

Toledo VM, Barrera-Bassols N (2009) A etnoecologia: uma ciência pós-normalque estuda as sabedorias tradicionais. Desenvolv Meio Amb 20:31–45. doi: http://dx.doi.org/10.5380/dma.v20i0.14519

Torres RA, Feitosa RB, Carvalho DC et al (2013) DNA barcoding approaches for fishing authentication of exploited grouper species including the endangered and legally protected Goliath Grouper *Epinephelus itajara*. Sci Mar 77(3):409–418

Tzadik OE, Goddard EA, Hollander DJ et al (2015) Non-lethal approach identifies variability of δ^{15}N values in the fin rays of Atlantic Goliath Grouper, *Epinephelus itajara*. Peer J 3:e1010. doi:10.7717/peerj.1010

Vander Zanden MJ, Hulshof M, Ridgway MS et al (1998) Application of stable isotope techniques to trophic studies of age-0 smallmouth bass. T Am Fish Soc 127:729–739

Vander Zanden MJ, Shuter BJ, Lester NP et al (2000) Within- and among-population variability in the trophic position of a pelagic predator, lake trout (*Salvelinus namaycush*). Can J Fish Aquat Sci 57:725–731

Vasconcelos RP, Reis-Santos P, Fonseca V et al (2009) Juvenile fish condition in estuarine nurseries along the Portuguese coast. Estuar Coast Shelf Sci 82:128–138

Vecsei P, Litvak MK, Noakes DLG et al (2003) A noninvasive technique for determining sex of live adult North American sturgeons. Environ Biol Fish 68:333–338

Veinott GI, Evans RD (1999) An examination of elemental stability in the Fin Ray of the White Sturgeon with Laser Ablation Sampling–Inductively Coupled Plasma–Mass Spectrometry (LAS-ICP-MS). Trans Am Fish Soc 128(2):352–361

Voegeli FA, Smale MJ, Webber DM et al (2001) Ultrasonic telemetry, tracking and automated monitoring technology for sharks. Environ Biol Fish 60:267–281

Wang X, Wang Q, Xie Z et al (2016) The complete mitochondrial genome of the *Epinephelus lanceolatus* (Perciformes: Serranidae). Mitochondrial DNA Part A 27(3):1738–1739

Welch DW, Boehlert GW, Ward BR (2003) POST-the Pacific Ocean salmon tracking project. Oceanol Acta 25:243–258

Willette DA, Allendorf FW, Barber PH et al (2014) So, you want to use next-generation sequencing in marine systems? Insight from the Pan-Pacific Advanced Studies Institute. Bull Mar Sci 90(1):79–122

Willmes M, Glessner JCG, Carleton SA et al (2016) $^{87}Sr/^{86}Sr$ isotope ratio analysis by laser ablation MC-ICP-MS in scales, spines, and fin rays as a nonlethal alternative to otoliths for reconstructing fish life history. Can J Fish Aquat Sci. doi:10.1139/cjfas-2016-0103

Wu X, Xie Z, Yang L et al (2015) The complete mitochondrial genome of the dusky tail grouper *Epinephelus bleekeri* (Serranidae: Epinephelinae). Mitochondrial DNA 26(5):722–723

Yang H, Xie Z, Li S et al (2016) The complete mitochondrial genome of the orange-spotted grouper *Epinephelus coioides* (Perciformes, Serranidae). Mitochondrial DNA Part A 27(3):1674–1676

Yu H, You X, Li J et al (2016) Genome-wide mapping of growth-related quantitative trait loci in orange-spotted grouper (*Epinephelus coioides*) using double digest restriction-site associated DNA sequencing (ddRADseq). Int J Mol Sci 17(4):501–512

Zatcoff MS, Ball AO, Sedberry GR (2004) Population genetic analysis of red grouper, *Epinephelus morio*, and scamp, *Mycteroperca phenax*, from the southeastern US Atlantic and Gulf of Mexico. Mar Biol 144:769–777

Zhuang X, Qu M, Zhang X et al (2013) Comprehensive description and evolutionary analysis of 22 grouper (Perciformes, Epinephelidae) mitochondrial genomes with emphasis on two novel genome organizations. PLoS One 8(8):e73561

Chapter 15
The Spreading Message in a Noisy World: Song Behavior and Acoustic Ecology in Humpback Whales (*Megaptera novaeangliae*), from the Southwestern Atlantic Ocean

Marcos R. Rossi-Santos

Abstract In this chapter, the song behavior and acoustic ecology of humpback whales (*Megaptera novaeangliae*) in the Brazilian breeding ground are documented, between 2005 and 2010 (July to October). Behavioral and ecological aspects such as group composition, spatial distribution, depth, tides and moon were analyzed, through boat surveys. We sighted 370 whale groups in 123 days, during a total of 912 h of sampling effort. In 36 days, 29% of the groups (n = 44; 82 individuals) were identified containing at least one singer male. Such groups were sighted along all the study area, mostly like alone individuals (n = 21 groups, 48%), followed by groups with two (n = 8; 18%), three (n = 3, 7%), four (n = 2, 4, 5%) individuals and also groups with female-calf pairs with one escort male (n = 4, 9%) or female-calf pairs plus two males (n = 1, 2%). Groups with singer males were sighted in a depth range of 16 and 173 m (mean = 54,6 ± 34,2 SD). The narrow continental shelf associated with the coastal distribution of the whales in the study area may influence in the group compositions containing singer males. So on, this work bring for the first time an ecological perspective of this recognized important singing behavior in this species, and also show the expansion of the acoustic activity, during the Brazilian breeding season, going far beyond the core area of Abrolhos Bank. Understanding the acoustic ecology of humpback whales, in which the song is performed, is crucial to provide insights into conservation measurements aiming to protect marine vertebrate and their habitat.

M.R. Rossi-Santos (✉)
Instituto Baleia Jubarte, Salvador, Brazil

Universidade Federal do Rio Grande do Norte, Natal, Brazil

Laboratório de Ecologia Acústica e Comportamento Animal, Centro de Ciências Agrárias, Ambientais e Biológicas, Universidade Federal do Recôncavo da Bahia, Rua Rui Barbosa, 710, Centro, 43800-000 Cruz das Almas, Bahia, Brazil
e-mail: marcos.rossi@ufrb.edu.br

M.R. Rossi-Santos, C.W. Finkl (eds.), *Advances in Marine Vertebrate Research in Latin America*, Coastal Research Library 22,
DOI 10.1007/978-3-319-56985-7_15

Keywords Humpback whales • Acoustic ecology • Song behavior • Southwestern Atlantic Ocean

15.1 Introduction

15.1.1 Ecology and Animal Communication

The most of the acoustic signals are modified when the pass by the environment since the sender to the receiver, which transforms and depredates this signal, but even so, results in information to the receiver, as some bird species who can estimate the distance from a sender individual through the signal degradation (McGregor and Krebs 1984; Naguib 1998).

Furthermore, the environment may shape evolutionary changes which may result in specific vocal adaptations in close related vertebrate species, and even inside the same species, leading to differences in sound repertoire which reflect their ecological habits (Podos 2001; Podos et al. 2004).

Dolphins and bats track their environment listening to the echoes which themselves produces, a process called echolocation (eg: Tyack 2000). Thus, in the field of cetacean communication there are some basic questions that still could be answered, such as what are the features that make this signal unique. Those questions are crucial to understand whether these signals are elaborated and have their specific usage. Part of this subject is included in the behavioral ecology context of the present work.

In this work, the behavioral ecology involving an acoustic display, the song, and the singer males in humpback whales (*Megaptera novaeangliae*) is explored in their breeding ground at northeastern Brazilian coast, discussing diverse parameters, such as singers' group structure, behavior, distribution and their possible relations with oceanographic features as depth, tide and moon phases influences on the humpback whale song inside the breeding season.

15.1.2 Species Characterization

The humpback whale *Megaptera novaeangliae* (Cetacea, Balaenopteridae) (Fig. 15.1), is a cosmopolitan species distributed along all the oceans worldwide (Clapham and Mead 1999), moving every year from high latitude feeding areas, staying during the autumn and summer, to the breeding areas in the tropics, staying during the spring and summer (Clapham and Mead 1999). These breeding areas are typically between islands and/or associated with coral systems (Whitehead 1981; Whitehead and Moore 1982).

In the feeding and breeding area, the humpback whale present a social organization characterized by unstable and small groups (2–3 animals). However, larger groups can be found during the feeding behavior or related to the aggressive competition between males during the breeding season (Clapham 1996).

Fig. 15.1 The humpback whale (*Megaptera novaeangliae*) is the most acrobatic of the large whales, exhibiting diverse surface behavior during their breeding season in Brazilian waters. The species is also known as the singer whale, due to the high complex acoustic behavior shown by males, as part of the mating process (Photo: Marcos Rossi-Santos/IBJ)

15.1.3 Occurrence and Distribution

Nowadays, there are seven humpback whale sub-populations (or stocks) in the southern hemisphere (IWC 1998), one of those, named by IWC "Breeding Stock A/BSA" to migrate to the Brazilian coast, where they breed from July to November.

Despite their occurrence along a large range in Brazil, from Rio Grande do Sul state, southern Brazil, to the Fernando de Noronha Archipelago, northeastern Brazil, (Towsend 1935; Lodi 1994), its core breeding area is the Abrolhos Bank (Fig. 15.2), Bahia state (Wedekin et al. 2010). However, the increase of humpback whale sightings northwards from the Abrolhos Bank, including the Bahia state capital, Salvador and the north coast of the state (Fig. 15.3), suggest the population recovery in this historical area, occupied by the whales prior the whaling period (Tavares 1916; Tollenare 1961; Rossi-Santos et al. 2008).

15.1.4 Ecology and Acoustic Behavior

The humpback whale is also known as singer whale because its unique characteristic of to exhibit a singing behavior, performed only by males, during their breeding season. Since the 1970ths, many studies described the physic structure of the songs (eg. Payne and McVay 1971; Helweg et al. 1998; Maeda et al. 2000; Arraut and Vielliard 2004) and even their probable functions at the population ecology level, such as female attraction, male-male competition and cultural exchange (eg.

Fig. 15.2 The Abrolhos Archipelago names this important coral reef Bank, where humpback whales are more abundant along the entire Brazilian coast. The shallow and warm waters in this place make it ideal to constitute a breeding ground for the species in the Southwestern Atlantic Ocean (Photo: Marcos Rossi-Santos/IBJ)

Fig. 15.3 The humpback whale may be sighted today along all the coast of Bahia state, including areas with a narrow continental shelf, such as those close to the capital Salvador, where intense humane activities take place (Photo: Marcos Rossi-Santos/IBJ)

McSweeney et al. 1989; Dawbin and Eyre 1991; Darling and Sousa-Lima 2005; Eriksen et al. 2005), however many other aspects of the acoustic ecology of the species during the singing behavior is still unknown, such as the socio-ecological context in which the males sing, bringing spatial information (eg. where do they sing? what depth?) and variations of this context inside the breeding season.

Humpback whale song was described as the most elaborated and spectacular single display of any animal species (Wilson 1975). The song was initially described in details by Payne and McVay (1971) and Winn et al. (1971). Extensive analyses of this unique acoustic behavior have revealed that: (1) males produce stereotyped, ordered sequences of sounds for long periods, (2) at any given time, sequences produced by individuals within a breeding area are highly similar (Winn et al. 1981; Payne and Guinee 1983), and (3) whales continuously modify these sequences over time (Payne et al. 1983; Payne and Payne 1985).

Explanations for the structural and acoustic features of humpback whale songs remain speculative. Because most researchers assume that songs are used primarily for long distance communication (Payne and McVay 1971; Winn and Winn 1978; Tyack 1981; Helweg et al. 1992; Frankel et al. 1995; Cerchio et al. 2001; Darling and Bérubé 2001), song features often are interpreted in terms of how they might facilitate the transfer of information among whales (Winn and Winn 1978; Frankel 1995). Additionally, song properties could reflect strategies for collecting environmental information, if components of songs are used as echolocation signals (Winn and Perkins 1976; Winn and Winn 1978; Frazer and Mercado 2000; Au et al. 2001).

Singers mostly appear as alone males (Winn and Winn 1978; Tyack 1981), yet some have been documented as singing in groups (Baker and Herman 1984) and while moving (Frankel et al. 1995). Singing appears to be most prevalent on the winter breeding grounds, despite also recorded on the Alaskan (McSweeney et al. 1989) and Gulf of Maine (Mattila et al. 1987) feeding grounds and during migration (Clapham 1996, 2000).

Humpback whale songs show clear structure based on aural analysis, with smaller repetitive units called "phrases" organized into larger "themes" which tend to occur in specific sequence (Payne and McVay, 1971, Guinee et al. 1983) (Fig. 15.4). The structure of song changes over the course of a winter season, yet at any given time all singers appear to be singing the same version of a song (Guinee et al. 1983; Payne et al. 1983; Payne and Payne 1985).

The relationship between singing and seasonal gonadal activity suggests that song production plays a role in the mating system. While humpback whale song appears to be the result of strong sexual selection (Tyack 1981; Smith et al. 2008), its function in the mating system remains unclear. Helweg et al. (1992) presented a detailed summary of current hypotheses regarding the role of humpback whale song, but its role is still far from certain. A number of possible functions of song have been proposed, including sexual advertisement to females (Payne and McVay 1971; Winn and Winn 1978; Tyack 1981, 2000), maintenance of spacing and territorial defense among singing males (Winn and Winn 1978; Tyack 1981; Frankel et al. 1995), inducement or synchronization of ovulation in females (Baker and Herman 1984), and a navigational "beacon" for migrating whales (Winn and Winn 1978).

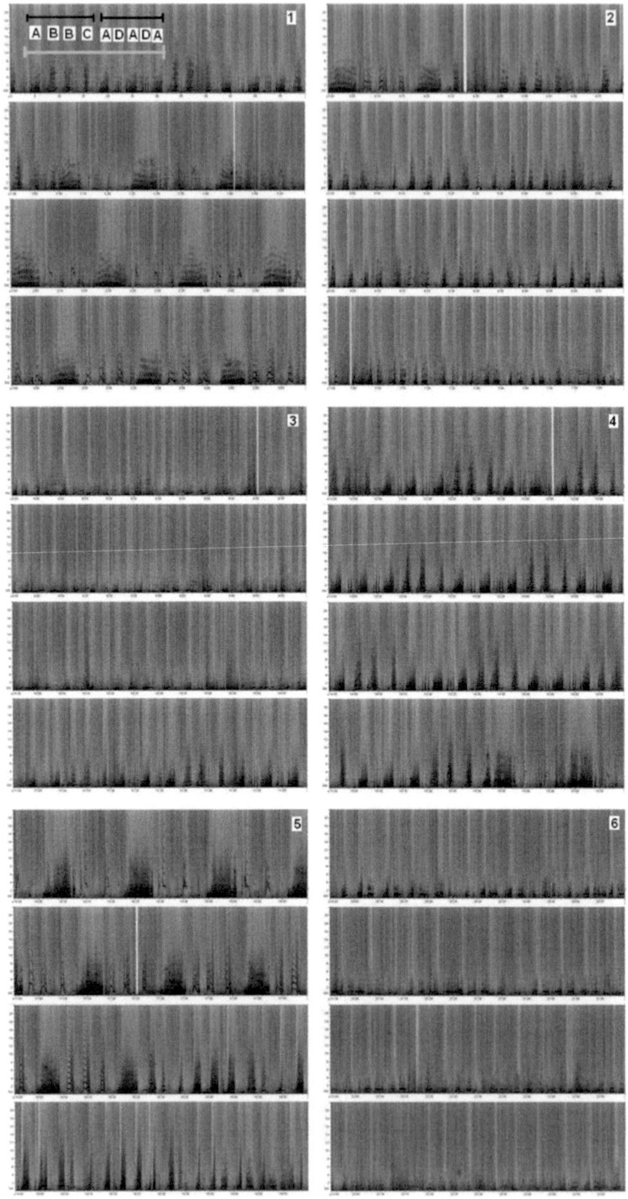

Fig. 15.4 A spectrogram is a graphical representation of the sound in a time x frequency acoustic domain. Here is a representation of the types of structural components typically present in sequences of sounds produced by singing humpback whales (*Megaptera novaeangliae*), that is called "song". Each letter represents one sound. Each individual sound is called a unit (different letters correspond to aurally distinctive sound units). Repeated groups of units are called phrases (*black lines*). A theme is a set of phrases (*gray line*). "Songs" consist of repeated theme sequences within a song session (1–4 complete the cycle, thereafter song session starts again)

Songs of humpback whales are typically recorded close to a whale in order to maximize the signal-to-noise ratio. However, when sounds are recorded at large distances from any whale, a number of whales singing may sound as in a chorus, although they do not sing in unison on the same portion of a song (Au et al. 2000).

The most studies bring acoustic information only by spectrograms (Fig. 15.4) (eg. Payne and McVay 1971; Winn et al. 1981; Arraut and Velliard 2004). The evolution in the study of the humpback whale frequency range is proportional with the advance in technology. The initial studies reported song range between up to 5–8 kHz (Payne and McVay 1971; Winn et al. 1971), and thereafter it goes up to 15–24 kHz (Helweg et al. 1992; Au et al. 2000, 2001, 2006; Fristrup et al. 2003). Mercado et al. (2003) measured acoustic patterns in humpback whale songs recorded in Hawaiian waters during four consecutive years (1992–1995) to determine whether subcomponents of songs differ in their detectability after long-range propagation. Such analyses can provide important insights into the functions of humpback whale songs.

Singing whales most often are reported to be alone (Winn and Winn 1978; Tyack 1981) although there are exceptions (Baker and Herman 1984). It is generally accepted that singing is performed by males in tropical breeding areas, where calves are being born and seasonal gonadal activity is high (Chittleborough 1955; Payne and McVay 1971).

This work aims to describe humpback whale songs in a wide geographic scale in the Brazilian breeding ground which includes the core area of Abrolhos Bank and the adjacent North Coast, comparing song form in different marine landscapes, adapting the Acoustic Ecology concept of Truax (1999), who define it as the study of relations between biological organisms and their sonic environment, or soundscape.

15.2 Material and Methods

15.2.1 Study Area

This study was conducted from 2005 to 2010, at the Northeastern Brazilian coast during the breeding season (July to October) of the Southwestern Atlantic Ocean humpback whale population, known as Breeding Stock A (BSA) by the International Whaling Commission (IWC 1998). Humpback whale songs were recorded along the study area, grouped in two main regions: Abrolhos Bank (16–19° S, 37–39° W) and North Coast (11–14° S, 37–38° W) (Fig. 15.5).

The Abrolhos Bank is very known as the oceanographic environment with the broadest extension of the Brazilian continental shelf, providing shallow and warm waters, and when the largest coral reef aggregation of the entire Southwestern Atlantic Ocean is located (Ekau and Knoppers 1999), making this place a hotspot for marine conservation and a reference for the humpback whale research as the core area for the species distribution along the Brazilian coast (Andriolo et al. 2006; Wedekin et al. 2010).

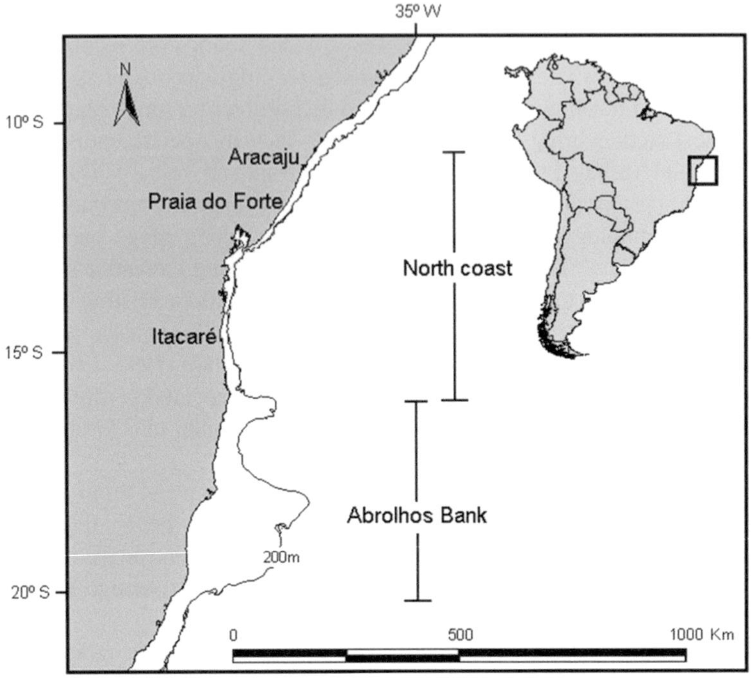

Fig. 15.5 Study area, named North coast, located between Itacaré and Aracaju, Northeastern Brazil. For reference, southwards it is located the Abrolhos Bank, core area of humpback whale (*Megaptera novaeangliae*) distribution in Brazil. The line represents the 200 m isobath, demarking the Brazilian continental shelf

Another stretch of coast, named North Coast, encompassed the northeast states of Bahia and Sergipe, from the touristic and fisherman village Itacaré (14° S, 38° W) to Aracaju (11° S, 37° W), capital of Sergipe, with main departing location at Praia do Forte, (13° S, 38° W) placed in the center of this area (Fig. 15.6). The main oceanographic characteristic for this area is a narrow continental shelf, with approximately 15 km of extension, with average depth of 40 m and the tide amplitude varying between 0.1 and 2.6 m (DHN 1995; Ekau and Knoppers 1999).

This region is characterized by two remarkable environments, being the presence of large bays and estuaries from Itacaré to Salvador, capital of Bahia state, and by long sand beaches, with large dune formation fringing the coast line, between Praia do Forte and Aracaju (Hatge and Andrade 2009).

15.2.2 Data Collection

We carried out boat based surveys (Fig. 15.7) searching for whales in a determined route, approaching animals when sighted to collect data (Fig. 15.8) such as number of individuals, composition of the groups, behavior observed before and after

Fig. 15.6 Praia do Forte, placed in the center of the study area is an strategic departing point for boat based surveys. Due to the narrow continental shelf, with approximately 15 km of extension, and average depth of 40 m, the humpback whales approach the coast, making this place ideal for whale-watching activities to spread the marine conservation message to the large touristic public (Photo: Marcos Rossi-Santos/IBJ)

approach, dive patterns, time of beginning and end of the sightings, photoidentification and bioacoustical recordings.

15.2.2.1 Behavioral Observations

Behavioral observations followed a combined *Focal Group* and *Ad libitum* methods (see Mann et al. 2000) registering a sequence of a certain behavior during the observation time of each group, broadly employed in the cetacean studies. Behavior and group structure nomenclature and definitions followed extensive previous works worldwide (Clapham 1996; Clapham and Mead 1999; Clapham 2000) and locally (Engel 1996; Lunardi et al. 2010).

15.2.2.2 Bioacoustics

Acoustic recordings were made by deploying a hydrophone from a small boat positioned less than 50 m from the singing whale (Fig. 15.9). During this period diverse recording equipment was utilized, according to the increase of technology available in Brazil. Therefore, sound data collected from 2005 to 2008 were made using an

Fig. 15.7 Boat based survey are being utilized in long term cetacean monitoring since the 1970s. In this research a 13 m wood trawler boat was adapted to perform data collection cruises, during the study of the humpback whale (*Megaptera novaeangliae*) acoustic ecology in Brazil (Photo: Marcos Rossi-Santos/IBJ)

Fig. 15.8 Data collection on the behavioral ecology of humpback whale singer males. Approaching the whales to investigate social parameters, such as number of individuals and group composition is an important step to see how and where the males are performing their song (Photo: IBJ Image Bank)

Fig. 15.9 Acoustic recordings were made by deploying a hydrophone from a small boat positioned less than 50 m from the singing whale, with engine turned off or simply padded out (Photo: Marcos Rossi-Santos/IBJ)

uncalibrated HTI – SSQ 96 hydrophone (sensitive to 20 kHz) that was connected to a Sony TCD5-M cassette recorder, frequency response up to 17 kHz). From 2008 to 2010 we utilized a CR 54 hydrophone connected to a digital recorder (M-Audio Microtrack Pro-II), increasing the frequency response up to 96 kHz.

Analyzes were first made both by listening and by visually comparing spectrograms, which are sound graphics with x-axis on time (seconds) and y-axis on frequency (Hertz). Spectrographic analyzes were performed using software RAVEN 3.1, selecting the following frequency parameters (Hertz), relative sound intensity (dB re SNR – referenced to the Signal to Noise Ratio) and time (seconds): minimum, maximum, central frequency, frequency amplitude, energy and duration. Whenever found, the harmonics and their intervals were also measured, as well as the visual contour shape from each note.

15.2.2.3 Sex Determination of Individuals

Genetic studies of humpback whales worldwide have shown that singing and escorting are behavioral roles performed only by males and that a whale closely associated with a calf is its mother, an adult female (eg. Palsboll et al. 1997; Darling and Berube 2001).

An 'escort' is defined as a whale that is associating with another adult whale known to be female or with a mother-calf pair. When two or more males are engaged in dispute behavior in the presence of a female, all are considered escorts (Herman and Tavolga 1980), forming a competitive group (Fig. 15.10).

In this study, the sex of observed animals was assumed following the convictions on these different behavioral roles of males and females. The term 'calf' refers to calves of that year (Fig. 15.11). It was not possible to visually differentiate between immature or sub-adult and adult whales, therefore the term 'whale' refers to all non-calf whales (Fig. 15.12).

Fig. 15.10 A competitive group of humpback whales can be sighted during their breeding season. This group composition is characterized by a central female and multiples escorts, engaged in dispute behaviors, even including a ruff fight for a mate partner (Photo: Marcos Rossi-Santos/IBJ)

Fig. 15.11 Genetic studies of humpback whales worldwide have shown that a whale closely associated with a calf is its mother, an adult female, as the same way that singing and escorting are behavioral roles performed only by males

15.2.2.4 Geographic Information System (GIS)

For the record of the geographic position of sighted animals, with reference to the boat position, a GPS Map 76 and a GPS Map 60c (Garmin®) were utilized, in the waypoint function. For posterior analysis and studies of the humpback whale distribution along the time in the study area, a GIS environment modeling was created, including important features such as the Brazilian coastline, bathymetry and the plotted geographic coordinates of each sighting, in a digital nautical charter, using the software Arc View GIS (version 3.1).

Fig. 15.12 Many times in the field it is not possible to differentiate the sex of an adult whale, assuming the sex of observed animals when matched with previously recognized behavioral roles for females and males, such as, respectively, nursing and singing. In this picture, a singer male in the group (in the *right*) may be escorting a female who slaps her tail as a receptive behavior (Photo: Marcos Rossi-Santos/IBJ)

15.2.2.5 Dive Time

Dive time was systematically registered during 2008 and 2009 breeding seasons, utilizing a digital chronometer, in order to measure and compare breathing behavior of small groups (less than four individuals), to date, Alone Males (solitary), Couples, Female with calf pairs, Female with Calf plus one escort male.

The dive time data was organized into the following definitions: (D.O.- minutes) = Direct Observation (the time spent observing humpback whale groups); (D.P. range – seconds) = Dive time – the time whales spent underwater; (Mean range D.P > 100 s) = mean number of Dive time larger than 100 s; (# of measur.) = number of dive measurements; (# measure. > 100 s) = number of dive measurements larger than 100 s; (Fe-Ca pairs) = female and calf pairs; (FeCa-E1) = female and calf pairs plus one escort male.

15.2.2.6 Tide and Moon Analysis

Tidal states were divided into eight different classes, looking for a similar time division and fine scale during the tidal regimen. Significant differences in tide and moon phases categories were analyzed utilizing the Chi-Square test (5%).

15.3 Results

Between 2005 and 2009, a total of 912 h of sampling effort was performed, distributed in 123 days, sighting 370 groups during our systematic fieldwork (Table 15.1). About 29% of the groups (n = 44; 82 individuals), in 36 days (41 h of direct observations), were identified containing at least one singer male in the group, with singing behavior confirmed by underwater recording.

15.3.1 Distribution

Humpback whale groups with singer males were sighted along the study area, from the southern sighting in Itacaré/Bahia state (14° S, 38° W) to the northern sighting in Aracaju/Sergipe state (11° S, 37° W) (Fig. 15.13), despite a small gap noted North of Praia do Forte towards Aracaju.

15.3.2 Group Structure

Regarding group structure, singer whales were sighted mostly alone (n = 21 groups, 48%), followed by groups containing two (n = 8, 18%), three (n = 3,7%), four (n = 2, 4,5%) adult individuals and groups with mother-calf pairs plus the singer male (n = 4,9%) or mother-calf pairs plus two more males (n = 1,2%) (Fig. 15.14).

During 2005 and 2006, singer males were observed in all mentioned categories, excepting three or four adults. In 2008 and 2009 these groups occurred. That may indicate a variation in singer male group composition along the time, with more singers in a certain area, maximizing the singing behavior with diverse potential groups for mating.

Table 15.1 Number of boat surveys, sampling effort (hours), covered distance (nautical miles), number of sighted whales, number of groups and direct observation (hours) of humpback whales in the northeastern Brazilian coast, between 2005 and 2009

Seasons	Survey n	Sampling effort (hs)	Direct observation (hs)	Covered distance (Nm)	Whales observed	Groups
2005	17	118,5	34,7	699	104	53
2006	37	266,6	109,4	1667	317	151
2007	7	51,7	18,8	339	58	26
2008	32	235,7	96,5	1463	291	21
2009	30	240	77	981	275	119
Total	123	912,3	336,4	5179	1045	370

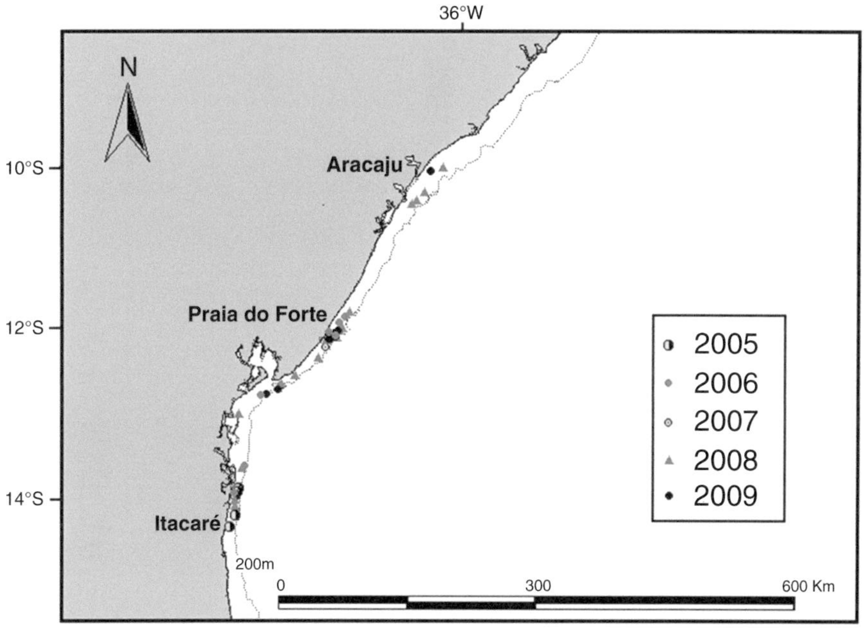

Fig. 15.13 The broad distribution of humpback whale (*Megaptera novaeangliae*) groups with singer males, in the study area, Brazilian breeding ground, between 2005 and 2009, indicates that the acoustic territory that is being utilized by the whales is much largest than previously though, and include a challenging situation of dealing with an endangered species in stock recuperation at the same time than an unprecedent industrial increase and its consequent degradation of oceanic environments

15.3.3 Behavioral Observations

Alone individuals (n = 21) are generally observed floating on the surface, with slow movements and performing resting behavioral events related to diving, such as tail-up and dorsum arching, with minor frequency of snaking (tail and peduncle shaking before dive) and breaching, both noted only for three individuals. Couples exhibited more active behaviors than alone animals, being observed breaching (three individuals), tail slapping (two individuals), diving and resting (two individuals).

Competitive groups including singer males were registered on four occasions, exhibiting active behaviors. In Oct, 1, 2005, during 40 min of direct observation we sighted four animals performing behaviors of consecutive tail-up for short dives, dorsum arching, belly exposition, breaching and bubble exhalation on the sea surface.

In Jul, 31, 2008, three animals were sighted, during 41 min of direct observation, showing head slapping, dorsum arching, rolling, breaching (n = 13), tail-up for diving (n = 12) and tail slapping (n = 2). In this sighting, a whale-watching zodiac boat was observed (about 9 m long with outboard engine of 250 Hp) near to the humpback group.

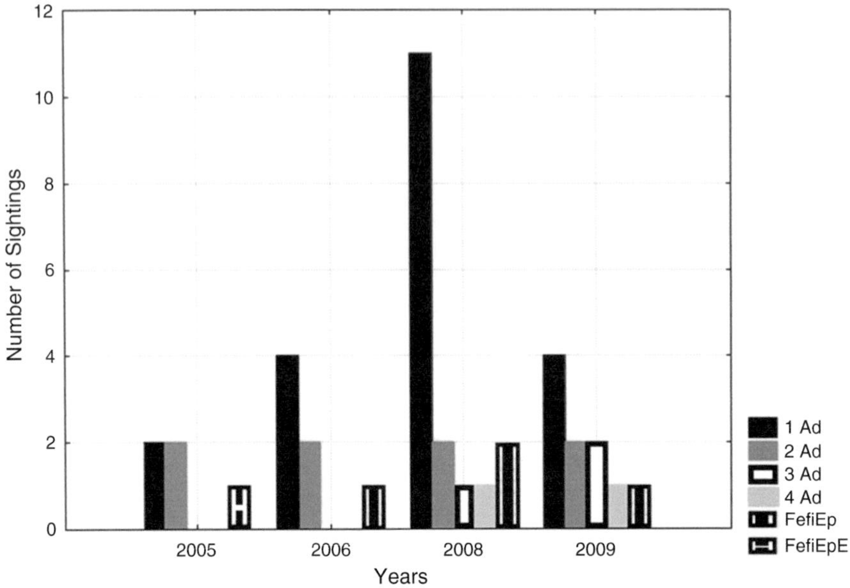

Fig. 15.14 Frequency of different group structure of humpback whales with singer males, between 2005 and 2009, in the Northeastern Brazilian coast. *1 Ad* 1 adult individual, *2 Ad* 2 adult individuals, *3 Ad* 3 adult individuals, *4 Ad* 4 adult individuals, *FefiEp* female and calf pair plus one escort male, *FefiEpE* female and calf pair plus two escort males

In Aug, 7, 2008, four animals were observed, for 85 min, with behavior such as breaching (n = 4) and tail-up during short dives (n = 37) and in Aug, 30, 2008, during 70 min, 4 animals were sighted performing head exposition (n = 2), tail-up during dives (n = 11) and pectoral fin slapping (n = 2).

15.3.4 Dive Time

To determine dive times, 956 measurements were obtained for 52 humpback whale observed groups (Table 15.2). The time at the sea surface, indicated a general pattern of few (2–5) very short breathings (less than 30 s) and one longer dive (>30 to 780 s).

Four groups of alone males were singing while measured, confirmed by underwater recordings. In Aug, 10, 2008, the singer male was at 51 m deep, showing long dive patterns and performed fluke exhibition in travelling behavior. Another sighting, in Aug, 30, 2008, a sequence of long dives (7 measurements >200 s) of the male at 40 m deep was registered, showing 5 fluke exhibitions in displacement behavior. In Aug, 28, a singer male was observed in active behavior with breaching and singing while showed a dive pattern of 4 dives less than 60 s and then one long (>100 s). Depth was not registered. In the last observation, during the research cruise of Jul, 14, 2009, a singer male was in rest behavior at 36 m deep, and the song caption was

Table 15.2 Summary of humpback whale dive time from the study area in the Brazilian coast between 2008 and 2009*

Group structure	# Groups	Depth range (m)	(D.O.) (min)	(D.P) range (s)	Mean range D.P > 100 s	# measur.	# measur. > 100 s
Alone males	21	8,7–126	977	2–780	134,3–780	278	59
Couples	18	49–126	810	2–660	125,6–540	389	88
FeCa pairs	3	–	120	5–132	116–166,3	59	7
FeCa-E1	10	66–155	500	2–349	122–479,5	230	31
Total	52	8,7–155	2407	2–780		956	185

*Definitions: (D.O.- minutes) = Direct Observation (the time spent observing humpback whale groups); (D.P. range – seconds) = Dive Pattern – the time whales spent underwater; (Mean range D.P > 100 s) = mean number of Dive Patterns larger than 100 s; (# of measur.) = number of dive measurements; (# measure. > 100 s) = number of dive measurements larger than 100 s; (Fe-Ca pairs) = female and calf pairs; (FeCa-E1) = female and calf pairs plus one escort male

intense, suggesting the proximity to our boat. This animal presented long dives, being the maximum value registered for the total sample of 780 s (13 min), and performed the same fluke exhibition than the other observed males.

The smallest dive range was registered for FeCa pairs (maximum of 132 s–2,2 min). Couples were the second category in long dive patterns (2–660 s) and the first in number of measurements larger than 100 s (n = 88) (Table 15.2).

15.3.5 Bioacoustics

15.3.5.1 Acoustic Parameters

Including the year 2010 in the acoustic sample, a 6 years (2005–2010) recording period was obtained, in effort of more than 50 h. From these raw data, we selected “type” songs from each sampled year, for both areas of Abrolhos Bank and the North Coast, chosen by containing all the characteristic phrases of the song from that year, as well by presents the best signal to noise ratio. In some years, like 2007, more than one song was selected, produced at different time during the breeding season, due to its high quality to extract additional acoustic information.

We measured a sample of 2868 component notes to describe the acoustic parameters of the humpback whale song in the Brazilian breeding ground. Data showed a wide frequency variation, typical from the long and complex humpback whale song, ranging from 20 Hz to the top of sampling rate (24,000 Hz), short duration notes (mean = 1,06 s) and sound energy from 60 to 130 dB re SNR (Table 15.3). However some features are remarkable, such as the mean central frequency at low frequency (596 Hz) (Fig. 15.15) and the intense sound energy putted into sound transmission (mean = 105 dB).

Table 15.3 Acoustic parameters from the humpback whale (*Megaptera novaeangliae*) song (sample = 2868 notes), in the study area, Northeastern Brazilian coast

Sample = 2868	Minimum frequency (Hz)	Maximum frequency (Hz)	Central frequency (Hz)	Frequency amplitude (Hz)	Duration (s)	Energy (dB)
Minimum	20	110	21,5	80,2	0,13	62
Maximum	5245	24,000	6421,9	24,000	6,55	130,4
Mean	224,4	4422	596	4198	1,06	105
St. Dev.	391	4695	625	4658	736	8,7

Harmonics were registered in 2098, 73% of the measured notes, varying in number of 1–85 (mean = 10,4). Harmonic intervals ranged from 10,3 to 4309 Hz (mean = 405).

15.3.5.2 The Humpback Whale Song Form

The song form, with descriptive information on the total number of notes, number of different notes and number of themes was represented by some best quality recording of each year for the Abrolhos Bank (Table 15.4) and for the North Coast (Table 15.5). Additionally, the number of singer males that could be heard at the background noise without confuse the main singer was also recorded.

The diversity in our recording systems evidenced that the humpback whale song has a wider frequency range than previously commented, with some notes exceeding the frequency response in some occasions (Fig. 15.16).

Behavioral observations of 50 min, simultaneous to the song recording of FV_ ITA2006 sequential files, made at 34 m deep, indicated that the singer male stops singing once blowing at the sea surface in short dive times (less than 30 s), when silent moments were registered (Fig. 15.17).

15.3.5.3 Chorus Behavior

During many times (66% = 7 from the 12 files presented in Tables 15.2 and 15.3) we could register the presence of more than one simultaneous singer male at the recordings, identified by the different intensities in sound caption, plus overlay of different song parts in the same file. More than two singer males, and even more than 3, were registered only in the Abrolhos Bank in two occasions (33% of 6 files).

After sound analyzes it was interestingly remarkable that when more than one male was singing, forming a chorus, the predominant song structure was much simpler song part than during singing of only one male, and composed, for example, by a long (4 s) whistle and a short hiccup, or even a long moan (Fig. 15.18).

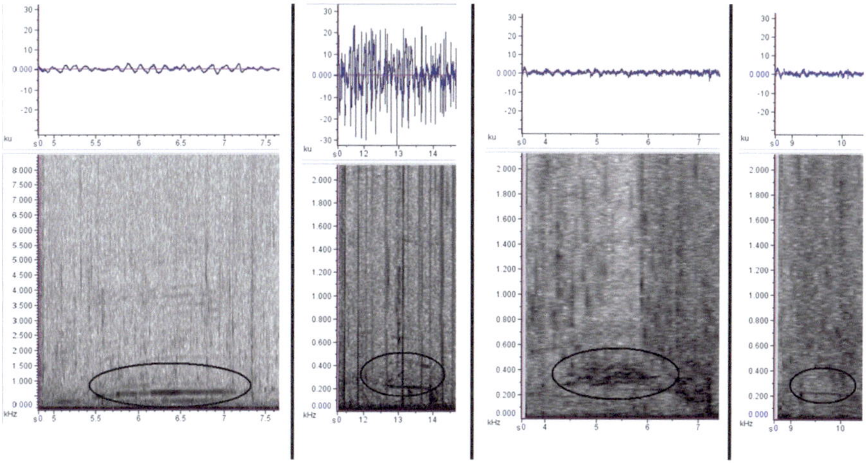

Fig. 15.15 Combination of spectrograms, a sound graphic with x-axis on time (seconds) and y-axis on frequency (kiloHertz), exemplifying some of the different humpback whale (*Megaptera novaeangliae*) notes with its remarkable mean central frequency around 500 Hz (indicated by *black circles*) *Blue line* graphs above represent the respective oscilograms or waveforms, sound graphics with x-axis on time (seconds) and y-axis on pressure (ku)

Table 15.4 Humpback whale (*Megaptera novaeangliae*) song characteristics from the northeastern Brazil (Abrolhos Bank), between 2005 and 2008, showing recording information –file identification and day-, file duration (minutes:seconds), total number of notes, different note types, number of themes and number of singers registered in each recording file

Song#	Day	Duration (m:s)	Total notes	Note types	N themes	N singers
ABL 2005	26 July 05	12:20	160	14	8	>3
ABL2006a		11:20	200	16	10	2
ABL2006b		24:12	290	21	8	2
ABL2007a	10 July 07	20:00	151	39	12	1
ABL2007b		16:25	220	40	15	3
ABL2008	21 Oct 08	56:14	188	24	7	1

Table 15.5 Humpback whale (*Megaptera novaeangliae*) song characteristics from the northeastern Brazil (North Coast), between 2005 and 2010, showing recording information –file identification and day-, file duration (minutes:seconds), total number of notes, different note types, number of themes and number of singers registered in each recording file

Song #	Day	Duration (m:s)	Total notes	Note types	N themes	N singers
ITA2005	20Ago05	50:00	212	17	8	2
F_ITA2006	06Ago06	35:00	131	21	8	1
FtVD-2008		20:00	78	14	6	1
ITA2009	04Ago09	35:00	101	12	6	2
PF2009	18Ago09	35:00	180	15	7	2
PF2010	Ago10	30:00	85	10	6	1

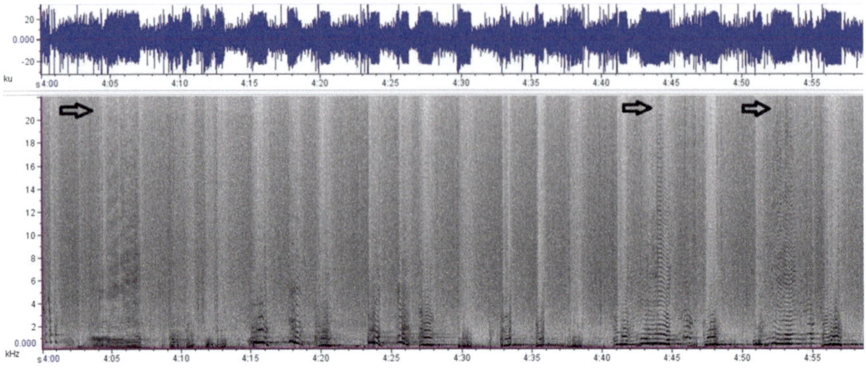

Fig. 15.16 Spectrogram showing humpback whale (*Megaptera novaeangliae*) notes surpassing the frequency response (24 kHz) of recording equipment (some evident notes indicated by *arrows*) utilized in the acoustic study off Northeastern Brazil

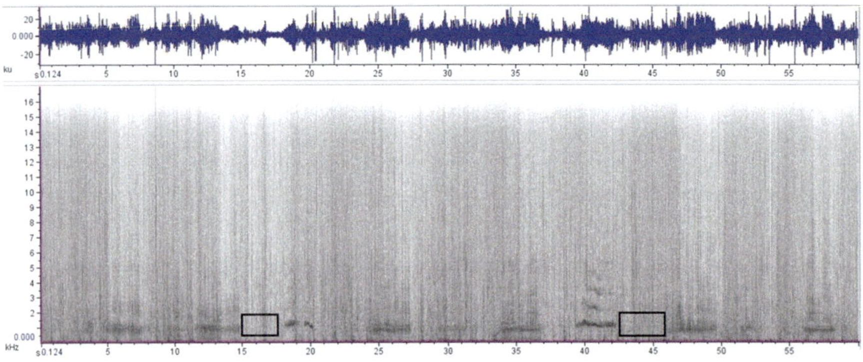

Fig. 15.17 Spectrogram exemplifying 1 min window from the FV_ITA2006 recording, indicating silent moments (*black rectangles*) once blowing at the sea surface during short dives (less than 30 s) of the humpback whale (*Megaptera novaeangliae*) in Northeastern Brazil

15.3.6 Environmental Features

15.3.6.1 Depth

Sightings of humpback groups with singer males occurred between 16 and 173 m (mean = 54,6; ±34,2 SD). Depth values were grouped into different classes (Fig. 15.19). For depth range 1 (0–20 m), only one competitive group of 4 animals was found at a 16 m location. For depth range 2 (21–30 m) 3 whale groups were sighted, for depth range 3 (31–40 m) 5 groups were registered, for depth range 4 (41–50 m) 7 whale groups were sighted, while for depth range 5 (51–70 m) 3 groups were registered. For depth range 6 (71–100 m) 3 whale groups were observed and, for depth range 7 (>100 m) 2 groups were sighted, being one with a alone individual at 126 m and a competitive group of 4 animals at 173 m deep.

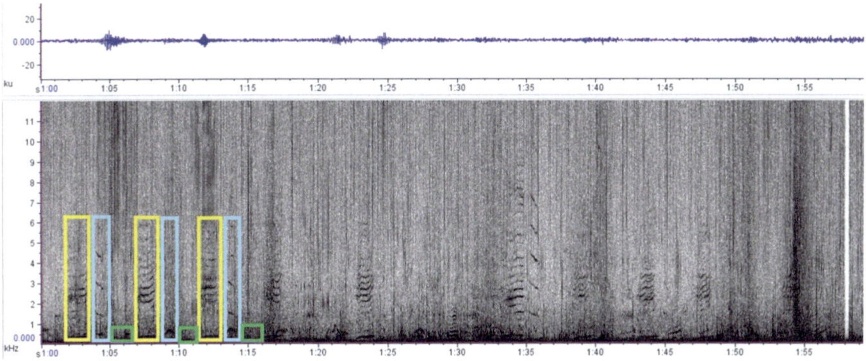

Fig. 15.18 Spectrogram exemplifying multiple male singing of the humpback whale (*Megaptera novaeangliae*), characterized by simpler parts of the song, in their breeding area off Northeastern Brazil (different colors exemplify different singers)

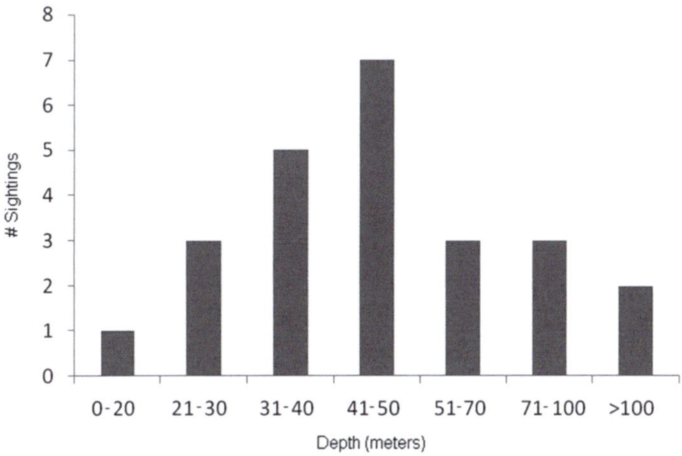

Fig. 15.19 Frequency (number of sightings) of humpback whale groups with singer males in different depth range, between 2005 and 2009, in the North Coast, Brazilian breeding ground

15.3.6.2 Tide

Even dividing the tidal states into time similar categories, and apparently diverse in the frequency distribution (Fig. 15.20), the Chi-Square test showed no statistical differences between singer male occurrence in the tidal classes (Chi-Square 5% = 11,55,556; df = 7; p = 0,12).

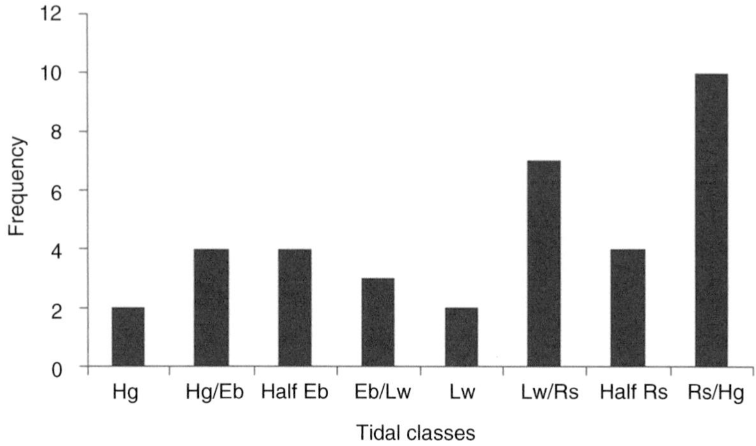

Fig. 15.20 Frequency of humpback whale groups with singer males in different tidal classes, between 2005 and 2009, in the North Coast, Brazilian breeding ground. (*Hg* High, *Hg/Eb* High/Ebbing, *Half Eb* Half Ebbing, *Eb/Lw* Ebbing/Low, *Lw* Low, *Lw/Rs* Low/Rising, *Half Rs* Half Rising, *Rs/Hg* Rising/High)

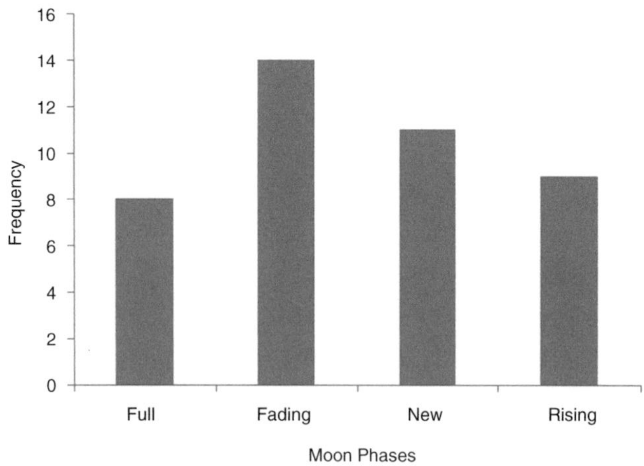

Fig. 15.21 Frequency of humpback whale (*Megaptera novaeangliae*) groups with singer males in different moon phases, between 2005 and 2009, in the North Coast, Brazilian breeding ground

15.3.6.3 Moon Phases

Singer males were sighted in all moon phases, but despite small differences (Fig. 15.21) there was no statistical difference between the four classes (Chi-Square 5% = 2,0; df = 3 p = 0,57), indicating no evident influences of this environmental feature in the singing behavior along the study area.

15.4 Discussion

During 5 years we found humpback whale groups containing singer males in all the extension of our study area, being about 520 km of coast. This confirm the large distribution of the humpbacks in the Northeastern region of Brazil, going farther than the core area of Abrolhos Bank, as previously commented (Rossi-Santos et al. 2008; Wedekin et al. 2010).

The lack of sightings in the stretch of coast between north of Praia do Forte to south of Aracaju may be probably due to the sampling route, always passing there during the night, because navigating conditions in those cruises, which depart from Praia do Forte to reach Aracaju during the early morning of the next day. However, the sightings north and southwards of this place attest the presence of the humpbacks, and because its high mobility one may assume they do occur along all that part of coast.

Despite the singer animals being typically described as alone males (eg. Payne and McVay 1971; Clapham 1996; Darling and Berube 2001), singing behavior may occur in other group composition, as the occasional escort males (eg. Baker and Herman 1984; Frankel et al. 1995; Darling and Berube 2001), whose may sing while escorting females with or without calves. Darling and Berube (2001) also reported about 14 occasions where another male approached singer males.

In the present work, singer males were mostly found as alone individuals. However singer males also occurred in groups varying from two to four individuals, besides males accompanying females and calves. In one occasion, two singers in the same group were confirmed by acoustic recordings. This show a large behavioral flexibility from humpback whales in increase the chances to find a mate partner during the breeding season.

During many occasions we noted singer males typically smaller (referenced to the boat as 8–9 m long animals), suggesting that young animals are using the study area as a way to search for new territories even less competitive than the core area of Abrolhos Bank, where traditionally large and old singer males utilize during the breeding season. We suggest the employ of the photogrammetry method (eg. Spitz et al. 2002) to access this hypothesis in next studies concerning body size estimation.

Humpback whale distribution in the study area may be related to ecological features such as oceanographic parameters of depth, continental shelf extension, while tide and moon seems do not influence singer male distribution.

Meanwhile, behavioral factors such as female distribution are theorized in the literature and may be determinant in the singer male occurrence during the breeding season. As humpback whale apparently do not feed in the breeding areas, females tend to move intensely, avoiding to meet with other female while, simultaneously, looking for increase their mate chances. Post-partum ovulation, common in this species, also benefits this high mobile female behavior and these factors together result in an irregular female distribution (Clapham 1996; Clapham and Mead 1999).

The fact of the study area be located in a continental region, different from the most of the described breeding areas, such as the Hawaiian Archipelago, (Au et al.

2006), South Pacific Ocean islands (Helweg et al. 1998), Panama Gulf (Olviedo et al. 2008), may result on differences in their acoustic signals.

As the oceanic dynamic between continental areas and islands is different, the influence of environmental parameters in continental areas may be different and provide the development of different acoustic signals that may be more efficient in a certain environment.

Further studies should develop efforts in ecological comparisons between coastal and oceanic areas which may be affecting male distribution and singing behavior in a fine scale.

We registered an increase in behavioral activity according to the increase in the group size and composition, with larger groups presenting a broader behavioral repertoire, in accordance with literature descriptions to other breeding areas (eg. Clapham 1996) and to the same breeding stock A, in the Abrolhos Bank (Morete et al. 2007), which may contribute with the complex group structure during the breeding season and may reflex in complexity of communicative skills (Freeberg et al. 2012).

The largest dive, of 13 min, was found for a singer male, very similar to the results from a recent study in the Abrolhos Bank, which analyzed the same 4 group categories and registered the largest time of 13,13 min, coming from a singer male (Bezamat et al. 2010), indicating a probable general pattern for areas as distant as 500 km, in the same breeding stock.

Humpback whales sometimes show avoidance to boat approaching increasing their dive patterns (Hoyt 2001). During many years researchers from both Abrolhos Bank and North Coast of Bahia state, noted that the whales in North Coast are more difficult to approach than in the core area of Abrolhos. The present data on the mean percentage of time spend by diving whales was 85% in relation to the direct observation time. This value was more than double than previous studies from the Abrolhos Bank, 42% in Peres (2006) and 30,1% in Benzamat et al. (2010).

The north coast do not present genetic differences from the Abrolhos Bank (Cypriano-Souza et al. 2010) but is being considered as an area of more dynamic movements of whales (Wedekin et al. 2010; Baracho-Neto et al. 2012), and this could lead, in consequence, in more travelling behavior and even boat avoidance through less time in the surface.

The narrow continental shelf in the North Coast also can play a role in the dive patterns, making individuals travel more in a north-south direction to keep in shallow waters characteristic of their distribution, while in the Abrolhos Bank, the core area in the Brazilian coast, more than double of density of whales (Andriolo et al. 2006, 2010; Wedekin et al. 2010).

As the whole breathing behavior of humpbacks do include those very short intervals, we decided to keep them all in the analyzes, different from Benzamat et al. (2010) which only included intervals larger than 60 s, even considering that smaller intervals prevailed, being 68% of their sampling.

Concluding, singer males are longer divers than other analyzed group structure, presenting similar values, of 13 min, to the Abrolhos Bank, distant about 500 km from our study area in the North Coast of Bahia state. Humpback whales in North Coast spent about 85% of the direct observation in underwater activities, more than the double of time found for the Abrolhos Bank, which could indicate longer underwater occurrence due by distinct habitat use, once the north coast is seen as an area of greater movement than the Abrolhos Bank. However, even with distinct environmental features as the deeper areas of the North Coast, it does not seem to present a strong influence on dive patterns of the humpback whale in the Brazilian Breeding ground.

15.4.1 Acoustic Parameters

Despite the long list of studies in more than 40 years of research about the singing behavior of humpback whales (eg. Payne and McVay 1971; Tyack 1981; Payne and Payne 1985; Helweg et al. 1998; Mercado et al. 2003; Au et al. 2006), still there are few descriptive studies that present numeric values for frequency, duration and sound intensity. This work brings the first song description in the Brazilian coast, outside the Abrolhos Bank, showing that humpbacks are utilizing a broader acoustic environment for their breeding activities than previously reported.

Frequency parameters varied from 20 Hz (minimum) to 24,000 Hz (maximum), being the interval between mean frequencies of 224,4 a 4422 Hz, duration of 0,13 to 6,5 s (mean = 1,06) and sound energy ranging from 62 to 130,4 dB (re SNR) (mean = 105).

Olviedo et al. (2008), studying at Las Perlas Archipelago, Panama Gulf, reported frequency parameters for year 2006, such as the minimum frequency (129,2–575,7 Hz), maximum (1151,7–2879,6 Hz) and note duration (0,79–5,50 s) and, for 2007, the minimum frequency (129–1076 Hz), maximum (2368–4866 Hz) and note duration (0,66–11,4 s). Number of notes varied from 13 to 15. Same authors also describe the environment as being calm and shallow waters, contributing to the whale breeding cycle, similar to the oceanographic conditions we observe in the Abrolhos Archipelago.

Arraut and Vielliard (2004), in Abrolhos Bank, reported 24 notes and 5 themes as the humpback whale song structure, during the breeding season of 2000, bringing only spectrograms as a physical reference for the acoustic parameters, with frequency axis ranging from 500 to 2000 Hz.

Thus the present work show similar frequency values from other geographical areas, such as the Panama Gulf (Olviedo et al. 2008), however, apparently by the spectrograms reported by Arraut and Vielliard (2004), song varied temporally inside the same breeding area off Brazil.

Another relevant finding from this study is the high frequency harmonic occurrence reaching the top of the frequency response (24 kHz). Au et al. (2006) reported high frequency harmonics for the humpbacks in the Hawaiian waters, suggesting the species may show an upper hearing limit as high or higher than this value.

The humpback whale song in this work seems to be more diverse and complex than other reported areas, such as in Hawaii, where Au et al. (2006) described 9 notes compounding the song of 2002. For the Brazilian coast, Arraut and Vielliard (2004) identified a similar number of notes, grouped into five themes, in the breeding season of 2000. This may represent the gradual modification described in literature (eg. Payne et al. 1983; Payne and Payne 1985) e o aporte de dados dos anos subsequentes poderá contribuir com uma melhor interpretação deste resultado.

The values for maximum frequency were larger than previously reported, which could indicate a better signal adaptation to the specific Brazilian acoustic environment. In this case, there is a big difference between the depth gradient utilized by the whales in places like the Abrolhos Bank and the environment of the narrow continental shelf from Itacaré to Aracaju. This consideration is strengthen by the harmonic values reported for Hawaii (Au et al. 2006), where the depth is similar to part of our study area, which include deeper waters. Biological implication of this would be a maximization of the reproductive signal travelling across the acoustic environment to increase the mate opportunities during the time-limited breeding season.

Thus, beyond the descriptive analyzes of the acoustic parameters for the song, this work bring information of the relations between song frequency and duration with the acoustic environment in which it is produced, stating that differences may occur even in a smaller geographic area, such as inside the same breeding area, where environmental differences may result in distinct song components, as shown with the maximum frequency and note duration.

Also, seems to be obvious that the advance in technology in the last decades have shown a broader frequency spectra utilized by humpback whales. In this way, the first decade in the "boom" of humpback song research was the 1970s, where notes were registered below 5 kHz (eg. Payne and McVay 1971; Winn and Winn 1978) as well as during the 1980s (eg. Payne and Payne 1985). The digital recording "Era" increased more than twice the previous frequency response, bringing the higher frequencies identifiable to the humpback repertoire (eg. Au et al. 2006, Mercado et al. 2003, present work).

15.4.2 Song Form

Diverse and complex song structure along the time was found for humpback whales off Brazilian waters, resulting in remarkable differences within a 6 years period of study (2005–2010). Such differences are broader than previously commented, even considering the differences inside the same breeding area, where environmental characteristics may contribute with signal diversity and specificity.

Payne and Payne (1985) analyzed and compared humpback whales songs in Bermuda for 13 years, between 1957 and 1975. They found that song changed conspicuously and progressively with time, being the songs separated by a number of years were very different in content. Furthermore, all the songs showed basic structural similarities making possible to define a song form, which characterizes songs from many years. Such analyses demonstrated high inter and intra-individual variability, none of which is as great as the variation between songs of consecutive years. Our study is the first long term acoustic description to both Atlantic and

Pacific coasts of South America, bringing important information on temporal variation in the humpback whale songs to this area.

Winn and Winn (1978) suggested that because low frequencies within humpback whale songs are attenuated less than high frequencies during propagation, higher frequency units might be used for transmission over short ranges whereas lower frequency units might be used for longer ranges. Mercado and Frazer (1999) add to this notion that it also depends on the depth where whales sing, because in shallow water environments in Hawaii, the lowest frequencies they produce do not propagate as far as the higher frequencies they produce). Mercado et al. (2003) also pointed out that no single frequency will propagate optimally to all positions within the water column. Thus, increasing the range of produced frequencies could increase the number of positions within a shallow water environment from which the sound could be detected.

Another possible argument to the high variability found in the form of humpback whale song in the Brazilian coast (present study) could be related to the high capacity of copying in humpbacks, controlling changes over time that can abruptly be modified (Noad et al. 2000), which suggest that individual whales select patterns and units to include in songs based on their recent experience, or kept in use during next seasons (Payne and Payne 1985; Cato 1991; Mercado et al. 2005). Furthermore, Mercado et al. (2005) discuss on song copying by humpback whales as being an open process in the spectral domain in that specific frequencies could be utilized by each different individual, within or across years, and that species-specific constraints determine song form and regionalization. These affirmations could, in part, explain the present results on the diversity of song structure between the two regions, the Abrolhos Bank and the North Coast of our study area.

Sequences of alternating tonal and pulsed units in humpback whale songs also could be possibly related to internalized 'inhale' and 'exhale' of air during sound production (Mercado et al. 2003). Further studies involving size estimation in the two regions of our study area could help to elucidate the matter of body size and song diversity, comparing song production among individuals of different sizes.

Despite the high variability within and between seasons, some song components are said to recur for over decades (Payne and Payne 1985; Mercado et al. 2003) indicating important features that may be maintained during time to build an effective communication system.

15.4.2.1 Song Form Diversity and Function

It is still debated the complete role that songs serve for (revised in Parsons et al. 2008). The behavior that accompanies singing indicates that it probably plays a part in courtship analogous to the role of singing in birds (Tyack 1981). Some authors say that it is primarily for inter- and intra-sexual advertisement, informing females about the location and reproductive fitness of singers, and informing other males about the location of singers to produce spacing among multiple singers (reviewed by Helweg et al. 1992; Clapham 1996), or advertising the fitness of males (Darling and Bérubé

2001). A different interpretation of the function of the song involving a cooperative male behavior during group formation in discussed by Darling et al. (2006).

15.4.3 Chorus Behavior

The chorus behavior is described as a part of the complex leking, where males tend to defend resources to which females are attracted, in this last case gain from stimulus pooling by displaying together and providing greater attraction for females (Krebs and Davies 1993). This strategy has been largely studied for ungulate, bats and frogs (revised in Krebs and Davies 1993).

While much attention has been given to the characteristics of songs, few studies go into the chorus behavior by multiple humpback whales (Au et al. 2000). Thompson and Friedl (1982) were the first to mention chorusing for the humpbacks in Hawaii, finding seasonality and daytime preference for chorusing in a breeding area.

Au et al. (2000) described spectral features as well temporal utilization of the chorus behavior in the humpbacks during the 1998 Hawaiian season, showing spectral peaks at 315 and 630 Hz. Their data also indicated a diurnal pattern in the sound pressure level, with levels at night significantly louder than the daytime levels, attributing different mating strategies to the species during a night and day scale, in which competitive groups are more frequent during the day, requiring more visual contact, while chorusing are more frequent during the night time.

Our study brought for the first time the chorus perspective to the humpback whale acoustic ecology in the Brazilian waters. The higher occurrence of chorus found in the Abrolhos Bank could indicate that this behavior should act as a trigger to the female choice as described for the terrestrial leks (Alcock 1993; Krebs and Davies 1993). Listening to an intense chorus, females would be guided to a good area concerning the search for a mate. Additionally, we are in accordance to the Au et al. (2000) suggestion of the viability of using different levels of chorusing to monitor and estimate the relative abundance of male humpback whales in the Brazilian breeding ground.

Summarizing, there are numerous hypotheses on the implications of the humpback whale song, many of which are not mutually exclusive. The fact is that many aspects of the humpback whale song components remain unknown, such as what are the reasons to keep for a longer time some components instead other and what kind of significance it has on reproductive success over time.

The humpback whale, as a migratory animal, requires a broad geographic scale to develop essential activities such as feeding and breeding, for consequence implying on a huge logistical structure to access their habits in accordance to such large scale.

The Ocean is a challenging environment to man, in the path of an ecological comprehension and global vision. Therefore, we still know very few about the whole implication of the humpback whale role in this broad oceanic scale. Perhaps in a near future, besides with the necessary technology to the understanding of oceanic processes in a deeper approach, we will find in the humpback whale song a key to last a long time with respect to this planet.

15.4.4 Mating System

Polygyny is a mating system in which one male mates with several females during a breeding season. "Lek" is defined as a traditional display site where males gather to defend small territories that lack resources useful to females, which nevertheless visit the site to mate. Therefore, lek polygyny occurs when a polygynous male use displays to attract several females to them at small display sites (Alcock 1993).

It is well studied for other mammals such those closely related to cetaceans, the ungulates, beyond 35 bird species, frogs and even insects (see Krebs and Davies 1993; Alcock 1993). Clapham (1996) suggested the term 'floating lek' to the humpback whale mating system, emphasizing the high mobility of the singer males, in contrast with the traditional lek which present a fixed territoriality.

In the present study, spatial (presence of multiple group composition) and behavioral (diverse repertoire while singing) features were found that may support the "lek polygyny" as the mating system for the humpback whale. The singer male distribution along the study area indicate that a broad area in the breeding ground is being utilized for displaying, refusing lek influenced by "hotspot", in which males compete for the most attractive site to display, while fitting better to lek by "hotshot" males, which may choose any place to sing and then attract females to mate and more "satellite" males, coming to maximize their mate chances by staying as close as the hotshot male (Krebs and Davies 1993; Alcock 1993).

Connor et al. (2000) discussing on male reproductive strategies in cetaceans, state that humpback whale seems to follow the restrictive definition of lekking: (1) no male parental care, (2) an arena where males gather and females come to mate, (3) resource-free display sites, and (4) opportunities for females to exercise mate choice (Clapham 1996). However, instead to emphasize only the male-male dominance and territoriality, Connor et al. (2000) also include a broader view in which leking behavior is found in association with other male mating strategies, as the example of male-male combat in competitive alternatively to displaying. Tyack (2000) suggest that humpback whale song plays a role in both male-male competitions and female choice, showing attributes associated with both intra and intersexual selection.

To determine a precise mating system for the humpbacks is a complex task, once we still do not know many basic issues as the ideal breeding area size for any population. Thinking broader, the Brazilian population increase (Ward et al. 2006; Zerbini et al. 2006; Andriolo et al. 2006) could lead to dispersion of young females, whose would attract different males, then forming the floating lek, either by hotspot or hotshot, and colonizing favorable adjacent areas.

At the same time that in the core area, the Abrolhos Bank, male arenas would be much more competitive, then causing younger males to move northwards and search for new areas to aggregate, through singing, and then maximize mate.

Furthermore, reflecting on the multi faced hypothesis of a mating system, would be important to analyze that the humpback whale population using the Brazilian breeding ground have a sex-ratio of 1.2:1, similar to the expected rate of 1:1, determined genetically by Cypriano-Souza et al. (2010). Baracho-Neto et al. (2012) also support the site fidelity similarity between genders through a long term photoidentification study in Praia do Forte, same than the present study.

All these pieces together may help to create a scenario suggesting a mating system more complex than previously reported, with more fluidity of individuals, females and males, along a broad area in the Brazilian coast and ideal breeding sites possibly influenced by some environmental features as the extent of the continental shelf, consequently depth gradient.

Furthermore, the diversity in singer male occurrence and behavior between different group compositions could be related to the complex communicative system in humpback whales, as recently described by Freeberg et al. (2012). Future studies should include environmental and whale occurrence data in a broader temporal scale to test specifically each one of these factors and their influence in the acoustic ecology and behavior of humpback whales.

15.5 Conclusions

This study contributed in the understanding of ecological mechanisms involved in the humpback whale song behavior. In a social context, about one third of the sighted groups presented at least one singer male, but also identifying groups with two to four individuals, plus groups with female-calf pairs with one to two escort males.

The narrow continental shelf associated with the coastal distribution of the whales in the study area may influence in the group compositions containing singer males.

We present an ecological perspective, including environmental features such as depth, tides and moon, of the singing behavior in this species, and also show the expansion of the acoustic activity, during the Brazilian breeding season.

References

Alcock J (1993) Animal behavior: an evolutionary approach. Sinauer Associates, Sunderland, 625 p

Andriolo A, Martins CCA, Engel MH, Pizzorno JLA, Más-Rosa S, Freitas A, Morete ME, Kinas PG (2006) The first aerial survey to estimate abundance of humpback whale (*Megaptera novaeangliae*) in the breeding ground off Brazil. J Cet Res Manag 8(3):307–311

Andriolo A, Kinas PG, Engel MH, Martins CCA, Rufino AM (2010) Humpback whale population estimates and distribution along the Brazilian breeding ground. Endanger Spec Res 11:233–243

Arraut EM, Vielliard JME (2004) The song of the Brazilian population of humpback whale *Megaptera novaeangliae* in the year 2000: individual song variations and possible implications. An Acad Bras Cienc 76(2):373–380
Au WWL, Mobley J, Burgess W, Lammers M, Nachtigall PE (2000) Seasonal and diurnal trends of chorusing humpback whales wintering in waters off western Maui. Mar Mamm Sci 16(3):530–544
Au WWL, Darling J, Andrews K (2001) High-frequency harmonics and source level of humpback whale songs. J Acoust Soc Am 110:2770
Au WWL, Pack A, Lammers ML, Herman LM, Deakos MH, Andrews K (2006) Acoustic properties of the humpback whale song. J Acoust Soc Am 120(2):1103–1110
Baker CS, Herman LM (1984) Aggressive behavior between humpback whales (*Megaptera novaeangliae*) wintering in Hawaiian waters. Can J Zool 62:1922–1937
Baracho-Neto CG, Santos Neto E, Rossi-Santos MR, Wedekin LL, Neves MC, Lima F, Faria D (2012) Site fidelity and residence times of humpback whales (*Megaptera novaeangliae*) on the Brazilian coast. J Mar Biol Ass UK, p 1 of 9, doi:10.1017/S0025315411002074
Bezamat C, Wedekin LL, Marcondes MCC, Valentin JL (2010) Padrão de mergulho da baleia jubarte na concentração reprodutiva do Banco de Abrolhos, Brasil. In: III Congresso Brasileiro de Oceanografia, Rio Grande, Brazil, pp 2308–2310
Cato DH (1991) Songs of humpback whales: the Australian perspective. Mem Queen Mus 30:277–290
Cerchio S, Jacobsen JK, Norris TF (2001) Temporal and geographical variation in songs of humpback whales, *Megaptera novaeangliae*: synchronous change in Hawaiian and Mexican breeding assemblages. Anim Behav 62:313–329
Chittleborough G (1955) Puberty, physical maturity, and relative growth of the female humpback whale, *Megaptera nodosa* (Bonnaterre) on the Western Australian coast. Aust J Mar Freshw 6:315–327
Clapham PJ (1996) The social and reproductive biology of Humpback Whales: an ecological perspective. Mamm Rev 26:27–49
Clapham PJ (2000) The humpback whale: seasonal feeding and breeding in a baleen whale
Clapham PJ, Mead JG (1999) Megaptera novaeangliae. Mamm Spec 604:1–9
Connor RC, Read AJ, Wrangham R (2000) Reproductive strategies and social bonds. In: Mann J, Connor RC, Tyack PL, Whitehead H (eds) Cetacean societies – field studies of dolphins and whales. Chicago Press, Chicago, pp 247–269
Cypriano-Souza AL, Gabriela P, Fernandez CA, Lima-Rosa V, Engel MH, Bonatto SL (2010) Microsatellite genetic characterization of the humpback whale (*Megaptera novaeangliae*) breeding ground off Brazil (Breeding Stock A). J Hered 101(2):189–200
Darling JD, Berube M (2001) Interactions of singing humpback whales with other males. Mar Mamm Sci 17:570–584
Darling JD, Sousa-Lima RS (2005) Songs indicate interaction between humpback whale (*Megaptera novaeangliae*) populations in the western and eastern South Atlantic Ocean. Mar Mamm Sci 21:557–566
Darling JD, Jones ME, Nicklin CP (2006) Humpback whale songs: do they organize males during the breeding season? Behaviour 143(9):1051–1101
Dawbin WH, Eyre EJ (1991) Humpback whale songs along the coast of Western Australia and some comparison with east coast songs. Mem Queen Mus 30(2):249–254
DHN-Departamento de Hidrologia e Navegação da Marinha do Brasil (1995) Tábua de Marés
Ekau W, Knoppers B (1999) An introduction to the pelagic system of the North-East and East Brazilian shelf. Arch Fishw Mar Res 47:113–132
Engel M (1996) Comportamento reprodutivo da Baleia Jubarte (*Megaptera novaeangliae*) em Abrolhos. Anim Etol 14:275–284
Eriksen N, Millar LA, Tougaard J, Helweg DA (2005) Cultural change in the songs of humpback whales (*Megaptera novaeangliae*) from Tonga. Behaviour 142:305–328

Frankel AS (1995) Interindividual variation in the songs of humpback whales. J Acoust Soc Am 99:2556
Frankel AS, Clark CW, Herman LM, Gabriele CM (1995) Spatial distribution, habitat utilization, and social interactions of humpback whales, *Megaptera novaeangliae*, off Hawaii, determined using acoustic and visual techniques. Can J Zool 73:1134–1146
Frazer LN, Mercado E III (2000) A sonar model for humpback whale song. IEEE J Ocean Eng 25:160–182
Freeberg TM, Ord TJ, Dunbar RIM (2012) The social network and communicative complexity: preface to theme issue. Phil Trans R Soc A 367:1782–1784
Fristrup KM, Hatch LT, Clark CW (2003) Variation in humpback whale (*Megaptera novaeangliae*) song length in relation to low-frequency sound broadcasts. J Acoust Soc Am 113:3411–3424
Guinee LN, Chua K, Dorsey EM (1983) Changes over time in the songs of known individual humpback whales (*Megaptera novaengliae*). In: Payne R (ed) Communication and behavior of whales. Westview Press, Boulder, pp 59–80
Hatge V, Andrade JB (2009) Baía de Todos os Santos: Aspectos Oceanográficos. Editora da Universidade Federal da Bahia, 264 p
Helweg DA, Farnkel AS, Mobley J, Herman LM (1992) Humpback whale song: our current understanding. In: Thomas JA, Kastelein RA, Supin AS (eds) Marine mammal sensory systems. Plenum, New York, pp 459–483
Helweg DA, Cato DH, Jenkins DF, Garrigue C, McCauley RD (1998) Geographic variation in South Pacific humpback whale songs. Behaviour 135:1–17
Herman LM, Tavolga WN (1980) The communication systems of cetaceans. In: Herman LM (ed) Cetacean behavior: mechanisms and functions. J. Wiley, New York, pp 149–209
Hoyt E (2001) Whale watching 2001: worldwide tourism numbers, expenditures and expanding socioeconomic benefits. IFAW Yarmouth Port.
IWC (1998) Report of the Scientific Committee. Rep Int Whal Comm 48:55–302
Krebs JR, Davies NB (1993) An introduction to behavioral ecology. Blackwell Scientific Publications, Oxford
Lodi L (1994) Ocorrências de baleias-jubarte, *Megaptera novaeangliae*, no Arquipélago de Fernando de Noronha, incluindo um resumo de registros de capturas no Nordeste do Brasil. Biotemas 7(1):116–123
Lunardi DG, Engel MH, Marciano JP, Macedo RH (2010) Behavioural strategies in humpback whales (*Megaptera novaeangliae*) in a coastal region of Brazil. J Mar Biol Ass UK 90(8):1693–1699
Maeda H, Higashi N, Uchida S, Sato F, Yamaguchi M, Koido T, Takemura A (2000) Songs of humpback whales, *Megaptera novaeangliae*, in the Ryukyu and Bonin regions. Mammal Study 25:59–73
Mann J, Connor RC, Tyack PL, Whitehead H (2000) Cetacean Societies: field studies of dolphins and whales. Chicago Press, Chicago
Mattila D, Guinee K, Mayo CA (1987) Humpback whale songs on a North Atlantic feeding ground. J Mammal 68:880–883
McGregor PK, Krebs JR (1984) Song learning and deceptive mimicry. Anim Behav 32:280–287
McSweeney DJ, Chu KC, Dolphin WF, Guinee LN (1989) North Pacific humpback whale songs: a comparison of southeast Alaskan feeding ground songs with Hawaiian wintering ground songs. Mar Mamm Sci 5:139–148
Mercado E III, Frazer LN (1999) Environmental constraints on sound transmission by humpback whales. J Acoust Soc Am 106:3004–3016
Mercado E III, Herman LM, Pack AA (2003) Stereotypical sound patterns in humpback whale songs: usage and function. Aquat Mamm 29:37–52
Mercado E III, Herman LM, Adam M, Pack A (2005) Song copying by humpback whales: themes and variations. Anim Cogn 8:93–102

Morete ME, Bisi T, Rosso S (2007) Temporal pattern of humpback whale (*Megaptera novaeangliae*) group structure around Abrolhos Archipelago breeding region, Bahia, Brazil. J Mar Biol Assoc U K 87:87–92
Naguib M (1998) Perpection of degradation in acoustic signals and its implications for ranging. Behav Ecol Sociobiol 42:139–142
Noad MJ, Catom DH, Brydenm M, Jenner N, Jener CS (2000) Cultural revolution in whale songs. Nature 408:537
Olviedo L, Guzman HM, Floréz-Ganzalés L, Capella Alzueta J, Mair JM (2008) The song of the southeast Pacific humpback whale (*Megaptera novaeangliae*) off Las Perlas Archipelago, Panama: preliminary characterization. Aquat Mamm 34(4):458–463
Palsboll PJ, Allen J, Berube M, Clapham PJ, Feddersen TP, Hammond PS, Hudson RR, Jorgensen H, Katona S, Larsen AH, Larsen F, Lien J, Mattila DK, Sigurjonsson J, Sears R, Smith T, Sponer R, Stevick P, Oien N (1997) Genetic tagging of humpback whales. Nature 388:767–769
Parsons ECM, Wright AJ, Gore MA (2008) The Nature of humpback whale (*Megaptera novaeangliae*) song. J Mar Anim Ecol 1(1):22–31
Payne S, Guinee LN (1983) Humpback whale (*Megaptera novaeangliae*) songs as an indicator of "stocks". In: Payne R (ed) Communication and behavior of whales. Westview Press, Boulder, pp 333–358
Payne S, McVay S (1971) Songs of humpback whales. Science 173:585–597
Payne K, Payne R (1985) Large scale changes over 19 years in songs of humpback whales in Bermuda. Zeitschrift fur Tierpsychologie 68(89–1):14
Payne KP, Tyack DR, Payne R (1983) Progressive changes in the songs of humpback whales (*Megaptera novaeangliae*): a detailed analysis of two seasons in Hawaii. In: Payne R (ed) Communication and behavior of whales. Westview Press, Boulder, pp 9–57
Peres GB (2006) Caracterização do tempo de mergulho em baleia jubarte (*Megaptera novaeangliae*) no Arquipélago de Abrolhos, Bahia – Brasil. In: Trabalho de Conclusão do Curso de Graduação em Ciências Biológicas, Universidade Federal de Santa Catarina
Podos J (2001) Correlated evolution of morphology and vocal signal structure in Darwin's finches. Nature 409:185–188
Podos JE, Southall JA, Rossi-Santos MR (2004) Vocal mechanics in Darwin's finches: correlations of beak gape and song frequency. J Exp Biol 207:607–619
Rossi-Santos MR, Neto ES, Baracho CG, Cipolotti SR, Marcovaldi E, Engel MH (2008) Occurrence and distribution of humpback whales (*Megaptera novaeangliae)* on the north coast of the State of Bahia, Brazil, 2000–2006. ICES J Mar Sci 65:667–673
Smith JN, Goldizen AW, Dunlop RA, Noad MJ (2008) Songs of male humpback whales, *Megaptera novaeangliae*, are involved in intersexual interactions. Anim Behav 76:467–477
Spitz SS, Herman LM, Pack AA, Deakos MH (2002) The relation of body size of male humpback whales to their social roles on the Hawaiian winter grounds. Can J Zool 80:1938–1947
Tavares JS (1916) A pesca da Baleia no Brasil. Brotéria, Revista de Sciencias Naturaes: Série de Vulgarização Scientífica 14:69–80
Thompson P, O. & Friedl, W. A. (1982) A long term study of low frequency sounds from several species of whales off Oahu, Hawaii. Cetology 45:1–19
Tollenare LF (1961) A pesca da baleia (pelo Recôncavo Baiano – 1817) (org Bruno ES, Riedel). In: Histórias e Paisagens do Brasil – Coqueirais e Chapadões – Bahia e Sergipe. Editora Cultrix, São Paulo, pp 63–79
Towsend CD (1935) The distribution of certain whales as shown by logbooks of American Whaleships. Zoologica 19(1):2–50
Truax B (1999) Handbook for acoustic ecology. Street Publishing, Burnaby. CD-Room version
Tyack P (1981) Interactions between singing Hawaiian humpback whales and conspecifics nearby. Behav Ecol Sociobiol 8:105–116
Tyack P (2000) Functional aspects of cetacean communication. In: Mann J, Connor RC, Tyack PL, Whitehead H (eds) Cetacean Societies: Field Studies of Dolphins and Whales. Chicago Press, Chicago, pp 270–307

Ward E, Zerbini AN, Kinas PG, Engel MH, Andriolo A (2006) Estimates of population growth rates of humpback whales (*Megaptera novaeangliae*) in the wintering grounds off the coast of Brazil (Breeding Stock A). Paper SC/58/SH14 presented to the IWC Scientific Committee

Wedekin LL, Neves M, Marcondes M, Baracho C, Rossi-Santos MR, Engel M, Simões-Lopes PC (2010) Site fidelity and movements of humpback whales (*Megaptera novaeangliae*) in the Brazilian breeding ground, southwestern Atlantic. Mar Mamm Sci 26(4):787–802

Whitehead H (1981) The behaviour and ecology of the Northwest Atlantic humpback whales. PhD thesis, University of Cambridge, Cambridge, 256 pp

Whitehead H, Moore MJ (1982) Distribution and movements of West Indian humpback whale in winter. Can J Zool 60:2203–2211

Wilson EO (1975) Sociobiology: the new synthesis. Belknap Press, Cambridge, MA

Winn HE, Perkins P (1976) Sounds of the minke whale, within a review of misticete sounds. Cetology 19:1–12

Winn HE, Perkins P, Poulter T (1971) Sounds of the humpback whale. In: Proceedings of the 7th annual conference on biological sonar and diving mammals. Menlo Park, CA, pp 39–52

Winn HE, Winn LK (1978) The song of the humpback whale *Megaptera novaeangliae* in the West Indies. Mar Biol 47:97–114

Winn H, Thompson E, Cummings JT, Hans WC, Hudnall J, Hays J, Steiner W (1981) Song of the humpback whale: population comparison. Behav Ecol Sociobiol 8:41–46

Zerbini AN, Ward E, Engel MH, Andriolo A, Kinas PG (2006) A Bayesian assessment of the conservation status of humpback whales (*Megaptera novaeangliae*) in the western South Atlantic Ocean (Breeding stock A). Paper SC/58/SH2 presented to the IWC Scientific Committee

Chapter 16
Conservation Advances for the Southern Right Whales in Brazil

Karina Groch

Abstract During the austral winter/spring time southern right whales occur in coastal waters of Brazil. The species was historically distributed between the southern and northeastern coast of Brazil but was severely depleted by commercial whaling until 1973. In the early 1980s, whales were 'rediscovered' in this region and have been studied there since then by the "Right Whale Project", a research and conservation program created with the aim of ensure the survival and recovery of the right whale population in Brazilian waters. Their main aggregation area is in the shallow and protected bays of the central-southern coast of Santa Catarina State. This area is marine protected area – Right Whale Environmental Protection Area (RWEPA). This chapter provides information on the species research techniques and conservation advances in Brazil.

16.1 Southern Right Whales: Where Are They Found?

The southern right whale (*Eubalaena australis*) (Desmoulins 1822) has circumpolar distribution in the Southern Hemisphere occurring between 20° and 60° latitude (Townsend 1935; Cummings 1985). The species spend the summer and early autumn in offshore high latitude feeding grounds and migrate to coastal waters of continents and oceanic islands during the austral winter/spring time seeking for shallow and protected areas to give birth, nurse their calves and mate (Payne 1986; Best 1988; Bannister 1990; Patenaude and Baker 2001).

Feeding grounds are regions between the subtropical and Antarctic convergences (IWC 2001). The evidence comes mainly from whaling records (Townsend 1935). One direct link between a nursery and feeding ground comes from five whales identified at Península Valdés, Argentina, that were resighted on feeding grounds off South Georgia and Shag Rocks (ca 53°S) (Best et al. 1993; Rowntree et al. 2001). Another direct link was found in Australia, where two right whales identified in the

K. Groch (✉)
Projeto Baleia Franca/Instituto Australis, CP 201, Imbituba 88780-000, SC, Brazil
e-mail: karinagroch@gmail.com; karina@baleiafranca.org.br

M.R. Rossi-Santos, C.W. Finkl (eds.), *Advances in Marine Vertebrate Research in Latin America*, Coastal Research Library 22,
DOI 10.1007/978-3-319-56985-7_16

Australian coast were resighted on feeding grounds offshore (Bannister et al. 1997). Other evidences for migratory patterns and locations of feeding grounds have been found from stable isotope studies (Rowntree et al. 2008; Valenzuela et al. 2009; Vighi et al. 2014). Recent information on migratory routes and feeding grounds of whales off Argentina was obtained from a satellite tagging project carried out between 2014 and 2015. Whales were tracked through the outer Patagonian shelf, (north of about 54°S and between about 57°–60°W) and the Scotia Sea (north of around 57°S and between about 28–42°W), and appeared to show feeding behavior (IWC 2016). There also is evidence that right whales can feed in their breeding grounds, both from studies on stable isotopes (Vighi et al. 2014; Best and Schell 1996) and from individuals repeatedly observed feeding on plankton blooms, particularly at the end of the season, in Argentina (Hoffmeyer et al. 2010).

During the austral winter/spring time southern right whales occur in coastal waters of South America, South Africa, Australia, New Zealand and some oceanic islands. In the South Atlantic five main breeding grounds are recognized: Argentina, Brazil, Tristan da Cunha, Namibia and South Africa (IWC 2001). Studies have also demonstrated the occurrence of southern right whales in Uruguay (Costa et al. 2007).

Coastal areas are the preferred habitat of mother/calf pairs, mainly because they enable energy conservation by the calves (Lockyer 1981; Thomas and Taber 1984; Corkeron and Connor 1999). Shallow waters and gently sloping beaches are preferred for being areas free from underwater obstacles where whales can drift freely following the tides (Thomas 1987). There is evidence that female select sheltered, inshore habitats during the early stages of nursing while their calves are presumably at their most vulnerable (Rayment et al. 2015). Shallow waters may also offer protection against killer whales (Thomas and Taber 1984; Thomas 1987; Corkeron and Connor 1999; Sironi et al. 2008a, b). Habitat use may also be determined by environmental factors that favor calf survival (Rowntree et al. 2001; Elwen and Best 2004a; Seyboth et al. 2015).

Southern right whales can be flexible in many aspects of their habitat use (Rowntree et al. 2001; Sironi et al. 2008a, b) but displays preference by certain aggregation areas as well as the segregation of concentration areas between mother/calf pairs and mating groups (Payne 1986; Bannister 1990; Best 1990; Burnell and Bryden 1997; Rowntree et al. 2001; Groch 2005; Rayment et al. 2015). Mother-calf pairs spend longer coastal residence periods than adults unaccompanied by calves (Bannister 1990; Burnell and Bryden 1997; Groch 2005) and present less movement along the coast (e.g. Bannister 1990; Best 1990; Rowntree et al. 2001; Groch 2005).

Southern right whales, especially mother-calf pairs, exhibit levels of site fidelity for breeding areas (Bannister 1990; Best 1990, 2000; Burnell and Bryden 1997; Rowntree et al. 2001; Groch 2005). Right whales have also shown site fidelity to feeding grounds (Valenzuela et al. 2009).

16.2 Southern Right Whales in Brazil

Southern right whales were severely depleted by commercial whaling from the 18th through the early twentieth centuries. In 1935 the species received international protection, however in the mid-twentieth century there was unreported Soviet catches of right whales on their pelagic summer grounds (Tormosov et al. 1998). The original Southern Hemisphere right whale population was estimated between 70.000 and 90.000 right whales (IWC 2001, 2013). By the time substantial catches have ceased, the population probably reached as few as 400 individuals.

In Brazil Right whales were historically distributed between the southern and northeastern coast of Brazil, between Rio Grande do Sul and Bahia states (~33–16oS) (Ellis 1969). Hunting in Brazil occurred illegally until 1973, and the population size declined drastically (Palazzo and Carter 1983). Whaling references revealed that the population was especially abundant off Santa Catarina State. In the early 1980s, whales were 'rediscovered' in this region and have been studied there since then (Câmara and Palazzo 1986). In 1982 the "Right Whale Project", a research and conservation program, was officially created with the aim of ensure the survival and recovery of the right whale population in Brazilian waters.

The species occurs in winter/spring and the time of peak abundance is September (Groch et al. 2005; Seyboth et al. 2015). Their main aggregation area is in the shallow and protected bays of the central-southern coast of Santa Catarina State, from the south of Ilha de Santa Catarina (27°25′S, 48°30′W) to Cabo de Santa Marta, Laguna (28°36′S, 48°48′W) (Fig. 16.1) (Simões-Lopes et al. 1992; Groch et al. 2005; Renault-Braga 2014). In the year 2000 this area was declared a marine protected area – Right Whale Environmental Protection Area (RWEPA) (Federal Decree of 14 September 2000) (Fig. 16.1). The RWEPA is managed by an Environmental Agency of the Government of Brazil, the Instituto Chico Mendes de Conservação da Biodiversidade (ICMBIO). The main purpose of the RWEPA was to give special protection to habitat crucial to the life cycle of the species, managing and regulating the human activities in the area. Most groups sighted in this region are mother-calf pairs with a few sightings of juveniles, single adults and mating groups (Simões-Lopes et al. 1992; Groch 2005; Seyboth et al. 2015). Right whale sightings have also been reported along the southeastern Brazilian coast along with a number of strandings (Lodi et al. 1996; Santos et al. 2001; 2010; Lodi and Rodrigues 2007; Figueiredo et al. 2013).

Possible factors determining distribution an occurrence of right whales off Brazil have been studied (Renault-Braga 2014; Seyboth et al. 2015). Whale numbers are higher in shallow areas with gentle slope, and areas less than 10 m depth and closer to river mouths are preferred (Renault-Braga 2014). There is also evidence that both mother-calf pairs and adults unaccompanied by calves prefer bays with dissipative beaches, and that cow-calf pairs apparently avoid bays facing southeast during days of strong east-west winds (Seyboth et al. 2015).

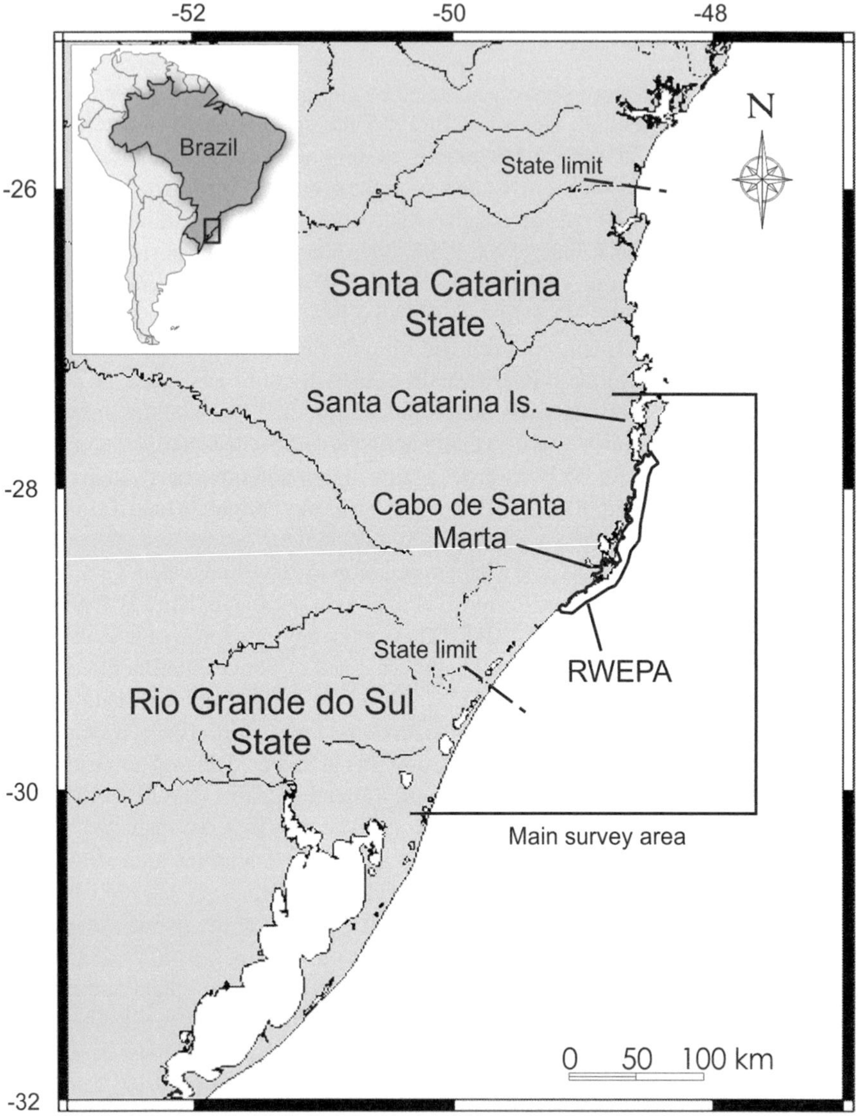

Fig. 16.1 Map showing the main survey area in southern Brazil, States limits, the Right Whale Environmental Protection Area (RWEPA), and the location reference of the whales main aggregation area between Cabo de Santa Marta and south of Ilha de Santa Catarina

16.3 Recognizing Right Whales and Learning About Their Lives

Individual recognition of animals in the wild is used since the first studies of ethology and ecology. The detailed investigation of activity patterns, associations and social interactions, as well as movements and habitat use, requires the individual identification of animals. Individual recognition also allow monitoring populations over time. Some species exhibit individually specific natural markings that facilitate identification in the field. For species that lack such distinctive markings, artificial marks, or tags, can be attached if individual identifications are desired (Wells 2009). Photographic identification of naturally marked animals or photo-identification, is a powerful, non-intrusive technique to obtain several information in natural populations (Hammond 1990).

Cetaceans exhibit a variety of individually distinct natural characteristics. In general, the features as head, back, dorsal fin and tail, that appear above the water surface during the respiratory cycle are the most useful, for ease of viewing and recording (Wells 2009). Variations occur in the skin color patterns, shape of dorsal fins, spots and scars along the body (Hammond et al. 1990).

The first systematic studies of photo-identification of cetaceans began in the 70s, after the discovery of the possibility of individually distinguish bottlenose dolphins by their dorsal fins (Wursig and Wursig 1977), and right whales (*Eubalaena* spp) from aerial photographs of the callosity patterns found on their heads Payne et al. (1983).

The callosities are raised patches of roughened skin on the head of right whales, colonized by thousands of cyamids (amphipod crustaceans: *Cyamus* spp) or "whale lice", whose light bodies clearly distinguish callosity tissue from intervening areas of smooth black skin (Fig. 16.2) (Payne et al. 1983). Cyamids are ectoparasites, which feed on the epidermal layer of whale skin and pass all live stages on cetaceans (Rowntree 1996). There are mostly three species of cyamids occurring on southern right whales: (1) *C. ovalis* (white, living in great quantities on the callosities); (2) *C. gracilis* (light yellow, restricted on the callosities, found, in jumbled masses); (3) *C. erraticus* (orange, found usually in isolated disc-shaped clusters on the base of callosities, on smooth skin between callosities, in the axes of flippers, in genital-anal grove and in wounds). In southern right whales the callosities are also inhabited by barnacles (Best 1991; Kenney 2009).

Callosities on a right whale head have similar distribution to the hairs on human face. They occur on the anterior and posterior margin of the blowholes, along the rostrum, on the upper margin of the lower lip, on eyebrows, on the chin, and along the mandible (Fig. 16.3) The number, shape and location of the calluses is unique to each whale. A calf's initial infestations arrive by vertical transmission from its mother, and the pattern is established in the first months of calves' life. There is a succession of cyamids on calves callosities, which are first "infested" by orange cyamids (*C. erraticus*), on the base of the callosities, on the sides of the head and the axes of flippers. As the calf grows, orange cyamids are replaced by white and yellow

Fig. 16.2 The callosities are raised patches of roughened skin on the head of right whales, colonized by thousands of cyamids (amphipod crustaceans: *Cyamus* spp.) or "whale lice" (**a**) Callosity pattern of an adult right whale (**b**) Cyamids

cyamids (*C. ovalis* and *C. gracilis*) (Fig. 16.4). Minor changes can occur with time, but the overall pattern remains constant from birth throughout the life of the whale (Fig. 16.5). Callosities that occur on the rostrum are called rostral islands.

Sometimes these islands can be connected becoming one uninterrupted patch from the blowholes to the anterior tip of the rostrum (called the "bonnet" by early whalers, who fancied it resembled a lady's bonnet), forming a peninsula (Fig. 16.6) (Payne and Dorsey 1983; Payne et al. 1983; Hammond et al. 1990). This pattern is rare in southern right whales but common in North Atlantic right whales (Hamilton et al. 2009).

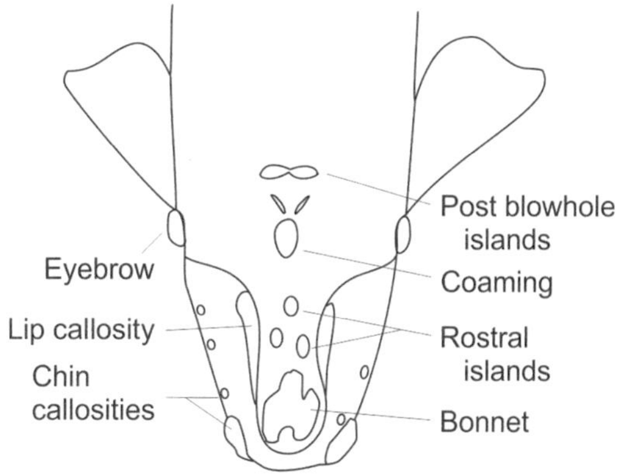

Fig. 16.3 Schematic drawing with location and name of callosities on a right whale head. They occur on the anterior and posterior margin of the blowholes, along the rostrum, on the upper margin of the lower lip, on eyebrows, on the chin, and along the mandible

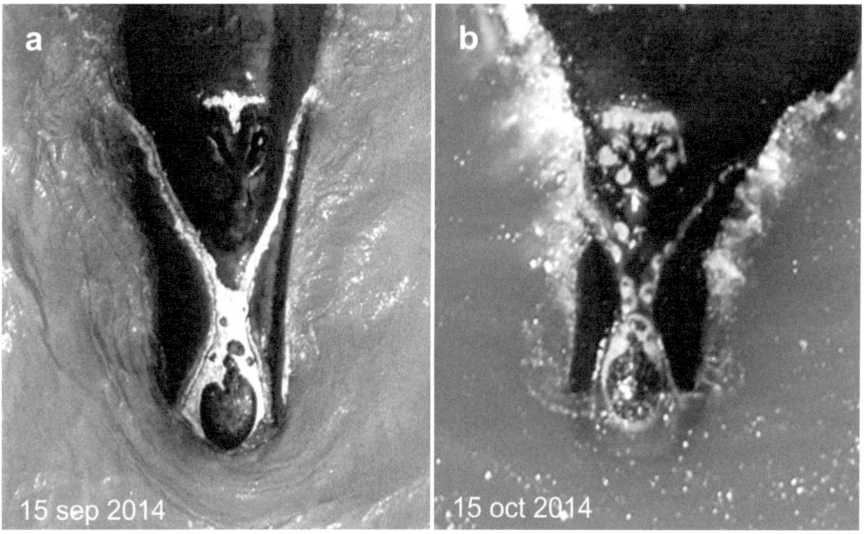

Fig. 16.4 Callosity pattern emerging on calf B118-04 (**a**) The base of the callosities is "infested" by *orange* cyamids (*C. erraticus*) (**b**) One month later orange cyamids begin to be replaced by *white* and *yellow* cyamids (*C. ovalis* and *C. gracilis*)

Despite the subtle difference, males have callosities in greater quantity and size than females. It is believed that the callosities are used for recognition among individuals and, based on the higher amount of scars and temporary scratches on the back of males compared to females, it is suggested that the callosities function as a tool for intraspecific aggression (Payne and Dorsey 1983).

In addition to the callosities on their head, right whales may have natural markings or spots on the back and belly (Fig. 16.7). These markings are white or grayish and can also be used for individual identification. These marks can occur throughout the body, resulting in partial albino individuals, also called grey-morph animals (Fig. 16.8) (Best 1990; Payne et al. 1983; Schaeff et al. 1999). A few anomalously

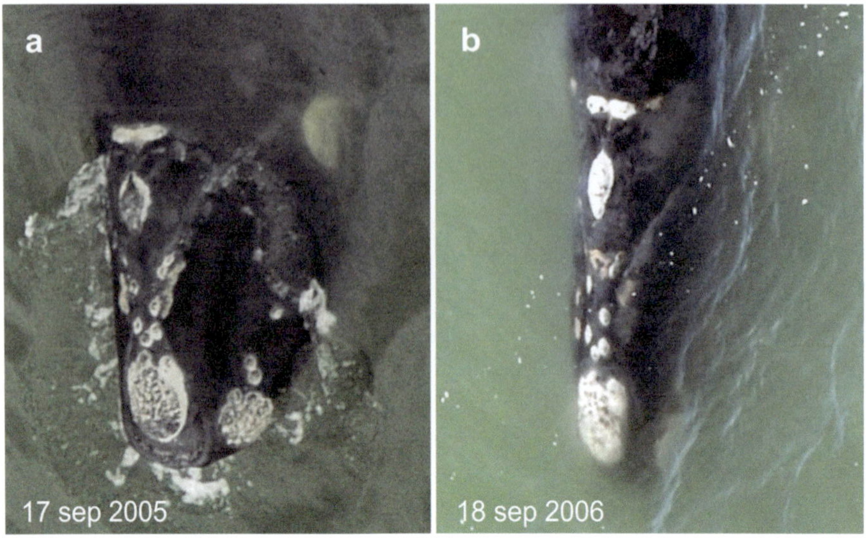

Fig. 16.5 Development of the callosity pattern through one year on calf B226-05 (**a**) First sighting in September 2005 and (**b**) One year later, in September 2006. The whale still carries loose *orange* cyamids on some portions of the head

Fig. 16.6 Rostral islands can be connected forming a peninsula, a pattern rare in southern right whale. The whale B410 was cataloged in 2006 not accompanied by calf. It has not been resighted up to 2010. There are two whales with this type of pattern in the BRWPC, however continuous callosities are common in North Atlantic right whales

Fig. 16.7 Typical southern right whale dorsal markings. The calf B558-10 was cataloged in 2010. The *white* dorsal markings were used for individual identification

white individuals have been noted for several species of large and small cetaceans (Fertl and Rosel 2009; Wells 2009), but southern right whales are the only species where such individuals are relatively common (Schaeff et al. 1999). In southern right whales, the partial albino individual body is mostly white with small black spots as a calf, but darkens in the first years of life becoming brown or gray (Fig. 16.8) (Payne et al. 1983; Best 1990; Schaeff et al. 1999). Individuals sexed genetically were males but at least three individuals have been photographed with calves in South Africa (Schaeff et al. 1999).

Right whales with natural markings as well as partial albino individuals have been recorded in Argentina, South Africa, Australia and New Zealand (Patenaude et al. 1998; Schaeff et al. 1999) as well as in Brazil (see below). North Atlantic right whales (*Eubalaena glacialis*) do not exhibit any of the dorsal skin color markings, only ventral markings (Schaeff and Hamilton 1999). North Pacific right whales (*Eubalaena japonica*) present dorsal white marks (Rowntree et al. 1980).

Callosities have allowed to follow the lives of individual right whales documenting basic features of the species' life history, such as age-at-first-reproduction, calving interval (e.g. Payne 1986; Whitehead et al. 1986; Best 1990; Rowntree et al. 2001) and estimate demographic parameters of right whale populations worldwide (e.g. Kraus et al. 1986; Caswell et al. 1999; Payne et al. 1990; Bannister 1990; Best et al. 2001; Cooke et al. 2001). From the individual recognition, it is possible to obtain information on reproductive aspects, movements, habitat use, estimates of abundance and population trends (Hammond 1986).

Fig. 16.8 Partial albino individuals are also called grey-morph animals and are relatively common in southern right whales (**a**) The body of the whale B190-05 was mostly *white* with *small black spots* when born (**b**) It began to grow *dark* in the first year of life (**c**) It become *gray* at 3 years old

16.4 Aerial Survey Methods

Aerial surveys for censuses and photographing individuals have been used for many species, and can be done using aircraft or helicopters, depending on the survey area and funding available.

Right whales have been surveyed along the southern coast of Brazil since 1987 by the Right Whale Project. The first aerial surveys were conducted from a single-engine aircraft (Simões-Lopes et al. 1992). Since 1998 a helicopter has been used (Groch et al. 2005). Despite the higher cost, the helicopter has proved to be a better platform because of its safety, maneuverability and more panoramic view of the whales. In regions where the water is turbid, the use of helicopter turns out to be crucial for the efficiency of the survey, because you cannot see the animals under water and the approach maneuver to register the sighting is facilitated.

Surveys have been concentrated along the central-southern coast of Santa Catarina State, and the North of Rio Grande do Sul State (Fig. 16.1). In five opportunities, surveys were extended, covering a larger area between Chui-RS (33°40′ S, 53°28′ W) and Porto Seguro-BA (16.27.15.2 S, 39.03.40.9 W), with the aim of gather information on occurrence and distribution of right whales along most of the species´ historical range. Surveys have been conducted during or close to the time of peak whale abundance, in September, however in Southern Brazil monthly surveys were also conducted from 2002.

For optimal results, the requirement is to conduct surveys during days with favorable sighting conditions, i.e. low wind, good visibility, sea state below 3 in Beaufort scale and adequate lighting. This condition is necessary to easy whales´ detection. Because the survey area is extensive, usually this task became a challenge for the time period programed for the survey. The challenge is to conciliate the best weather forecast with the availability of the aircraft, and a similar time window every year.

Surveys have been conducted at low altitude, about 300 m and a speed of 111–167 km/h), depending on the aircraft, following a parallel trajectory at a distance of approximately 500 m from shore. The search for individuals is usually restricted to a stripe of up to 1500 m from shore, where mother-calf pairs concentrate. Up to 1997 a single engine aircraft was used. Since 1998 a helicopter has been used. In most of the surveys (from 2002 to the present) a Bell Jet Ranger helicopter (Fig. 16.9) with capacity for the pilot and up to four passengers has been used. Therefore two to three observers can be involved in the search duty and data recording. The main observer sits to the left of the pilot and searches for whales ahead of the aircraft. The second observer sits behind the pilot and scan to the right of the aircraft and is responsible for taking photographs of the sighted animals. The third observer sit next to the photographer and monitor the area to the left of the aircraft. This side was also observed by the main observer when only two observers are onboard.

Whenever a whale or group of whales is spotted they are approached at a minimum height of (100 m) for determination of the number of whales and group composition. The helicopter hover over the group and as the callosity patterns of individuals

Fig. 16.9 Survey team on board a helicopter used for the aerial surveys for locate and photographing right whales off Southern Brazil. (**a, b**) Main observer searching for whales and taking notes, (**c**) Second observer taking photographs of whales sighted

became visible, they are photographed (Groch et al. 2005). From the airplane, the whales were circled until all animals were photographed. The methodology for taking and analyzing photographs was developed by Payne et al. (1983). An approach is halted if it appear to change the whales' behavior, even if it results in not photographing all of the counted whales. Number of individuals, group composition and

Fig. 16.10 A female right whale accompanied by a calf photographed during aerial surveys off southern Brazil

their location are recorded for each sighting. Photographs taken from all aerial surveys have been used to create a catalogue of right whales identified off southern Brazil.

To classify group composition and obtain life-history parameters the following definitions of different age/sex categories are usually applied in right whale studies: (1) females are whales accompanied by calves; (2) calves are whales between 1/3 to just over 1/2 of their mother's length (Fig. 16.10); (3) yearlings are one year old whales (recognized if photo-identified in their birth year); (4) juveniles or 'sub-adults' are immature whales of unknown age classified based on relative length and morphology of the head. Juveniles are mid-size, bigger than a calf, but smaller than an adult, however there are difficult to distinguish if not accompanied by an adult whale; (5) adults are individuals with adult size, not accompanied by a calf, with undetermined sex and age (Fig. 16.11). Males can usually only be distinguished if they have been morphologically or genetically sexed. Behavior and consecutive resightings can give a clue to maleness (Payne and Dorsey 1983). For example, if an adult size whale is seen four consecutive years without a calf, or involved in a mating group, it is classified as male.

Fig. 16.11 Adult size right whales photographed during aerial surveys off southern Brazil

16.5 The Matching of the Photographs

Before the digital photography age, the process of analyzing photographs looking for matches of right whales was done manually, from the categorization of photographs according to specific characteristics of the identification pattern, to the comparison with individuals already catalogued. So each new photograph was compared by eye to the existing catalogue, which was as selection of slides or prints organized as a photo album. For catalogues with a large number of identified individuals the by eye comparison becomes a laborious and time consuming task. The advent of digital images (obtained through scanning of slides, negatives or prints) enabled the development of computer-assisted automated systems. About 15 years ago digital cameras have largely replaced film cameras for photo-identification work (Markowitz et al. 2003), enabling to obtain images in real-time, facilitating manipulation and storage. The development of high-resolution digital cameras also improved the quality of photographs.

Manipulate large amounts of photographs is time consuming and prone to errors. The use of computers for analysis and comparison of individuals can limit the number of photographs that must be visually inspected to verify that two observations are of the same individual.

In parallel, many whale populations have been recovering, including southern right whales (IWC 2013) and so the number of sightings and consequently individuals photographed in each aerial survey. This demanded for the development of some automated system which could assist the comparisons. In 1998, Hiby and Lovel (2001) developed a computer program called "Right Whale Photo-identification Software", to aid on the analysis of aerial photographs taken from right whales.

The "Right Whale Photo-identification Software" makes use of the variation in shape, number and location of callosities on the rostrum and lips, transforming the three-dimensional shape of the head of right whales in a two-dimensional picture, where image pixels define the presence of callosities, generating a library of patterns for comparison with new images analyzed. The objective of the software is to suggest the order in which the existing catalogue be searched for matches to a new photograph, and to eliminate from the search those photographs in the library that cannot possibly be potential matches of the new one (Hiby and Lovell 2001).

The method consists in extract from each new photograph a simplified map of the callosity pattern. The process generate similarity scores with extracts from photographs already in the catalogue (Hiby and Lovell 2001). The process use a three-dimensional surface model on the head of right whales, allowing to compensate differences in the angle from which the photograph was taken (Fig. 16.12). The orientation and scaling of the model to the whale in the photograph is made by identifying four reference points: the front margin of the bonnet, the rear margin of the coaming and the upper margin of the eyebrow callosities. The next step is to identify samples of callosity and skin from the digital image, which the program then uses to generate an extract of the callosity pattern (Fig. 16.12). The operator can extend or reduce the selected areas, and also manually paint the callosity area. The act of painting is almost an artistic work, and must be done with the aid of a graphic tablet. In addition, the operator has to identify which of the callosities lie along the lip margins so that the entire image is classified into areas of rostral callosity, lip callosity and skin. Areas with uncertainty can also be highlighted (Hiby and Lovell 2001). Information on the date of the photograph, if the lip callosities are present or absence, as well as if the year of birth are required at the end of this step. The extracts are initially stored in a pending directory and then automatically compared. The digital images from which the extracts with the best similarity scores were obtained can be displayed next to the image corresponding to the pending file, to confirm or reject a match (Fig. 16.13). Each file has a unique name identifying the date when the photo was taken and a temporary number of the whale. In case a match is confirmed the file receive the whale number in the library, or a new number if not a match.

The numbering of the files in the catalog consists of adding a letter (corresponding to the country, e.g. A for Argentina, B for Brazil) and a sequential number identifying the whale. If cataloguing a calf, it receives the mother's number, with the two last digits of the year of birth added at the end of the mother's number. For example B101-03 is the calf of the whale B101, born in 2003. All the information associated to the whale in the photograph, e.g. date, sighting location, accompanied whales, presence of calf, is stored in a database associated to the automated system. The

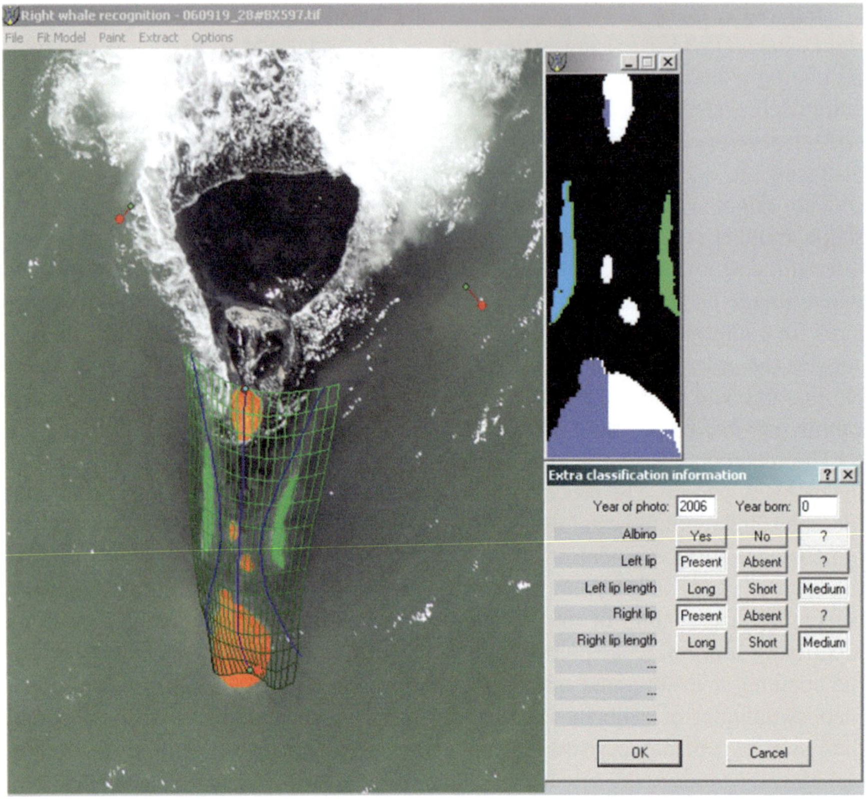

Fig. 16.12 The "Right Whale Photo-identification Software" (Hiby and Lovell 2001) use a three-dimensional surface model on the head of right whales. The orientation and scaling of the model to a whale photograph taken from distinct angles is made by identifying four reference points marked in *green* and *red* on this figure

uniqueness for each photo being analyzed is given by the date of the sighting and a key code generated in the database. The process take a several minutes to complete for each new photograph, depending on the experience of the operator. However for each new whale to be compared the interpretation of the pattern by the operator is made only once. As more experienced the operator, the better the potential of the system.

Other software have been developed, incorporating photos from different perspectives. One is called DIGITS (Digital Image Gathering and Information Tracking System), developed by the New England Aquarium (http://rwcatalog.neaq.org) to incorporate imagens of North Atlantic right whales. DIGITS is a data management system designed to accommodate aerial and shipboard images, providing a platform to include images from different angles of a right whale (Fig. 16.14). The software rely on a detailed coding system, rather than an automated interpretation of the image as the Hiby and lovel (2001) system, to allow for whale-matching searches

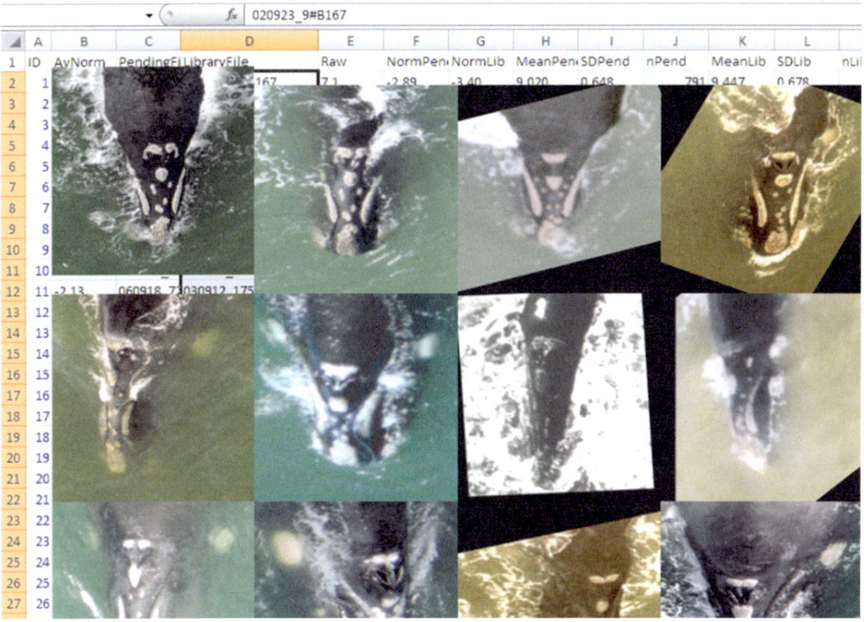

Fig. 16.13 The digital images of similar whales can be displayed next to the image corresponding to the pending file, to confirm or reject a match

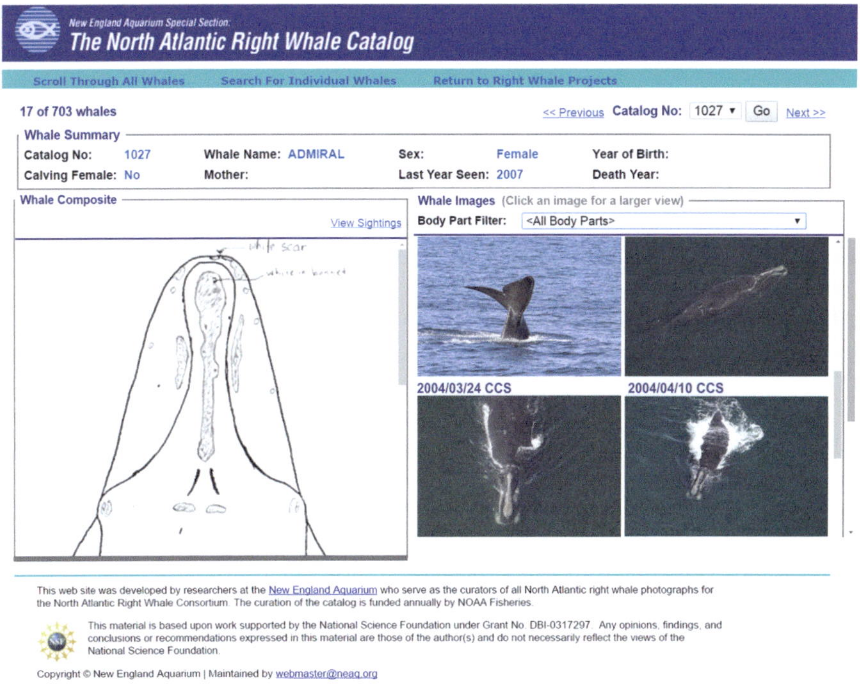

Fig. 16.14 Screen of the DIGITS data management system (New England Aquarium http://rwcatalog.neaq.org). The software is server-based and is accessible through the Internet

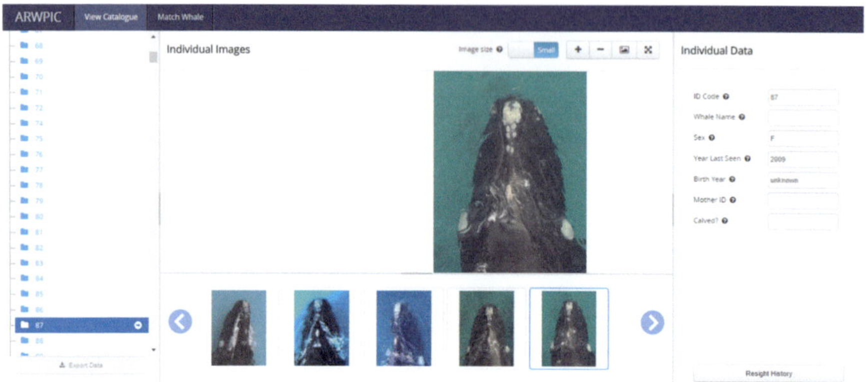

Fig. 16.15 Screen of the ARWPIC data management system (https://data.marinemammals.gov.au/arwpic/index.php?route=catalog/individual)

(Hamilton et al. 2009). The software was designated to be server-based and is accessible through the Internet. This allows data and images to be processed and integrated from field stations, as well as provides a framework to allow public assess. The database can also include photographs of skin lesions and general body condition, allowing for health assessment studies.

A similar data management system was developed by Australian researchers, as a need for a broader-based catalogue that provides for integration of data and linkages between projects (Australian Government 2009; Pirzl and Watson 2011). The Australasian Right Whale Photo-identification Catalogue (ARWPIC) integrate southern right whale photo-identification collections and related data of Australian and New Zealand so that data are archived and accessible from one web-based portal as a collaborative research and management initiative.

ARWPIC provides an online platform for researchers to share sightings, image and other related information for photo-identified southern right whales with each other, other interested professionals, and the public (Fig. 16.15) (Pirzl and Watson 2011). ARWPIC is similar to DIGITS, and uses a computer-assisted matching system called *BigFish* to match uploaded images to other whales in the catalogue (http://www.skadia.com.au/resources.html). *BigFish* is a computer-assisted matching program that works by comparing coded values of identification features of the whale being matched, to values for whales already catalogued. The user enter specific identifying characteristics and callosity pattern features of right whales, and then query the data for matching individuals. The Search Results form allows the user to scan through the selected whales, accessing further images if required. *BigFish* also support for dynamic analysis of calving cycles and other sighting information through filtering and calculations (Pirzl 2010).

The three systems use different approaches and the choice between them depend first on the angle of image available for the analysis and the need to be accessible by different researchers from different places and the general public. No matter the choice the use of this type of technology is a fundamental tool aiding and optimizing the effort of collection, maintenance and analysis of large amounts of photo-identification data.

16.6 The Brazilian Right Whale Photo-Identification Catalog

The Brazilian Right Whale Photo-identification (BRWP) Catalog was created after the first aerial surveys conducted in Brazil, in 1987. Each whale photographed in a new aerial survey is compared to the existing catalogue to look for matches. If a whale is not found in the catalog, it is given a new number and added to the catalogue. Photographs are analyzed with the aid of the "Right Whale Photo-identification Software" (Hiby and Lovel 2001), and the information associated to each whale sighting is stored in the database associated to the system.

The BRWP Catalog has over 2300 photos from 670 individual whales identified up to 2010. From the total, 38.2% (n = 256) are mature females (whales sighted at least once accompanied by a calf), 18.5% (n = 124) are calves identified in the year of birth, 3.6% are (n = 24) juveniles, and 39.7% (n = 266) are adult whales never seen accompanied by calf. The catalog includes 8.4% (n = 56) of whales presenting dorsal marks and 1.6% (n = 11) of whales with a partial albinism.

16.7 What Have We Learned from the Catalog?

The number of right whales on the Brazilian breeding ground was low in the first years of aerial surveys with a mean of 11 new whales added to the catalog by year up to 2001. However a rapid increase in the population was observed in 2002 (Groch et al. 2005), and so the number of right whales in the BRWP Catalog. The year with the highest number of new whales added to the catalog was 2002 (n = 124 whales) (Fig. 16.16). From 2002 and 2010 a mean of 57 new whales were added to the catalog each year. Resightings of individuals begun to occur in 1994 but 95% were recorded from 2003, when a pattern emerged. From 2003 to 2010, an average of 39% of whales identified each year were resightings. The year with the highest number of resightings in relation to the total identified individuals was 2007, when 56.5% (n = 35) of individuals sighted were known whales.

The greater number of new whales added to the catalog every year in relation to the number of resightings indicate that this population is being ´discovered´. The number of whales added to the catalog vary between years, but some individual whales have been providing valuable information on patterns of occurrence and site fidelity to this area. Twenty-five percent of the catalogued whales up to 2010 were seen in two or more years. The longest sighting interval is 18 years, shown by the whale B016, first sighted in 1988 and resighted in 2006 (Fig. 16.17). In both opportunities she was sighted with a calf. This whale had been previously sighted in Argentina (PV256) (Best et al. 1993), where she was also sighted with a calf, in

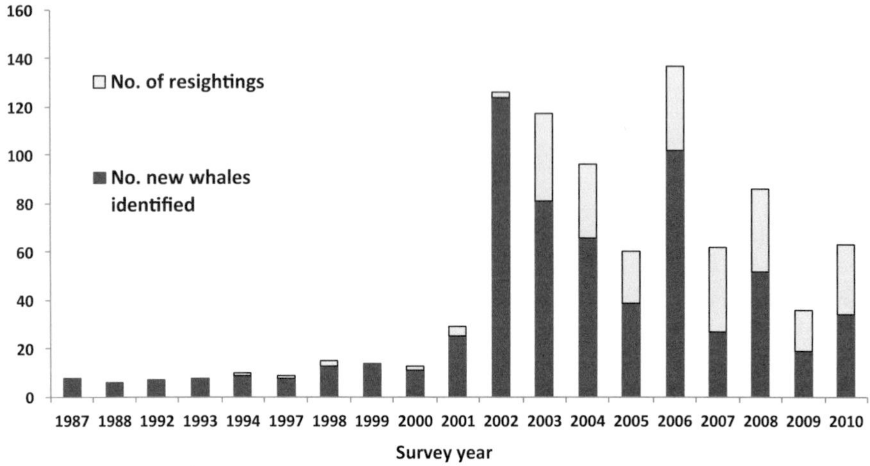

Fig. 16.16 Number of right whales photo-identified by year during aerial surveys off Southern Brazil. Note the rapid increase in the population observed in 2002 and the variation in the total number of whales each year

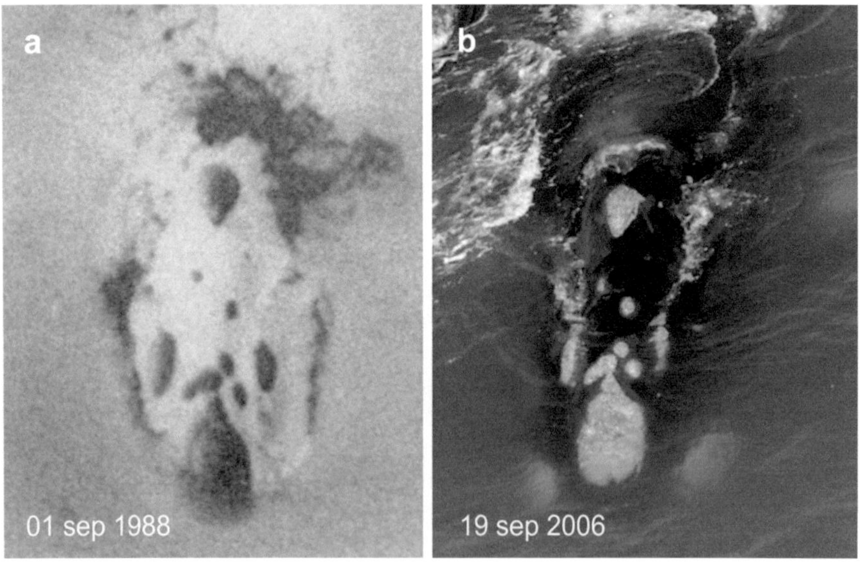

Fig. 16.17 The female B016, (**a**) First sighted in 1988, (**b**) Resighted in 2006, 18 years later. She was previously sighted in Argentina in 1971 (PV256) (Best et al. 1993)

1971. Female southern right whales sexual maturity is reached at 6 years old and the modal age of first reproduction is 9 years (Payne 1986; Payne et al. 1990; Best et al. 2001; Cooke et al. 2001). Therefore, whale B016 was at least 44 years old when she was last sighted in Brazil. This whale represented one of the first direct links between the two breeding areas.

Differently, 60 females have shown some level of site fidelity, returning to Brazil in three-year intervals for calving and nursing, consistent with successful reproduction (Groch et al. 2005). One example is whale B082, the female with the highest frequency of return (Fig. 16.18). She was added to the catalog in 1999, and resighted in four opportunities, all with a newborn calf. She has a series of parallel propeller scars on the caudal peduncle, resulting from a ship collision, which made her easy recognizable from shore. For this reason, she has become a very familiar whale to the Right Whale Project team, and received a name: Felicia. The name is a result of the combination of "fe", from "female" and the word "hélice" (portuguese translation for "propeller") as she was referred when sighted during land-based surveys. In 2007 Felicia was honored with a replica of 1 m in length, which is used for educational purposes. Whales in the catalog can also receive names based on distinct dorsal marks or callosity pattern that looks like some object, therefore becoming familiar whales.

Eight whales identified in the year of birth were resighted one to five years later. Despite the great number of calves added to the catalog, the number of resightings is low. Usually calves are difficult to catalog. The establishment of the callosity pattern in right whales occurs during the first months of life (Payne et al. 1983). The pattern is well established mostly from the seventh month of life (Hamilton et al. 2009). Calves were added to the catalog only if the callosity pattern was relatively well-established, if they presented distinct dorsal marks or partial albinism. The low number of resightings can indicate that the waters in southern Brazil are not being used by young whales. Most calves was added to the catalog from 2002 onwards. If whales are using this area mainly as a nursery area, the individuals most likely to return are females that reached sexual maturity and is producing their first offspring. Therefore, calves identified in southern Brazil have a better chance of resighting after 2010. Cataloging calves in its year of birth enables the construction of family trees and monitoring the life history of whale families.

The resighting of individuals between monthly surveys provide information on how the whales use an area of occurrence. Considering the interval between flights conducted in southern Brazil, and that during the interval the whales remain in the area, right whales spend between 25 and 37 days in the region (Groch 2005). The longest sighting interval was observed for two females with calves sighted in three consecutive monthly surveys, and may have spent up to 64 days in the region. Females with calves sighted during shore and boat-based observations in Santa Catarina have been reported to stay at least 20–25 days in the same bay and as long as 69 days moving between different bays (Palazzo et al. 1999; Groch 2000, Groch et al. 2003). There is a tendency for mother-calf pairs to stay longer in Southern Brazil than unaccompanied whales (Groch 2005; Seyboth et al. 2015), a pattern also observed in other breeding grounds (Payne 1986; Bannister 1990; Burnell and Bryden 1997; Best 2000; Rowntree et al. 2001).

One important result obtained from the BRWP Catalog was the finding about the relationship of right whales that frequent Brazil with other areas of reproductive concentration. Resighting of calving females in Santa Catarina waters that were previously photoidentified on the breeding area off Península Valdés, Argentina, (Best et al. 1993, Groch and Rowntree, ongoing research) indicates some degree of rela-

Fig. 16.18 (**a**) Southern right whale B082 is the female with the highest frequency of return, known since 1999, (**b**) She has a series of parallel propeller scars on the caudal peduncle, resulting from a ship collision, which made her easy recognizable from shore

tionship between these right whale populations. Genetic studies have shown that the Argentinean and Brazilian individuals are part of a single genetic population due to the high gene flow among these breeding grounds (Ott 2002). However, there is also evidence from stable isotope studies that individuals from southern Brazil are isotopically distinct from those from central Argentina which means that individuals from the two areas exploit different feeding grounds (Vighi et al. 2014). Right whales feed in at least four feeding grounds in the South Atlantic, whose location they learn from their mothers (Valenzuela et al. 2009). It is suggested some degree of population substructure within the western South Atlantic right whales (Vighi et al. 2014).

The ongoing comparison with individuals cataloged in Argentina, as well as the inclusion of individuals cataloged in Uruguay, are fundamental to the understanding of aspects of the ecology of the species in the western South Atlantic Ocean. Photo-identification research on right whales have also been used to assess body condition and health from an external perspective providing a tool to monitor the status of both individual whales and the population as a whole (Rolland et al. 2009). From the stories of individual whales one can understand the ecological needs of whales at the population level, identify possible threats, generating invaluable information to its conservation and management.

16.8 The Brazilian Right Whale Population

The growth rate and other demographic parameters have been estimated from the number of whales photo-identified each year during aerial surveys off their breeding grounds (e.g. Bannister 1990; Best et al. 2001; Cooke et al. 2001). The most recent global estimate is 13,611 southern right whales in 2009, which means about 20% of its pre-exploitation abundance (IWC 2013). The species is currently listed as least concern (LC) in the IUCN Red List of Threatened Species (Reilly et al. 2013).

The recovery of right whales in Brazil was apparently slow in the first years of aerial surveys, but a rapid increase in numbers was observed in early 2000s (Groch et al. 2005). The number of right whales off Brazil was estimated to have increased at a rate of 14% per year up to 2003, a rate considered outside the range of biological plausibility for a natural rate of increase for southern right whales (Best et al. 2001). A possible explanation for the increase was immigration from other breeding grounds such as Peninsula Valdes, Argentina, where the whale numbers have been increasing at a rate of 6.5% (Cooke et al. 2013). An update analysis for southern right whales off Brazil estimated an average 12.0% (CI 8.514.2%) rate of increase over the period 1987–2010. Population size in 2010 was estimated at 197 mature females (CI 146–234) (IWC 2013).

With the growing numbers of right whales off Brazil it is expect the whales to expand their range as they have off Argentina and South Africa (Best 1990; Rowntree et al. 2001). Whaling records indicate that right whales were found up to the north-eastern coast of Brazil (~16°S) (Ellis 1969). In five dedicated aerial surveys conducted along the southeast (2001, 2002, 2004 and 2008) and northeast (2001–2002) cost of Brazil during the peak of the right whale season, 8 mother-calf pairs were recorded in 2001 and 2002, and no sightings were recorded in 2004 and 2008 (Groch et al. unpublished). A possible reduction in sightings along the southeast of Brazil was also detected reviewing data of strandings and sightings available from various sources (Figueiredo et al. 2013). Between 2000 and 2011 a total of 36 records was found (n = 59 individuals), between May and October. Mother and calf pairs represented 76% of all sightings. Annual sightings were significantly reduced when compared previously to available data from the period between 1981 and 1999 (Santos et al. 2001). Much of this area is now developed which will increase the potential of conflicts between right whales and human activities as the whales reoccupy or expand their range (Groch et al. 2005).

16.9 Potential Threats

The National Action Plan for the conservation of aquatic mammals list the potential threats for the recovery and conservation of right whales in Brazil (Rocha-Campos and Camara, I. G. 2011). Currently, the main threats faced by right whales are thought to be ship collisions and entanglement in fishing gear.

Collisions with vessels are a recognized source of severe injuries and mortality of whales. Between 1999 and 2014 four cases of southern right whales with injuries or scars compatible with vessel collisions, including 3 live animals and one calf washed ashore dead, are reported along Santa Catarina coast (Groch et al. 2015). The live animals consisted of 2 cataloged females and one calf, all resighted later with the lesions completely healed. The most severe case was the female recognized as B128, cataloged since 2002 (Fig. 16.19). Despite the severity of the lesions she was resighted in apparently good general condition, spanning 57 days since first sighted with the open wounds. The whale was monitored the whole period and was accompanied by a calf.

Vessel collisions have been proposed to cause the death of right whales mostly south of Santa Catarina State (Greig et al. 2001; Gomes 2005). A review on strandings along the Brazilian coast between 1936 and 2004 reported the death of 6 whales caused by ship strikes, out of 45 stranding records (Gomes 2005). Most of the cases occurred in Rio Grande do Sul State, south of the species main concentration area in Santa Catarina. Marine traffic in the area is extensive due to the presence of the Port of Rio Grande, the only fishing and cargo ships facility on the state coast, which my increase the probability of collisions (Greig et al. 2001).

Strandings have been reported for Santa Catarina, however in most of the cases the causes were not determined (Groch et al. 2013). During the period of 2002–2013, a total of 23 strandings were registered, including all classes of whales. A rate of 1.9 strandings *per* year is observed, and only one case of anthropogenic cause of mortality was confirmed, due to vessel collision. Most strandings were of dead individuals except for four live animals, from which one calf died shortly after stranded, two juvenile were successfully rescued and one adult had to be euthanized. Calves constitute the majority of events and there is a prevalence of strandings in the central and northern part of the state coast. In the northern state coast the ship traffic is high due to the 2 large Ports located in the area, and vessel traffic might be a cause of

Fig. 16.19 (**a**) The southern right whale B128 was found in 2012 with three large open cuts on the dorsal peduncle compatible with a vessel collision, (**b**) The whale was resighted in apparently good general condition, spanning 57 days since first sighted with the open wounds. She was accompanied by a calf the whole period of monitoring

mortality, even though necropsy fails to identify clear evidence. However, this is an area with low number of right whale sightings. Compared to Rio Grande do Sul State, it's interesting to note that, including the two Ports mentioned above, Santa Catarina State has three Ports, the other located in the central area of the RWEPA, and the number of deaths related to vessel collisions are much lower. Since 2007 a stranding network has been established with a protocol to improve notification and response to strandings along the RWEPA.

Entanglement of right whales have been recorded in Southern Brazil. Between 1999 and 2014 a total of 38 cases are reported (Groch et al. 2015). All records were of live animals, mostly adult females accompanied by a calf (74%) The majority of the entanglements consisted of parts of monofilament gillnet, which attach to the whales callosities when they swim through a fishing net (Fig. 16.20). No deaths were reported due to these cases and photo-identification enabled a follow up of 16 whales, from which nine (56%) were free of fishing gear when resighted. Strandings associated to entanglement in fishing gear, have been reported mostly in southeastern Brazil, where a total of 7 cases have been recorded (Gomes 2005; Lodi and Rodrigues 2007). In Santa Catarina artisanal fisheries are an important income source for local traditional communities, resulting in the use of gillnets close to shore as well as whale entanglements (Zappes et al. 2013).

In September 2016 a disentanglement training workshop was conducted by an expert of the International Whaling Commission, in the SRWEPA. The aim was training teams to rescue whales in cases when entanglement may poses risk to the whale survival, using the best technique available. While entanglements reported for right whales off southern Brazil so far have not been at a survival risk level, there is a concern on the number of cases reported. However it has been emphasized that a priority must be given to work on prevention measures.

Whalewatching activity is another potential threat to the species, because of the coastal habits of mother-calf pairs (Palazzo and Groch 2011). In Santa Catarina, the activity started in 1999 stimulated by the regular presence of right whales. In this area boats must operate in agreement to the national legislation. In Brazilian waters, Southern right whales are protected from whaling and harassment since 1987 (Federal Decree No. 7643), and were declared a State Natural Monument of Santa Catarina in 1995. In 1996, considering the need to regulate the recently emerged whalewatching tourism, the Ordinance IBAMA no. 117/1996 (amended by n° 24/2002) established several regulations for the activity. From 1999 to 2008, the number of whalewatching cruises increased 13 times, and the number of passengers engaged in the cruises increased 14.6 times (Groch et al. 2013). There was one boat operating in 1999 and a total of six boats were registered in the RWEPA by 2008, though not all of them operating regularly.

Preliminary results from an analysis of movement patterns of right whales in relation to boat approaches collected in 2002 found no clear evidence of immediate disturbance to this right whale population (Groch et al. 2005). Analysis of respiration rates sampled in 2006 found no significant changes related to boat approaches (Corrêa and Groch 2007). Despite the increasing overall interest by the whalewatching activity in the RWEPA, the number of boats operating in the region never exceeded six boats and usually two boats operated regularly. In 2013 the activity was suspended by a judicial decision as a precautionary measure after complaints of harassment resulting from the whalewatching activity. A control plan was developed by the RWEPA and should be implemented if the activity resumes. Because right whales approach the coast, shore-based whalewatching has also been con-

Fig. 16.20 The majority of the southern right whale entanglements off Santa Catarina consist of parts of monofilament gillnet, which attach to the whales callosities when they swim through a fishing net, (**a**) A female with calve, (**b**) A juvenile

Fig. 16.21 A southern right whale sighted in the Port of Imbituba bay. The Port is located in the central area of the RWEPA, adjacent to the main right whales aggregation bay

ducted, and have been the alternative as a commercial activity. Whalewatching industry has considerable economic and educational potential. With proper regulation and effective enforcement the activity can bring socioeconomic benefits especially for coastal communities of developing countries, often in areas where other economic options are scarce.

Other potential threats are habitat loss, degradation, and pollution. In the southern portion of the RWEPA, the discharge of coal mining and burning waste for thermoelectric purposes have been resulting in habitat degradation whose potential effects on the species are unknown (Palazzo and Groch 2011). The expansion of the Port of Imbituba represent an additional threat to the species, due both to the risk of collisions and underwater noise caused by the increase in vessel traffic. The Port is located in the central area of the RWEPA (Fig. 16.21), adjacent to the main right whales aggregation bay (Renault-Braga 2014). During the expansion, concluded in 2011, a “Right Whale Monitoring Program” was implemented. The program included theodolite tracking of whales from a shore-based station for monitoring the occurrence and movements of the whales in the port area, and aerial surveys for population monitoring along the RWEPA and adjacent areas (Groch, unpublished). After completion of the expansion, the program became a permanent monitoring program included in the Environmental Management Plan of the Port.

Facing the potential threats of vessel traffic to right whales, the repertoire of vocalizations produced by southern right whale mother-calf pairs in southern Brazil have been characterized using passive acoustic monitoring providing baseline information on the vocal behavior of this species as a tool for conservation and research of the species off its main concentration area (Dombroski et al. 2016).

Despite increase in the right whale population in southern Brazil, climate variability may currently constitute the main potential threat. A recent study suggest that global climate indices influence southern right whale breeding success in the area (Seyboth et al. 2016). There is positive correlation between the number of southern right whale calves in southern Brazil and krill densities near South Georgia, considered an important feeding ground for this whale population, suggesting that the reproductive success of southern right whale is directly influenced by food availability during the early months of gestation. The study analyzed data from 1997 to 2013. In 2016 only seventeen calves were counted in Santa Catarina, the lowest number since 2001. A low number was also recorded in 2015, when 29 calves were recorded (Groch unpublished). A likely explanation is that these years were also influenced by climate anomalies. If the frequency of years with reduced krill abundance continues, is likely to reduce the current rate of recovery of southern right whales and jeopardize its expansion to historical range (Seyboth et al. 2016).

The southern right whale breeding population of Península Valdés, Argentina, have also found to produce fewer calves than expected following years of higher sea surface temperatures around South Georgia (Leaper et al. 2006). Southern right whales can be flexible in many aspects of their habitat use, however they have shown site fidelity both to breeding and feeding grounds. There is also evidence of culturally inherited site fidelity to feeding grounds (Valenzuela et al. 2009).

More than six hundred southern right whale calves died in their breeding ground off Penınsula Valdes, Argentina, from 2003 through 2011. There is a varying levels and patterns of calf mortality suggesting that many of the deaths have an unusual cause or causes (Rowntree et al. 2013). The mortality of calves may be a consequence of: (1) cow nutritional stress; (2) exposure to HAB- and/or bacteria-associated biotoxins in (a) the feeding ground resulting in in utero exposure of the calf or (b) the calving/nursery ground; (3) infectious disease (viral, bacterial, protozoal, etc.); (4) kelp gull parasitism; (5) density-dependent processes; and (6) a decline in food availability (IWC 2016).

In 2009, the southern right whale breeding ground in Argentina was estimated to include 3373 whales (IWC 2013). As of 2010, the southern right whale subpopulation that calves off Península Valdés was estimated to be less than 20% of its initial size before whaling (Rowntree et al. 2013). It is expected that the ongoing high mortality of calves will significantly affect its recovery.

The severe harassment on right whales by kelp gulls (Rowntree et al. 1998; Marón et al. 2015), has been suggested to contribute to right whales immigration to Brazil (Groch et al. 2005). The percentage of living mothers and calves with gull lesions in Argentina increased from an average of 2% in the 1970s to 99% in the 2000s (Marón et al. 2015), when the rapid increase in the Brazilian population started. Resightings in Brazil of whales previously photographed off Argentina (Best et al. 1993, Groch and Rowntree ongoing research), as well observations of whales with kelp gulls attack marks (Groch, unpublished) support the immigration hypothesis. Kelp gulls attacks on right whales was first reported in Brazil in early 2000s (Groch 2001), consisting of occasional pecking bouts with no obvious lesions on the whale's back. No evidence of increase have been observed so far.

On the other hand, it has been found that right whales from southern Brazil are isotopically distinct from those from central Argentina, meaning that individuals from the two areas exploit different feeding grounds (Vighi et al. 2014). The isotopic difference also might indicate that intermingling between the two areas is limited or non-existing, supporting the authors hypothesis of isolation between the right whales off central Argentina and southern Brazil. The results contradict the resightings of individuals between these areas, and indicate the occurrence of some level of structure in the Southwest Atlantic right whale population. The authors suggest the use of two management units for the southern right whale in the Southwest Atlantic Ocean and independently focused conservation regulations, calculations of their biological parameters and demographic trends.

The ongoing comparison of the photo-identification catalogs from Brazil and Argentina will help define the extent of the overlap between both populations, the direction of movement between these breeding areas, and if the high mortalities off Argentina can affect the recovery of the right whale population off Brazil.

16.10 The Future

Regardless of the recent rapid increase, the low number of mature females and the potential threat of coastal anthropogenic activities and habitat degradation right whales are listed as "endangered" in the List of the Brazilian Threatened Fauna (Ott et al. 2008; Rocha-Campos and Camara 2011).

The National Action Plan for the Conservation of Aquatic Mammals in Brazil (Rocha-Campos and Camara 2011) emphasize that more effective measures must be taken to ensure compliance with legislation, such as adequate supervision of protected areas and periodic review of legal instruments with the aim to their improvement.

A Conservation Management Plan (CMP) for the southern right whale Southwest Atlantic population was developed by experts of range states from the International Whaling Commission (IWC 2013). The overall objective of the CMP is to protect southern right whale habitat and minimize anthropogenic threats to maximize the likelihood that southern right whale will recover to healthy levels and recolonize their historical range.

Understanding the influence of climate anomalies and variation of food supply on right whale reproductive success may be important for species conservation strategies as they can support decisions about protective measures taken place both in its breeding area off Brazil and in its feeding ground in the Southern Ocean (Seyboth et al. 2016). Conservation and management actions must be determined to ensure availability of krill species such as southern right whales that depend on it as a food resource, in face of the krill fishing industry in the south Atlantic sector of the Southern Ocean.

Critical data such as abundance estimates, population trends, movements and habitat use, generated from the systematic aerial surveys have been used for the formulation of conservation policies for the southern right whale. Although it is a major effort of collecting, maintaining and analyzing data, these studies highlight the importance of continuing the aerial surveys and the use of photo-identification research for the conservation and recovery of southern right whales in southern Brazil, the second most important breeding ground of the species in the western South Atlantic.

References

Australian Government (2009) Report of the National Southern right whale workshop. Hobart

Bannister JL (1990) Southern right whales of western Australia. Rep Inter Whal Comm 12(Special Issue):279–288

Bannister J, Burnell S, Burton C, Kato H (1997) Rep Intern Whal Comm 47:441–444

Best PB (1988) Right whales *Eubalaena australis* at Tristan da Cunha – a clue to the 'non-recovery' of depleted stocks? Biol Conserv 46:23–51

Best PB (1990) Trends in the inshore right whale population off South Africa, 1969–1987. Mar Mamm Sci 6(2):93–108

Best PB (1991) The presence of Coronuline barnacles on a southern right whale *Eubalaena australis*. S Afr J Mar Sci 11:585–587

Best PB (2000) Coastal distribution, movements and site fidelity of right whales *Eubalaena australis* off South Africa, 1969–1998. S Afr J Sci 22:43–55

Best PB, Schell DM (1996) Stable isotopes in southern right whale (*Eubalaena australis*) baleen as indicators of seasonal movements, feeding and growth. Mar Biol 124:483–494

Best PB, Payne R, Rowntree V, Palazzo JT Jr, Both MC (1993) Long-range movements of South Atlantic right whales *Eubalaena australis*. Mar Mamm Sci 9(3):227–234

Best PB, Brandão A, Butterworth DS, (2001) Demographic parameters of southern right whales off South Africa. J Cetacean Res Manage 2(Special Issue) 161–169

Burnell SR, Bryden MM (1997) Coastal residence periods and reproductive timing in southern right whales, *Eubalaena australis*. J Zool (Lond) 241:613–621

Câmara IG, Palazzo Jr JT (1986) Novas informações sobre a presença de *Eubalaena australis* no sul do Brasil. In: Actas de la Primera Reunion de Trabajo de Expertos en MamiferosAcuaticos de America del Sur, Buenos Aires, pp 35–41

Caswell J, Fujiwara M, Brault S (1999) Declining survival probability threatens in the North Atlantic right whale. Proc Nat Acad Sci (USA) 96(6):3308–3313

Cooke JG, Rowntree V, Payne R (2001) Estimates of demographic parameters for southern right whales (*Eubalaena australis*) observed off Península Valdés, Argentina. J Cetacean Res Manag (special issue) 2:125–132

Cooke JG, Rowntree V, Sironi M (2013) Southwest Atlantic right whales: interim updated population assessment from photo-id collected at Península Valdéz, Argentina. J Cetacean Res Manag 14(Suppl):188

Corkeron PJ, Connor RC (1999) Why do baleen whales migrate? Mar Mamm Sci 15(4):1228–1245
Corrêa AA, Groch KR (2007) Respiration patterns of right whales in Southern Brazil – are they affected by whalewatching boats? Abstracts: 17th biennial conference on the biology of marine mammals, 28 November – 03 December 2007, Cape Town
Costa PM, Piedra PF, Paez E (2007) Distribution and habitat use patterns of southern right whales, *Eubalaena australis*, off Uruguay. J Cet Res Manage 9:45–51
Cummings WC (1985) Right Whales, *Eubalaena glacialis* (Muller, 1776) and *Eubalaena australis* (Desmoulins, 1822). In Ridway SH, Harrison SR (eds) Handbook of marine mammals 3: The Sirenians and Baleen Whales. p 275–304
Dombroski JRG, Parks SE, Groch KR et al (2016) Vocalizations produced by southern right whale (*Eubalaena australis*) mother-calf pairs in a calving ground off Brazil. J Acoust Soc Am 140:1850–1857. doi: http://dx.doi.org/10.1121/1.4962231
Ellis M (1969) A baleia no brasil colonial. Editora da Universidade de Sao Paulo, Sao Paulo
Elwen SH, Best PB (2004) Environmental factors influencing the distribution of southern right whales (*Eubalaena australis*) on the south coast of South Africa I: broad scale patterns. Mar Mamm Sci 20:567–582
Fertl D, Rosel PE (2009) Albinism. In: Perrin WF, Wursig B, Thewissen JGM (eds) Encyclopedia of marine mammals, 2nd edn. Academic Press, San Diego, pp 24–26
Figueiredo GC, Santos MC de O, Siciliano S et al (2013) Southern right whale, *Eubalaena australis*, in southeastern Brazil: are we losing an illustrious visitor? Scientific Committee of International Whaling Commission. Available via International Whaling Commission Secretariat, Cambridge
Gomes VSS (2005) Encalhes de baleias-francas-do-sul *Eubalaena australis* (Desmoulins, 1822), na costa brasileira: síntese do conhecimento. Dissertation, Faculdades Integradas Maria Thereza, Rio de Janeiro
Greig AB, ER Secchi, Zerbini AN et al (2001) Stranding events of southern right whales, *Eubalaena australis*, in southern Brazil. J Cet Res Manage 2(Special Issue):157–160
Groch KR (2000) Ocupação preferencial de áreas de concentração pela Baleia Franca Austral, *Eubalaena australis* (Desmoulins, 1822), CETACEA, MYSTICETI, no litoral sul do Brasil. Dissertation, Universidade Federal do Rio Grande do Sul, Porto Alegre
Groch K (2001) Cases of harassment by kelp gulls (*Larus dominicanus*) on right whales (*Eubalaena australis*) of Southern Brazil. Biotemas 14:147–156. doi:10.1111/j.1748-7692.1998.tb00693.x
Groch KR (2005) Biologia populacional e ecologia comportamental da baleia franca austral, *Eubalaena australis* (Desmoulins, 1822), Cetacea, Mysticeti, no litoral sul do Brasil. Dissertation, Universidade Federal do Rio Grande do Sul
Groch KR, Palazzo JT, Flores PAC et al (2005) Recent rapid increases in the right whale (*Eubalaena australis*) population off southern Brazil. LAJAM 4:41–47
Groch KR, Correa AA, Rocha MEC (2013) Development of whalewatching activities in Southern Brazil: conservation implications for right whales. Unpublished document, Brazil
Groch KR, Flores PAC, Kolesnikovas CKM et al (2015) Collisions between vessels and southern right whales (*Eubalaena australis*) in the Brazilian coast. In: Abstracts of the 21st biennial society for marine mammalogy conference, San Francisco, 13–18
Hamilton PK, Knowlton AR, Marx MK (2009) Right whales tell their own stories: the photo-identification catalog. In: Kraus SD, Rolland RM (eds) The urban whale: North Atlantic right whales at the crossroads. Harvard University Press, Cambridge, pp 75–104
Hammond PS (1986) Estimating the size of naturally marked whale populations using capture-recapture techniques. Intern Whal Comm 8(Special Issue):253–281
Hammond P (1990) Capturing whales on film – estimating cetacean population parameters from individual recognition data. Mammal Rev 20(1):17–22
Hammond PS, Mizroch SA, Donovan GPE (1990) Individual recognition of cetaceans: use of photo-identification and other techniques to estimate population parameters. International Whaling Comission, Cambridge
Hiby L, Lovell P (2001) Trials of a computerized aid for matching right whale callosity patterns. J Cet Res Manage 2(Special Issue):291–295

Hoffmeyer MS, Lindner MS, Carribero A, Fulco VK, Menendez MC et al (2010) Planktonic food and foraging of *Eubalaena australis* on Peninsula Valdes (Argentina) nursery ground. Rev Biol Mar Oceanogr 45(1):131–139. doi:10.4067/s0718-19572010000100013
International Whaling Commission (2001) Report of the workshop on the comprehensive assessment of right whales: a worldwide comparison. J Cet Res Manage 2(Special Issue):1–60
International Whaling Commission (2013) Report of the workshop on the assessment of southern right whales. J Cet Res Manage 14(Supplement):439–462
International Whaling Commission (2016) Report of the Scientific Committee of International Whaling Commission IWC/66/Rep01 Bled, Slovenia, 7–19
Kenney R (2009) Right whales *Eubalaena glacialis*, *E. japonica*, and *E. australis*. In: Perrin WF, Wursig B, Thewissen JGM (eds) Encyclopedia of marine mammals, 2nd edn. Academic, San Diego, pp 962–972
Kraus SD, Moore KE, Price CA et al (1986) The use of photographs to identify individual North Atlantic right whales (*Eubalaena glacialis*). Rep Int Whal Commn 10(Special Issue):145–151
Leaper R, Cooke J, Trathan P, Reid K, Rowntree VJ, Payne R (2006) Global climate drives southern right whale (*Eubalaena australis*) population dynamics. Biol Lett 2(2):289–292. doi: 10.1098/rsbl.2005.0431
Lockyer C (1981) Growth and energy budgets of large baleen whales from the Southern Hemisphere. FAO Fisheries Series III(5):379–487
Lodi L, Rodrigues MT (2007) Southern right whale on the coast of Rio de Janeiro State, Brazil: conflict between conservation and human activity. J Mar Biol Assoc UK 87:105–107
Lodi L, Siciliano S, Bellini C (1996) Ocorrências e conservação de baleias-francas-do-sul, *Eubalaena australis*, no litoral do Brasil. Pap Avul Zool 39(17):307–328
Markowitz TM, April DH, Wursig B (2003) Digital photography improves efficiency of individual dolphin identification. Mar Mammal Sci 19(1):217–223 (Notes)
Marón CF, Beltramino L, Di Martino M et al (2015) Increased wounding of southern right whale (*Eubalaena australis*) calves by kelp gulls (*Larus dominicanus*) at Península Valdés, Argentina. PLoS One 10(10):e0139291. doi:10.1371/journal.pone.0139291
Ott PH (2002) Diversidade genética e estrutura populacional e duas espécies de cetáceos do Atlântico Sul Ocidental: *Pontoporia blainvillei* e Eubalalaena australis. Dissertation, Pontifícia Universidade Católica do Rio Grande do Sul, Porto Alegre
Ott PH, Groch KR, Danilewicz D (2008) *Eubalaena australis* (Desmoulins, 1922). In: Machado ABM, Drummond GM, Paglia AP (eds) Livro Vermelho das espécies da fauna brasileira ameaçadas de extinção. Ministério do Meio Ambiente, Brasília, pp 802–804
Palazzo JT, Carter L (1983) A caça de baleias no Brasil. Associação Gaúcha Proteção Ambiental à Natureza, Porto Alegre, p 25
Palazzo Jr JT, Groch KR (2011) Baleia franca austral. In: Rocha-Campos C, Camara IG (eds) Plano de ação nacional para conservação dos mamíferos aquáticos: grandes cetáceos e pinípedes: versão III – Brasília: Instituto Chico Mendes de Conservação da Biodiversidade p 30–33
Palazzo Jr JT, Flores PAC, Groch KR (1999) First resighting of a southern right whale (*Eubalaena australis*) in Brazilian waters and an indicative of a three-year return and calving interval. In: Abstracts of the 13th biennial conference on the biology of marine mammals, Maui, 28 November – 3 December 1999
Patenaude NJ, Baker CS (2001) Population status and habitat use of southern right whales in the sub-Antartic Auckland Island of New Zealand. J Cet Res Manage 2(Special Issue):111–116
Patenaude NJ, Baker CS, Gales N (1998) Observations of southern right whales on New Zealand's sub Antarctic wintering grounds. Mar Mamm Sci 14(2):350–355
Payne (1986) Long term behavioral studies of the southern right whale (*Eubalaena australis*). Rep Int Whal Commn 10:161–168
Payne R, Dorsey E (1983) Sexual dimorphism and aggressive use of callosities in right whales (*Eubalaena australis*). In: Payne R (ed) Communication and behavior of whales, AAAS selected symposia series 76. Westview Press, Boulder, pp 295–328

Payne R, Brazier O, Dorsey E et al (1983) In: Payne R (ed) External features in southern right whales (*Eubalaena australis*) and their use in identifying individuals, Communication and behavior of whales, AAAS selected symposia series 76. Westview Press, Boulder, pp 371–445

Payne R, Rowntree VJ, Perkins JS (1990) Population size, trends and reproductive parameters of right whales (*Eubalaena australis*) off Peninsula Valdes, Argentina. Rep Int Whal Commn 12(Special Issue):271–278

Pirzl R (2010) BigFishv4.1 Manual. South Australian Department for Environmenet and Heritage. 7p

Pirzl R, Watson M (2011) Australasian right whale photo-identification catalogue (ARWPIC) technical outline. Skadia Pty Ltd, Victoria

Rayment W, Dawson S, Webster T (2015) Breeding status affects fine-scale habitat selection of southern right whales on their wintering grounds. J Biogeogr 42:463–474. doi:10.1111/jbi.12443

Reilly SB, Bannister JL, Best PB et al (2013) *Eubalaena australis*. In: The IUCN red list of threatened species 2013. Accessed 24 Aug 2016

Renault-Braga EP (2014) Distribuição espaço-temporal da baleia-franca-austral – *Eubalaena australis* (desmoulins 1822) – no sul do Brasil. Dissertation, Universidade Federal do Rio Grande, Rio Grande

Rocha-Campos C, Camara IG (eds) (2011) Plano de ação nacional para conservação dos mamíferos aquáticos: grandes cetáceos e pinípedes: versão III – Brasília: Instituto Chico Mendes de Conservação da Biodiversidade p 30–33

Rolland RM, Hamilton PK, Marx MK, Pettis HM et al (2009) External perspectives on tight whale health. In: Kraus SD, Rolland RM (eds) The urban whale: North Atlantic right whales at the crossroads. Harvard University Press, Cambridge, pp 273–309

Rowntree V (1996) Feeding, distribution, and reproductive behavior of cyamids (Crustacea: Amphipoda) living on humpback and right whales. Can J Zool 74:103–109

Rowntree V, Darling J, Silber G et al (1980) Rare sighting of a right whale (*Eubalaena glacialis*) in Hawaii. Can J Zool 58(2):309–312

Rowntree VJ, McGuinness P, Marshall K et al (1998) Increased harassment of right whales (*Eubalaena australis*) by kelp gulls (*Larus dominicanus*) at Península Valdés, Argentina. Mar Mamm Sci 14:99–115. doi:10.1111/j.1748-7692.1998.tb00693.x

Rowntree VJ, Payne R, Schell D (2001) Changing patterns of habitat use by southern right whales (*Eubalaena australis*) on their nursery ground at Península Valdés, Argentina, and in their long-range movements. J Cet Res Manage 2:133–143

Rowntree VJ, Valenzuela LO, Fraguas PF et al (2008) Foraging behaviour of southern right whales (*Eubalaena australis*) inferred from variation of carbon stable isotope ratios in their baleen. Scientific Committee of International Whaling Commission SC/60/BRG23

Rowntree V, Uhart M, Sironi M et al (2013) Unexplained recurring high mortality of southern right whale *Eubalaena australis* calves at Península Valdés. Argentina Mar Ecol Prog Ser Inter-Research 493:275–289. doi:10.3354/meps10506

Santos MC d O, Siciliano S, De Souza SP, Pizzorno JLA (2001) Occurrence of southern right whales (*Eubalaena australis*) along southeastern. Brazil J Cet Res Manage 2:153–156

Santos MCO, Siciliano S, Vicente AFDC et al (2010) Cetacean records along São Paulo State coast, southeastern Brazil. Braz J Oceanogr 58:123–142

Schaeff CM, Hamilton PK (1999) Genetic basis and evolutionary significance of ventral skin color markings in North Atlantic right whales (*Eubalaena glacialis*). Mar Mamm Sci 15(3):701–711

Schaeff CM, Best PB, Rowntree VJ et al (1999) Dorsal skin color pattern among southern right whales (*Eubalaena australis*): genetic basis and evolutionary significance. J HeredJ Heredity 90(4):464–471

Seyboth E, Groch KR, Secchi ER et al (2015) Habitat use by southern right whales, *Eubalaena australis* (Desmoulins, 1822), in their main northernmost calving area in the western South Atlantic. Mar Mamm Sci 31:1521–1537

Seyboth E, Groch KR, Dalla Rosa L et al (2016) Southern right whale (*Eubalaena australis*) reproductive success is influenced by krill (*Euphausia superba*) density and climate. Sci Rep 6:28205. doi:10.1038/srep28205

Simões-Lopes PC, Palazzo JT, Both MC et al (1992) Identificação, movimentos e aspectos biológicos da baleia franca-austral (*Eubalaena australis*) na costa sul do Brazil. Proocedings of III Reunión de Trabajo de Expertos en Mamíferos Acuáticos de América del Sur Montevideo 1988:62–65

Sironi M, López JC, Bubas R (2008a) Predation by killer whales (*Orcinus orca*) on southern right whales (*Eubalaena australis*) off Patagonia, Argentina: effects on behavior and habitat choice. Scientific Committee of International Whaling Commission SC/60/BRG29

Sironi M, Rowntree V, Snowdon C (2008b) Kelp gulls (*Larus dominicanus*) feeding on southern right whales (*Eubalaena australis*) at Península Valdés, Argentina: updated estimates and conservation implications. Scientific Committee of International Whaling Commission SC/61/BRG19

Thomas PO (1987) Social behavior, habitat use, and interspecific interactions of southern right whales (*Eubalaena australis*) mother-calf pairs. Dissertation, University of California, Santa Cruz

Thomas PO, Taber SM (1984) Mother-infant interaction and behavioral development in southern right whales, *Eubalaena australis*. Behaviour 88:42–60. doi:10.1163/156853984x00470

Tormosov DD, Mikhaliev BYA, Best PB et al (1998) Soviet catches of southern right whales *Eubalaena australis*, 1951–1971. Biological data and conservation implications. Biol Conserv 86:185–197

Townsend CH (1935) The distribution of certain whales as shown by logbook records of American whaleships. Zoologica. Sci Contr New York Zool Soc 19:1–50

Valenzuela LO, Sironi M, Rowntree VJ, Seger J (2009) Isotopic and genetic evidence for culturally inherited site fidelity to feeding grounds in southern right whales (*Eubalaena australis*). Mol Ecol 18:782–791

Vighi M, Borrell A, Crespo EA, Oliveira LR, Simões-Lopes PC, Flores PAC, García NA, Aguilar A (2014) Stable isotopes indicate population structuring in the Southwest Atlantic population of right whales (*Eubalaena australis*). PLoS One 9:e90489

Wells (2009) Identification methods. In: Perrin WF, Wursig B, Thewissen JGM (eds) Encyclopedia of marine mammals, 2nd edn. Academic Press, San Diego, pp 593–599

Whitehead H, Payne R, Payne M (1986) Population estimate for the right whales off Peninsula Valdes, Argentina, 1971–1976. Rep Int Whal Commn 10(Special Issue):169–171

Wursig B, Wursig M (1977) The photographic determination of group size, composition, and stability of coastal porpoises (*Tursiops truncatus*). Science 198(4318):755–756. doi:10.1126/science.198.4318.755

Zappes CA, da Silva CV, Pontalti M et al (2013) The conflict between the southern right whale and coastal fisheries on the southern coast of Brazil Mar. Policy 38:428–437

Chapter 17
Behavior and Ecology of Endangered Species Living Together: Long-Term Monitoring of Resident Sympatric Dolphin Populations

Marta J. Cremer, Annelise C. Holz, Camila M. Sartori, Beatriz Schulze, Renan L. Paitach, and Paulo C. Simões-Lopes

Abstract Sympatric species may exploit the same resources, potentially acting as competitors when resources are limited. For cetaceans, ecological separation between sympatric species is based on differences in diet and habitat use. The franciscana dolphin *Pontoporia blainvillei* (Pontoporiidae) and the Guiana dolphin *Sotalia guianesis* (Delphinidae) live in sympatry along the southern and southeastern Brazilian coast, but Babitonga Bay is the only estuarine region in which these species occur throughout the year and live in direct sympatry. Twenty years of data on these populations give some indication of how this ecologically similar species share the same habitat. That coexistence is facilitated through differences in diet, distribution, and habitat use patterns. The Guiana dolphin is probably dominant in

M.J. Cremer (✉) • B. Schulze • R.L. Paitach
Laboratório de Nectologia – Centro de Ciências Biológicas, Universidade da Região de Joinville (UNIVILLE) – Unidade São Francisco do Sul,
Rodovia Duque de Caxias, 6365, Iperoba, 89240-000 São Francisco do Sul, Santa Catarina, Brazil

Programa de Pós-Graduação em Ecologia – Centro de Ciências Biológicas, Universidade Federal de Santa Catarina (UFSC) – Campus Universitário, s/n, Departamento MIP – Córrego Grande, 88040-970 Florianópolis, Santa Catarina, Brazil
e-mail: mjc2209@yahoo.com.br

A.C. Holz • C.M. Sartori
Laboratório de Nectologia – Centro de Ciências Biológicas, Universidade da Região de Joinville (UNIVILLE) – Unidade São Francisco do Sul,
Rodovia Duque de Caxias, 6365, Iperoba, 89240-000 São Francisco do Sul, Santa Catarina, Brazil

P.C. Simões-Lopes
Programa de Pós-Graduação em Ecologia – Centro de Ciências Biológicas, Universidade Federal de Santa Catarina (UFSC) – Campus Universitário, s/n, Departamento MIP – Córrego Grande, 88040-970 Florianópolis, Santa Catarina, Brazil

Laboratório de Mamíferos Aquáticos (LAMAQ) – Centro de Ciências Biológicas, Universidade Federal de Santa Catarina (UFSC) – Campus Universitário, s/n, Departamento de Ecologia e Zoologia – Córrego Grande, 88040-970 Florianópolis, Santa Catarina, Brazil

M.R. Rossi-Santos, C.W. Finkl (eds.), *Advances in Marine Vertebrate Research in Latin America*, Coastal Research Library 22,
DOI 10.1007/978-3-319-56985-7_17

the niche partitioning system in Babitonga Bay. This fact, combined with small franciscana population size, makes the latter highly vulnerable to local extinction.

Keywords *Pontoporia blainvillei* • *Sotalia guianensis* • Population • Southern Brazil • Conservation

17.1 Introduction

Long-term monitoring of habitat use patterns is important for understanding population trends and identifying potential threats for the species (Krebs 2008). Monitoring programs enhance our knowledge of species life history traits such as population size (Wilson et al. 1999), migration patterns (Rock et al. 2006), distribution (Williams et al. 1993), critical habitats (Ingram and Rogan 2002), and social structure (Whitehead 2008), and help us to identify changes in home range, residence patterns, and behavior, among others.

Recognition of individual animals in a population allows researchers to analyze individual residence levels, providing critical information on temporal investment in specific geographic areas (Wells and Scott 1990). This information can help us estimate the relevance of specific areas for survival of the population. Photoidentification is a non-invasive technique used for recognition of individual animals (Würsig and Jefferson 1990). Natural body marks are reliable distinguishing features among cetaceans, and are thus commonly used in ecological studies and to inform species conservation and management plans (Gomez-Salazar et al. 2011).

Sympatry describes the scenario in which two ecologically similar species share the same habitat (Ricklefs 1996). Sympatric species may exploit the same resources, potentially acting as competitors when resources are limited (Pianka 1983). Niche separation and specialization help species to avoid direct interaction, and many authors have suggested that ecologically similar species exhibit stronger differences when they co-occur geographically than when they occur in non-overlapping parts of a territory (Pulliam 2000; Bonesi et al. 2004). For cetaceans, ecological separation between sympatric species is based on differences in diet and habitat use (Bearzi 2005). Species distribution is directly related to competition, dispersal, niche size, resource availability, and habitat stability over time and space (Pulliam 2000). Depending on the spatial scale of analysis, populations tend to show patterns of aggregation that depend on specific habitat characteristics. Resources are generally not randomly or regularly distributed, but rather concentrated in certain areas. This aggregated resource distribution results in aggregated distribution of dolphin populations (Begon et al. 1996).

Many small cetacean species occur in broad sympatry (sensu Bearzi 2005), in which two or more species co-occur over a wide geographic area. One example is *Pontoporia blainvillei* (Pontoporiidae), known as franciscana dolphin, and *Sotalia guianesis,* the Guiana dolphin (Delphinidae) in the southwestern Atlantic Ocean. These species are sympatric over much their ranges along the Brazilian coast, which extends from the southern limit of Guiana dolphin distribution in Florianópolis, Santa Catarina (27°23′S; 48°33′W) (Simões-Lopes 1988), to the northern limit of distribution of the franciscana dolphin in Itaúnas, Espírito Santo (18°25′S; 30°42′W) (Moreira and Siciliano 1991). Guiana dolphins commonly occupy bays and inlets throughout their species range (Pizzorno 1999; Flores 1999; Geise et al. 1999; Edwards and Schnell 2001; Lodi 2003), which extends north to Nicaragua (14°35′N, 83°14′W) (Carr and Bonde 2000), and includes Babitonga Bay (Hardt et al. 2010). Franciscana dolphin distribution extends southward to the Golfo Nuevo in Argentina (42°35′S; 64°48′W) (Crespo et al. 1998), and Babitonga Bay is the only estuarine bay in which this species occurs throughout the year (Cremer and Simões-Lopes 2005). These two dolphin species live in direct sympatry in Babitonga Bay (sensu Bearzi 2005), i.e., they co-occur in the same immediate habitat.

The first cetacean surveys in Babitonga Bay took place in 1995. Guiana dolphins are well-known to occupy bays and estuaries in southern Brazil, and this knowledge combined with reports from the local community led to a consensus that they were likely the only species in Babitonga Bay. However, in late 1996 researchers discovered a franciscana dolphin population in the bay (IBAMA 1998). Local fishermen also had prior knowledge of franciscana dolphins in the area, but there is no available information on how long either species have been residing there.

Various efforts have been made to better understand the ecology and behavior of both dolphin species. Following is a review of the knowledge of franciscana and Guiana dolphins in Babitonga Bay, derived from abstracts, papers, books, monographs, dissertations, theses, and reports published over the last 20 years.

17.2 Babitonga Bay

Babitonga Bay lies on the northern coast of Santa Catarina state in southern Brazil (26°02′ – 26°28′S, 48°28′ – 48°50′W). The bay contains the last remaining major mangrove habitat in the southern hemisphere (IBAMA 1998) (Fig. 17.1). This subtropical estuary has an area of 160 km^2, with many islands, rivers, sand banks, and rock formations. The tide shows a maximum range of 2.3 m. Mean bay depth is roughly 6 m, with a deeper entrance channel that can reach 25 m. The bay margins are composed mainly of mangrove forest, but also contain sandy-muddy beaches and rocks.

Despite the importance of this ecosystem and high species richness, the area has been suffering anthropogenic effects of extensive human occupation of surrounding areas. In 2010, there were 620.572 inhabitants among the six municipalities surrounding the bay (IBGE 2010). Detrimental human activities include water pollution, intense boat traffic, port activity, and illegal occupation of mangrove areas, among others; these issues present major challenges for conservation. The bay has two large harbors, and also hosts various economic activities such as fishing, sand

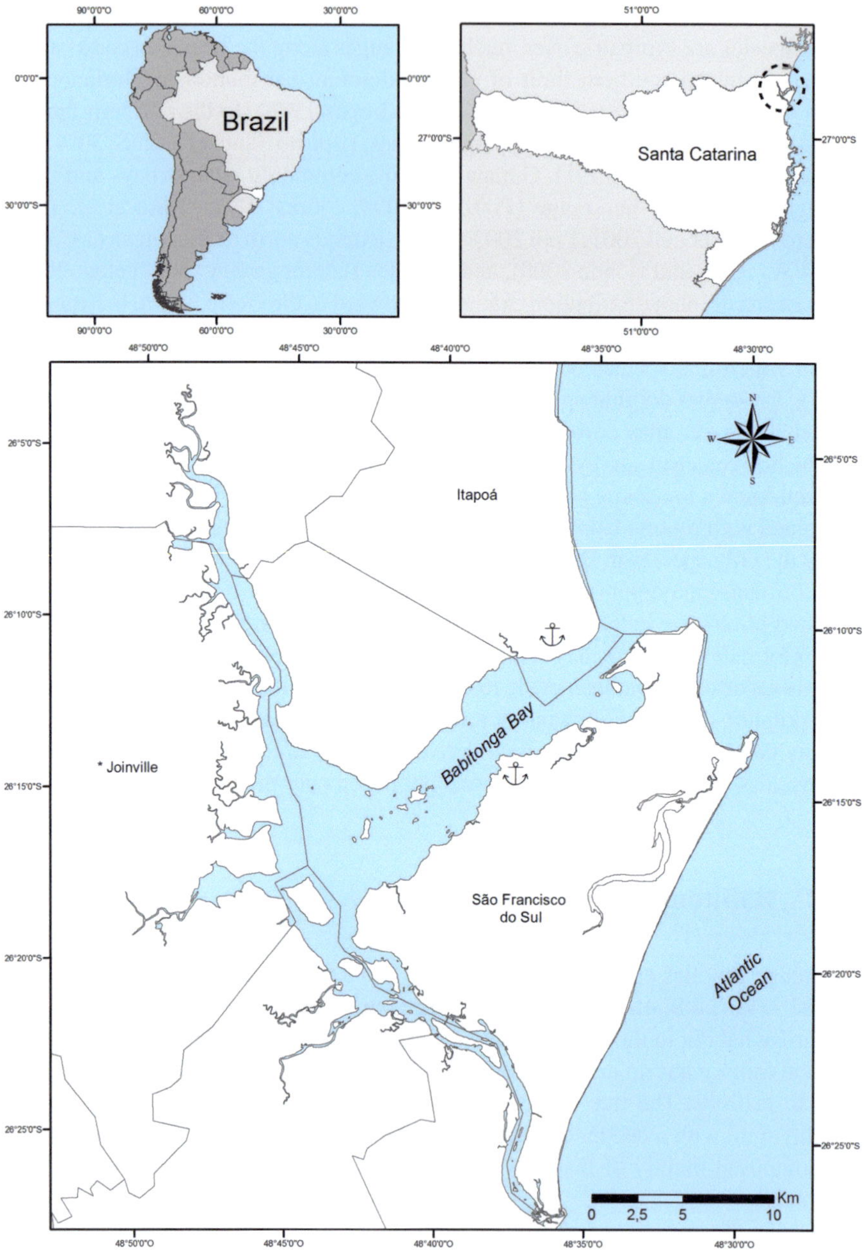

Fig. 17.1 Babitonga Bay, in Santa Catarina state, southern Brazil

mining, tourism, and aquaculture. Despite its economic importance, the area has been designated as a "priority area for conservation, sustainable use and sharing of benefits from Brazilian biodiversity", by the Brazilian government, in the "extremely high" priority category (PROBIO 2007).

17.3 Sympatric Species in Babitonga Bay

17.3.1 The Guiana Dolphin, Sotalia guianensis

Among the small cetacean species of the Brazilian coast, the Guiana dolphin is undoubtedly, the best studied. Along its coastal distribution it is often found associated with bays, estuaries, and coves, and these areas are amenable to the observation of populations. The maximum body length recorded for this species is 2.20 m, with a weight of 121 kg (Rosas and Monteiro-Filho 2002). The Guiana dolphin has gray coloration on the back that becomes lighter in the belly region; this area can be pinkish in calves. The rostrum is medium in size, and the species has about 140 teeth. There are no accurate estimates of maximum age for the Guiana dolphin, but the oldest known animal was a 30 year-old female (Rosas and Monteiro-Filho 2002) (Fig. 17.2).

Some studies indicate that the Guiana dolphin has a promiscuous mating system with sperm competition (Rosas and Monteiro-Filho 2002; Santos and Rosso 2008). Body length at birth is estimated to be between 92 and 106 cm, and the gestation period is about 11 months (Rosas and Monteiro-Filho 2002). The number of calves seems to be higher in the summer months in Babitonga Bay (Cremer 2015).

Fig. 17.2 A Guiana dolphin (*Sotalia guianensis*) pair traveling. We can observe the medium size rostrum, typical triangular dorsal fin and gray color

This species has been considered under threat of extinction in Brazil since 2014 (MMA 2014), however the IUCN classifies it as "data deficient" (Secchi 2012). Only two populations live in Santa Catarina state, one in the southern limit (North Bay in Florianópolis), and the other in Babitonga Bay. The main conservation threat to the species is accidental catch by fishing nets (Secchi 2012).

17.3.2 The Franciscana Dolphin, Pontoporia blainvillei

The franciscana dolphin is one of the smallest cetaceans in the world. Females are larger and can reach a body length of 1.8 m, while male maximum body length is around 1.6 m. Individuals in the southern portion of the species distribution are larger than those living further north. Maximum weight recorded for the species is 53 kg, and maximum age recorded was 21 years. Calves are born with a body length of about 71 cm (Rosas and Monteiro-Filho 2001). Coloration varies from greyish brown on the back to light brown or light gray in the belly region. A distinguishing characteristic of this species is its rostrum, which is proportionally the longest among cetaceans. This species has more than 200 teeth (Fig. 17.3).

Franciscana dolphins are thought to be a monogamous species (Danilewicz et al. 2004; Rosas and Monteiro-Filho 2001; Wells et al. 2013), with groups forming functional social units (Mendez et al. 2010). The main threat for this species is the accidental capture in fishing nets, a phenomenon that is currently leading to population declines. The IUCN classifies this species as "vulnerable" (Reeves et al. 2012), however the Brazilian government classification is "critically endangered" (MMA 2014).

Fig. 17.3 Franciscana dolphin (*Pontoporia blainvillei*) showing her long and thin rostrum, the main characteristic of the specie

17.4 Group Size and Behavior

Franciscana dolphin group size varies along the species distribution, driven mainly by differences in prey abundances and habitat quality (Cremer et al. 2011). These differences in group size are related to several variables, including physical characteristics, prey concentration and distribution, and population social organization. Data on the Babitonga Bay franciscana population indicate group sizes ranging from one to 22 individuals (mean = 7.1 ± 5.42, n = 79) (Cremer and Simões-Lopes 2005). Solitary individuals are uncommon (3.8% of total observations). Groups with two or three individuals are most common (35.4%), and 59.5% of the groups have more than four individuals. Calves were observed in all seasons, and were present in 30.4% of the groups. Groups composed of one adult (likely a female) accompanied by a calf and one or two adults are often observed (Fig. 17.4). Along the entire geographical distribution of the species, group sizes range from 2 to 15 individuals (Crespo 2009). It is common to find many groups in close proximity, forming “groupings” with up to 40 individuals. The majority of the population tends to remain aggregated in this manner and concentrated within the same region, at least for part of the time.

Franciscana dolphin is a discrete species with rare aerial events and reduced dorsal exposure during emersion (Cremer and Simões-Lopes 2005). After one dive, in general the long rostrum first projects out of the water, followed quickly by the back which moves with only a slight curvature, restricting above-water exposure to a small portion of the body (Fig. 17.5). The back and dorsal fin have few marks. The dorsal fin barely cuts the water line, emerging and submerging vertically.

Fig. 17.4 Mother and calf of franciscana dolphin swimming together, side by side

Fig. 17.5 Dive behavior of franciscana dolphin in Babitonga Bay. The long rostrum is first projected out of the water, followed quickly by the back

Fig. 17.6 Guiana dolphin group in Babitonga Bay

Guiana dolphin mean group sizes range from 5.3 to 6.5 individuals (Cremer 2000). Group size across the species geographical distribution ranges from two to 29 individuals in most areas (e.g., Edwards and Schnell 2001; Daura-Jorge et al. 2005; Nascimento et al. 2008) (Fig. 17.6).

Mixed-species groups have not been observed. Large groupings of both Guiana (Cremer 2000) and franciscana dolphins (Cremer and Simões-Lopes 2005) have been observed interacting with seabirds while hunting in the bay area. Franciscana dolphins are known to interact with terns (*Sterna* spp), cormorants (*Phalacrocorax brasilianus*), and brown boobies (*Sula leucogaster*) (Cremer and Simões-Lopes 2005). Guiana dolphins interact with the same species, as well as frigate birds (*Fregata magnifiscens*) and gulls (*Larus dominicanus*) (Cremer et al. 2004) (Figs. 17.7 and 17.8).

Fig. 17.7 Guiana dolphin (*Sotalia guianensis*) interacting whit birds during feeding behavior in Babitonga Bay

Fig. 17.8 Franciscana dolphin (*Pontoporia blainvillei*) interacting whit birds during feeding behavior in Babitonga Bay

17.5 Species Distributions and Habitat Use Patterns

Many variables contribute to species distribution and habitat use patterns in cetaceans, including water temperature (Gaskin 1968; Au and Perryman 1985), distance from the shore (Karczmarski et al. 2000; Edwards and Schnell 2001), depth (Würsig and Würsig 1979, 1980; Shane 1990), tide (Würsig and Würsig 1979; Shane 1990; Félix 1994; Bordino et al. 1999; Edwards and Schnell 2001), time of day (Geise

1991), current speed (Irvine et al. 1981) and bottom topography (Selzer and Payne 1988; Cañadas et al. 2002; Ingram and Rogan 2002). Environmental parameters are considered indirect measures because they influence primarily prey distribution, which in turn affects dolphin spatial and behavioral patterns (Wells et al. 1980; Shane et al. 1986; Selzer and Payne 1988; Baumgartner 1997; Acevedo-Gutiérrez and Parker 2000; Hastie et al. 2004).

Both species show an aggregated pattern of distribution with core areas, similar to that of other coastal dolphins such as *Tursiops truncatus* (Würsig and Würsig 1979; Ingram and Rogan 2002), *Souza chinensis* (Karczmarski et al. 2000), *Lagenorhynchus obscurus* (Würsig and Würsig 1980), and *Lagenorhynchus acutus* (Selzer and Payne 1988). The area occupied by both species has substantially reduced in the last 20 years. Although the species distribution strongly overlap, there is no overlap of core areas (Cremer 2007).

Guiana dolphins were recorded in the innermost area of the bay between 1997 and 1999, including the Palmital channel and the harbor inlet (Cremer 2000). The harbor inlet in São Francisco do Sul was intensively used by Guiana dolphins, mainly for hunting and feeding (Cremer et al. 2009). Dolphins in this area have been observed using the ships as a barrier to trap fish prey. However, in 2000 the harbor was enlarged and underwent repairs required the use of dredges, pile-drivers, and other heavy machinery, causing intense disturbance in the area. Guiana dolphins were rarely seen in the São Francisco do Sul vicinity from 2000 to the 2006, and were not found at all in the harbor inlet (Cremer et al. 2004, 2009). Seasonal changes in distribution have been analyzed in recent years, and revealed the use of a larger area in winter and a smaller one in the summer. Guiana dolphin core areas are located central and south of the islands, and do not change throughout the year (Fig. 17.9).

A similar trend was observed for franciscana dolphins, however changes in distribution encompassed smaller areas. In 2006, the population was well distributed along the central region of the bay. Since then, the species has preferentially occupied the north coast of the bay, and greatly reduced habitat area within the bay (Cremer and Simões-Lopes 2005; Sartori 2014) (Fig. 17.10).

Patterns of dolphin habitat use in Babitonga Bay vary by daily tidal cycles, but in different ways for the two species. Both species utilize areas closer to the continental shore at the end of flood tide (Paitach et al. 2017). This is the period when tidal flats overflow, causing fish to migrate to these important feeding areas (Reis-Filho et al. 2011). Dolphins use these areas opportunistically, as fish are more vulnerable to predation in these locations (Reis-Filho et al. 2011).

Franciscanas move toward the mouth of the bay during ebb tide and away from the mouth during flood tide, following the flow of the current (Paitach et al. 2017). This behavior is similar to that of franciscanas in Anegada Bay (Bordino 2002). Guiana dolphins do not exhibit strong patterns in relation to tidal current, typically occupying inner areas at early ebb stage and remaining near the mouth of the bay in late flood stage. Although a counter-current foraging strategy has already been described for *T. truncatus* (Shane 1990), movement following the direction of

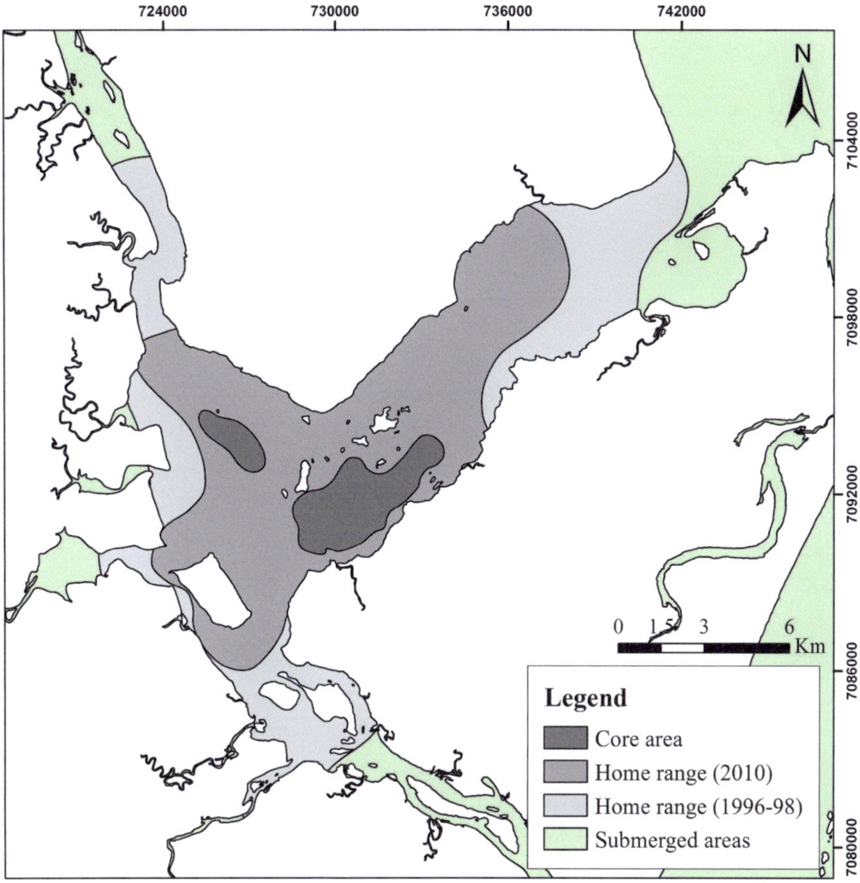

Fig. 17.9 Changes in the distribution of Guiana dolphins (*Sotalia guianensis*) inside Babitonga Bay

the current is more common (e.g., Bordino 2002; Mendes et al. 2002; Fury and Harrison 2011).

Patterns of habitat use in Babitonga Bay dolphins are a reflection of prey availability (Cremer 2007). Fish abundance in estuaries tends to be higher in warmer seasons (Gratwicke and Speight 2005). The dolphins used larger areas in winter and smaller areas in summer, showing more dispersed patterns when prey availability is lower (Paitach et al. 2017). The central region of Babitonga Bay hosts a cluster of islands around which there is a strong convergence of currents. This causes nutrient accumulation and attracts fish, making this area the primary zone of activity for franciscana and Guiana dolphins (Cremer and Simões-Lopes 2008; Cremer et al. 2011).

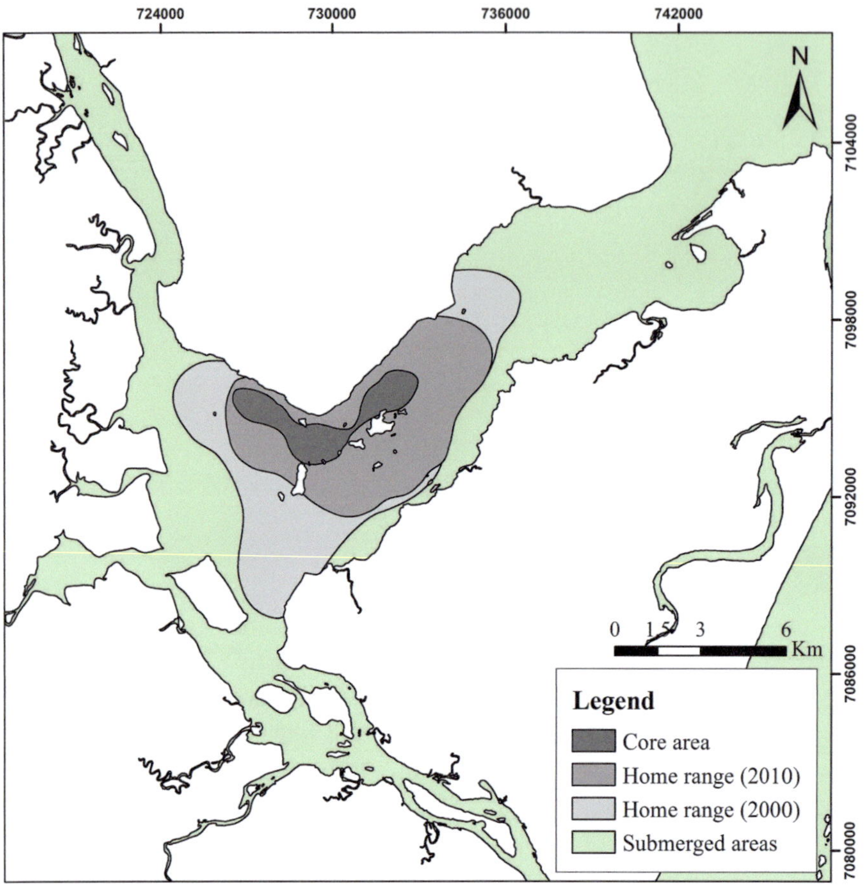

Fig. 17.10 Changes in the distribution of franciscana dolphins (*Pontoporia blainvillei*) inside Babitonga Bay

17.6 Density, Abundance, and Residence

Estimates of population size are important for assessment of conservation status, and identification of temporal or spatial trends in abundance is an integral component of conservation and management strategies (Heppell et al. 2000; Bearzi et al. 1997).

Recognition of individuals in a population permits to understand patterns of residence. Photoidentification is the primary technique used to estimate abundances of small cetaceans, and consists of identifying individuals by natural or artificial markings (e.g., Hammond et al. 1990) (Figs. 17.11 and 17.12). Color patterns, scars, scratches, nicks, and mutilations are markings commonly used for recognition; dorsal fin and dorsal surface markings are primarily used for small cetaceans (Würsig and Jefferson 1990). These markings are typically acquired over time and can be

Fig. 17.11 Staff in field work, finding a Guiana dolphin group during photoidentification survey

Fig. 17.12 Staff in field work, finding a franciscana dolphin group during photoidentification survey

temporary (e.g., scratches) or permanent (e.g., nicks and mutilations), caused by a variety of factors. Natural markings can be caused by bites from conspecifics, or abrasions from the ground (Würsig and Jefferson 1990; Dufault and Whitehead 1998). Generally, only a portion of the population can be identified through natural marks (Gowans and Whitehead 2001).

Recent abundance estimates in Babitonga Bay were acquired using mark-recapture methods, but prior to these studies distance sampling technique surveys over linear transects were employed. Guiana dolphin abundance in Babitonga Bay in 2010 and 2011 was estimated at 209 individuals (95% CI: 174–252) by Schulze (2012) using photoidentification techniques and mark-recapture methods. Cremer et al. (2011) generated similar estimates about 10 years prior using distance sampling method over linear transects, with values of 245 (95% CI: 142–422) individuals for 2001, 186 (95% CI: 93–374) for 2002, and 179 (95% CI: 93–144) for 2003. Guiana dolphin abundance in Babitonga Bay is intermediate compared to other populations, however the population is small enough to require precautionary management (Thompson et al. 2000). For franciscana dolphins, distance sampling produced a population estimate of 50 individuals (95% CI: 28–89) from 2001 to 2003 (Cremer and Simões-Lopes 2008) and 55 individuals in 2011 (Zerbini et al. 2011). Photoidentification techniques with mark-recapture methods obtained similar estimates (i.e., between 52 and 82 individuals from 2011 to 2013) (Sartori 2014). Population size varies slightly with birth and death rates in the area, but remains stable over time.

Twenty-three franciscana dolphins and 78 Guiana dolphins have been identified in Babitonga Bay through natural markings (Schulze 2012; Sartori 2014) (Figs. 17.13 and 17.14). For both species, nicks on dorsal fin were the main natural marks. Small nicks were the only identified markings on franciscana dolphins, with the exception of a scratch on one individual. Individual residence patterns varied from 5.26% to

Fig. 17.13 Sample of the franciscana dolphin photoidentification catalog. The total of identified individuals is 23 in Babitonga Bay

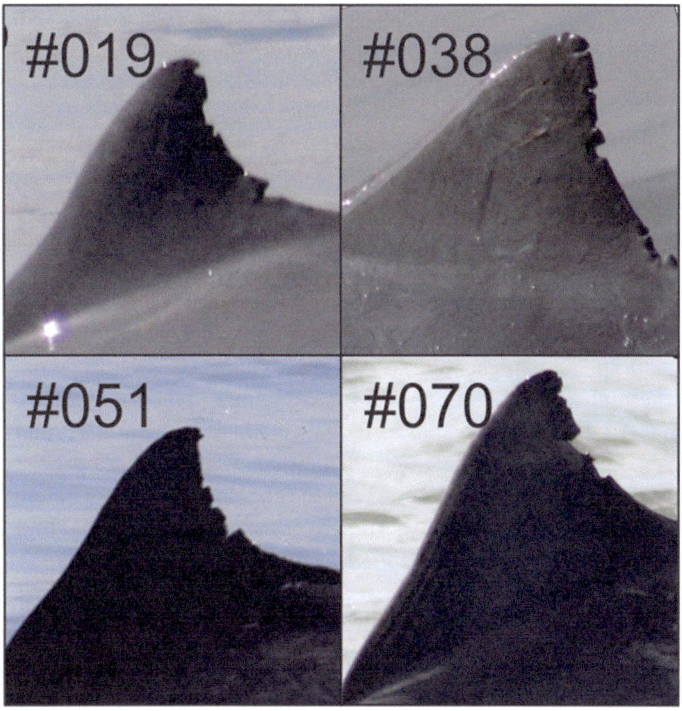

Fig. 17.14 Sample of the Guiana dolphin photoidentification catalog. The total of identified individuals is 78 in Babitonga Bay

78.95%, and 40% of the population had a residence rate greater than 50% (Sartori 2014). For Guiana dolphins, larger nicks and scratches were common. Residence patterns varied from 5% to 60%, with 79.5% of the population considered resident and 20.5% considered transient (Schulze 2012).

Studies of mitochondrial diversity (Dias et al. 2013), skull morphometry (Alves 2013), and intestinal parasite burden (Alves et al. 2017) indicate that franciscana dolphins in Babitonga Bay may be an isolated population. However, the species is widely distributed along the coast. Guiana dolphins seem to occur outside of the bay, but always near the estuary (Vianna et al. 2016). This species seems to be highly dependent on estuaries in the southern and southeastern Brazilian coast, and differences in skull morphometry suggest differences among the three populations living closer to the southern limit of the species range (Deon 2015).

17.7 Diet and Fish Availability

Studies of feeding ecology contribute to our general understanding of the biology and ecology of predators and prey, as well as trophic structure, energy flow, and ecosystem function (Katona and Whitehead 1988; Bowen and Siniff 1999). Niche overlap can change several components of behavior, including feeding behavior, and may reduce competition (Begon et al. 1996). According to the definition of Pauly et al. (1998), franciscana dolphins and Guiana dolphins occupy a similar trophic level.

Direct observation of feeding behavior in marine mammals can be difficult, thus analysis of stomach contents is the method typically used for diet studies (Barros and Odell 1990). Analysis of stomach contents of franciscana and Guiana dolphins in Babitonga Bay revealed that teleosts are the most common prey item for both species (Cremer et al. 2011; Paitach 2015). The squid *Lolliguncula brevis* was also found in the stomach contents of both species.

Twenty-one fish species belonging to six families were identified in the diet of franciscana dolphins, representing 69.4% of the numerical frequency of prey. Fish also had the highest frequency of occurrence among prey, accounting for 91.3%. Cephalopod frequency of occurrence was estimated at 47.8%, and numeric frequency was estimated at 30.6% (Paitach 2015). The number of prey species and the numerical dominance of fish compared to cephalopods agree with other studies throughout the species' geographical ranges (Bassoi 1997; Di Beneditto 2000; Di Beneditto and Ramos 2001; Rodríguez et al. 2002; Oliveira 2003; Silva 2011; Paso-Viola et al. 2014).

Among the prey species, the 'cangoá' (*Stellifer rastrifer*) is the most important in the franciscana diet, with a 70% index of relative importance. When the second most important prey is added (the sardine *Pellona harroweri*), the value is closer to 90%. This result indicates that only a few prey species sustain franciscana population, as observed in other studies (Oliveira 2003; Silva 2011). Physical characteristics of this species (e.g., long rostrum and numerous small teeth) favor small-sized prey like the sardine and the cangoá (Pinedo 1982; Oliveira 2003; Bassoi 2005; Silva 2011; Cremer et al. 2012).

Twenty-eight fish species from 13 families were identified in the Guiana dolphin diet. Teleosts also had the highest frequency of occurrence (88.2%), but cephalopods had higher numerical frequency (53.1%) than teleosts (46.9%). The number of prey species for the Guiana dolphin varies from 10 to 36 throughout its geographical range (Di Beneditto 2000; Zanelatto 2001; Santos et al. 2002; Oliveira 2003; Di Beneditto and Siciliano 2006; Daura-Jorge et al. 2011). Guiana dolphin population sizes, behavior, habitat use, and social dynamics vary geographically, and variation in prey richness may reflect the different ways that populations explore their respective environments (Daura-Jorge et al. 2007; Wedekin et al. 2007; Cantor et al. 2012; Schulze 2012).

For Guiana dolphins, five prey species together had a 70% relative importance value: *Mugil curema, Pellona harroweri, Stellifer rastrifer, Micropogonias furnieri* and *Trichiurus lepturus*. The Guiana dolphin diet is composed of many prey species with similar importance values and gradual reduction, including up to six species with a combined importance value of 75%; this pattern agrees with other studies (Zanelatto 2001; Oliveira 2003; Daura-Jorge et al. 2011).

The availability and distribution of food resources are considered key factors influencing habitat use for predators in both marine and terrestrial habitats (Acevedo-Gutiérrez and Parker 2000; Castro 2003; Hastie et al. 2004). Fish were sampled in Babitonga Bay to evaluate potential relationships between diet (prey items) and dolphin distributions (Fig. 17.15). Ninety-five teleost species were collected between 2004 and 2005, of which 25 were considered part of the dolphin diet. The central region of the bay accumulates the highest prey abundances, especially species of high importance for the franciscana dolphin that are abundant throughout the year. These data support a hypothesis of generalist and opportunistic feeding behavior in franciscana dolphins, which appear to utilize whichever resource is most abundant. Guiana dolphins also have generalist feeding behavior but select prey with lower abundances in the bay, possibly because these species produce higher energetic returns (Cremer 2007). Paitach (2015) found that prey were more abundant in low tide and that this factor constrains local spatial distributions.

Thirteen prey species were shared between franciscana and Guiana dolphins, representing 62% of total franciscana prey species and 46% of Guiana dolphin prey species (Cremer et al. 2011; Paitach 2015). Stable isotope analysis of $\delta^{13}C$ and $\delta^{15}N$ in prey and predator tissue suggests that there is an overlap in the diet of these species, especially with respect to some preys (*L. brevis, Diapterus rhombeus* and *S. rastrifer*) (Hardt et al. 2013).

17.8 Conservation Threats and Mortality

Coastal cetacean species are more vulnerable to impacts of human activities, primarily because coastal environments are intensively used by human populations. Dolphin populations with restricted geographic ranges and strong dependence on specific environmental characteristics are more vulnerable to habitat loss (Reeves et al. 2003). Human activities such as ecotourism and associated noise pollution and environmental degradation can promote changes in patterns of cetacean distribution and habitat use (Richardson et al. 1995; Aguilar et al. 2000; Forcada 2002; Lusseau 2003). This is certainly the case for franciscana and Guiana dolphins in Babitonga Bay (see section on Species Distribution and Habitat Use Patterns). Babitonga Bay dolphins are exposed to many threats, and the consequences are not well understood. The negative impacts of human activities may occur directly (e.g., by causing mortality or diseases), or indirectly when affect dolphin habitat.

Habitat loss will be one of the greatest threats to conservation in the coming decades (Simberloff 1998; Groom and Vynne 2006). The reduction of the area of occurrence for both dolphin species in Babitonga Bay is a clear indication of habitat loss, caused mainly by impacts of human activities. Santos and Lacerda (1987) also demonstrated abandonment of the Sado estuary in Portugal by *T. truncatus*, and attributed it to significant habitat changes. The decline of *T. truncatus* populations

Fig. 17.15 Fish sampling using trawl nets: (a) beach trawl and (b) bottom trawl

in the western Adriatic Sea is thought to be a result of the high levels of pollution and other disturbances (Bearzi et al. 1997). Richardson et al. (1995) also cite numerous examples of cetacean populations that have modified habitat use and reduced total living area due to noise disturbance.

Mortality records show that 43 Guiana dolphin and 36 franciscana dolphin carcasses were recovered in Babitonga Bay over the last 17 years. The number of dead animals fluctuates among years, with annual mean values of 2.9 individuals/year for Guiana dolphins and 2.5 individuals/year for franciscana dolphins; these values represent at the least 1.5% and 5% of the populations, respectively. Mortality is likely underestimated, as some carcasses may sink or be carried out of the bay with currents rather than being washed ashore.

17.8.1 Interactions with Fisheries

The cause of death could not be determined for many individuals because carcasses were often found in advanced state of decomposition. Markings on the head indicated that 38% of the dolphins died as a consequence of entanglement in fishing nets (Figs. 17.16 and 17.17). Accidental capture in fishing nets is a threat throughout the year for Babitonga Bay dolphins (Pinheiro and Cremer 2004). Cremer et al. (2012) reported that the highest number of carcasses was recovered in August, September and October, when drift nets are used in the area. Births occur between the months of October and January, and 11.4% of the carcasses found were either

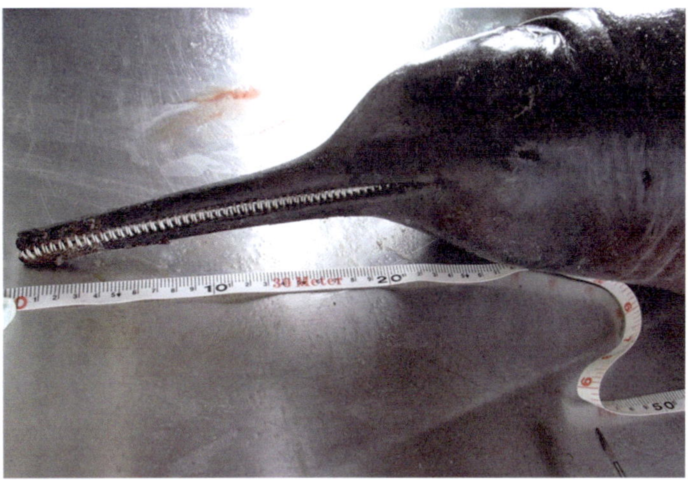

Fig. 17.16 Stranded franciscana dolphin found with evidences of entanglement in fishing net

Fig. 17.17 Guiana dolphin carcass found entangled in fishing net

calves or fetuses (less than 80 cm in length) (Cremer et al. 2013). There is no evidence of intentional capture of dolphins inside Babitonga Bay, or any other location along the southern and southeastern Brazilian coast. However, local fishermen are reluctant to report dead animals due to fear of reprisal. Reports indicate that these uncounted carcasses typically sink, reducing the reliability of mortality estimates (Simões-Lopes and Paula 1997). Accidental entangling in fishing nets is the major cause of dolphin mortality worldwide (Currey et al. 1990; Siciliano 1994; Rosas et al. 2002; Reeves et al. 2003).

17.8.2 Noise Disturbance

Underwater noise stems from numerous sources and varies both temporally and spatially. The increase in anthropogenic underwater noise over the past few decades coincides with intensified human activity in association with the aquatic environment (Weilgart 2007; Wright et al. 2007; Hildebrand 2009). According to Hatch and Wright (2007), low-level frequency noise has increased by 10–15 dB over the past 50 years due to boat traffic. High levels of underwater noise can inhibit vital activities for species that use sound as a means of communication and environmental recognition (Richardson et al. 1995; Janik and Slater 1998; Weilgart 2007).

Underwater noise in Babitonga Bay is mainly due to harbor activities and maritime traffic (Holz 2014), as is the case in other coastal areas around the world (Richardson et al. 1995, McQuinn et al. 2011; Cruz 2012; Bittencourt et al. 2014). Predictive maps of sound energy show that the innermost region of Babitonga Bay has lower noise intensity (Fig. 17.18), and that noise intensity increases towards the mouth of the bay with increasing proximity to the two harbors. High underwater noise levels can cause a range of problems for cetaceans. One example is behavioral

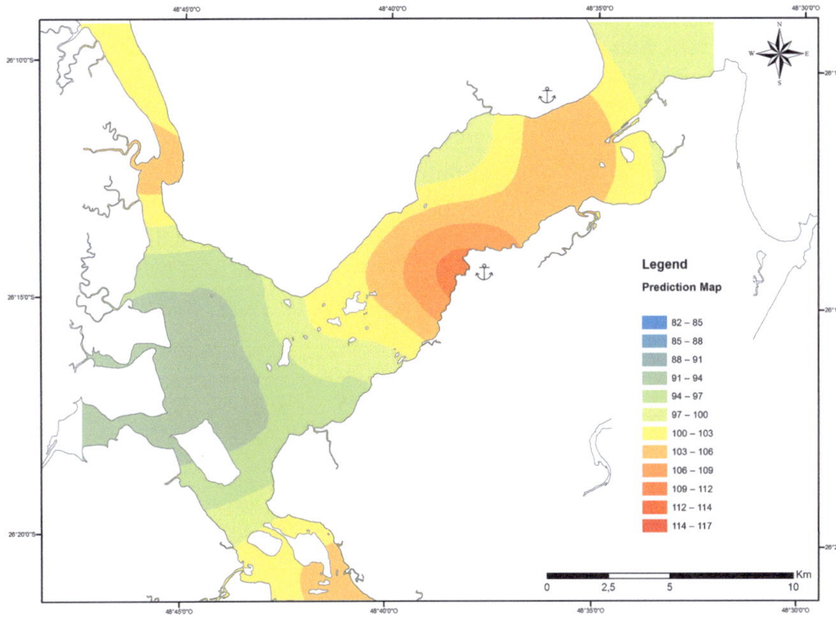

Fig. 17.18 Predictive map of sound energy showing the innermost region of Babitonga Bay that has lower noise intensity

and vocal changes due to noise from ships. Holz et al. (2010) found that the Guiana dolphin increased the frequency and reduced the duration of whistles during the passing of small boats in Babitonga Bay. Studies of *P. blainvillei* distribution patterns in Babitonga Bay for over 10 years show that this species does not approach the harbors (Cremer and Simões-Lopes 2008), which may be partially due to high noise levels in these areas. Data for *S. guianensis* indicate that in the past, this species intensively occupied the area near the Port of São Francisco do Sul (Cremer et al. 2009). However, after the beginning of harbor extension project the species abandoned the site, likely due to noise. The permanence of Guiana dolphins in areas with intense noise, even for hunting purposes, is detrimental for the species due to the potential for resulting physiological problems (Richardson et al. 1995) and contamination.

17.8.3 Boat Traffic

Boat traffic can generate two specific threats: collision and noise disturbance. Boats with erratic trajectories are one of the main culprits, and are associated mainly with human leisure activities and water sports (Simões-Lopes and Paula 1997). Wells and Scott (1997) highlighted cases of dead or injured cetaceans with cut wounds on the back, clearly caused by boat propellers. Flach (2006) reported a high number of estuarine dolphins with injuries related to human activities in Sepetiba Bay in Rio de Janeiro, Brazil. Areas with high dolphin concentration in the bay tend to coincide with tourist routes, and there are few reports of animals with cut wounds, probably due to collision with boats (Cremer 2000).

Although there is only one case of injury caused by boat propeller in Babitonga Bay, boat traffic can change animal behavior (Figs. 17.19 and 17.20). As vessels approach, dolphin emersions become gradually more discreet, and dives become longer in duration. Dolphins may also move away from the area temporarily, and groups may divide as a consequence of frequent boat passage.

Fig. 17.19 A case of injury caused by boat propeller in Babitonga Bay

Fig. 17.20 Boats pass in high speed inside the core areas of franciscana dolphins population in Babitonga Bay

17.8.4 Contaminants

Deposit of waste into the environment, especially in the water, is a growing problem worldwide. Garbage has been found in the stomach contents of cetaceans found dead on beaches, indicating that dolphins can and do swallow plastics, possibly confusing the material with prey (e.g., squid) (Geise and Gomes 1988). In Argentina, Denuncio et al. (2011) noted that franciscana dolphins in an estuary region had more plastic debris in the stomach than those in the adjacent coastal region. Cunha et al. (2014) also found plastic debris in franciscana dolphin carcasses recovered from the coastal region outside of Babitonga Bay.

Contamination of the marine environment is partially due to the large contribution of domestic and industrial effluents. Water contamination by chemical compounds is a major concern, mainly due to cumulative effects. Organochlorines are one of the major groups of pollutants that affect ecosystems due to persistence in the environment and toxicity to organisms (Clark 2001). For cetaceans, organochlorines are linked to reproductive problems such as miscarriages, fetal malformations, hormonal changes, and calf mortality, as well as skin problems. One study tested for organobromine compounds in liver samples from franciscana dolphins stranded along the Brazilian coast, and found the highest polybrominated diphenyl ethers (PBDEs) concentrations in individuals from Babitonga Bay (Alonso et al. 2012).

17.9 Conclusion

Babitonga Bay is likely the only region of direct sympatry along the geographical distribution of franciscana and Guiana dolphins. Although these species share the same habitat and occupy a similar trophic level, minor differences in habitat use and resource abundance probably contribute to the coexistence of these potentially competing species (Bearzi 2005). For example, Guiana dolphins are around twice the size of franciscana dolphins, and this difference should produce higher energetic demands. Although the two species share many prey in the area, their main prey differ, and Guiana dolphins tend to prey upon larger species. The prey species that are shared by dolphins are highly abundant in Babitonga Bay. Opportunistic feeding behavior was also identified for sympatric populations of *Stenella coeruleoalba* and *Delphinus delphis*, which prey on species that are most abundant (Das et al. 2000). In Patagônia, Argentina, *D. delphis* and *Lagenorynchus obscurus* feed mainly on the most abundant local fish species, which supports mixed-species aggregations in the sympatric area (Romero et al. 2012). Differences in feeding strategy may also explain the direct sympatry of *D. delphis* and *T. truncatus* in the Ionian Sea (Bruno et al. 2004).

The sharing of prey species is minimized by different patterns of habitat use. Although the home ranges of these dolphin populations strongly overlap, the core areas differ and they never use the same areas at the same time. Effects of tidal cycles on dolphin distribution is another factor contributing to differences in habitat use. Mixed groups were never observed, and is probable that these species avoid physical interaction due to potential for aggression (Bearzi 2005). Observational data suggest that when Guiana dolphins arrive in the core areas of franciscana dolphins, they temporarily leave the area.

The Babitonga Bay Guiana dolphin population is roughly four times larger than the franciscana population. Both show a high degree of residency inside the bay, however the distribution of Guiana dolphins extends to the adjacent coast for hunting while franciscana dolphins are more dependent on the estuary.

Physical and behavioral characteristics of the Guiana dolphin indicate that it is probably dominant in the niche partitioning system in Babitonga Bay (Bonesi et al. 2004). In fact, resident populations of Guiana dolphins in southern and southeastern Brazilian coast live in a broad sympatry with franciscana dolphins; it is only in Babitonga Bay the two species live in direct sympatry. In the southern portion of the Guiana dolphin range, another coastal species, the bottlenose dolphin (*T. truncatus*), occupies bays and estuaries (Simões-Lopes and Fabian 1999). In this areas franciscana dolphins do not live in direct sympatry with bottlenose dolphins. In Argentina and Uruguay it is uncommon to find *T. truncatus* near the coast, and in these areas franciscana dolphin populations reside in bays and estuaries (Bordino et al. 1999; Rodríguez and Bastida 2002; Failla et al. 2012; Negri et al. 2012).

There is no clear explanation for the resident and potentially isolated franciscana dolphin population in Babitonga Bay, but one interesting characteristic of this bay is the deep channel that connects the bay to the ocean. Unlike other estuaries in which the entrance channels are narrow and sand banks are common, Babitonga Bay has a wide, deep, natural channel (25 m depth, 1.8 km width) without buildings and no necessity for dredging to maintain it.

The home range area of both dolphin populations has declined over the last 20 years, probably as a consequence of human disturbances like sound pollution. This situation increases species overlap, potentially increasing competition for habitat and other resources. When food is abundant, competition is reduced or absent (Begon et al. 1996), but habitat degradation and overfishing can alter prey abundances in Babitonga Bay. Reduction of prey may then drive changes in the diets of sympatric species (Bonesi et al. 2004).

Populations seems to be stable at this time, but the absence of abundance estimates in the beginning of the monitoring period makes analysis of long-term trends difficult. Further, different methods were used among studies to estimate abundances over the history of monitoring in the area. Based on current data we estimate that if disturbance pressure continues to increase, coexistence could be unsustainable in the future. Considering that the Guiana dolphin is the dominant species in estuarine areas throughout its range in southern Brazil, is possible that franciscana dolphins will not persist in Babitonga Bay.

Although both species vulnerable to the same threats, franciscana dolphins seems to be more sensitive than Guiana dolphins to accidental capture in fishing nets and effects of habitat loss. Franciscana dolphins do not occur near very impacted areas like harbors and cities; this is in stark contrast to Guiana dolphins, which are known to hunt near anchored ships at the harbor and in highly degraded areas near cities. Despite evidence that Guiana dolphins are more robust under disturbance scenarios, these populations (like other small cetaceans) are also susceptible to anthropogenic threats such as noise pollution and environmental contamination. Although we do not have specific information for Babitonga Bay, the literature indicates that franciscana and Guiana dolphins have different social organization systems. Guiana dolphins are thought to be promiscuous while franciscana dolphins are likely monogamous. Coastal cetacean species are more vulnerable to human-driven impacts because coastal environments are intensively used by human populations, and dolphin populations with restricted geographic ranges and strong dependence on specific environmental characteristics are more vulnerable to resulting habitat loss (Reeves et al. 2003). Population viability analysis conducted with coastal dolphin species indicated that populations with less than 100 individuals are at high risk of extinction, even if mortality rates are relatively low (Thompson et al. 2000; Slooten 2007).

The long-term presence of franciscana and Guiana dolphins in Babitonga Bay highlight its importance for feeding and reproduction for both species. The high degree of residency by both species increases exposure to local impacts, and urgent measures are needed to control the impacts of anthropogenic activities in the region. This may include restrictions in the use of gillnets and controlling habitat degradation resulting from expansion of port areas, among other strategies. Long-term monitoring will be a powerful tool for continued detection of population trends and determination of conservation status, which is especially important for endangered species.

Acknowledgements The authors are grateful to the staff members at the Laboratório de Nectologia for assistance with field studies. Financial support was provided by the Universidade da Região de Joinville (UNIVILLE), the Fundação O Boticário de Proteção à Natureza, CNPq, Capes, and FAPESC. Studies between 2010–2015 were supported by the Projeto Toninhas/UNIVILLE, sponsored by Petrobras through the Programa Petrobras Ambiental.

References

Acevedo-Gutiérrez A, Parker N (2000) Surface behavior of bottlenose dolphins is related to spatial arrangement of prey. Mar Mamm Sci 16:287–298

Aguilar N, Díaz F, Carrillo M, Brito A, Barquín J, Alayón P, Falcón J, González G (2000) Evidence of disturbance of protected cetacean populations in the Canary Islands. Presented to the IWC Scientific Committee, London (SC/53/WW1)

Alonso MB, Eljarrat E, Gorga M, Secchi ER, Massoi M, Barbosa L, Bertozzi CP, Marigo J, Cremer MJ, Domit C, Azevedo AF, Dorneles PR, Torres JPM, Lailson-Brito J, Malm O, Barceló D (2012) Natural and antropogenically produced brominated compounds in endemic dolphins from Western South Atlantic: another risk to a vulnerable species. Environ Pollut 170:152–160

Alves AKM (2013) Variação morfológica e morfométrica dos crânios de *Pontoporia blainvillei* no litoral norte de Santa Catarina. Dissertation, Universidade da Região de Joinville

Alves AKM, Souza EAL, Marigo J, Cremer MJ (2017) Intestinal helminths of Franciscana (Pontoporia blainvillei) and Guiana dolphins (Sotalia guianensis) from the north coast of Santa Catarina state, Brazil. Oecol Aust 21(01):83–87

Au DWK, Perryman WL (1985) Dolphin habitats in the eastern tropical Pacific. Fish Bull 83(4):623–643

Barros NB, Odell DK (1990) Food habits of bottlenose dolphins in the southeastern United States. In: Leatherwood S, Reeves RR (eds) The bottlenose dolphin. Academic, San Diego, pp 309–328

Bassoi M (1997) Avaliação da dieta alimentar de toninha, *Pontoporia blainvillei* (Gervais& D' Orbigny, 1844), capturadas acidentalmente na pesca costeira de emalhe no sul do Rio Grande do Sul. Dissertation, Federal University of Rio Grande

Bassoi M (2005) Feeding ecology of franciscana dolphin, *Pontoporia blainvillei* (Cetacea: Pontoporiidae), and oceanographic processes on the Southern Brazilian coast. Dissertation, Graduate School of the National Oceanography Centre (USA), University of Southampton

Baumgartner MF (1997) The distribution of Risso's dolphin (*Grampus griseus*) with respect to the physiography of the northern Gulf of Mexico. Mar Mamm Sci 13:614–638

Bearzi M (2005) Aspects of the ecology and behaviour of bottlenose dolphins (*Tursiops truncatus*) in Santa Monica Bay, California. J Cetacean Res Manag 7:75–83

Bearzi G, Notarbartolo di Sciara G, Politi E (1997) Social ecology of bottlenose dolphins in the Kvarneric (northern Adriatic Sea). Mar Mamm Sci 13:650–668

Begon M, Harper JL, Townsend CR (1996) Ecology - individuals, populations and communities, 3rd edn. Blackwell Science, Oxford

Bittencourt L, Carvalho RR, Lailson-Brito J, Azevedo AF (2014) Underwater noise pollution in a coastal tropical environment. Mar Pollut Bull 83:331–336

Bonesi L, Chanin P, MacDonald DW (2004) Competition between Eurasian otter *Lutra lutra* and American mink *Mustela vison* probed by niche shift. Oikos 106:19–26

Bordino P (2002) Movement pattern of franciscana dolphins (*Pontoporia blainivillei*) in Bahia Anegada, Argentina. Lat Am J Aquat Mamm 1(1):71–76

Bordino P, Thompson G, Iñiguez M (1999) Ecology and behaviour of the franciscana dolphin *Pontoporia blainvillei* in Bahía Anegada, Argentina. J Cetacean Res Manag 1(2):213–222

Bowen WD, Siniff DB (1999) Distribution, population biology, and feeding ecology of marine mammals. In: Reynolds JE III, Rommel SA (eds) Biology of marine mammals. Smithsonian Institution Press, Washington, pp 423–484

Bruno S, Politi E, Bearzi G (2004) Social organization of a common dolphin community in the eastern Ionian Sea: evidence of a fluid fission-fusion society. European Research on Cetaceans 15:49–51

Cañadas A, Sagarminaga R, García-Tiscar S (2002) Cetacean distribution related with depth and slope in the Mediterranean waters off southern Spain. Deep-Sea Res I 49:2053–2073

Cantor M, Wedekin LL, Daura-Jorge FG, Rossi-Santos MR, Simões-Lopes PC (2012) Assessing population parameters and trends of Guiana dolphins (*Sotalia guianensis*): an eught year mark-recapture study. Mar Mamm Sci 28:63–83

Carr T, Bonde RK (2000) Tucuxi (*Sotalia fluviatilis*) occurs in Nicaragua, 800 km north of its previously known range. Mar Mamm Sci 16:447–452

Castro CSS (2003) Tamanho da área de vida e padrão de uso do espaço em grupos de sagüis, *Callithrix jacchus* (Linnaeus) (Primates, Callitrichidae). Rev Brasil Zool 20(1):91–96

Clark RB (2001) Marine pollution, 5th edn. Oxford, New York

Cremer MJ (2000) Ecologia e Conservação de *Sotalia fluviatilis guianensis* (Cetacea, Delphinidae) naBaía de Babitonga, litoral norte de Santa Catarina. Dissertation, Federal University of São Carlos

Cremer MJ (2007) Ecologia e conservação de populações simpátricas de pequenos cetáceos em ambiente estuarino no sul do Brasil. Dissertation, Federal University of Paraná

Cremer MJ (2015) Os mamíferos marinhos da Baía da Babitonga, 1st edn. Editora da Univille, Joinville, p 104

Cremer MJ, Simões-Lopes PC (2005) The occurrence of *Pontoporia blainvillei* (Gervais&d'Orbigny) (Cetacea, Pontoporiidae) in an estuarine area in southern Brazil. Rev Bras Zool 22:717–723

Cremer MJ, Simões-Lopes PC (2008) Distribution, abundance and density estimates of Franciscanas, *Pontoporia blainvillei* (Cetacea: Pontoporiidae), in Babitonga Bay, southern Brazil. Rev Bras Zool 25:397–402
Cremer MJ, Hardt FAS, Tonello AJ Jr, Simões-Lopes PCA, Pires JS (2004) Core areas changes in *Sotalia guianensis* (Cetacea, Delphinidae) population in Babitonga Bay, Santa Catarina. Revta Univille 9:12–16
Cremer MJ, Simões-Lopes PC, Pires JSR (2009) Occupation pattern of a harbor inlet by the estuarine dolphin, *Sotalia guianensis* (P. J. Van Bénéden, 1864) (Cetacea, Delphinidae). Braz Arch Biol Technol 52(3):765–774
Cremer MJ, Hardt FAS, Tonello AJ Jr, Simões-Lopes PC (2011) Distribution and status of the Guiana dolphin *Sotalia guianensis* (Cetacea, Delphinidae) population in Babitonga Bay, southern Brazil. Zool Stud 50:327–337
Cremer MJ, Pinheiro PC, Simões-Lopes PC (2012) Prey consumed by Guiana dolphin *Sotalia guianensis* (Cetacea, Delphinidae) and franciscana dolphin *Pontoporia blainvillei* (Cetacea, Pontoporiidae) in an estuarine environment in southern Brazil. Iheringia (Sér Zool) 102(2):131–137. doi:10.1590/S0073-47212012000200003
Cremer MJ, Sartori CM, Holz AC, Schulze B, Dos Santos NZ, Alves AKM, Paitach RL (2013) Franciscana strandings on the north coast of Santa Catarina state and insights into birth period. Biotemas 26(4):133–139. doi:10.5007/2175-7925.2013v26n4p133
Crespo EA (2009) Franciscana – *Pontoporia blainvillei*. In: Perrin WF, Würsig B, Thewissen JGM (eds) Encyclopedia of marine mammals, 2nd edn. Academic, Amsterdam, pp 466–469
Crespo EA, Harris G, González R (1998) Group size and distributional range of the franciscana, *Pontoporia blainvillei*. Mar Mamm Sci 14(4):845–849
Cruz EM (2012) Caracterização do ruído subaquático produzido pelo tráfego marítimo no estuário do sado e potenciais impactos sobre a população residente de *Tursiops truncatus* (Montagu, 1821). Dissertation, University of Lisboa
Cunha HA, Medeiros BV, Barbosa LA, Cremer MJ, Marigo J, Lailson-Brito J, Azevedo AF, Solé-Cava AM (2014) Population structure of the endangered franciscana dolphin (*Pontoporia blainvillei*): reassessing management units. PLoS One 9:e85633. doi:10.1371/journal.pone.0085633
Currey D, Lonsdale J, Thornton A, Reeves R (1990) The global war against small cetaceans. Environmental Agency, London
Danilewicz D, Claver JÁ, Perez Carrera AL, Secchi ER, Fontoura NF (2004) Reproductive biology of male franciscanas (*Pontoporia blainvillei*) (Mammalia: Cetacea) from Rio Grande do Sul, southern Brazil. Fish Bull 102:581–592
Das K, Lepoint G, Loizeau V, Debacker V, Dauby P, Bouquegneau JM (2000) Tuna and dolphin associations in the northeast Atlantic: evidence of different ecological niches from stable isotope and heavy metal measurements. Mar Pollut Bull 40:102–109
Daura-Jorge FG, Wedekin LL, Piacentini V, Simões-Lopes PC (2005) Seasonal and daily patterns of group size, cohesion and activity of the estuarine dolphin, *Sotalia guianensis* (P. J. van Bénéden) (Cetacea, Delphinidae), in southern Brazil. Rev Bras Zool 22:1014–1021
Daura-Jorge FG, Rossi-Santos MR, Wedekin LL, Simões-Lopes PC (2007) Behavioral patterns and movement intensity of *Sotalia guianensis* (P. J. van Bénéden) (Cetacea, Delphinidae) in two different areas on the Brazilian coast. Rev Bras Zool 24(2):265–270
Daura-Jorge FG, Wedekin LL, Simões-Lopes PC (2011) Feeding habits of the Guiana dolphin, *Sotalia guianensis* (Cetacea: Delphinidae), in Norte Bay, southern Brazil. Sci Mar 75:163–169
Denuncio P, Bastida R, Dassis M, Giardino G, Gerpe M, Rodríguez D (2011) Plastic ingestion in Franciscana dolphins, *Pontoporia blainvillei* (Gervais and d'Orbigny, 1844), from Argentina. Mar Pollut Bull 62(8):1836–1841. doi:10.1016/j.marpolbul.2011.05.003
Deon B (2015) Análise da morfometria craniana do boto-cinza, *Sotalia guianensis* (Delphinidae) como ferramenta de diferenciação de duas unidades populacionais no limite sul de sua distribuição. Dissertation, Universidade da Região de Joinville
Di Beneditto APM (2000) Ecologia alimentar de *Pontoporia blainvillei* e *Sotalia fluviatilis* (Cetacea) na costa norte do Estado do Rio de Janeiro, Brasil. Dissertation, State University of North Fluminense

Di Beneditto APM, Ramos RMA (2001) Biology and conservation of the franciscana (*Pontoporia blainvillei*) in the north of Rio de Janeiro State, Brazil. J Cetacean Res Manag 3(2):185–192

Di Beneditto APM, Siciliano S (2006) Stomach contents of the marine tucuxi dolphin (*Sotalia guianensis*) from Rio de Janeiro, south-eastern Brazil. J Mar Biol Assoc UK. in press

Dias CP, Cremer MJ, Azevedo AF, Cunha HA (2013) Baixa diversidade mitocondrial sugere filopatria das toninhas *Pontoporia blainvillei* (Gervais& d'Orbigny 1844) da Baía da Babitonga, SC. In: Abstracts of the II Encontro de Mastozóologos do Rio de Janeiro, 30 november – 3 december 2013, Parque Nacional da Serra dos Órgãos, Teresópolis

Dufault S, Whitehead H (1998) Regional and group-level differences in fluke markings and notches of sperm whale. J Mammal 79:514–520

Edwards HH, Schnell GD (2001) Status and ecology of *Sotalia fluviatilis* in the Cayos Miskito Reserve, Nicaragua. Mar Mamm Sci 17:445–472

Failla M, Seijas VA, Espósito R, Iníguez MI (2012) Franciscana dolphins, *Pontoporia blainvillei*, of the Río Negro Estuary, Patagonia Argentina. Mar Biodivers Rec 5:e102. doi:10.1017/S1755267212000875

Félix F (1994) A case of epimeletic behaviour in a wild bottlenose dolphin *Tursiops truncatus* in the Gulf of Guayaquil, Ecuador. Investig Cetac 25:227–234

Fernando Augusto Sliva Hardt, Marta Jussara Cremer, Antonio José Tonello Junior, Antonio Bellante, Gaspare Buffa, Giuseppa Buscaino, Salvatore Mazzola, André Silva Barreto, Luiz Antonio Martinelli, Giovanni Maria Zuppi, (2013) Use of carbon and nitrogen stable isotopes to study the feeding ecology of small coastal cetacean populations in southern Brazil. Biota Neotropica 13 (4):90–98

Flach L (2006) Photo-identification study reveals human threats towards estuarine dolphins in Southeast Brazil. In: Siciliano S, Borobia M, Barros N, Trujillo MF, Flores PAC (eds) Workshop on research and conservation of genus *Sotalia*, Armação dos Búzios, Rio de Janeiro, Brasil

Flores PAC (1999) Preliminary results of a photoidentification study of the marine tucuxi *Sotalia fluviatilis* in southern Brazil. Mar Mamm Sci 15(3):840–847

Forcada J (2002) Distribution. In: Perrin WF, Wursig B, Thewissen JGM (eds) Encyclopedia of marine mammals. Academic, San Diego, pp 327–333

Fury CA, Harrison PL (2011) Seasonal variation and tidal influences on the use of two Australian estuaries by bottlenose dolphins (*Tursiops aduncus*). Estuar Coast Mar Sci 93:389–395

Gaskin DE (1968) Distribution of Delphinidae (Cetacea) in relation to sea surface temperatures off eastern e southern New Zealand. NZ J Mar Fresh Res 2:527–534

Geise L (1991) *Sotalia guianensis* (Cetacea, Delphinidae) population in the Guanabara Bay, Rio de Janeiro, Brazil. Mammalia 55(3):371–379

Geise L, Gomes NMB (1988) Ocorrência de plástico no estômago de um golfinho do gênero *Sotalia* Gray 1886 (Cetácea, Delphinidae). In: Abstracts of the Tercera reunión de trabajo de especialistas em mamíferos acuáticos de América del Sur, Montevideo

Geise L, Gomes N, Cerqueira R (1999) Behaviour, habitat use and population size of *Sotalia fluviatilis* (Gervais, 1853) in the Cananéia estuary region, São Paulo, Brasil. Rev Bras Biol 59:183–194

Gomez-Salazar C, Trujillo F, Whitehead H (2011) Photo-identification: a reliable and noninvasive tool for studying pink river dolphins (*Inia geoffrensis*). Aquat Mamm 37(4):472–485

Gowans S, Whitehead H (2001) Photographic identification of northern bottlenose whales (*Hyperoodon ampullatus*): sources of heterogeneity from natural marks. Mar Mamm Sci 17:76–93

Gratwicke B, Speight M (2005) The relationship between fish species richness, abundance and habitat complexity in a range of shallow tropical marine habitats. J Fish Biol 66:650–667

Groom MJ, Vynne CH (2006) Habitat degradation and loss. In: Groom MJ, Meffe GK, Carroll CR (eds) Principles of conservation biology. Sinauer Associates, Sunderland, pp 173–212

Hammond PS, Mizroch SA, Donvan GP (1990) Individual recognition of cetaceans: use of photo-identification and other techniques to estimate population parameters. Reports of the IWC (Special Issue):12

Hardt FAS, Cremer MJ, Tonello Junior AJ, Bellante A, Buffa G, Buscaino G, Mazzola S, Barreto AS, Martinelli LA, Zuppi GM (2013) Use of carbon and nitrogen stable isotopes to study the feeding ecology of small coastal cetacean populations in southern Brazil. Biota Neotrop 13(4):90–98

Hardt FAS, Cremer MJ, Tonello AJ Jr, Simões-Lopes PCA (2010) Residence patterns of the Guiana dolphin *Sotalia guianensis* in Babitonga Bay, Southern coast of Brazil. LAJAM 8:117–121

Hastie GD, Wilson B, Wilson LJ, Parsons KM, Thompson PM (2004) Functional mechanisms underlying cetacean distribution patterns: hotspots for bottlenose dolphins are linked to foraging. Mar Biol 144:397–403

Hatch LT, Wright AJA (2007) Brief review of anthropogenic sound in the oceans. Int J Comp Psychol 20:121–133

Heppell S, Pfister C, de Kroon H (2000) Elasticity analysis in population biology: methods and applications. Ecology 81(3):605–606

Hildebrand J (2009) Anthropogenic and natural sources of ambient noise in the ocean. Mar Ecol Prog Ser 395:5–20. doi:10.3354/meps08353

Holz AC (2014) Caracterização do ambiente acústico subaquático e uso de habitat pela toninha, *Pontoporia blainvillei*, na Baía da Babitonga, litoral norte de Santa Catarina, Brasil. Dissertation, Universidade da Região de Joinville

Holz AC, Dias CP, Cremer MJ (2010) Efeito da poluição sonora sobre o boto-cinza (*Sotalia guianensis*) na Baía da Babitonga. In: Abstracts of the 6° SeminárioIntegrado de Ensino, Pesquisa e Extensão. University of the Region of Joinville, Joinville

IBAMA (1998) Proteção e controle de ecossistemas costeiros: Manguezal da Baía de Babitonga. Instituto Brasileiro do Meio Ambiente e dos Recursos Naturais Renováveis, Brasília

IBGE (2010) Censo Demográfico 2010 (Census 2010), Brazilian Institute of Geography and Statistics, Rio de Janeiro, http://censo2010.ibge.gov.br/

Ingram S, Rogan E (2002) Identifying critical areas and habitat preferences of bottlenose dolphins *Tursiops truncatus*. Mar Ecol Prog Ser 244:247–255

Irvine AB, Scott MD, Wells RS, Kaufmann JH (1981) Movements and activities of the atlantic bottlenose dolphin, *Tursiops truncatus*, near Sarasota. Florida Fish Bull 79(4):671–688

Janik V, Slater P (1998) Context-specific use suggests that bottlenose dolphin signature whistles are cohesion calls. Anim Behav 56(4):829–838

Karczmarski L, Cockcroft VG, Mclachlan A (2000) Habitat use and preferences of Indo-Pacific humpback dolphins *Sousa chinensis* in Algoa Bay, South Africa. Mar Mamm Sci 16(1):65–79

Katona S, Whitehead H (1988) Are Cetacea ecologically important? Oceanogr Mar Biol 26:553–568

Krebs CJ (2008) Ecology: the experimental analysis of distribution and abundance. Benjamin Cummings Press, San Francisco, California, p 688

Lodi LF (2003) Seleção e uso do habitat pelo boto-cinza, *Sotalia guianensis* (van Bénéden, 1864) (Cetacea, Delphinidae), na Baía de Paraty, Estado do Rio de Janeiro. Bioikos 17(1/2):5–20

Lusseau D (2003) The emergent properties of a dolphin social network. Proc R Soc Lond B Biol Sci 270:S186–S188

María Alejandra Romero, Silvana L. Dans, Nestor García, Guillermo M. Svendsen, Raúl González, Enrique A. Crespo, (2012) Feeding habits of two sympatric dolphin species off North Patagonia, Argentina. Marine Mammal Science 28 (2):364–377

Marcos César de Oliveira Santos, Sergio Rosso (2008) Social organization of marine tucuxi dolphins, Sotalia guianensis, in the Cananéia estuary of Southeastern Brazil. J Mammal 89(2):347–355

Mcquinn IH, Lesage V, Carrier D, Larrivée G, Samson Y, Chartrand S, Michaud R, Theriault J (2011) A threatened beluga (*Delphinapterus leucas*) population in the traffic lane: vessel-generated noise characteristics of the Saguenay-St, Lawrence Marine Park, Canada. J Acoust Soc Am 130:3661–3673

Mendes S, Turrell W, Lütkebohle T, Thompson P (2002) Influence of the tidal cycle and a tidal intrusion front on the spatio-temporal distribution of coastal bottlenose dolphins. Mar Ecol Prog Ser 239:221–229

Mendez M, Rosenbaum H, Subramaniam A, Yackulic C, Bordino P (2010) Isolation by environmental distance in mobile marine species: molecular ecology of franciscana dolphins at their southern range. Mol Ecol 19:2212–2228

MMA (2014) Lista Nacional Oficial de Espécies da Fauna Ameaçadas de Extinção (Portarias MMA no 444/2014), Ministry of the Environment, Brazil

Moreira LM, Siciliano S (1991) Northward extension range for *Pontoporia blainvillei*. In: Abstracts of the IX Biennial Conference on the Biology of Marine Mammals, 5–9 December, Chicago

Nascimento LF, Spinelli LHP, Santos E, Queiroz REM, Pansard KCA, Medeiros PIAP, Gondim MA, Jesus AH, Silva FJL, Yamamoto EM (2008) Atypical coloration in a specimen of estuarine dolphin, *Sotalia guianensis*, on the litoral of the state of Rio Grande do Norte, north-east Brazil. Mar Biodiv Rec 1:1–2

Negri MF, Denuncio P, Panebianco MV, Cappozzo HL (2012) Bycatch of franciscana dolphins *Pontoporia blainvillei* and the dynamic of artisanal fisheries in the species' southernmost area of distribution. Braz J Oceanogr 60:149–158

Oliveira MR (2003) Ecologia alimentar de *Sotalia guianensis* e *Pontoporia blainvillei* (Cetacea, Delphinidade e Pontoporiidae) no litoral sul do Estado de São Paulo e litoral do Paraná. Dissertation, Federal University of Paraná

Paitach RL (2015) Ecologia alimentar e espacial da toninha (*Pontoporia blainvillei*) e do boto-cinza (*Sotalia guianensis*) na Baía da Babitonga, Sul do Brasil. Dissertation, Federal University of Santa Catarina

Paitach RL, Simões-Lopes PC, Cremer MJ (2017) Tidal and seasonal influences in dolphin habitat use in a southern Brazilian estuary. Sci Mar 81(1):49

Paso-Viola MN, Denuncio P, Negri MF, Rodríguez D, Bastida R, Cappozzo HL (2014) Diet composition of franciscana dolphin *Pontoporia blainvillei* from southern Buenos Aires, Argentina and its interaction with fisheries. Rev Biol Mar Oceanogr 49(2):393–400

Pauly D, Trites AW, Capuli E, Christensen V (1998) Diet composition and trophic levels of marine mammals. J Mar Sci 55:467–481

Pianka ER (1983) Ecología evolutiva. Ediciones Omega, Barcelona

Pinedo MC (1982) Análise dos conteúdos estomacais de *Pontoporia blainvillei* (Gevais e D'Orbigny, 1844) e *Tursiops gephyreus* (Lahille,1908) (Cetacea, Platanistidae e Delphinidae) na zona estuarina e costeira de Rio Grande, RS, Brasil. 1982. Dissertation, Fundação Universidade do Rio Grandeio Grande do Sul

Pinheiro L, Cremer MJ (2004) Etnoecologia e captura acidental de golfinhos (Cetacea: Pontoporiidae e Delphinidae) na Baia da Babitonga, Santa Catarina. Desenvolvimento e Meio Ambiente 8:69–76

Pizzorno JLA (1999) Estimativa populacional do boto-cinza, *Sotalia fluviatilis*, na Baía de Guanabara, por meio de catálogo de fotoidentificação. Dissertation, Federal Rural University of Rio de Janeiro

PROBIO (2007) Levantamento da Cobertura Vegetal Nativa do Bioma Mata Atlântica. In: IESB: Instituto de estudos Socioambientais do Sul da Bahia, Rio de Janeiro

Pulliam HR (2000) On the relationship between niche and distribution. Ecol Lett 3:349–361

Reeves RR, Smith BD, Crespo E, Notarbartolo di Sciara G (2003) Dolphins, whales and porpoises 2002–2010, Conservation action plan for the world's cetaceans. IUCN – International Union for Conservation of Nature, Gland, p 139

Reeves R, Dalebout M, Jefferson TA, Karkzmarski L, Laidre K, O'corry-Crowe G, Rojas-Bracho L, Secchi E, Slooten E, Smith BD, Wang JY, Zerbini AN, Zhou K (2012) *Pontoporia blainvillei*. The IUCN Red List of Threatened Species. Version 2014.3

Reis-Filho JÁ, Barros F, Nunes JACC, Sampaio CLS, Souza GBG (2011) Moon and tide effects on fish capture in a tropical tidal flat. J Mar Biol Assoc UK 91(3):735–743

Renan Lopes Paitach, Paulo César Simões-Lopes, Marta Jussara Cremer, (2017) Tidal and seasonal influences in dolphin habitat use in a southern Brazilian estuary. Scientia Marina 81 (1):49

Richardson WJ, Greene CR Jr, Malme CI, Thomson DH (1995) Marine mammals and noise. Academic, San Diego, p 576

Ricklefs RE (1996) A economia da Natureza. 3.ed, Guanabara Koogan, Rio de Janeiro, p 470
Rock J, Pastene LA, Kaufman G, Forestell P, Matsuoka K, Allen J (2006) A note on East Australia group V stock humpback whale movement between feeding and breeding areas based on photo-identification. J Cet Res Manage 8:301–305
Rodríguez D, Bastida R (2002) From the conquest to ecotourism: environmental consequences of human activities in coastal Argentina. In: UNU-Iwate-UNESCO joint International Conference on 'Conserving our coastal Environment'. United Nations University, Tokyo, pp 109–128
Rodríguez D, Rivero L, Bastida R (2002) Feeding ecology of the franciscana (*Pontoporia blainvillei*) in marine and estuarine waters of Argentina. Lat Am J Aquat Mamm 1:77–94
Romero MA, Dans SL, García N, Svendsen GM, González R, Crespo EA (2012) Feeding habits of two sympatric dolphin species off North Patagonia, Argentina. Mar Mamm Sci 28(2):364–377
Rosas FCW, Monteiro-Filho ELA (2001) Reproductive parameters of *Pontoporia blainvillei* (Cetacea, Pontoporiidae), on the coast of São Paulo and Paraná states, Brazil. Mammalia 66:231–245
Rosas FCW, Monteiro-Filho ELA (2002) Reproduction of the estuarine dolphin (*Sotalia guianensis*) on the coast of Paraná, southern Brazil. J Mamm 83:507–515
Rosas FCW, Monteiro-Filho ELA, Oliveira MR (2002) Incidental catches of franciscana (*Pontoporia blainvillei*) on the Southern coast of São Paulo State and the coast of Paraná State, Brazil. Lat Am J Aquat Mamm 1:161–167
Santos ME, Lacerda M (1987) Preliminary observations of the bottlenose dolphin (*Tursiops truncatus*) in the Sado estuary (Portugal). Aquat Mamm 13:65–80
Santos COM, Rosso S, Santos RA, Lucato SHB, Bassoi M (2002) Insights on small cetacean feeding habits in southeastern Brazil. Aquat Mamm 28:38–45
Sartori CM (2014) Fotoidentificação, abundância e área de vida de toninhas, *Pontoporia blainvillei*, na Baía da Babitonga, Santa Catarina, Brasil. Dissertation, Universidade da Região de Joinville
Schulze B (2012) Estimativa populacional e área de vida do boto-cinza *Sotalia guianensis*, (Cetacea, Delphinidae) na Baía da Babitonga, Santa Catarina, Brasil. Dissertation, Federal University of Santa Catarina
Secchi E (2012) *Sotalia guianensis*. In: IUCN 2012. IUCN Red List of Threatened Species, Version 2012.2.
Selzer LA, Payne PM (1988) The distribution of white-sided (*Lagenorhynchus acutus*) and commom dolphins (*Delphinus delphis*) vs. environmental features of the continental shelf of de northeastern United States. Mar Mamm Sci 4:141–153
Shane SH (1990) Behavior and ecology of the bottlenose dolphin at Sanibel Island, Florida. In: Leatherwood S, Reeves RR (eds) The bottlenose dolphin. Academic, San Diego, pp 369–386
Shane SH, Wells RS, Würsig B (1986) Ecology, behavior, and social organization of the bottlenose dolphin: a review. Mar Mamm Sci 2:34–63
Siciliano S (1994) Review of small cetaceans and fishery interactions in coastal waters of Brazil. In: Abstracts of the 46ª meeting of the Scientific Committee of the International Whale Commission (IWC). Puerto Vallarta 15:241–250
Silva BH (2011) Ecologia alimentar da toninha *Pontoporia blainvillei* (Cetacea). 2011a. Dissertation, University of Sao Paulo
Simberloff D (1998) Flagships, umbrellas, and keystones: is single-species management passe in the landscape era? Biol Conserv 83:247–257
Simões-Lopes PC (1988) Ocorrência de uma população de *Sotalia fluviatilis*, Gervais, 1853, (Cetacea, Delphinidae) no limite sul de sua distribuição, Santa Catarina, Brasil. Biotemas 1:57–62
Simões-Lopes PC, Fabian ME (1999) Residence patterns and site fidelity in bottlenose dolphins, *Tursiops truncatus* (Montagu) (Cetacea, Delphinidae) off Southern Brazil. Rev Bras Zool 16:1017–1024
Simões-Lopes PC, Paula GS (1997) Mamíferos aquáticos e impacto humano: diretrizes para conservação e "utilização não letal". Aquitaine Ocean 3:69–78

Slooten E (2007) Conservation management in the face of uncertainty: effectiveness of four options for managing Hector's dolphin bycatch. Endang Species Res 3:169–179

Thompson PM, Wilson B, Grellier K, Hammond PS (2000) Combining power analysis and population viability analysis to compare traditional and precautionary approaches to conservation of coastal cetaceans. Biol Conserv 14:1253–1263

Vianna TS, Loch C, Castilho PV, Gaidzinski MC, Cremer MJ, Simões-Lopes PC (2016) Review of thirty-two years of toothed whale strandings in Santa Catarina, southern Brazil (Cetacea: Odontoceti). Fortschr Zool 33:e20160089

Wedekin LL, Daura-Jorge FG, Piacentini VQ, Simões-Lopes PC (2007) Seasonal variations in spatial usage by the estuarine dolphin, *Sotalia guianensis* (van Bénéden, 1864) (Cetacea; Delphinidae) at its southern limit of distribution. Braz J Biol 67(1):1–8

Weilgart LS (2007) The impacts of anthropogenic ocean noise on cetaceans and implications for management. Can J Zool 85:1091–1116

Wells RS, Scott MD (1990) Estimating bottlenose dolphin population parameters from individual identification and capture-release techniques. Rep Int Whal Comm 12:407–415

Wells RS, Scott MD (1997) Seasonal incidence of boat strikes on bottlenose dolphins near Sarasota, Florida. Mar Mamm Sci 13:475–480

Wells RS, Irvine AB, Scott MD (1980) The social ecology of inshore odontocets. In: Herman LM (ed) Cetacean behavior: mechanisms and functions. John Wiley, New York, pp 263–317

Wells RS, Bordino P, Douglas DC (2013) Patterns of social association in the franciscana, *Pontoporia blainvillei*. Mar Mamm Sci 29:520–528

Whitehead H (2008) Analyzing animal societies: quantitative methods for vertebrate social analysis. Chicago University Press, Chicago

Williams J, Dawson S, Slooten E (1993) The abundance and distribution of bottlenose dolphins (*Tursiops truncatus*) in doubtful sound, New Zealand. Can J Zool 71:2080–2088

Wilson B, Hammond PS, Thompson PM (1999) Estimating size and assessing trends in a coastal bottlenose dolphin population. Ecol Appl 9:288–300

Wright AJ, Soto NA, Baldwin AL, Beteson M, Beale CM, Clark C, Deak T (2007) Do marine mammals experience stress related to anthropogenic noise? Int J Comp Psychol 20:274–316

Würsig B, Jefferson G (1990) Methods of photoidentification for small cetaceans. Rep Int Whal Comm 12:43–52

Würsig B, Würsig M (1979) Behavior and ecology of the bottlenose dolphin, *Tursiops truncatus*, in the south Atlantic. Fish Bull 77:399–412

Würsig B, Würsig M (1980) Behavior and ecology of dusky dolphins, *Lagenorhynchus obscurus*, in the South Atlantic. Fish Bull 77:871–890

Zanelatto RC (2001) Dieta do boto-cinza, *Sotalia fluviatilis* (Cetacea, Delphinidae) no complexo estuarino da Baía de Paranaguá e sua relação com a ictiofauna estuarina. Dissertation, Federal University of Paraná

Zerbini AN, Danilewicz D, Secchi ER, Andriolo A, Cremer M, Flores PAC, Ferreira E, de Alves LCP S, Sucunza F, de Castro FR, Pretto D, Sartori CM, Schulze B, Denuncio P, Laake F (2011) Assessing bias in abundance estimates from aerial surveys to improve conservation of threatened franciscana dolphins: preliminary results from a survey conducted in southern Brazil. Paper SC/63/SM9 presented to the IWC Scientific Committee, Trømso, Norway, June 2011, p 13

MIX
Papier aus verantwortungsvollen Quellen
Paper from responsible sources
FSC® C105338

If you have any concerns about our products,
you can contact us on
ProductSafety@springernature.com

In case Publisher is established outside the EU,
the EU authorized representative is:
Springer Nature Customer Service Center GmbH
Europaplatz 3, 69115 Heidelberg, Germany

Printed by Libri Plureos GmbH
in Hamburg, Germany